March 11, 1977

To Seymour Cohn

Thank you for the fine comment you make about my book. Your r

is graced with excellence. The

can be no better recommendation

Best regards

[signature]

Semiconductor Control

Semiconductor Control

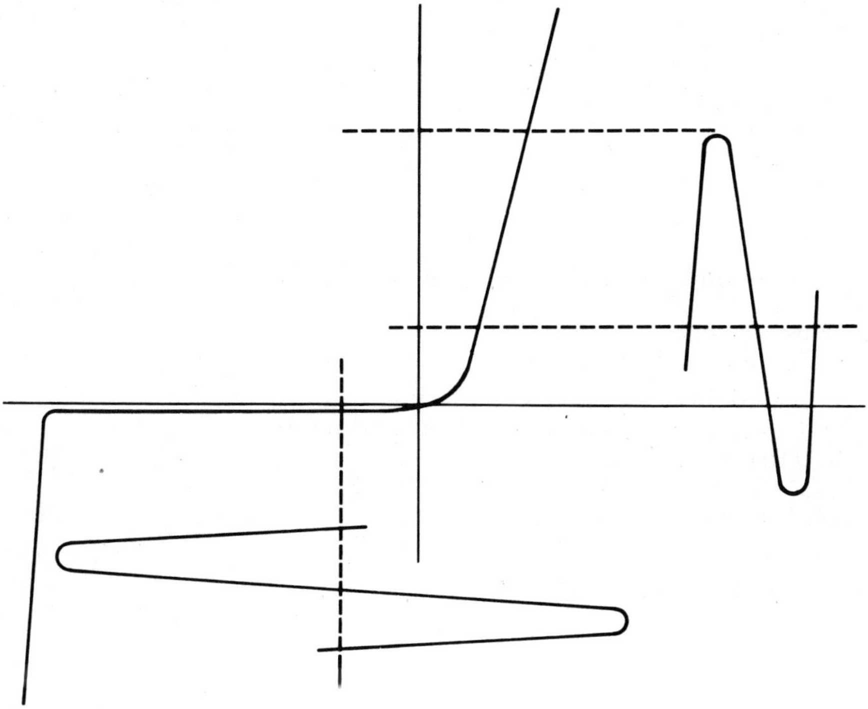

Joseph F. White Technical Director
Semiconductor Operations
Microwave Associates, Inc.
Burlington Massachusetts

Artech

Patricia Jean Collins
Art Director

Neal Charles Vitale
Editor

Maura Zeman
Typesetter

Library of Congress Cataloging in Publication Data

White, Joseph F 1938-
Semiconductor control.

Includes bibliographies and index.
1. Electronic control. 2. Semiconductors.
3. Microwave circuits. I. Title.
TK7881.2.W46 629.8'04'3 76-1793
ISBN 0-89006-028-2

ARTECH HOUSE, INC.
610 Washington Street
Dedham, Massachusetts 02026
Printed and bound in the United States of America

Library of Congress Catalog Card Number: 76-1793
International Standard Book Number: 0-89006-028-2

To Christopher, Patricia Jeanne, Catherine, and Elizabeth

Foreword

Joseph F. White has studied, worked, and taught in all aspects of microwave semiconductor materials, control diodes, and circuit applications. He is thoroughly grounded in the physics and mathematics of the field, but has primarily the engineer's viewpoint, combining basic knowledge with experience and ingenuity to generate practical designs under constraints of required performance and costs of development and production. As a result of his teaching experience and numerous technical papers and oral presentations, he has developed a clear, well-organized writing style that makes this book easy to use as a self-teaching text, a reference volume, and a design handbook.

Dr. White believes that an engineer must have a good understanding of semiconductor physics, a thorough knowledge of microwave circuit theory, at least an elementary acquaintance with transistor drivers, and the ability to check and refine a microwave circuit on a computer terminal to be qualified for modern, creative design of microwave semiconductor control components. These subjects are well covered in approximately the first half of the book; the second half treats the general and specific design of switches, attenuators, limiters, duplexers, and phase shifters, with many examples drawn from his experience and that of others.

Especially important is the chapter *Mathematical Techniques and Computer Aided Design;* it should appeal to all who are concerned with microwave circuits, whether or not these circuits include semiconductor elements. In particular, engineers seeking a good introduction to computer programming will find this book serves the function well; typical books and instruction manuals on FORTRAN are too general, too complex, and too confusing for those who wish to learn the programming skills needed for computer aided design of microwave circuits. Dr. White provides the necessary information for time-shared FORTRAN IV work, starting with the fundamentals and covering integer, real, and complex variables; operation symbols; real and complex functions; subroutines; simple READ, WRITE, and FORMAT statements; flow charts; and so on. He gives a number of program listings as examples, starting with simple ones, then moves on to subroutines for lossy line immittance transformation, complex ABCD matrix multiplication, response calculation, and a final flow chart and listing of a general network analysis program for the cascaded two-port building blocks commonly found in microwave circuits. Also, he reproduces a number of printouts from his terminal which illustrate the effect of the WRITE and FORMAT statements, thereby making their meanings unambiguous.

I believe that other microwave engineers will share my admiration for the broad value of this book, and will consult it frequently regarding various aspects of microwave circuits, either with or without semiconductors as elements.

Seymour B. Cohn
Tarzana, California

January 1977

Preface

Theodore Saad suggested I write this book; I am grateful that he did.

This book's publication, through the generous cooperation extended by Microwave Associates, Inc., is a demonstration of their commitment to professionalism.

Claire Alterio finished most of the manuscript and assisted Neal Vitale in its coordination. The initial chapter manuscripts, begun much earlier, were prepared by Jan Beland and Eleanor Walsh. Perhaps only one who has attempted a book knows how deeply I appreciate their skills, enthusiasm, and perseverance in the pursuit of the final manuscript.

As I thought of possible content, the things which I selected, in relative importance, were:

1) *The effects and methods of computer aided design.*
2) *A description of semiconductors from the microwave engineer's viewpoint.*
3) *A straightforward specification for PIN diodes.*
4) *A means for establishing PIN power ratings.*
5) *A convenient-to-use summary of microwave math techniques, including FORTRAN IV, matrix algebra, even and odd mode analysis, waveguide impedance definition (through Green's functions), and development of three-port network theory.*
6) *A text showing fundamental limits, circuit techniques, and practical results for semiconductor switches, duplexers, and phase shifters.*
7) *Collected formulas, constants, and design graphs.*

Only the reader can say how effectively these objectives have been realized. I hope that the book proves useful enough to warrant eventual revision; therefore, suggestions, comments, and noting of errors that are directed to me will be especially appreciated.

Joseph F. White
Lexington, Massachusetts

November 1976

Introduction

This book is intended to be an introductory text and reference for the design and analysis of microwave semiconductor control circuits, and is written to be read on either of two levels.

At the general insight level, the reader can skim this book, obtaining an introduction to the subject. One may be interested, for example, in determining what a semiconductor is, how lifetime is defined, why it is important, and what circuit performance it affects. From a circuit viewpoint, one may want to find out how switches, duplexers, or phase shifters function, and what need be specified about them. This reading requires only a general working understanding of the microwave field. All special terms and definitions are printed in italic type within the text.

At the design level, the reader seeks enough information to recap the state of the art on one or more detailed matters in semiconductor control, either because there is a need to analyze various approaches to a very demanding application or there is a desire to extend the art to realize performance not obtained previously. For this reader, the most fundamental exposition of the subject has been included, for which conversance with complex circuit analysis is needed. A bachelor's degree in Electrical Engineering or Physics should provide sufficient background, since most microwave analysis methods, including use of the Smith Chart, are derived directly in the text and the appendices.

Chapters I-III introduce the semiconductors themselves, evolving the physical principles behind semiconductors, the PN junction, rectification, the PIN diode, and the characteristics which affect microwave performance. The concepts of doping, hole and electron carriers, lifetime, mobility, charge, punchthrough, breakdown, depletion, C(V) law, dielectric relaxation frequency, microwave resistance determination through charge control modeling, transit time, microwave breakdown, parasitic reactances, and thermal time constant modeling are defined and illustrated with examples. This introductory treatment develops the physical reasoning and the measurement definitions and techniques behind semiconductor operation with which the designer must be familiar. It culminates with data for typically available diodes.

Intimately related areas, often overlooked in device treatments, receive special attention. *Chapter IV* is devoted to drivers for PIN devices, defining balanced and unbalanced TTL logic, showing how diode and driver are chosen for switching speed, and describing measurements, built-in fault detection (BITE), pulse leakage considerations, and complementary switching.

The fundamental limits of control circuitry are derived, based upon the general properties of three-port networks, in *Chapter V.* Not only are the power handling and insertion loss relationships useful to control circuitry, their derivations demonstrate practical uses of complex variable theory and the impedance matrix to establish general circuit results of practical importance.

No field is more open to mathematical analysis than is microwaves. Prior to the last two decades, nearly all scientific calculations were performed by hand; today, nearly all are, or can be, done by computer. The former method imparts more understanding of the physical behavior, but the latter can provide more speed, efficiency, and accuracy. The experienced scientist and engineer understands that reliance on both is necessary, and more importantly, that the effective mathematical approaches for each are different.

Chapter VI – Mathematical Techniques introduces, in terms readily assimilated by the electrical engineer or physicist, matrix algebra with definitions and a library of functions needed for most circuits; even and odd mode analyses demonstrated by an evaluation of the phenomenal "backward wave coupler;" FORTRAN IV programming, most practical for microwave engineering because of its capacity for direct complex number manipulation; and procedures for generating interactive microwave analysis computer programs, including a complete program example capable of analyzing general two-port networks. This program is then used throughout the remainder of the book to analyze most of the circuit examples.

The last three chapters describe the theory and practice of semiconductor control circuit design. *Chapter VII – Limiters and Duplexers* describes nonlinear limiting in terms of microwave frequency and semiconductor diode properties. A detailed description of avalanche limiting using bulk semiconductors is also presented. Waveguide diode limiter circuit evaluation is performed in a manner which demonstrates how to use the waveguide Green's functions. The bulk element analysis shows how the general three-port theory derived in Chapter V can be applied. Practical designs of actual limiters in coax, waveguide, and integrated circuit microstrip are given. These examples show tuning and its effectiveness using either fixed capacitance and inductance or distributed tuning to resonate the reactances of the diode and bulk semiconductor elements. Design data are presented in universal curves for easy determination of ultimate device frequency, loss, and isolation limitations.

Chapter VIII – Switches and Attenuators shows how semiconductor elements can be embedded within filters, thereby permitting a large body of filter theory to be useable for control circuit design. Practical results with an octave bandwidth, high power switch demonstrate not only the filter approach, but also the effectiveness of "quarter wavelength spacing" and stagger series tuning of diodes for increased isolation bandwidth. Waveguide, stripline, coaxial, and microstrip circuit media operating from 0.5 to 18 GHz are covered within the demonstration of switching circuit techniques. Packaged beam lead individual diode chips as well as bulk effect window devices are used in the examples given to illustrate the different semiconductors which are possible.

Chapter IX – Phase Shifters contains complete analysis of digital phase shifters using the switched line, loaded line, lumped element, and reflection circuits, with practical examples from 0.4 to 15 GHz. Techniques for increasing bandwidth and power handling capacity are shown. Continuous phase shifters are treated, including derivation of the maximum phase shift per decibel of loss, nonlinear power limits, and varactor selection.

Appendices A-J include charts and computer programs to evaluate the characteristics of various transmission lines and couplers, the frequency response of bias blocks and returns, material properties, often used constants and formulas, and a review of the basis and use of the Smith Chart.

Table of Contents

Chapter I

The PN Junction

A. The Need to Understand Low Frequency Diode Behavior

The use of PN and PIN diodes to switch microwaves is a technology that is over 20 years old. [1, 2] Within that period, many engineers involved in microwave switching have observed that a small dc forward bias current of 100 m/A is sufficient to turn on a PIN diode such that it is essentially a short circuit to microwave signals with current amplitudes of tens of amperes, and that a -100 V dc reverse bias is sufficient to hold the diode in a nonconducting capacitance state even though the peak RF voltage impressed upon it may be as large as 1000 V. This operation is shown diagrammatically in Figure I-1. It would be an understatement to say that such a situation is useful as well as remarkable. But the reasons underlying it often are given too little, even inaccurate, thought. Misleadingly, this important behavior is often explained by an inertial argument – "The diode conduction mechanism cannot respond rapidly enough to follow the RF waveform; hence, its impedance state is established by the dc bias."

The problem with this inertia model is that it is such a plausible half-truth that few microwave circuit engineers look further into the diode's conduction mechanisms. The inertia model for the diode physics is an oversimplification even on a qualitative basis. It relates only to the carrier diffusion and drift process within the diode. It cannot, therefore, be used even for the qualitative description of carrier generation by impact ionization which establishes the maximum RF voltage sustainable by the diode in reverse bias, above which even the short half-period duration of the microwave voltage can destroy a reverse biased diode.

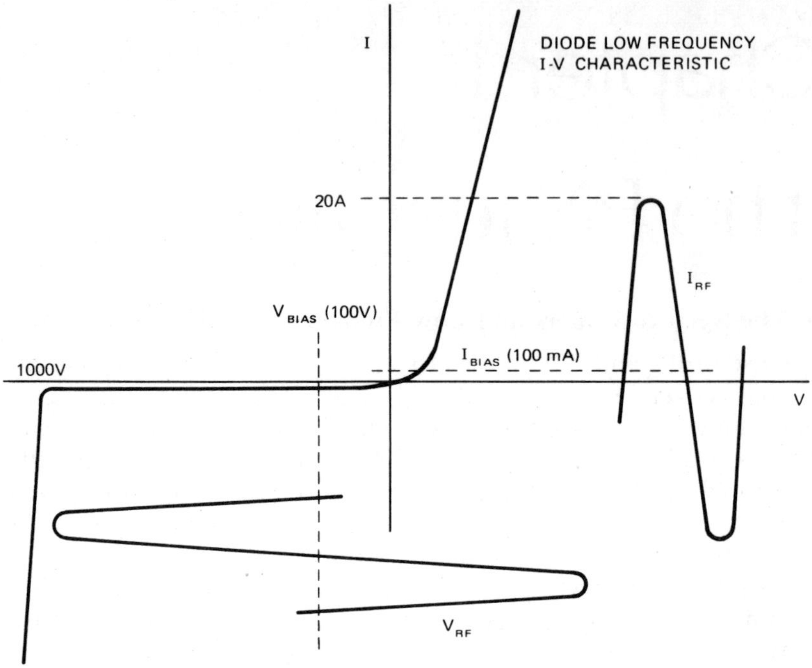

Figure I-1 **Diode Low Frequency Characteristic with Superimposed Bias and RF Excitations**

The inertia model can not be used to relate the actual microwave resistance of a PIN diode to the dc bias current, either. Indeed, when the need arises to estimate the microwave resistance at forward bias, one whose mental image of the diode is based simply on the inertia model encounters an irresistible temptation to say that "RF resistance can be approximated by the slope about the forward bias point of the static I-V curve" (Not shown in Figure I-1 to avoid possible proliferation of erroneous reasoning). This fallacy is common despite the fact that such an approximation is an implicit contradiction of the inertia model simplification.

How, then, is the diode's microwave resistance estimated? In Chapter II it is demonstrated that a "charge control" model is required. In this chapter, however, we first develop a basic appreciation of the diode conduction process (the treatment used to show the rectifying character of a PN semiconductor junction) supplemented by those semiconductor properties needed in Chapter II to describe the PIN. As such, the PN varactor description given in Chapter I leads to the PIN model described in Chapter II.

B. Silicon — The Semiconductor

1. Intrinsic Conduction

The purpose of this section is to provide a qualitative model for the conduction process in a PN junction. The circuit designer need not have a quantitative description [3, 4] of PN junction physics to appreciate both the nature of the rectifying junction and the time required for its conductivity to be modulated by the diffusion and drift of carriers.

To begin the description of the semiconductor PN junction, some theoretical starting point must be selected, usually the physical description of the atom with its surrounding electron orbitals. Here we shall try to pass rapidly over the atomic structure referring the reader to References 3 and 4 for a more detailed presentation. The object is to develop a good qualitative understanding of, first, the conduction process of the PN junction, and, later, of the PIN diode, which is a variation on this device that proves so useful in microwave switching.

Semiconductors such as germanium and silicon have two important characteristics which make the realization of a solid-state diode possible. First, Column 4 elements such as germanium and silicon have four electrons in their outermost occupied (valence) band, yet only a moderate activation energy is required to move an electron to the next outermost (conduction) band. As solids, adjacent atoms *share* electrons, forming covalent bonds. Eight electrons can occupy the valence band around each atom, but the four host atom electrons together with four shared neighboring electrons occupy all of the vacancies. A second important characteristic arising from the covalent bonds is that these elements form regular crystalline solids with atoms so close together (about 1 Å center-to-center) that neighboring atoms' electron "orbits" overlap. Thus, an electron, when once imparted with the necessary energy to reach the conduction band, is able to move from atom to atom through the regular crystalline lattice quite freely, thereby providing current flow.

So far, the picture of a semiconductor is one of a crystal in which the electron flow can occur readily, provided that the electrons can be somehow induced to occupy the normally unfilled outer energy (conduction) band about their host atoms. One method of achieving such electron activation is by heating the semiconductor. In this way, the average energy of electrons surrounding the atoms is increased causing some to be excited to higher energy states in the conduction band. Such a process is not useful to the operation of a silicon rectifier; in fact, it is a deleterious effect which occurs when the rectifier is used at high temperatures. Nevertheless, it is one of the methods by which electrons can be excited to the conduction state.

Another means of promoting electrons to the conduction band is by having photons of light energy absorbed by the crystal. The resulting increase in energy can raise electrons from the valence to the conduction band. Such photon stimulation of the electrons is also a means of increasing the conductivity of a semiconductor. It has, however, no practical utility in the operation of a silicon rectifier.

These two mechanisms result in *intrinsic* conduction, a diagram for which is shown in Figure I-2. The electron excited to the conduction band leaves behind a vacancy that could be occupied by a valence electron from a neighboring atom. Filling this vacancy (called a *hole*) creates a new vacancy in the neighboring atom. This moving vacancy results in a net current flow similar to that of a positive mobile charge. To distinguish it from the electron flow in the conduction band, it is called *hole flow*. Intrinsic conduction

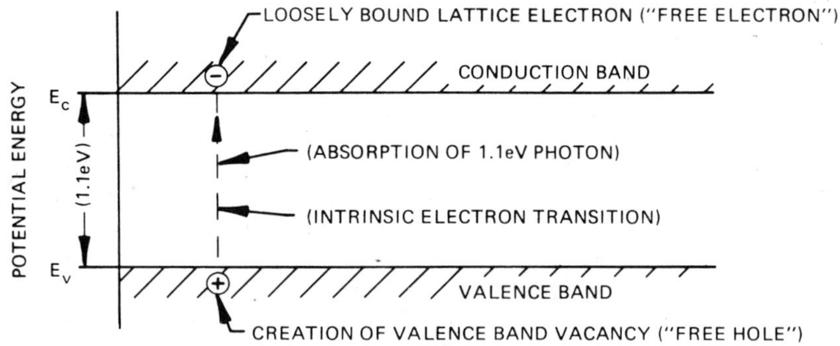

Figure I-2 **Energy Band Schematic Representation for Pure Silicon Showing (Intrinsic) Creation of an Electron Hole Pair of Free Carriers**

by such simultaneously produced electron-hole pairs is parasitic both in conventionally used microwave diodes and lower frequency rectifier diodes. An alternate method for creating separate electrons or hole mobile carriers is necessary for the PN junction device to be described shortly.

2. *Extrinsic Conduction*

The means for promoting an electron to the conduction band *extrinsically* represents an ingenious piece of engineering. By embedding atoms within the silicon host structure (*doping*) whose conduction band contains five electrons (Column 5 elements), a source of electrons requiring little additional energy for occupancy of the conduction band can be obtained on a static basis and without the simultaneous creation of a valence band hole. Ready availability of the electron is possible because the fifth electron is not needed in the covalent bond and is, therefore, bound weakly to its host atom. Strictly speaking, any of the Column 5 elements might be suited to this purpose. A good match to the semiconductor's crystalline lattice geometry and the resultant practical rates at which the impurity can be diffused into the silicon lattice make phosphorus a commonly used dopant to introduce this source of electron donors (called *donor* or *N-type* impurities) into silicon PN diodes. Donor doping is diagrammed in Figure I-3.

The corresponding introduction of an extrinsic hole is also possible (as shown in Figure I-4) by the addition of a suitable dopant; in this case, one whose valence band contains only three electrons. Thus, elements listed in Column 3 of the Periodic Table serve as *acceptor* dopants. Practically, boron* is a widely used acceptor dopant in silicon PN junctions.

The efficiency of these dopants in providing electrons to the conduction band (in the case of donor dopants) or accepting electrons from valence band (in the case of acceptor dopants) can be surmised by recognizing that the energy required for an electron transition from the valence band to the conduction band in silicon is approximately 1.1 eV. However, the energy required to stimulate an electron from a phosphorus donor atom into the conduction band of the silicon crystal lattice, creating a mobile electron,

*There is a handy mneumonic for remembering which is the donor and which is the acceptor. Donor materials are called N-type and acceptor materials are called P-type. Of the names boron and phosphorus, one contains an N and the other contains a P. The mneumonic is to remember that boron would be expected to be N-type and phosphorus to be P-type, but this conclusion is *exactly wrong!*

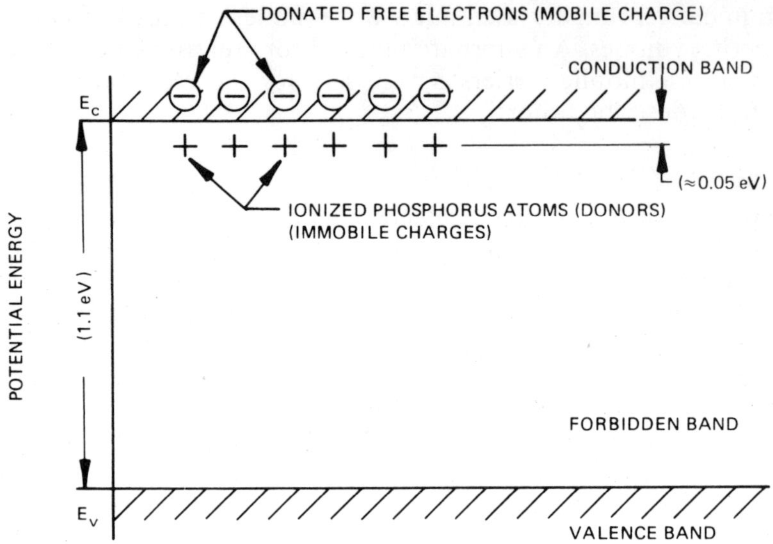

Figure I-3 **Energy Band Schematic for Phosphorus (N-Type) Donor Impurity in Silicon Showing Creation of Mobile Electron Charge Carriers in Conduction Band of Silicon Lattice**

is only about .05 eV. Similarly, only about .05 eV is required to promote an electron from the valence band of a host silicon atom, creating a mobile hole, into the conduction band of a boron dopant atom. This small amount of energy is readily present as thermal activation at normal ambient temperatures. Such activation of the dopant atoms is called *dopant ionization*, since the loss of an electron by a donor or the gaining of an electron by an acceptor leaves the immobile dopant atom electrically charged. Although the dopant atoms are not free to move within the silicon crystal, their loss of electrical neutrality produces an electric field within the crystal *(space charge),* described by Poisson's equation, the effect of which we see in the description of the rectifying junction. Because of the need for some thermal activation, dopants as contributors to conduction within silicon lose their effect if the silicon crystal is cooled down sufficiently. However, for boron and phosphorus in silicon, statistically over 95% of the dopant atoms are ionized (and hence operative) at temperatures above -40°C, the lowest operating temperature of most practical electronic applications.

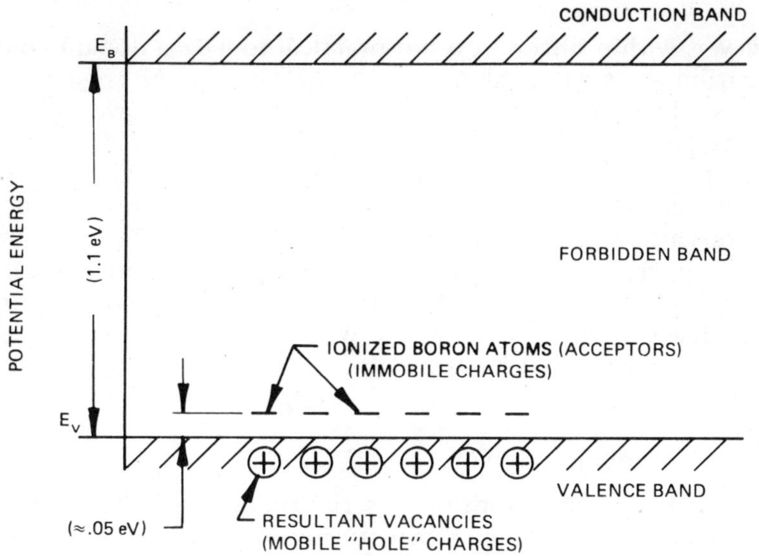

Figure I-4 **Energy Band Schematic for Boron (P-Type) Acceptor Impurity in Silicon Showing Creation of Mobile Vacancies (Holes) in Valence Band of Silicon Lattice**

Reviewing our model for the semiconductor building block from which silicon diodes are made, we have seen that silicon is substantially a non-conductor until the presence of some conduction contributing mechanism such as heat, photon absorption, or atomic doping is introduced. It is because of the inherent *non-conduction property which is easily converted to a conduction property* that silicon is called a *semiconductor*. Before advancing to the discussion of the PN junction, it is worthwhile to gain an engineering appreciation for the order of magnitude of doping needed to make silicon a conductor in a practical diode device. To make this evaluation, it is illustrative to compare silicon-the-semiconductor with copper-the-conductor.

A crystal of silicon contains an orderly three-dimensional arrangement of atoms located on approximately 2.3 Å (2.3×10^{-8} cm) centers. It follows that there are then approximately 10^{23} atoms/cm^3 in a silicon crystal. Within copper, the atomic density is comparable. Thus, if each copper atom can contribute one electron to current flow, there will be 10^{23} electrons/cm^3 available for conduction. The resistivity of copper is 1.7 x 10^{-6} Ω-cm.

However, within silicon, it is not practical to have a dopant concentration greater than about 10^{19} to 10^{20} before the crystal lattice is so disturbed that the charge *mobility* (i.e., the average drift velocity per unit of applied electric field) is so inhibited that further dopant density actually decreases conductivity. Thus, even in the most massively doped silicon material, the concentration of electrons available for conduction is less than 1/1000 of that present in copper; the lowest bulk resistivity is greater than .001 Ω-cm, typically 0.1-1.0 Ω-cm. Nonetheless, this carrier concentration makes device fabrication practical. With these properties of silicon and its conduction mechanisms in mind, we now consider the PN junction.

C. The PN Depletion Zone

This discussion begins with the description of an idealized silicon PN junction model, shown schematically in Figure I-5(a). The silicon host crystal represents an orderly array of atoms bound together by the covalent, electron-sharing bonds, involving the four outermost electrons of each silicon atom. The drawing shown in Figure I-5(a) is a schematic only and is not to be taken as numerically representative; this fact is evident in the discussion that follows. The presence of boron and phosphorus impurity atoms is also depicted in Figure I-5(a). These atoms are *substitutional* in that each replaces a silicon atom site in the crystal – a necessity if they are to be electrically active in contributing or accepting electrons. In practice, some dopant atoms are located *interstitially* in the crystal – they do not contribute carriers, and cause lattice imperfections which reduce electrical conductivity and carrier lifetime (described later). One liberty taken in Figure I-5(a) is the square array of atoms. The actual atomic arrangement in silicon has one silicon atom each at four of the eight corners of a cube with another silicon atom at the center. Each silicon atom shares one of its four electrons with four different neighbors and the resulting *diamond structure* imparts mechanical hardness (the same structure which gives diamond its hardness). But, for present purposes, our model is adequate.

A second liberty taken in the drawing of Figure I-5(a) is the distribution of impurity dopants. In order to show enough of the crystal in cross-section to reveal at least one of each kind of dopant in this PN junction example, we actually would have to show

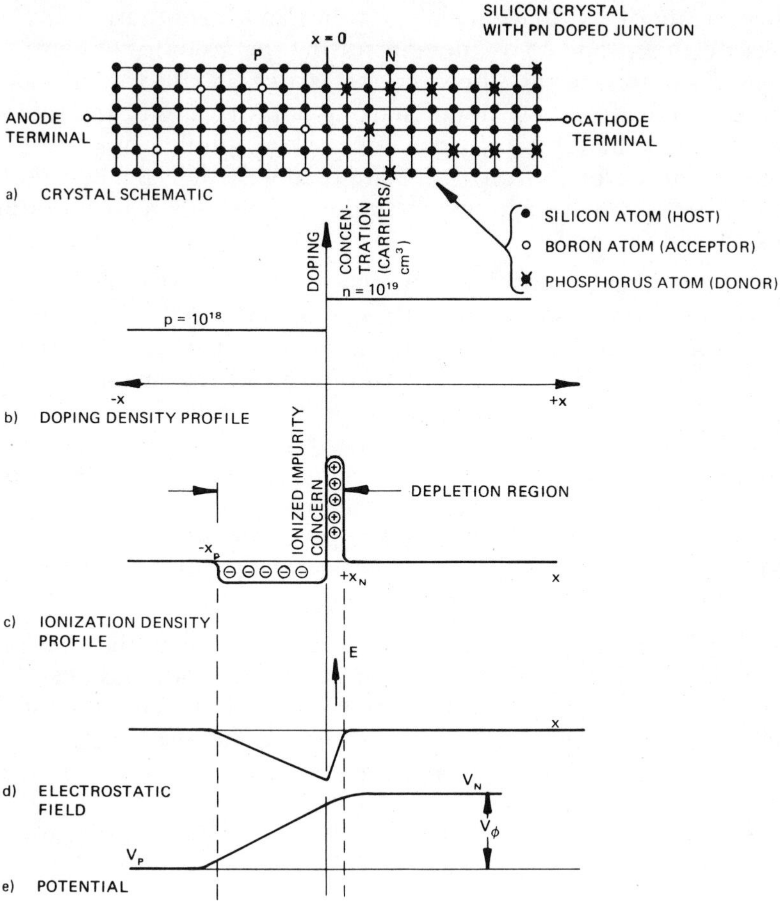

Figure I-5 **Ideal PN Junction Details**

about 100,000 atom sites for the P region (whose doping concentration is 10^{18}) and 10,000 of the atomic sites in the N region (with doping concentration of 10^{19} atoms/cm^3).

The linkages among atoms in the crystal are shown schematically as vertical and horizontal lines connecting the atoms which compose this crystalline lattice. An electron from a phosphorus donor atom could leave to occupy an energy orbital surrounding any one of the four adjacent silicon atoms (unless one of these sites is already occupied by another phosphorus impurity — a fairly unlikely situation). From this new silicon host site, the electron can then move to any of its adjacent silicon atoms and from there to another neighbor, and so on, demonstrating the mobility of this *donated electron* in the silicon crystal lattice. In this way an electron can move from any donor starting point to any of the silicon atoms within the crystal. By the same token, a hole can move from any acceptor (boron) atom site to any of the valence band energy states which surround every one of the silicon atoms in the crystal.

At room temperature, there is enough thermal energy present that any of the countless random paths one would care to describe from atom to atom in the crystal has some probability of actually being taken by a hole or an electron. In one of the smallest of diodes whose active silicon volume is approximately that of a cylinder 1 mil in diameter and 5 mils high there are more than 10^{18} atoms, and as many as 10^{15} mobile charges. Clearly, a *microscopic* evaluation describing the paths of *each* charge carrier of such a device is out of the question and therefore a shift to the *macroscopic* description (describing the overall effect of the sum of such carrier motions) of the device is necessary.

Figure I-5(b) shows a possible profile of the impurity distribution in an ideal PN junction. In this example the N region has been shown more heavily doped than the P region, but virtually any combination of doping densities for the two regions is possible. In all cases, a *rectifying junction* (i.e., *one which passes current in one direction with less applied voltage than required for the opposite current direction*) is obtained, but the capacitance versus voltage profile (to be described shortly) as well as the current density for a given applied voltage is characteristic of the particular doping concentrations in the two regions.

For purposes of illustration, another liberty has been taken in the drawing. The two impurity concentrations are shown with an abrupt dicontinuity both in impurity type and concentration at the "PN junction." In practice this situation is never realizable. The PN junction may be created by diffusing one impurity type into a silicon host which already contains the other type, or it may be formed by direct crystal growth of a doped layer on one of the opposite type. The latter, a crystalline formation method, is called *epitaxial growth*. In any case, the actual junction is formed at that region where, *on the average*, there are equal quantities of both impurities. In such a region, the electrons contributed by donors are virtually all captured by acceptors, the net number of mobile charges is zero, the impurities are said to be *compensated*, and there is no *net* contribution of mobile charge. This region of the semiconductor, from a conduction point of view, is similar to undoped silicon. The lifetime of a mobile charge passing through such a region, however, is less; more is said about lifetime in Chapter III.

Having cautioned with respect to the idealness of our diagram in Figure I-5, we can now discuss the first aspect of a PN junction; namely, its static potential, V_ϕ. To develop a physical picture for this situation, consider what would happen if the crystal idealized in Figure I-5(a) is first cooled to a temperature near absolute zero. Under this condition, all of the electrons associated with the donors remain with them since they do not have enough thermal activation to leave these parent atoms. Similarly, no electrons surrounding silicon host atoms are able to enter the valence bands of any neighboring acceptor dopant atoms. If an ohmmeter is then connected to the anode and cathode terminals of the crystal, a high resistance will be experienced regardless of with which polarity the ohmmeter hook-up is made. Next, imagine that the crystal is allowed to warm up to room temperature. At this time, electrons surrounding donor atoms obtain sufficient thermal energy to leave their parent atoms and to roam about (*diffuse*) through the silicon crystal. In like fashion, holes may also move through the crystal.

Consider the boundary or junction between the P and N regions. Mobile charges diffuse randomly in all directions but, *on the average*, holes will move from the P to the N region simply because there are fewer holes in the N region than there are in the P region. By the same token, electrons will move from the N region to the P region. Such diffusion motion is similar to that which occurs when an amount of gas is released into the atmosphere. Given enough time, the gas sample will distribute itself evenly throughout the limits of the atmosphere's boundaries. Gas molecules move in random directions; on the average, the net flow is from an area where the gas density is great to one where it is less dense.

Unlike gas equilibrium, however, a uniform distribution of holes and electrons throughout the whole crystal is inhibited by the *space charge* which such diffusion causes. Each time a negatively charged electron crosses the junction into the P region, it leaves behind a positively charged immobile donor atom. This atom is too massive to move through the lattice, and it couldn't move, even if it were less massive, because it is bound to its neighboring silicon atoms by the covalent bonds which hold the whole crystal together. Similarly, each positively charged hole which crosses into the N region leaves behind a negatively charged acceptor atom — likewise immobile within the lattice.

The recombination rate of a given concentration of carriers is proportional to the concentration itself; therefore, the resulting recombination is an exponential decay with time, in a manner similar to the discharge of a capacitor through a resistor. The time constant of this decay is defined as the *carrier lifetime,* τ, analogous to the R-C time constant of the capacitor-resistor. Since the average time in which a carrier is available for conduction is approximately equal to τ, the term "lifetime" is appropriate. Lifetime is also dependent upon the silicon crystalline condition and boundaries, the effect of which is to introduce other recombination and carrier storage sites (called *traps* because they prohibit conduction by immobilizing the carrier). Thus, although the lifetime concept is simple, its theoretical quantitative application is not. What happens to an electron that diffuses from the N into the P region? It combines either with a free hole or, more likely, with an ionized donor atom. Likewise, a hole which diffuses from the P to the N region recombines, usually with an ionized acceptor atom. *The recombination process annihilates electron and/or hole free carriers.* Due to the plentiful number of doping impurities in a PN junction, the recombination process takes place very rapidly (on the order of nanoseconds or microseconds). Thus, the "low-frequency" I-V characteristic of a PN junction with heavily doped P and N regions actually can apply in a frequency range in the tens of megahertz.

The charges on the ionized, immobile impurities on either side of the PN junction restrict the extent of mobile hole and electron diffusion. Consider the electric field which is developed by these opposing P and N region ion formations, shown schematically in Figure I-5(c). From Poisson's Equation, the resulting internal electrostatic field within the region of the ionized impurities (Figure I-5(d)) is determined. The number of ionized donors equals the number of ionized acceptors on the two sides of the junction, and the zone from $-x_P$ to $+x_N$ is called the *depletion zone* because few mobile charges can remain within it. Any minority holes or electrons which diffuse to its boundaries are rapidly conducted across the depletion zone. The boundaries, $-x_P$ and $+x_N$, are such that the net number of ionized donors equals the net number of ionized acceptors, a necessary definition of the depletion zone boundaries if the electrostatic field is to be zero outside of the depletion zone.

D. Junction Potential

The E field diagram and its potential function integral (Figure I-5(e)) create a paradox. The potential V_N is greater than V_P; for a silicon junction this difference is about 0.7 V. One might ask what happens if the two ends of the crystal are connected together by some conducting path, as diagrammed in Figure I-6. *Will a current flow indefinitely, producing a source of perpetual motion? Or will the junction be "shorted out" and the internal field collapse?* Neither of these events occurs. When contacts are made to the P and N doped ends of the crystal, static charge dipole layers are created at the contacts due to the differences in free carrier potentials between the contact materials and the doped silicon. The algebraic sum of the potentials at these contacts is exactly the negative of the PN junction potential (assuming uniform temperature throughout the circuit); hence, no net current results from the internal electrostatic field even when the external diode leads are "shorted."

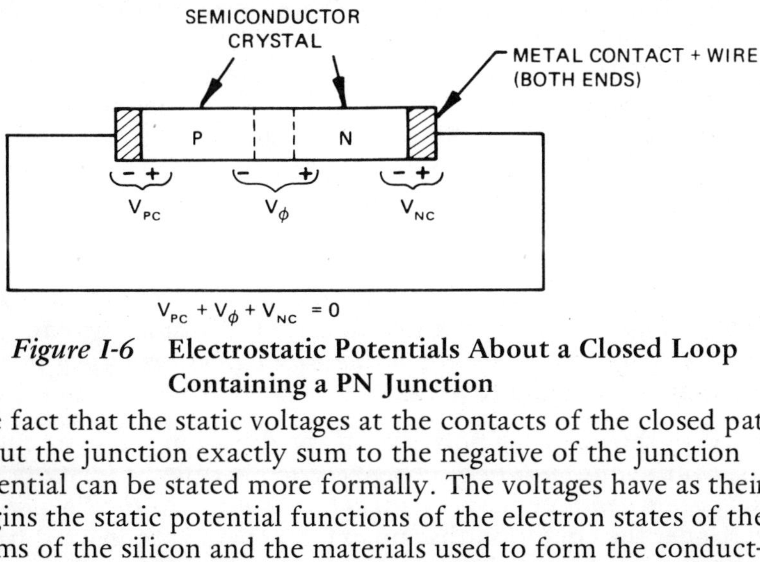

Figure I-6 **Electrostatic Potentials About a Closed Loop Containing a PN Junction**

The fact that the static voltages at the contacts of the closed path about the junction exactly sum to the negative of the junction potential can be stated more formally. The voltages have as their origins the static potential functions of the electron states of the atoms of the silicon and the materials used to form the conducting path. The ensemble of these potential functions must, by the Second Law of Thermodynamics, compose a conservative force field such that (neglecting frictional losses) the average charge carrier completing any closed path circuit within the system of atoms forming that path neither acquires nor loses energy.

E. Diffusion and Drift Currents

As described in the previous section, the electrostatic potential, V_ϕ, results from the diffusion, or migration, of holes and electrons on opposite sides of the junction from the respective regions where they are numerous to those where they are few; namely, holes diffuse from the P to the N zone and electrons diffuse from the N to the P. Such diffusion represents a *net current flow* and is referred to as the *diffusion current*, I_{DIFF}, as shown in Figure I-7. It results from the flow of *majority carriers* from the respective P and N zones from which the two types of diffusing carriers originate, where they represent respectively the majority charge carrier type. Were there no other mechanism of charge transfer across the junction between the P and N zones, this current would exist only as a transient during the physical formation of the junction. The junction potential created as a result of the diffusion would soon prohibit further diffusion, bringing the charge transfer and hence the magnitude of I_{DIFF}, to zero.

In actuality, however, there are some electrons in P-type material and some holes in N-type material. These charge carriers are appropriately referred to as *minority carriers.* Their presence arises because there is always some thermal activation of intrinsic electron-hole pairs, as described earlier in the discussion of silicon as a semiconductor material. Since their polarity in the respective P and N zones is opposite to that of the majority carriers (whose diffusion is responsible for the junction potential) it follows that any minority carriers in the vicinity of the junction will be swept across it by V_ϕ and the net current flow produced by this motion would be opposite I_{DIFF}. This current flow is referred to as *drift current,* I_{DRIFT}; since, in contrast to carrier diffusion current, *carriers which move under the influence of an electric field in a semiconductor are said to drift.*

Because of the opposing polarities of the diffusion and drift currents, *continuous motion of charge carriers across the junction exists even when no external connection is made to the PN junction.* Under the condition that no external path is connected to the diode (i.e., the zero bias condition), the external current must be zero; hence, *the net current flow across the junction must likewise equal zero.* Since this current is the sum of the oppositely

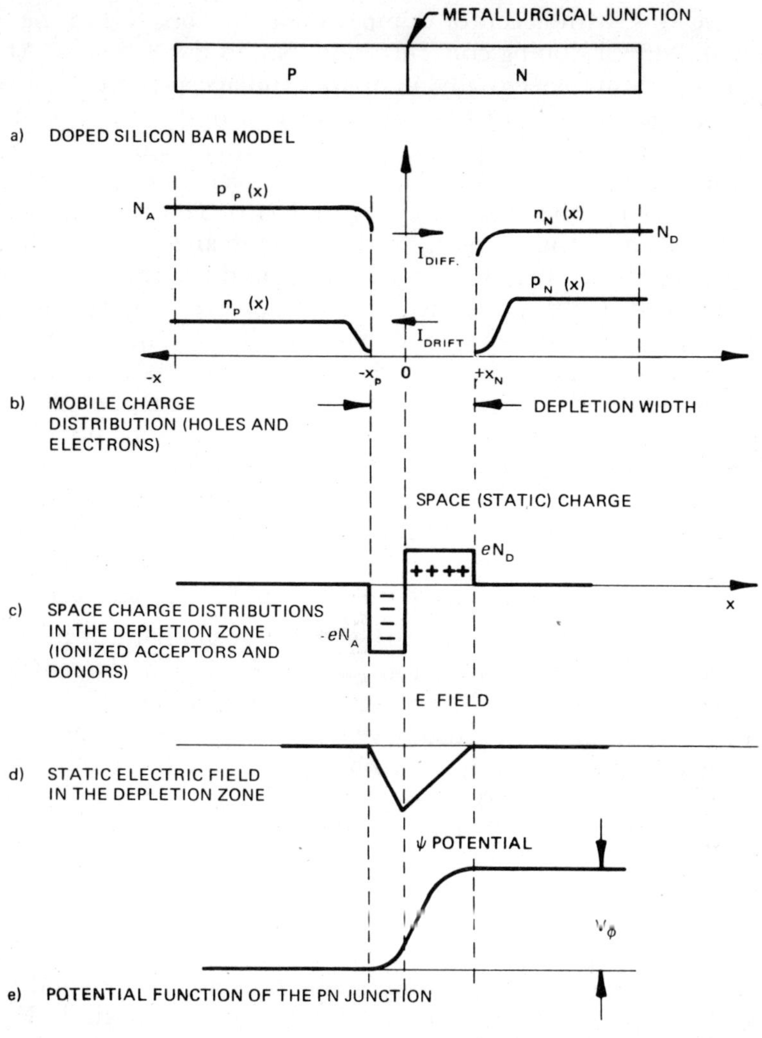

Figure I-7 **Unbiased Ideal PN Junction (After A.B. Phillips [3])**

directed diffusion and drift currents, it follows that they have equal magnitudes under zero bias.

The rectification properties of the PN junction can be described in terms of the change in relative magnitudes of these two oppositely directed currents under the application of forward and reverse bias voltages to the PN junction.

Figure I-7 shows an ideal PN junction consisting of a junction of silicon regions, with acceptor doping concentration, N_A, in the P region and donor doping concentration, N_D, in the N region. At room temperature, essentially all donors and acceptors are ionized; therefore, the majority carrier concentrations in the P and N regions, $p_P(x)$ and $n_N(x)$ respectively, are essentially equal to the acceptor and donor concentrations, except near the depletion zone, as shown in the diagram in Figure I-7(b). The minority concentrations, consisting of $n_P(x)$ in the P region and $p_N(x)$ in the N region are also shown. For this illustration the acceptor concentration exceeds the donor concentration; however, any relative donor and acceptor density combination will result in a PN rectifying junction. The minority carrier concentrations are also shown plotted in the diagram. Although not necessary for the qualitative understanding of the PN junction, it may be noted that the product of free electrons and free holes is a constant in a given silicon sample, the value of which depends on the semiconductor material and temperature.* At room temperature $p \cdot n \approx 10^{20}$ carriers/cm^3. Thus, the greater the majority doping concentration, the smaller the concentration of intrinsic (minority) carriers.

From the diagram it can be seen that the large gradient of majority carriers produces the diffusion current, I_{DIFF}. This current is made up of both holes and electrons. They travel in opposite directions but, because they also have opposite polarities, there is a net positive current flow, I_{DIFF}, from the P to the N region. Once these mobile charges cross the depletion zone they represent *minority carriers* and quickly combine with the opposite carrier type — holes with electrons and vice versa. They are thus rapidly annihilated (by recombination) as mobile charges. Within the depletion zone itself there are no holes or electrons** except for those which are participating in the diffusion and drift currents. The acceptor and donor impurity atoms in this zone are ionized and their static charge produces the electric field diagrammed again in Figure I-7(c). The value of this field, E(x), is determinable from Poisson's Equation (as is demonstrated later in the determination of the C(V) law at reverse bias); it is plotted in Figure I-7(d). Integrating this field magnitude as a function of distance through the depletion zone gives the potential difference between the P and N

*This important result is used in the quantitive derivation of the I-V characteristic for the PN junction.

**Some are thermally generated intrinsically, but their concentration is negligible at normal operating temperatures.

regions, V_ϕ, called the *junction potential*. In the description of rectification, it is helpful to keep in mind that the forward bias current is in the direction of I_{DIFF} (traveling from P to N) and reverse bias current is in the direction of I_{DRIFT}.

Diffusion current, resulting from majority carrier flow, has a magnitude which is *limited only by the junction potential.* If an external bias is applied which tends to reduce V_ϕ, majority carriers from both the P and the N regions will flood across the junction, analogous to the way in which sand pours out of the back of a truck when the gate is opened. Theoretically this current magnitude approaches infinity as $V_\phi \rightarrow 0$, but practically, ohmic losses prevent application of a bias sufficient to reduce V_ϕ to zero.

On the other hand, *the drift current is limited in magnitude by the quantity of minority carriers which diffuse close enough to the junction to come under the influence of the depletion zone field,* at which time they drift (i.e., are moved by the electric field) across it. The magnitude of this current is limited, regardless of how high a voltage is applied across the junction. For this reason this drift current, observed under reverse bias, is called the *reverse saturation current,* I_{SAT}.

Figure I-8 shows diagrammatically the current distributions under forward and reverse bias conditions. To summarize the current properties of the PN junction, the net current across the junction is made up of I_{DIFF} and I_{DRIFT}. In turn these two oppositely directed currents are themselves made up of two separate hole and electron current components, I_P and I_N. The hole and electric components of either I_{DIFF} or I_{DRIFT} result from the motion of holes and electrons which, because they have opposite charge polarities, cross the junction in opposite directions but result in a common current direction when they are composed entirely of either majority carriers, as in the case of I_{DIFF}, or minority carriers, as in the case of I_{DRIFT}. The directions of these current components are shown in Figure I-8.

F. Rectification and the I-V Law

1. Assumptions

The operation of the PN junction under bias is usually described under the following assumptions [3] :

1) The ohmic resistance of the P and N regions are negligible and hence the electric field is zero except in the depletion zone.
2) No recombination of carriers occurs within the depletion zone.

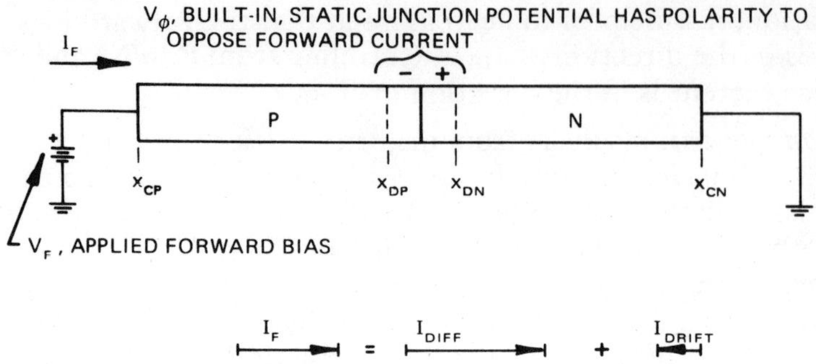

a) FORWARD BIAS DIODE CURRENT POLARITY SHOWING DIFFUSION AND (OPPOSITELY DIRECTED) DRIFT COMPONENTS

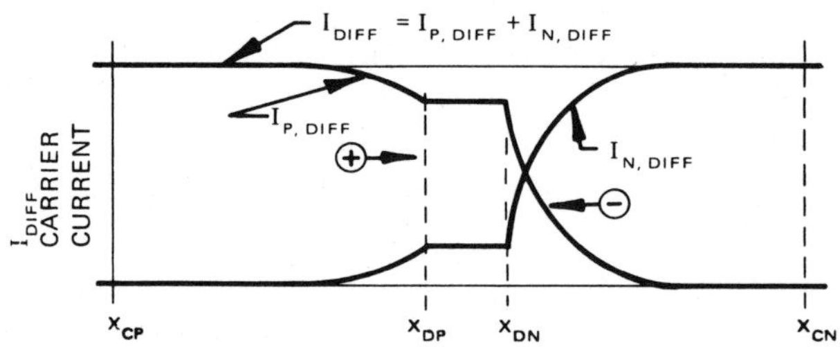

b) CARRIER CURRENT (I_{DIFF}) WITH HOLE ($I_{P, DIFF}$) AND ELECTRON ($I_{N, DIFF}$) COMPONENTS

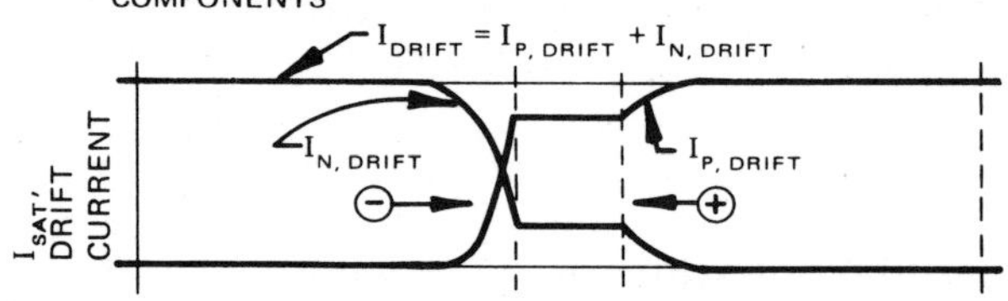

c) DRIFT CURRENT (I_{DRIFT}) WITH HOLE AND ELECTRON CURRENTS

Figure I-8 **Drift Current (I_{DRIFT}) with Hole and Electron Currents**

Recombination occurs when an electron in a conduction band drops to the valence band and combines with a hole, a process which eliminates the mobility of, and therefore annihilates the pair of charge carriers.

3) The minority carrier *diffusion lengths* (i.e., the distances travelled in a time equal to the lifetime) in the P and N region are

shorter than the lengths of these regions so that minority carriers injected into these regions after crossing the depletion zone recombine as necessary to establish equilibrium levels before the ohmic contacts are reached.

4) The concentration of diffusing carriers which cross the depletion zone (referred to as the injected minority carrier concentration) is small compared to the majority carrier concentration in that region. In other words, it is assumed that the P and N regions are so heavily doped that the forward current is carried through the depletion zone by a density of injected carriers which is small compared to the majority carrier concentrations of the P and N regions.

2. *Forward Bias*

Under the preceding assumptions, it is possible to discuss the current flow across the junction in simple terms. If, as shown in Figure I-8(a), an external voltage, V_F, is applied whose polarity is opposite the internal static potential of the PN junction, V_ϕ, the effect of the applied voltage is to reduce the internal field and to increase the diffusion current. The resulting net current, I_F, is the sum of the electron and hole majority carrier currents injected into the depletion zone, less the oppositely directed drift current, I_{SAT}. Referring to Figure I-8(b), the relative percentages of the total diffusion current $I_{DIFF} = I_{P,DIFF} + I_{N,DIFF}$ carried by holes and electrons respectively varies between the two contacts of the diode as shown in Figure I-8(b).

Suppose one begins at the P side contact, x_{CP}. At this contact, the current flow is made up entirely of holes $I_{DIFF} \approx I_{P,DIFF}(x_{PC})$. Moving then toward the depletion zone boundary x_{DP}, a portion of the hole current begins to recombine with electrons (minority carriers) which have been injected from the N region. The resulting recombination serves to absorb some of the hole carriers (moving left-to-right in Figure I-8). The remaining portion of these holes is injected into the depletion zone and crosses to the N region, x_{DN} — by assumption without undergoing recombination within the zone. Once in the N region, the injected concentration of holes, $p(x_{DN})$, represents a surplus of minority carriers in the N region, encounters abundant electrons, and recombines rapidly. In this region, there is (by assumption) negligible electric field, and motion of both holes and electrons is by diffusion. Because of the abundance of electrons in this region, the holes do not travel far before they recombine with electrons. The diffusion length of holes in the N region, L_{PN}, is very short.

Referring again to Figure I-8(b), we see that the fraction of I_{DRIFT} made up of holes in the N region drops rapidly to zero as one moves from the N edge, x_{DN}, of the depletion zone toward the N contact, corresponding to the distance x_{DN} to x_{CN}. An alternate way of describing this rapid recombination is to say that the *hole lifetime* in the N region, τ_{PN}, is very short. Of course the current through the entire diode must be continuous; therefore, at a given cross-section, whatever current is not carried by holes must be carried by electrons (which move in the opposite direction of current flow because of their negative charge).

This description of current under the forward bias condition was made in terms of majority* carrier diffusion current. Of course a minority carrier current in the opposite direction still takes place even under the forward bias shown in Figure I-8(c), but its magnitude is much smaller than the forward current for any appreciable level of forward bias. This reverse drift current, I_{SAT}, is always present.

Even with large applied forward bias the net voltage across the junction still has the polarity associated with V_ϕ. The astute reader might well ask, if V_ϕ is only equal to about 0.7 V for a silicon junction, why is it that an externally applied voltage V_F having a magnitude greater than 0.7 V could not be applied which would reverse the electric field direction at the junction? To answer this question, one must reexamine the assumptions. When a forward bias is applied, it reduces the junction potential. Majority carriers in both the P and N regions then diffuse across the depletion zone. The magnitude of the current which results approaches infinity as the magnitude of the electric field at the junction approaches zero. In the usual quantitive analysis of the PN junction, it is assumed that there is no ohmic voltage drop in the P and N regions outside of the depletion zone. This assumption, though useful for describing the relationship between diffusion and drift currents, is not valid under large forward currents — the *high injection level* case. There is considerable ohmic voltage drop in the P and N regions, and it is necessary to distinguish between the forward bias voltage

*Confusion can be minimized by observing that the carrier definition (in terms of *majority or minority*) changes in crossing the depletion zone. For example, majority carrier holes in the P region which diffuse across the depletion zone are called *injection minority carriers* since they enter the N region where the majority carrier type is the electron. It is common practice to refer to the PN junction as a *minority carrier device,* because the major current flow (i.e., the forward current, I_{DIFF}) causes the injection of minority carriers into the P and N regions.

applied to the ohmic contacts of the diode, V_F, and the actual *applied junction potential,* V_J, under the high current condition. In practice, for any practical current density with a forward bias the actual junction voltage, V (equal to $V_\phi - V_J$), always has the polarity of V_ϕ, opposite to the polarity of V_F.

3. Reverse Bias

When a reverse bias is applied (Figure I-8(b)), its effect is to reinforce the built-in junction potential, V_ϕ, which prevails before the bias is applied. This rapidly reduces the magnitude of the diffusion current into the depletion zone to near zero. The direction of the field has a polarity which would be expected to increase drift current across the depletion zone but, unlike the diffusion mechanism wherein a large supply of carriers is available, the minority carrier densities available in the P and N regions are small. For reverse current, the participating charge carriers are limited to those which diffuse near enough to the depletion zone boundaries within the N and P regions themselves to come under the influence of the depletion zone electric field and be swept across. The magnitude of this reverse current is limited to the small value, I_{SAT}, equal to the value of drift current which, at equilibrium with zero applied bias, balances the diffusion current across the depletion zone. Thus, it can be seen that the net reverse current, I_R, resulting from the application of a reverse bias voltage, V_R, actually comes about through the extinction of the competing (oppositely directed) equilibrium diffusion current.

4. The I-V Law

The wide disparity in diffusion and drift current magnitudes for a practical silicon PN junction produces an excellent low frequency rectifier. When the conduction mechanisms just described are analyzed quantitatively, it is found that the diffusion current into the PN junction diode increases exponentially with the applied bias, V, (neglecting the ohmic drop in the P and N regions). Reversing the polarity gives reverse bias, causing the diffusion current to decrease exponentially with the same exponential magnitude. The reverse directed drift current remains essentially unchanged with bias of either polarity and has a magnitude just sufficient to balance that of the diffusion current at zero bias. Quantitative analysis [3, 4] shows that the total diode current, I, consisting of the algebraic sum of the diffusion and drift currents, can be expressed by Equation (I-1) in terms of the actual junction voltage, V.

$$I = I_{SAT}\,(e^{eV/kT} - 1) \qquad \text{(I-1)}$$

This relation, known as the PN junction I-V Law, is shown graphically in Figure I-9. It must be emphasized that this relation is derived by neglecting the ohmic voltage drop in the P and N regions and, hence, implicitly assumes that the magnitude of I is small. Of course, the resistance, R, of these regions including the contact resistances can be determined experimentally and the net bias voltage applied to the junction can be estimated, according to Equation (I-2).

$$\text{Effective applied voltage} = V_J = V_F - IR \qquad \text{(I-2(a))}$$

$$\text{Actual net junction voltage} = V_\phi - V_J \qquad \text{(I-2(b))}$$

5. Summary of Low Frequency Conduction

To summarize the conduction behavior of the PN junction as discussed so far, the following observations can be made:

1) Current through the depletion zone consists of diffusion (having *forward* current polarity) and drift (having *reverse* current polarity). Diffusion current results from the spreading of an abundance of (majority) carriers in one region into another region having a lower density of that carrier. For purposes of analysis, *the presence of an electric field is ignored insofar as diffusion currents are concerned.* On the other hand, *drift currents result from the motion of holes and electrons under the influence of an electric field.*
2) Both diffusion and drift currents are, in turn, composed of hole and electron carrier flow.
3) Both hole and electron carriers are considered *majority* or *minority* carrier types depending upon the region being discussed. Holes are majority carriers and electrons are minority carriers in P-type material. Electrons are majority carriers and holes minority carriers in N-type material.
4) Current conduction through the PN junction, including the property of rectification, can be explained in terms of the *modulation* of the depletion zone by an external voltage with resultant change in only the diffusion current magnitude. The saturation current direction and magnitude remain unchanged with bias.

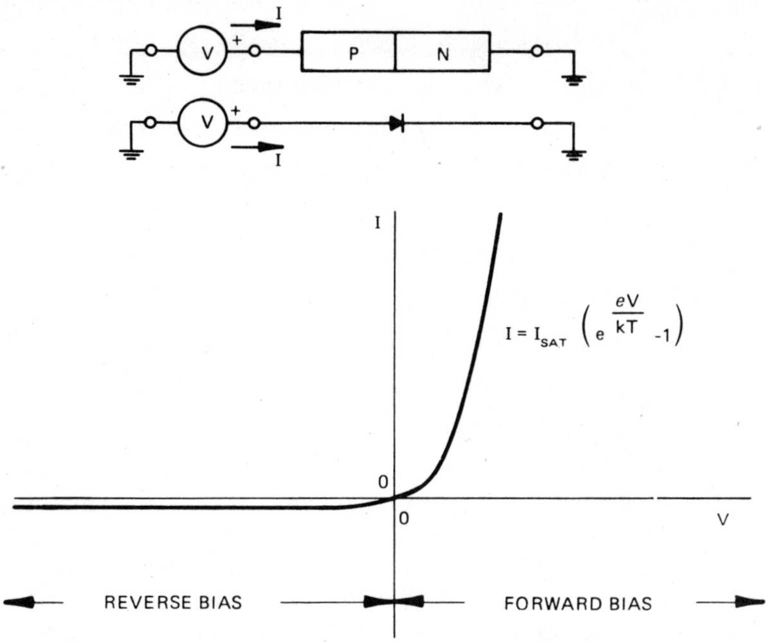

Figure I-9 **Low Frequency PN Junction I-V Law**

The modulation with bias of the charge density within the depletion zone is the basis for the description of microwave diode properties to be presented in Chapter II.

G. Depletion Zone Width Modulation — the Depletion C(V) Law

1. The Abrupt Junction

We have discussed how the external application of bias *modulates* the depletion zone to influence the diffusion current. The modulation is such as to vary the depletion zone width with applied voltage. At each boundary of the depletion zone there is a high conductivity P or N region. Thus the depletion zone is equivalent to a parallel plate capacitor whose plate separation and, accordingly, capacitance, C(V), are voltage variable. It turns out that for practical purposes, the value of capacitance is given by Equation (I-3), although, since the "plates" are moveable with applied voltage, this conclusion could not be assumed *a priori.*

$$C(V) = \epsilon_0 \epsilon_R \frac{A}{W(V)} \qquad \text{(I-3)}$$

where $C(V)$ = voltage dependent capacitance

ϵ_0 = free space dielectric constant (permittivity)

= 8.85×10^{-2} picofarad/centimeter

= 2.25×10^{-4} picofarad/mil

ϵ_R = relative dielectric constant

= 11.8 (silicon)

A = junction (depletion zone) area

$W(V)$ = voltage dependent depletion zone width

Equation (I-3) gives the junction capacitance due to the depletion zone itself (sometimes referred to as the *transition capacitance*). This equation gives a good approximation to the actual small signal junction capacitance for zero to reverse bias conditions, and for frequencies from a few hertz to the microwave spectrum of tens of gigahertz. Under forward bias conditions an additional parallel capacitance effect called *diffusion capacitance* is experienced when the recombination of carriers which cross the depletion zone is slow compared to the period of the ac signal. However, for microwave switches and phase shifters, the conductance of the junction under forward bias is usually small enough that the device can be modelled as a single resistor; therefore, diffusion capacitance is not considered here.

The voltage variable capacitance of PN junctions has several important practical microwave applications; for this reason, special diodes designed to exploit this feature have been termed *varactor diodes* to emphasize the variable alternating current reactance property resulting from the variable capacitance. Of the phase shifters to be discussed, only the *continuous phase shifter* type uses varactor diodes for controlling microwave signals. Most microwave switches and phase shifters employ the PIN diode whose capacitance at microwave frequencies is essentially independent of the microwave voltage. However, because the description of PIN follows so directly from that of the PN junction, the varactor diode is described here first. Furthermore, all semiconductor diodes exhibit a voltage variable capacitance to some extent, and it is important to keep this in mind from a microwave engineering viewpoint since this reactance nonlinearity does produce circuit performance nonlinearities in all practical devices to some level. Usually, with the PIN diode to be described, such effects are negligible. In some applications, however, the generation of harmonic signals – even though faint when compared to the strength of the fundamental operating frequency – is of significance.

The quantitive effect of depletion zone width spreading with applied voltage is determined using Poisson's Equation (Equation (I-4)) as follows:

$$\nabla^2 \psi(x,y,z) = \frac{-\rho(x,y,z)}{\epsilon_0 \epsilon_R} \tag{I-4}$$

where ψ (x, y, z) = the static potential function in a space (x, y, z)

ρ (x, y, z) = the distribution of charge in the (x, y, z) space

We analyze two idealized models for the varactor diode – the *abrupt* and the *linearly graded* doping profiles. No real diode is perfectly represented by either of these approximations because its impurity profile is a more complex function, depending upon the method of doping (diffusion from fixed or constant impurity source, alloy, epitaxial, etc.) and the specific parameters of the process (temperature, time, background doping, diffusion constants in the silicon host of the donor and acceptor impurities used, etc.). However, the resulting C(V) law of practical junctions usually falls somewhere between the inverse square and cube root functions which are predicted respectively from the *abrupt* and *linearly graded* models. This fact and the relative ease of calculating the C(V) law for these two idealized cases underlie the models' popularity.

The abrupt junction is defined as having constant impurity densities on either side of the junction. The depletion zone, under this condition, is as shown in Figure I-10(c), and analyzed under the following assumptions:

1) Thermal velocity effects are neglected so that the depletion is sharply defined.
2) No mobile carriers are found in the depletion zone.

Using a one dimensional model to represent the charge profile of the depletion zone, Poisson's Equation is written

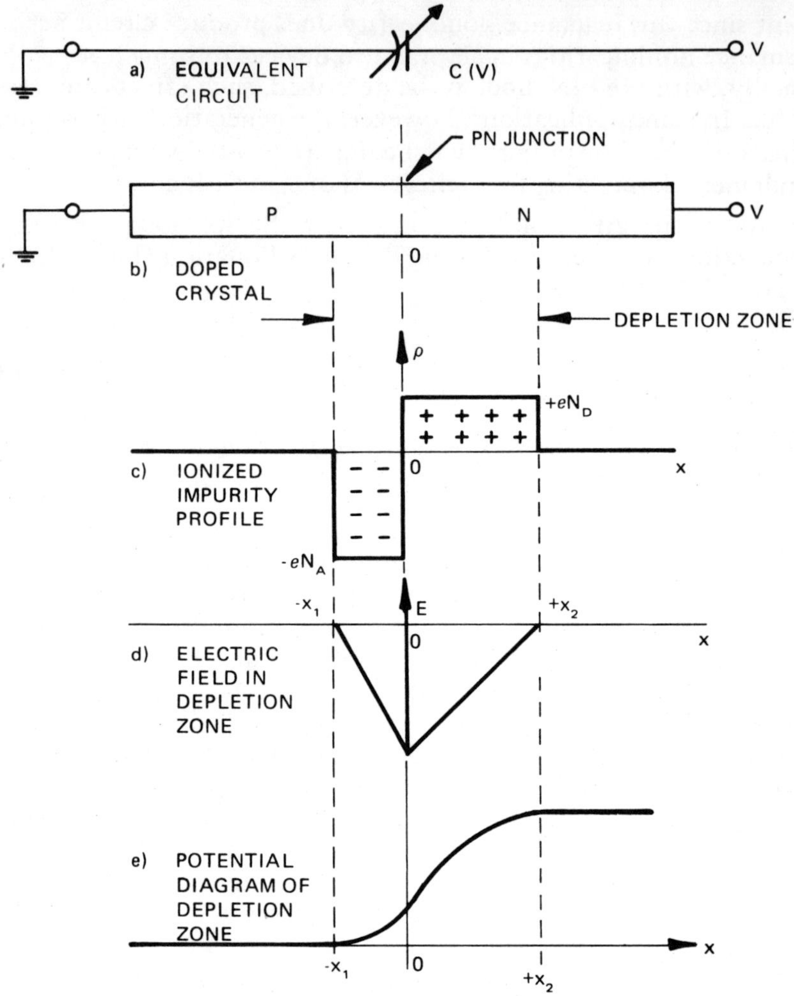

Figure I-10 **Charge, Field, and Voltage Plots for an Abrupt PN Junction**

$$\frac{d^2\psi}{dx^2} = \frac{eN_A}{\epsilon_0\epsilon_R} \quad \text{for } -x_1 < x < 0 \tag{I-4(a)}$$

$$\frac{d^2\psi}{dx^2} = \frac{-eN_D}{\epsilon_0\epsilon_R} \quad \text{for } 0 < x < x_2 \tag{I-4(b)}$$

Integrating with respect to x produces $d\psi/dx$, the gradient of the scalar potential, which is the negative of the electric field, E. The constant of integration is evaluated by imposing the boundary condition that the net charge within $-x_1$ to x_2 is zero; hence $E = 0$ for $x = x_1, x_2$.

$$E = \frac{-d\psi}{dx} = \frac{-eN_A}{\epsilon_0 \epsilon_R}(x + x_1) \quad \text{for } -x_1 \leqslant x \leqslant 0 \qquad \text{(I-5(a))}$$

$$E = \frac{-d\psi}{dx} = \frac{eN_D}{\epsilon_0 \epsilon_R}(x - x_2) \quad \text{for } 0 \leqslant x \leqslant x_2 \qquad \text{(I-5(b))}$$

A second integration produces expressions for ψ with integration constants, ψ_P and ψ_N, which are respectively the scalar potentials in the P and N regions (i.e., in the ohmic regions)

$$\psi = \frac{eN_A}{2\epsilon_0 \epsilon_R}(x + x_1)^2 + \psi_P \qquad -x_1 \leqslant x \leqslant 0 \qquad \text{(I-6(a))}$$

$$\psi = \frac{-eN_D}{2\epsilon_0 \epsilon_R}(x - x_2)^2 + \psi_N \qquad 0 \leqslant x \leqslant x_2 \qquad \text{(I-6(b))}$$

As sketched in Figure I-10(d and e), both the electric field, E, and scalar potential, ψ, are continuous functions of x. Hence the pairs in Equations (I-5) and (I-6) must give uniform results at $x = 0$ where they apply together and

$$\frac{-eN_A x_1}{\epsilon_0 \epsilon_R} = \frac{-eN_D x_2}{\epsilon_0 \epsilon_R} \qquad (x = 0)$$

from which $N_A x_1 = N_D x_2$ (I-7)

The net number of ionized acceptor atoms equals the net number of ionized donor atoms; this fact comes as no surprise because, under equilibrium conditions, for each acceptor atom which becomes ionized by gaining an electron, a donor is ionized by contributing that electron.

Notice that it is the products $N_A x$ and $N_D x_2$ which form the equality. Thus if the acceptor doping concentration, N_A, is much greater than the donor atom doping level, N_D, (as is suggested by Figure I-10(c)) the depletion zone *spreads* farther into the zone of lighter doping (and higher resistivity). *The fact that the depletion zone spreads easily through a high resistivity zone but progresses only a short distance into a heavily doped region underlies the principle of the PIN diode, discussed in Chapter II.*

Equating the expressions for $\psi(x)$ at $x = 0$ in Equation (I-6), we obtain

$$\frac{e}{2\epsilon_0 \epsilon_R} (N_A x_1^2 + N_D x_2^2) = \psi_N - \psi_P \tag{I-8}$$

But $\psi_N - \psi_P$ is the junction potential, V. Substituting into Equation (I-8) and eliminating N_A using the charge conservation $N_A = N_D (x_2/x_1)$

$$V = \frac{eN_D}{2\epsilon_0 \epsilon_R} (x_1 x_2 + x_2^2) \tag{I-9}$$

It is more useful to cast this result in terms of the total depletion layer width, $W(V) = x_1 + x_2$. Therefore, the quantity $N_D(x_1 x_2 + x_2^2)$ is rewritten to eliminate the variables x_1 and x_2 as follows

$$W = x_2 (1 + x_1/x_2) = x_2 (1 + N_D/N_A)$$

$$x_2 = \frac{WN_A}{N_A + N_D}$$

and the quantity to be recast is then

$$N_D (x_1 x_2 + x_2^2) = N_D W x_2 = N_D W^2 \frac{N_A}{N_A + N_D}$$

Substituting into Equation (I-9) and simplifying gives

$$W = \left[\frac{2\epsilon_0 \epsilon_R V}{e} \left(\frac{1}{N_A} + \frac{1}{N_D} \right) \right]^{1/2} \tag{I-10}$$

This expression indicates that the depletion layer width for an abrupt junction varies as the square root of the junction potential. Under reverse bias there is essentially no steady current through the diode; hence, the voltage applied to the diode terminals is equal to the junction voltage plus the built-in potential, V_ϕ. The voltage variable capacitance from zero to reverse bias for the abrupt PN junction is then

$$C(V_R - V_\phi) = \frac{\epsilon_0 \epsilon_R A}{W} = A\sqrt{\frac{e\epsilon_0\epsilon_R}{-2(V_R - V_\phi)\left(\frac{1}{N_A} + \frac{1}{N_D}\right)}} \tag{I-11}$$

where A = area of junction

V_R = applied voltage*

In practice either $N_A \gg N_D$ or $N_D \gg N_A$; therefore, the spreading occurs almost totally in either the P or the N zone respectively. Equation (I-11) can then be written approximately as either

$$C \approx A\sqrt{\frac{e\epsilon_0\epsilon_R N_A}{-2(V_R - V_\phi)}} \qquad \textit{P-type} \tag{I-12(a)}$$

or

$$C \approx A\sqrt{\frac{e\epsilon_0\epsilon_R N_D}{-2(V_R - V_\phi)}} \qquad \textit{N-type} \tag{I-12(b)}$$

In actual practice the area of the junction is often not known with sufficient precision to justify calculation of the absolute value of junction capacitance. The relations in Equation (I-12) are simplified by using the zero biased capacitance as the calibration constant, the C(V) law in terms of applied voltage then becomes

$$C(V) = \frac{C(V_R = 0)}{\sqrt{-(V_R - V_\phi)}} \qquad \textit{abrupt junction} \tag{I-13}$$

2. *The Graded Junction*

The model of the graded junction differs from the abrupt junction in that the impurity profile is represented as a linear function rather than a step discontinuity at the P/N interface, as is shown schematically in Figure I-11(c). Poisson's Equation for the one dimensional graded geometry then becomes

*The applied voltage is a negative number when applied in the non-conducting direction.

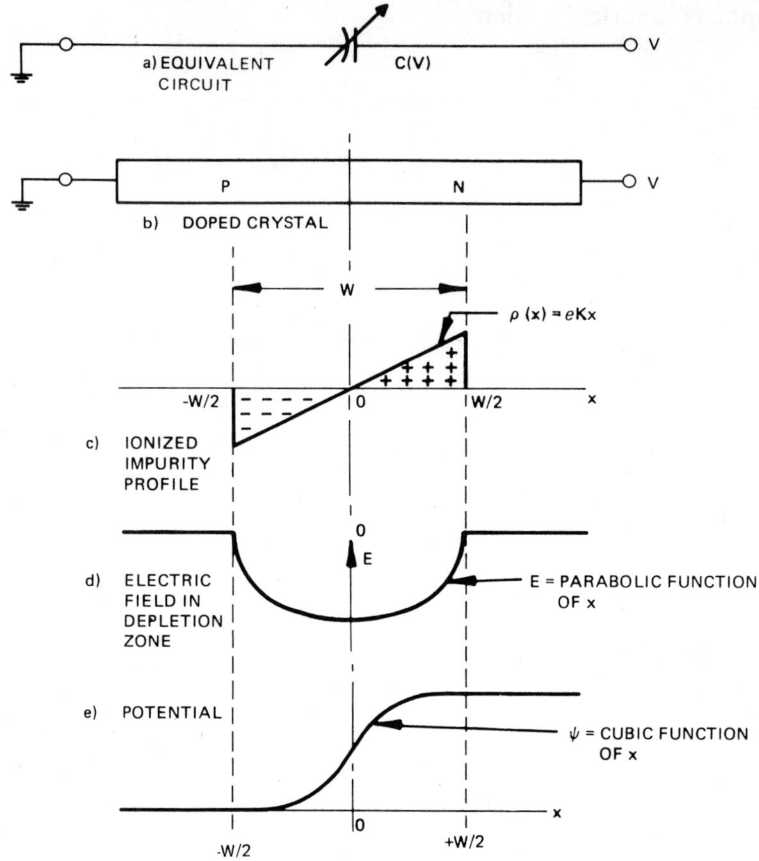

Figure I-11 **Charge, Field, and Voltage Plots for a Graded PN Junction**

$$\frac{d^2\psi}{dx^2} = \frac{\rho(x)}{\epsilon_0\epsilon_R} = +\frac{eKx}{\epsilon_0\epsilon_R} \qquad \text{(I-14)}$$

where $-\dfrac{W}{2} \leqslant x \leqslant \dfrac{W}{2}$

and K is a constant which describes the impurity doping gradient of the junction as shown in Figure I-10(c). A first integration produces electric field, E

$$E = -\frac{d\phi}{dx} = -\frac{eKx^2}{2\epsilon_0\epsilon_R} + A$$

The constant of integration, A, is evaluated by noting that the electric field goes to zero at the depletion zone boundaries. Due to symmetry, both boundaries are the same distance from the junction at x = 0. This distance is W/2 and the expression for electric field then becomes

$$E = -\frac{d\phi}{dx} = -\frac{eK}{2\epsilon_0\epsilon_R}\left[x^2 - \frac{w^2}{4}\right] \tag{I-15}$$

$$\left(-\frac{W}{2} \leqslant x \leqslant \frac{W}{2}\right)$$

The junction voltage, V, is obtained from the second integration.

$$\phi = \frac{eK}{6\epsilon_0\epsilon_R}\left[X^3 - \frac{W^2x}{4}\right] + B \tag{I-16}$$

The integration constant, B, need not be evaluated specifically because only the total potential difference across the junction, $V_R + V_\phi$, is of interest. Substituting x = ±W/2 into Equation (I-16) and subtracting

$$V_R + V_\phi = \psi\left(\frac{W}{2}\right) - \psi\left(\frac{-W}{2}\right) = \frac{eKW^3}{12\epsilon_0\epsilon_R} \tag{I-17}$$

and hence the junction width becomes

$$W = \left[\frac{12\epsilon_0\epsilon_R(V_R + V_\phi)}{eK}\right]^{1/3} \tag{I-18}$$

and the corresponding capacitance for the graded junction varies as the inverse cube root of voltage compared with the inverse square root for the abrupt junction.

$$C(V_R) = A\left[\frac{\epsilon_0{}^2\epsilon_R{}^2eK}{12(V_R + V_\phi)}\right]^{1/3} = \frac{C(V_R = 0)}{(V_R + V_\phi)^{1/3}} \quad \textit{graded junction} \tag{I-19}$$

3. Practical Varactor Diodes

Large *C-change* diodes – that is, those with large* ratio of $C(V_R = 0)/C(V_R = V_B)$ – are typically designed using an abrupt junction and therefore would be expected to have an inverse square law relationship between $C(V_R)$ and $V_R + V_\phi$. However, it is not possible to achieve an absolutely abrupt change in doping density in a practical PN junction. Thus, practical varactors can be expected to have a capacitance versus junction voltage slope somewhere between -0.5 and -0.33 when plotted on a log-log scale, favoring -0.5 as the doping profile approaches the abrupt case described by Figure I-10. Figure I-12 shows the measured characteristic for an abrupt junction tuning diode having a 90 V breakdown voltage. The C-change from zero applied volts ($V_R + V_\phi = 0.7$ V) to breakdown is nearly 10 to 1. The characteristic is closely described by a straight line with -0.46 slope closely approximating the 0.5 slope of an ideal square law diode.

H. Reverse Voltage Breakdown

The low frequency I-V characteristic shown in Figure I-9 implies that any magnitude of reverse voltage can be applied without producing a current greater than I_{SAT}. This point is true, of course, only insofar as the low field model is concerned. However, when sufficiently large reverse voltages are applied to cause the internal electric field in the depletion zone to reach about 300,000 V/cm, a small fraction of the electrons and holes may acquire sufficient kinetic energy to impart to valence band electrons with which they collide enough energy (1.1 eV) to reach the conduction band. Electron-hole pairs are thereby created.** The new free carriers so generated can likewise be accelerated by the high electric field to create new carriers by further impacts with valence band electrons.

The rapid increase of carriers produces an essentially infinite rate of increase of current with applied voltage, once the internal field

*"Hyper-abrupt" varactors are sometimes designed for extremely large capacitance var-actions; they also, however, have more complex doping profiles. They are used rarely for microwave control.

**Another form of breakdown consists of tunnelling directly through the depletion zone and is called *zener* [2 (Chapter 3)] breakdown, but it occurs only with PN junctions having much higher concentrations in *both* P and N regions than are used with practical microwave control diodes. Zener breakdown occurs practically in only very thin diodes, typically below 10 V breakdown, where high fields can be obtained. Varactor and PIN microwave control diodes have breakdown voltages greater than 10 V. Therefore, zener breakdown is not treated here.

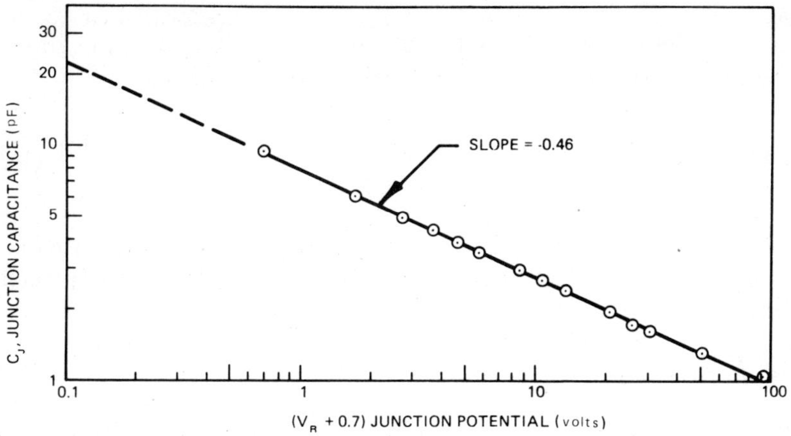

Figure I-12 **Measured C(V) Law for an Abrupt Junction Tuning Varactor**

is sufficiently high that, on the average, every free carrier produces at least one additional free carrier. For this reason the breakdown is called *avalanche breakdown.* The new carriers produce ionized silicon atoms when, through impact, they promote valence electrons to the conduction band. Accordingly this form of electron-hole pair production is sometimes called *impact ionization.* It is analogous to the ionization of gas molecules which occurs naturally in lightning. It also is the mechanism which is used usefully in microwave gas T-R tubes to separate high power transmitter signals from low power received signals in a radar antenna.

To describe ionization, McKay [5] has defined the rate of ionization, $\alpha(E)$, as the average number of electron-hole pairs produced per unit distance by one free carrier traveling under the influence of the applied electric field, E. Assuming $\alpha(E)$ is the same for electrons and holes, *either* an electron or a hole can produce N electron-hole pairs in traversing the depletion zone width, W, according to

$$N = \int_0^W \alpha(E)\, dx \qquad \text{(I-20)}$$

In turn, the N pairs generate N^2 new pairs and the initial reverse bias leakage current I_0 is increased by some factor, M, to

$$I = MI_0 = I_0 (1 + N + N^2 + \dots). \tag{I-21}$$

If $N < 1$, the infinite series $1 + N + N^2 + N^3 + \dots .N^n + \dots$ has the closed form sum $1/(1 - N)$ and so the total current is

$$I = MI_0 = \frac{I_0}{1 - N} \tag{I-22}$$

where $M = 1/(1-N)$ is defined to be the avalanche *multiplication factor.* As already described M approaches infinity as N approaches 1, the condition at which breakdown occurs is

$$N = \int_0^W \alpha(E)\, dx = 1 \tag{I-23}$$

On a purely analytical basis, Equation (I-23) is practically unsolvable because $\alpha(E)$ is not known analytically with sufficient accuracy. A good discussion of the semiconductor properties that influence $\alpha(E)$ is given by Phillips, [3 (pp. 116-125)] . Experimentally, the electric field, E_B, which produces avalanche has been determined. It varies weakly with impurity concentration, N_B, either donor or acceptor type) as shown in Figure I-13.

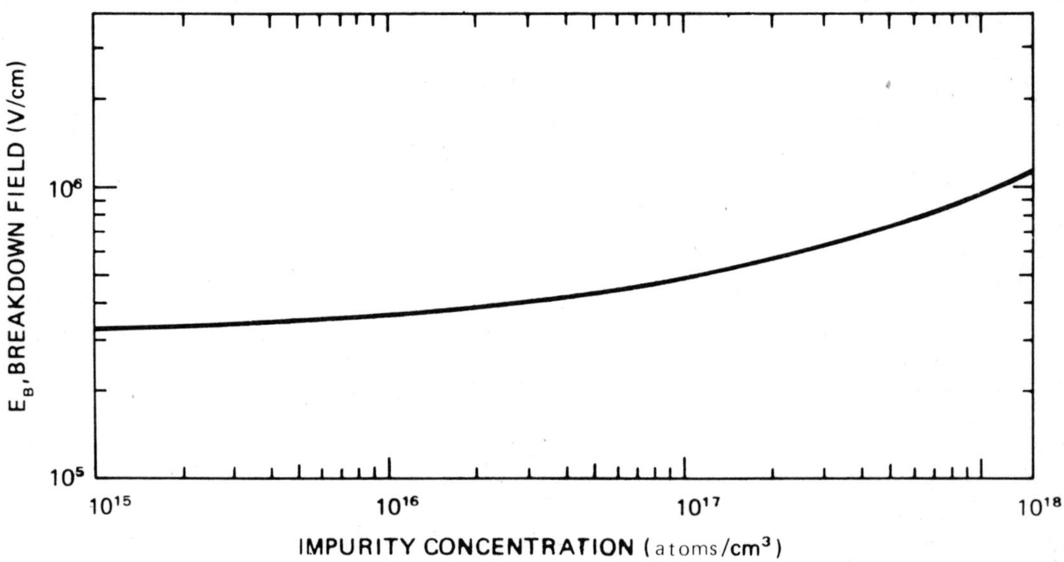

Figure I-13 Avalanche Breakdown in Doped Silicon (After Lee [4 (p. 124)])

For most practical diodes, because $\alpha(E)$ is a very rapidly increasing function near breakdown, avalanching typically occurs at about 300,000 V/cm. Actual devices always have crystalline defects at which lower average fields produce breakdown. Due to the small geometries of these defects the breakdown is often localized, at least initially, and the ionizations referred to as *microplasmas.*

The second consideration, neglecting non-uniformities and surface effects, is how the diode's voltage breakdown can be increased. For the abrupt junction described in Figure I-10, the electric field is maximum at $x = 0$. If we assume $N_A \gg N_D$, essentially all of the depletion zone spreading is in the N region. The breakdown voltage for most useful varactor diodes is greater than 30 V. Therefore, we may usually neglect the 0.5-0.7 V built-in junction potential and assume the applied reverse voltage, V, is equal to the total junction potential. Because of the linear slope of E(x) in the junction

$$V_B = \frac{E_B \cdot W}{2} \tag{I-24}$$

where E_B and V_B are the field and voltage at breakdown.

But, from Equation (I-10), $W \approx [2\epsilon_0\epsilon_R V/eN_D]^{1/2}$. Thus, at breakdown, $V = V_B$ and, for the abrupt junction

$$V_B = \frac{E_B{}^2 \epsilon_0 \epsilon_R}{2eN_D} \tag{I-25}$$

For the graded junction, E(x) is parabolic in the depletion zone and depends on the linear gradient, K, of impurity atoms. When this fact is taken into account the breakdown for the graded junction is [3 (p. 139)]

$$V_B = \frac{4\ E_B{}^{3/2}}{3} \left(\frac{2\ \epsilon_0 \epsilon_R}{eK}\right)^{1/2} \tag{I-26}$$

For present purposes it is only important for the reader to be aware of what influences diode breakdown. In Chapter II the breakdown of the PIN junction, most useful for high power microwave control, is discussed.

References

[1] Armstead, M.A.; Spencer, E.G.; and Hatcher, R.D.: "Microwave Semiconductor Switch," *Proceedings of the IRE, Vol. 44,* p. 1875, December 1956.

[2] Garver, R.V.; Spencer, E.G.; and Harper, M.A.: "Microwave Semiconductor Switching Techniques," *IRE Transactions on Microwave Theory and Techniques, Vol. MTT-6,* pp. 378-383, October 1958.

[3] Phillips, Alvin, B.: *Transistor Engineering* (Chapters 2-6), McGraw-Hill, Inc., New York, 1962.

[4] Watson, H.A., (ed.): *Microwave Semiconductor Devices and Their Circuit Applications* (Chapters 2-7), McGraw-Hill, Inc., New York, 1969.

[5] McKay, K.G.: "Avalanche Breakdown in Silicon," *Physical Review, Vol. 94,* pp. 877-884, May 1954.

Questions

1. Show that the conductivity of a material is given by

$$\sigma = e[(\mu_P p) + (\mu_N n)]$$

where e is the charge on an electron, p and n are the respective free carrier hole and electron densities, and μ_P and μ_N are the corresponding mobilities.

2. Use the above result to estimate the resistivity ($\rho = 1/\sigma$) for single crystal silicon at 50°C having a uniform doping density of fully ionized phosphorus atoms of 10^{16} atoms/cm^3. Use the graph in the next chapter in Figure II-8 to estimate mobility.

3. Suppose a sample of silicon material has uniformly distributed, fully ionized equal concentrations of boron and phosphorus doping atoms. Explain why the resistivity is very high, no matter what the concentration of these carriers. (This principle is that of *compensation.*)

4. Suppose a silicon sample has a *net* ionized donor concentration (after acceptor compensation) of 10^{14} atoms/cm^3. Using the mobility data in Figure II-7 and the conductivity expression in Question 1, estimate the resistivity of the sample when the

background concentrations of fully ionized, uniformly distributed acceptors are 0, 10, 10^2, 10^3, 10^4 and 10^5 atoms/cm^3.

5. If an ideal semiconductor diode (i.e., no surface leakage) has I_{SAT} = 1 nA (10^{-9} A) what forward voltage must be applied to produce currents of I = 0.1, 1, and 10 A if a) the internal resistance, R, is 0? b) R = 0.1 Ω? How much power is dissipated in the diode in each instance? (Boltzmann's Constant, k, equals 8.63 x 10^{-5} (eV/K), T = 300K).

Chapter II

PIN Diodes and the Theory of Microwave Operation

A. The PIN Diode — An Extension of the PN Junction

1. Structure

The PIN diode should not be thought of as something pysically different from the PN junction discussed in Chapter I, but rather different in a sense of degree. In Chapter I we saw that with the abrupt junction the width of the depletion zone is inversely proportional to the resistivity of the P or N region, whichever has the lesser impurity doping concentration. As the width of the depletion zone increases, the capacitance per unit area of the junction decreases. This effect is very beneficial for a diode which is intended for use as a microwave switch because the lower the capacitance the higher the impedance of the diode under reverse bias, and the more effective the device is as an "open circuit."

The limiting case of high resistivity material is undoped (or "intrinsic") *I* silicon. In practice, of course, no silicon material is without some impurities. A practical PIN diode, then, consists of an extremely high resistivity P or N zone between low resistivity (highly doped) P and N zones at its boundaries, as shown in Figure II-1. To distinguish unusually heavily or lightly doped material, special nomenclature has evolved. Heavily doped P and N materials are referred to as P+ and N+, respectively. To identify very lightly doped, high resistivity P and N material, the Greek letters are used; thus high resistivity P material is called *π-type* and high resistivity N material is called *ν-type.* Recognizing that perfectly intrinsic material is not practically obtainable, the I region of a PIN diode can consist of either ν- or π-type material. The result-

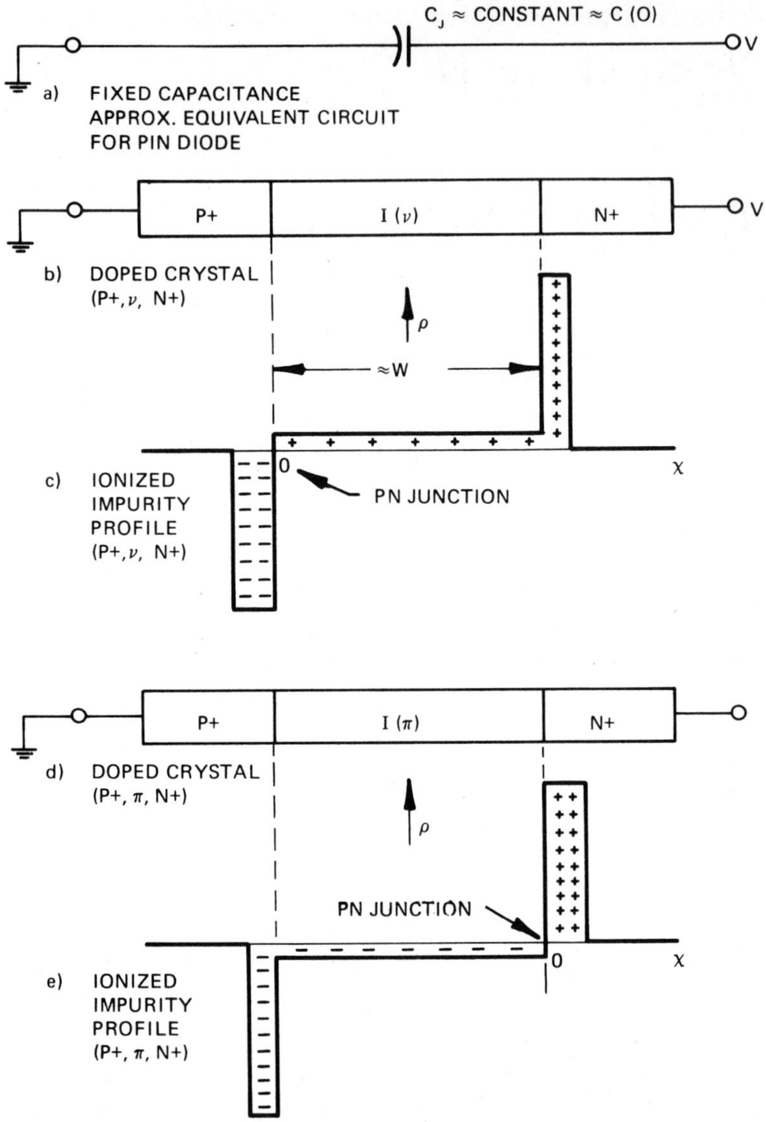

Figure II-1 Profiles for the Two PIN Diode Types

ing diodes are indistinguishable from a microwave point of view; however, the actual junction forms at opposite ends of the intrinsic zone depending on the choice. This distinction is diagrammed for both cases in Figure II-1.

The first type shown in Figure II-1(b) shows a P+, ν, N+ diode structure. If the I region is of sufficiently high resistivity, what few impurity atoms it has will be ionized and the depletion zone will extend throughout the I region and include a small penetration into both the P and N regions. Because of the heavy doping in the P+ and N+ zones the depletion zone will not extend very far into them, and the depletion zone will be essentially equal to the I layer width, W_I. The alternate diode structure, P+, π, N+ is shown schematically in Figure II-1(d). Here the depletion zone width is likewise approximately equal to the width of the intrinsic layer but the junction is formed at the N+ interface rather than that of P+. Controlling the location of the junction has important consequences from the standpoint of passivating the diode chip, but no impact on performance. Most PIN diodes use ν material for the I region and the junction is formed at the P+ interface.

2. *C(V) Law and Punchthrough Voltage*

In the preceding section it was assumed that the I layer is of such high resistivity that, even with no applied bias, the depletion zone extends across the I layer to the P+ and N+ zones. Under such circumstances C_J is practically independent of applied voltage. At zero voltage the depletion zone has already extended through the I region; as further reverse bias is applied to the diodes, little further widening of the depletion zone proceeds because of very high impurity concentrations and correspondingly large availability of ionizable donors and acceptors in the P+ and N+ regions.

The PIN diode which actually does have so high a resistivity I layer that it is depleted at zero bias is called a *zero punchthrough* diode, because the depletion zone has "punched through" to the high conductivity zones even before bias is applied.

Such a situation, however, represents an idealization. Not all practical diodes are zero punchthrough. A more *general definition of the PIN is a semiconductor diode which consists of two heavily doped P and N regions separated by a substantially higher resistivity P or N region.*

Figure II-2 shows schematically a practical PIN diode with ionized impurity profiles at zero bias and at punchthrough. At zero bias a

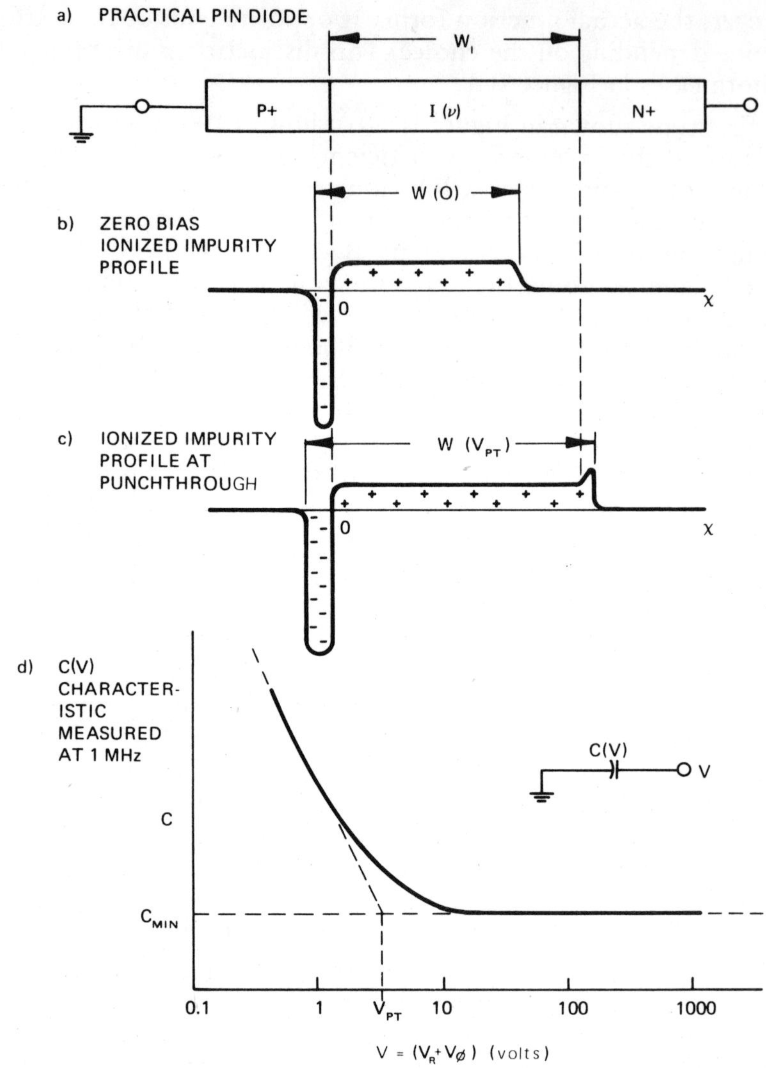

Figure II-2 **Practical PIN and Reversed Punchthrough Characteristics**

large portion, but not necessarily all, of the I region impurities have been ionized and the depletion zone, W(0), may be somewhat less than the I layer width, W. As reverse bias voltage is applied to this diode, depletion layer spreading occurs, and the capacitance, shown in Figure II-2(b), decreases until the depletion layer has spread definitely to the N+ region, as shown in Figure II-2(c). At this voltage the depletion layer width, $W(V_{PT})$, is approximately equal to W_I. Further spreading of the depletion layer into the low resistivity P+ and N+ regions is, for most applications, negligible. The voltage at which the depletion zone just reaches the N+ contact is the punchthrough voltage, V_{PT}.

Because in practice the resistivity levels in the P+, I, and N+ regions do not change abruptly, the resulting capacitance versus voltage characteristics have a *soft knee.* Therefore the punchthrough voltage is not directly measureable with precision. However, the practical diode usually does have two definable slopes in its C(V) characteristic, when plotted using semilog paper as shown in Figure II-2(d). *By convention, the voltage intersection of these two straight line projected slopes is called the punchthrough voltage.*

It is to be emphasized that this C(V) characteristic is what one obtains when the measurements are made at relatively low frequencies, typically 1 MHz. At microwave frequencies, the dielectric susceptibility of silicon is much larger than the conductivity of ν or π material; thus, the capacitance is effectively equal to the minimum capacitance for all values of reverse bias, as is shown in the following discussion of dielectric relaxation.

3. Capacitance Measurements and Dielectric Relaxation

If the capacitance of a PIN diode which does not punch through at zero bias is measured at zero bias, a larger value of capacitance will be measured at a low frequency (such as 1 MHz) than would be measured at microwave frequencies (such as with a slotted line measurement at 1 GHz). The reason is that silicon, in addition to being a variable conductor, also has a high dielectric constant. Therefore, its bulk differential equivalent circuit appears as a parallel combination of conductance and capacitance. The relative current division between these two equivalent circuit parameters varies with the frequency of the applied signal, higher frequency currents being carried mostly by the capacitive path.

To illustrate this point, consider Figure II-3 which shows a PIN diode below punchthrough. The portions of the P+ and the I regions which are depleted represent the depletion zone, or "swept region." The remainder of the I region is "unswept" and can be modelled, as shown in Figure II-3(c), as a parallel resistance-capacitance circuit, represented by the equivalent circuits elements, C_{US} and R_{US}.

The division of current through C_{US} and R_{US} depends upon the ratio of the susceptance of C_{US} to the conductance ($1/R_{US}$). This ratio in turn depends on the dielectric constant of silicon to its bulk resistivity. The frequency at which the current division between these two elements is equal (i.e., when the susceptance is equal to the conductance) is defined as the *dielectric relaxation frequency*, f_R, of the material.

When the operating frequency, f, is equal to or greater than 3 f_R, the total capacitance represented by the series combination of C_{SW} and C_{US} is approximately equal to C_J (within 10%), the parallel plate capacitance of the totally depleted I region. This value corresponds to the minimum capacity C_{MIN} measured beyond punchthrough at low frequency.

This point is a major one in the practical characterization of PIN diodes intended for microwave switching applications. It means that practical measurements of the capacitance of a PIN junction can be made at 1 MHz, and the values so attained will represent a good approximation to the actual capacitance applicable at microwave frequencies. This test only requires that sufficient bias voltage is used during the low frequency measurement to insure that the I region is fully depleted. A check to determine whether the I region is in fact fully depleted can be made simply by plotting the C(V) characteristic for a few representative diodes from the production lot to determine at what minimum bias voltage the measured capacitance reaches what is essentially its minimum value.

The remaining required quantity to determine the applicability of the low frequency C_{MIN} as a representation for the microwave capacitance, C_J, is an estimate of the relaxation frequency for the I region of the diodes being measured. High purity silicon material used to make PIN diodes typically has resistivity in the range of 500-10,000 Ω-cm prior to the diffusion and/or epitaxial growth steps used to achieve the low resistivity P+ and N+ regions. However, after the high temperature processing needed to realize these regions, the resistivity of the I region is always less than that of the starting crystal. Typical values for I region resistivity are in the range of 100-1000 Ω-cm. The dielectric relaxation frequency for the unswept portion of the I region can be written in terms of the equivalent circuit parameters, directly from the definition which

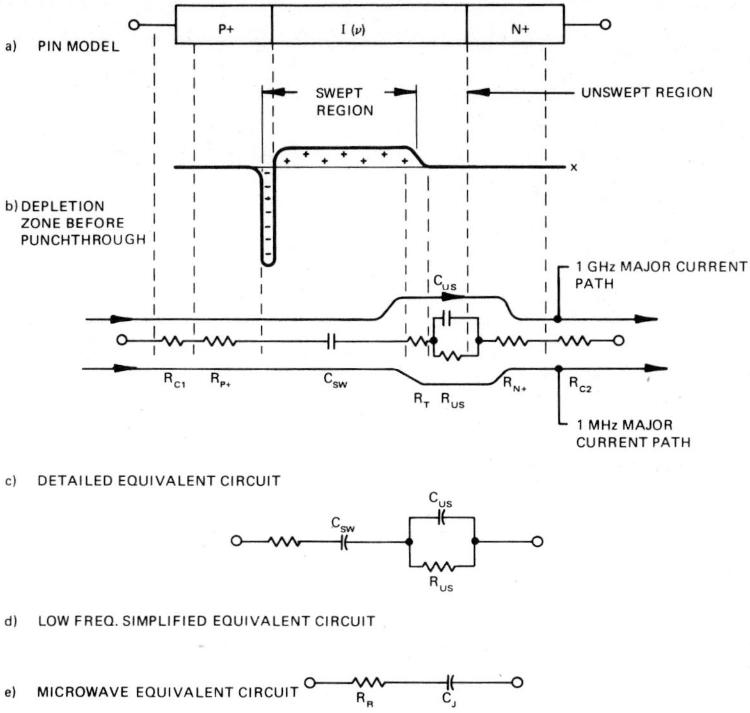

Figure II-3 **Reverse Bias PIN Equivalent Circuit**

requires that the conductance and capacitive susceptance be equal at f_R. The result is

$$f_R = \frac{1}{2\pi R_{US} C_{US}} \tag{II-1}$$

In turn, the specific values for R_{US} and C_{US} can be written in terms of the length, L, and the area, A, of this unswept region together with the bulk resistivity, ρ, and the absolute dielectric constant, $\epsilon_0 \epsilon_R$, as follows

$$R_{US} = \frac{\rho L}{A} \tag{II-2}$$

$$C_{US} = \frac{\epsilon_0 \epsilon_R A}{L} \qquad \text{(II-3)}$$

Substituting these expressions into Equation (II-1) together with the value $\epsilon_R = 11.8$ for silicon yields Equation (II-4), which gives the dielectric relaxation frequency directly in GHz when the resistivity, ρ, is known.

$$f_R = \frac{1}{2\pi \, \epsilon_0 \epsilon_R \rho}$$

$$f_R = \frac{153}{\rho(\text{ohm-centimeters})} \quad \text{(gigahertz)} \qquad \text{(II-4)}$$

This expression is shown graphically in Figure II-4.

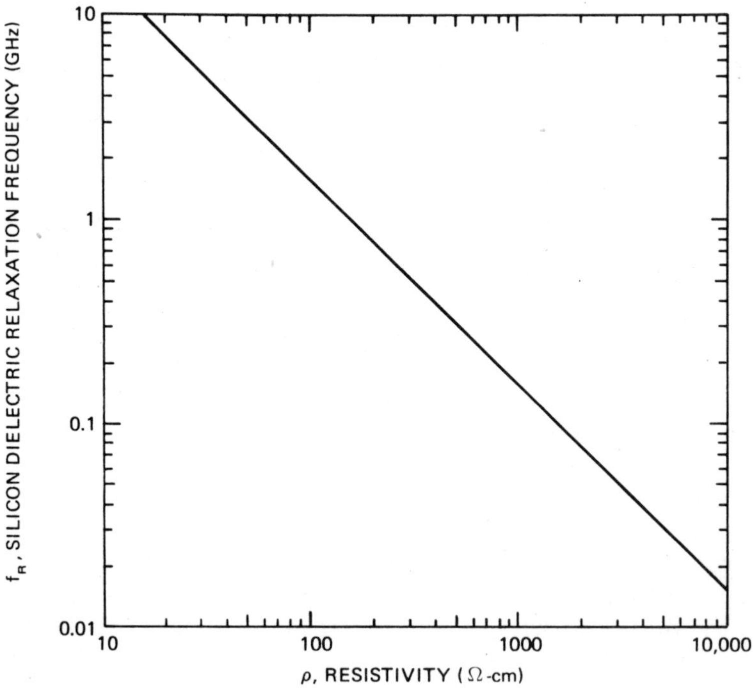

Figure II-4 **Dielectric Relaxation Frequency in Silicon of Various Resistivities**

Strictly speaking, since the final resistivity of the I layer of a practical diode depends upon the actual processing steps used to fabricate the diode, one could not know beforehand what dielectric relaxation frequency would apply for a particular diode unless a method for determining the magnitude of ρ as realized in a final device were available. Usually a PIN diode has an I layer resistivity of at least 100 Ω-cm, which corresponds to $f_R = 1.53$ GHz. Thus for operating frequencies of 5 GHz or more the simplified equivalent circuit in Figure II-3(e) can nearly always be applied.

An experimental method does exist for the determination of I layer resistivity through the measurement of the punchthrough voltage and knowledge of the I layer width, which usually is known with reasonable accuracy by the diode manufacturer. To make the calculation, Equation I-10 is solved for V_{PT} at which the depletion layer is equal to the I region width, W. Recognizing that for a PIN the impurity concentration of the N+ contact, N_A, is much larger than the impurity concentration in the I region, N_D, the result becomes

$$V_{PT} = \frac{eN_D W^2}{2\epsilon_0 \epsilon_R} \qquad \text{(II-5)}$$

But the resistivity of the I region is related to the donor impurity density according to

$$\rho = \frac{1}{N_D e \mu_N} \qquad \text{(II-6)}$$

where μ_N is the *electron mobility* (i.e., the effective drift velocity of electrons in the I region per unit applied electric field) and e is the charge of a single electron. Substituting this result in Equation (II-5) gives

$$\rho = \frac{W^2}{2V_{PT}\epsilon_0 \epsilon_R \mu_N} \qquad \text{(II-7)}$$

$$\rho = \frac{(2.4 \times 10^8)\, W^2}{V_{PT}} \text{(ohm-centimeters)}$$

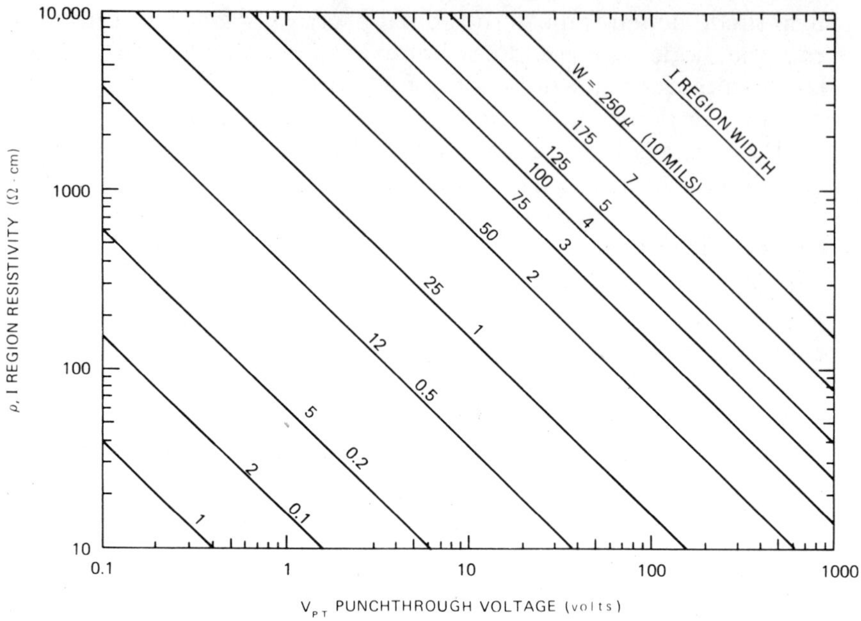

Figure II-5 **Punchthrough Voltage vs. I Region Resistivity for Various I Region Widths**

where ϵ_R = 11.8 (silicon)

μ_N = 2000 (centimeters^2/volt-second)*

W = I region width (centimeters)

V_{PT} = punchthrough voltage (volts)

Thus, for example, if a particular diode having an I region width of 0.0025 cm (1 mil) is found to have a punchthrough voltage of 10 V, the resultant average resistivity is 150 Ω-cm. Equation (II-7) is shown graphically in Figure II-5 for various values of I region width, W.

B. Microwave Equivalent Circuit

1. *Charge Control Model*

Transit Time Limit of the I-V Law

In Chapter I the I-V characteristic for a PN junction was given (Equation (I-1)). The same characteristic applies for the PIN *at*

*This mobility value is representative for electrons in high resistivity N type silicon, as shown in Figure II-8. Thus, this method of I region width determination is limited by the accuracy with which mobility can be estimated.

low frequencies, for which the RF period is long compared with the *transit time* of an electron or hole across the I region.

The discussion to follow using a simple carrier transit time model is only approximate. Real diodes have more complex carrier flow, which is non uniform, subject to applied voltages (i.e., nonlinear), and so forth. The approximation is useful, as it permits estimates of frequency behavior and switching speed. The transition between low and high frequency behavior occurs when this transit time is equal to the RF period. To estimate the transit time, recall that the injection of carriers into the depletion zone occurs under forward bias by diffusion. That is, once forward bias is applied it reduces the magnitude of the built-in junction potential, causing holes to diffuse from the P to the N region and electrons to diffuse in the opposite direction.

The mechanics of this diffusion charge transport are described by *diffusion constants* for holes and electrons, D_P and D_N respectively. Diffusion, being the flow of carriers from a region of high to lower density, is described in terms of a current density proportional to the spatial gradient of charge density according to Equations (II-8) and (II-9).

$$\text{(For holes)} \quad \vec{J}_P = -eD_P(\nabla p) \tag{II-8}$$

$$\text{(For electrons)} \quad \vec{J}_N = -eD_N(\nabla n) \tag{II-9}$$

where

J = current density

e = unit charge *magnitude* = $(+1.6 \times 10^{-19}$ coulomb)

$D_{P,N}$ = diffusion constants for holes and electrons respectively

∇p = spatial gradient of hole density

∇n = spatial gradient of electron density

To illustrate diffusion, let us estimate the approximate *transit time* for holes, the slower moving carrier, across the depletion zone of a PN junction of width W. A one dimensional analysis is used, and Equation (II-8) becomes

$$J_P = -eD_P \frac{dp}{dx} \tag{II-10}$$

The minus sign is required (D is defined as a positive constant) since current flow is opposite to the direction of increasing charge density. Figure II-6 shows a simplified model of the PIN and majority carrier profiles. The gradient dp/dx is abrupt at the P/I interface and an exact analysis would require an analytic representation of p(x). However, as an approximation, we use the *average gradient of the hole density across the I region,* or

$$\frac{dp}{dx} \approx \frac{P_P}{W}$$

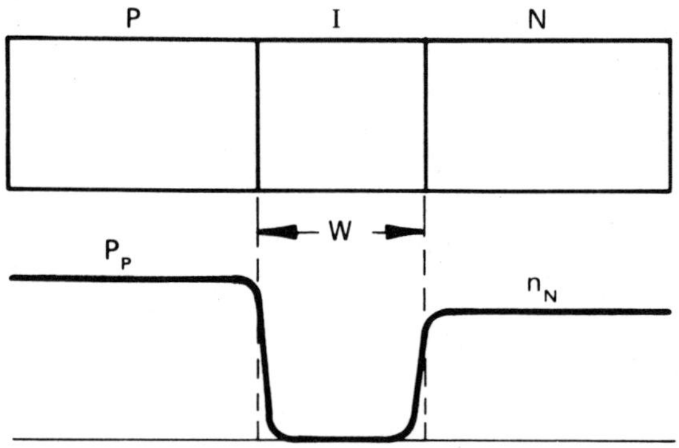

Figure II-6 **Depletion Zone Model Used to Estimate Transit Time Frequency**

Equation (II-10) then becomes

$$J_P = P_P e v_P \approx -eD_P \frac{P_P}{W}$$

where J_P has been written explicitly using carrier velocity, v_P, and the density of carriers participating in the hole current flow. But the hole transit time, T_P equals W/v_P; therefore

$$\text{Transit Time} = T_P \approx \frac{W^2}{D_P} \qquad \text{(II-11)}$$

Accordingly, we can expect that the low frequency I-V characteristic (Equation (I-1)) can no longer be used at frequencies for which the RF period is comparable to T_P. If a transition frequency, f_T, is defined for the PIN diode at which

$$f_T = \frac{1}{T_P}$$

then

$$f_T = \frac{D_P}{W^2}$$

Frequently, the mobility, μ, rather than the diffusion constant, D, is evaluated for semiconductor materials. These two constants are related according to the Einstein relationship

$$D = \mu \frac{kT}{e} \quad \text{(centimeters}^2\text{/second)} \qquad \text{(II-12)}$$

where D = diffusion constant (centimeters^2/second)

μ = mobility (average carrier drift velocity per unit applied electric field)

k = Boltzmann's Constant

T = absolute temperature (Kelvin)

At 300 K (near room temperature) kT/e = .026 V; thus

$$D = .026\,\mu \text{ (at 300 K)} \qquad \text{(II-13)}$$

The hole and electron mobilities vary both with impurity densities (see Figure II-7) and temperature (see Figure II-8). For the present example the hole mobility at 300 K in high resistivity silicon is about 500 cm^2/V-s and therefore

$$f_T = \frac{1300}{w^2} \text{ megahertz} \qquad \text{(II-14)}$$

where w is I region thickness in microns.

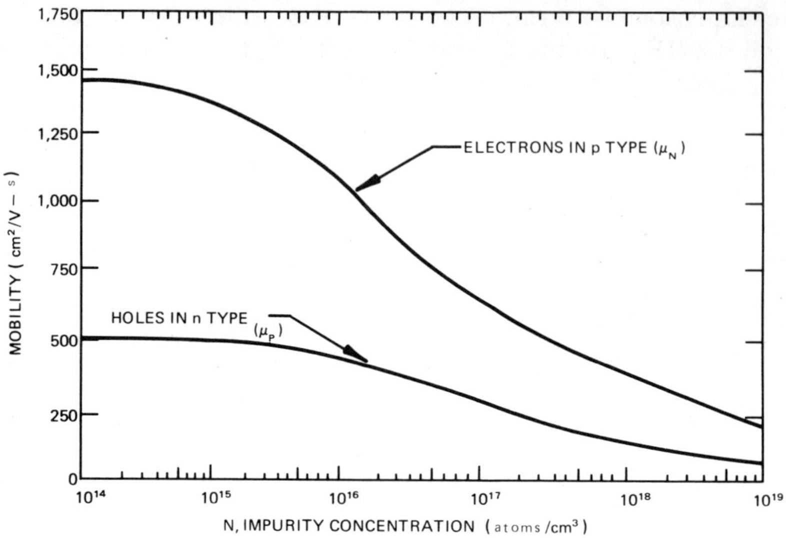

Figure II-7 **Hole and Electron Mobilities (at 300 K in silicon) vs. Impurity Density (After A.B. Phillips [1])**

Thus, even for a very thin base PIN diode having only a 2.5μ (0.1 mil) I region, f_T = 200 MHz. A graph showing how f_T varies with w is shown in Figure II-9. In practice, PIN diodes used for microwave switching have I region widths of 25-250 μm (1-10 mils) *and accordingly the low frequency I-V characteristic given in Equation (I-1) is useless for evaluating microwave resistance.*

I Region Charge and Carrier Lifetime

However, all of the concepts introduced so far to describe low frequency behavior are easily applied to determine the microwave resistance. We shall evaluate I region charge and use it to gauge resistance. From Figure II-9 it is evident that once charge, consisting of holes and electrons, has been injected into the I region under forward bias, *it cannot be removed in the brief duration of a half cycle of RF frequency* if that RF frequency is above a few hundred megahertz, even for the thinnest I region (or base width) diodes.

The *charge control model* for the PIN diode allows RF performance to be related to the net steady state hole and electron

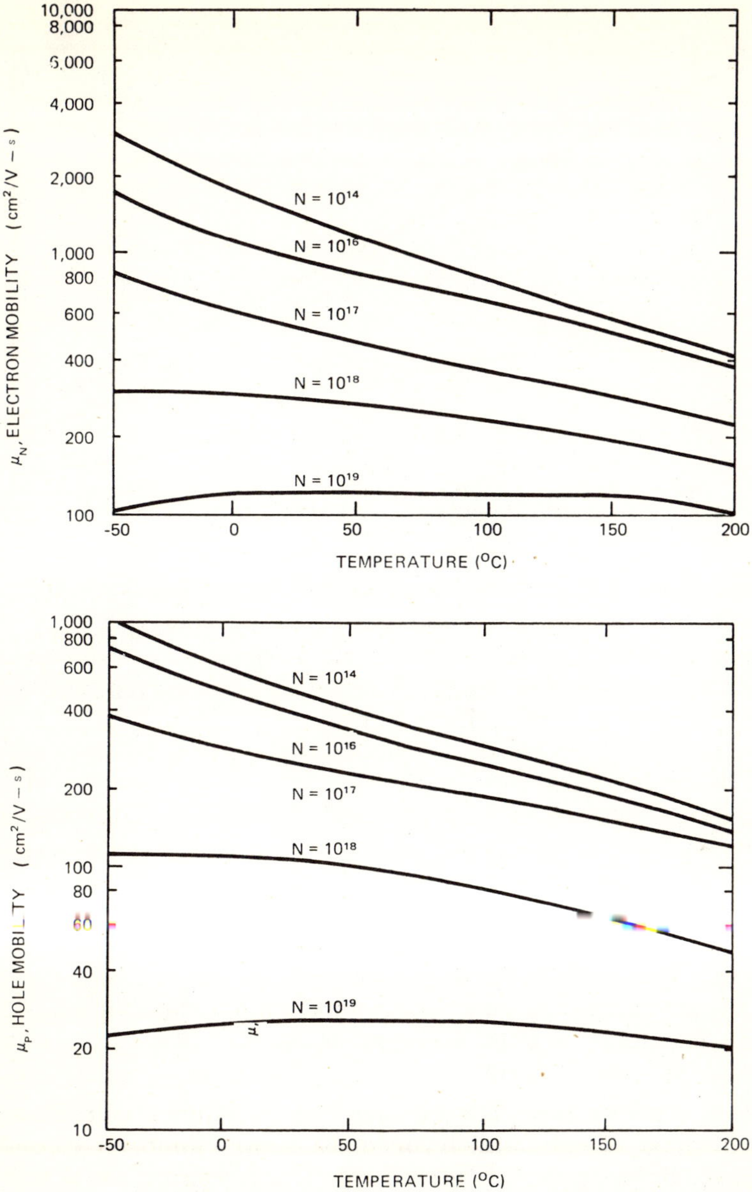

Figure II-8 Hole and Electron Mobilities vs. Temperature for Various Impurity Densities in Section (After A.B. Phillips [1])

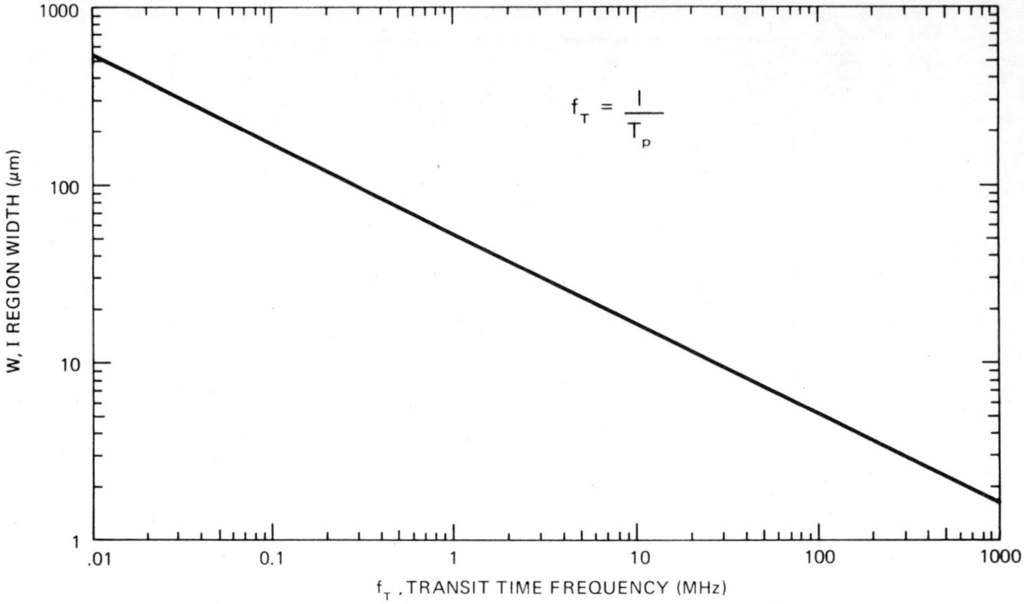

Figure II-9 **I Region Width vs. Transit Time Frequency for Silicon PIN Diodes at Room Temperature**

charges, Q_P and Q_N respectively, in the I region. These charges are equal to the product of the (low frequency) bias current and the respective average carrier lifetime, thus

$$Q_P = I_0 \cdot \tau_P \tag{II-15}$$

$$Q_N = I_0 \cdot \tau_N \tag{II-16}$$

That is, after the turn on transient during which the I region charge density is established, the bias current serves as a replenishment source for holes and electrons which have recombined. Referring to Figure II-10, the bias current at the P/I interface consists almost entirely of holes being injected into the I region. At the I/N interface the same bias current consists mainly of electron injection into the I region.

The longer the lifetime, the less bias current required to maintain a given charge density and, accordingly, a given microwave conductivity. Before proceeding further it is important to note that *long lifetime does not necessarily imply slow switching speed.* A properly designed driver can remove I region charge, and thereby reverse bias the diode, in a period shorter than the lifetime. Rather, long lifetime should be considered a measure of the crystalline perfection within the diode.

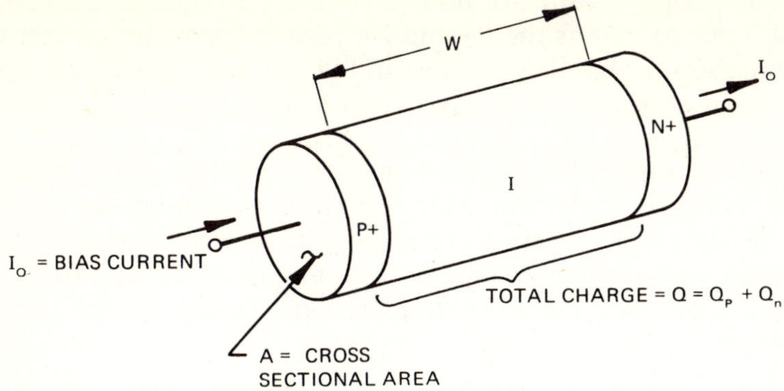

Figure II-10 **Cylindrical I Region Model Used to Estimate I Region Resistance of PIN Diode**

A pure intrinsic silicon crystal has a calculated carrier lifetime of 3.7 *seconds.* With impurity doping of 10^{15} cm^{-3} this figure drops to 0.11 ms. [1 (p. 80)] In actual diodes the lifetime typically ranges from 0.1 to 10 μs, orders of magnitude less than these theoretically attainable values. To appreciate the reason for this great disparity, it is necessary to review what lifetime represents.

Lifetime is proportional to the *improbability* that an electron and hole will recombine. Imperfections in the regular array of crystal atoms create energy states within the otherwise *disallowed* band gap of silicon. Such intermediate states provide a virtual energy "staircase" by which the recombination proceeds. In a very regular crystalline structure, energy must be given off in the transition of an electron from conduction to valence bands in the form of a 1.1 eV (light emitting) photon; the statistical probability of such an occurrence is low. But with crystalline irregularities, intermediate allowed energy states between these two bands permit a transition in a "staircase" of smaller energy transitions with corresponding low energy phonon (lattice vibrations) emissions, the overall probability of which is higher. Thus lifetime is reduced and recombination is enhanced by the presence of crystalline imperfections and/or impurities.

There are two categories of crystalline irregularities — boundary surfaces and bulk impurities. For a PIN diode the I region boundaries consisting of the highly doped P+ and N+ represent rapid recombination surfaces for carriers which diffuse into them. Likewise the peripheral surface boundary of the I region, although not to the same extent as the P+ and N+ regions, provides greater recombination probability than would be present for carriers were the silicon crystal of infinitely extended dimensions. Furthermore, from a bulk point of view, even the structure of an undoped silicon crystal is never ideal. There are stress lines and faults where the probability of electron-hole recombination increases. A doped crystal is all the more susceptible to such imperfections because of the temperature shocks, imperfect atomic fit of doping atoms within the silicon, and related crystal stress producing factors associated with diode manufacture.

This brief discussion of lifetime and its determining factors is qualitative. Even an approximate theoretical treatment of the effective lifetime for a real diode is impractical, although some bulk quantitive analytical treatments of semiconductor crystal lifetime have been made. [3] For the diode maker and user, resort must be made to experimental means by which *average* carrier lifetime can be measured. The conventional method for measuring PIN diode lifetime, τ, consists of injecting a known amount of charge, Q_0, into the I region and measuring the time, τ_S, required to extract it using a "constant" reverse bias current. [4,5] To appreciate this method, consider the equivalent circuit and charge versus time profiles shown in Figure II-11.

A forward bias current, I_F, is established and permitted to flow for a period long compared to the expected lifetime, thus storing a charge, Q_0 equal to $I_F \cdot \tau$ in the diode under test. The current supplies are chosen so that $R_R \ll R_F$. Thus, when the switch, S, closes, the diode current, I_D, reverses direction and reaches a magnitude, I_R-I_F. The stored charge is removed by this current until it is fully depleted. If the discharge period, τ_S, is short compared to the lifetime ($\tau_S \ll \tau$), then negligible recombination occurs during the turnoff and the total stored charge is recovered. In this case, $Q_0 = I_F \cdot \tau = (I_R - I_F)\tau_S$ and the lifetime is found from

$$\tau \approx \tau_S \left(\frac{I_R}{I_F} - 1\right) \quad \text{where } \tau_S \ll \tau \tag{II-17}$$

This same expression gives the approximate switching time, τ_S, of a driver which switches from forward bias, I_F, to reverse bias and has a reverse bias transient current switching capability of $I_R - I_F$

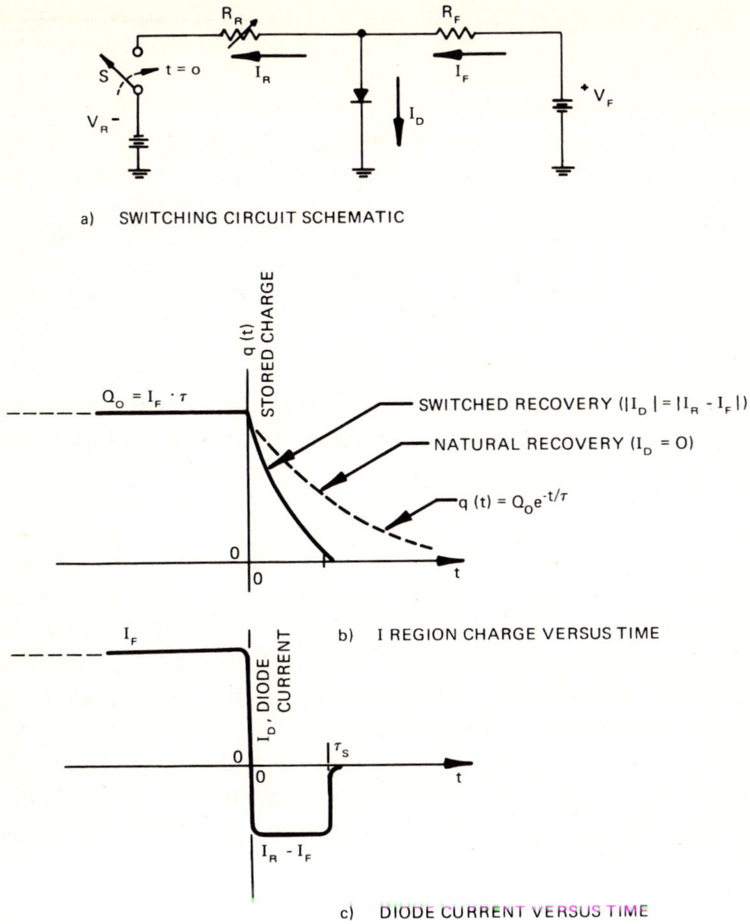

Figure II-11 **Lifetime Measuring Method**

amperes. Of course, in a practical driver circuit the forward current supply, I_F, would be switched off during reverse bias.

Practically, however, Equation (II-17) is not always directly useable because it may be difficult to switch the diode off in a time short compared with the lifetime. Typical PIN diode lifetimes may range from 0.1-10 μs, requiring extremely fast switches to satisfy the requirements that τ_S be small, say one-tenth, of the expected value of τ. To overcome this problem, a test setup is

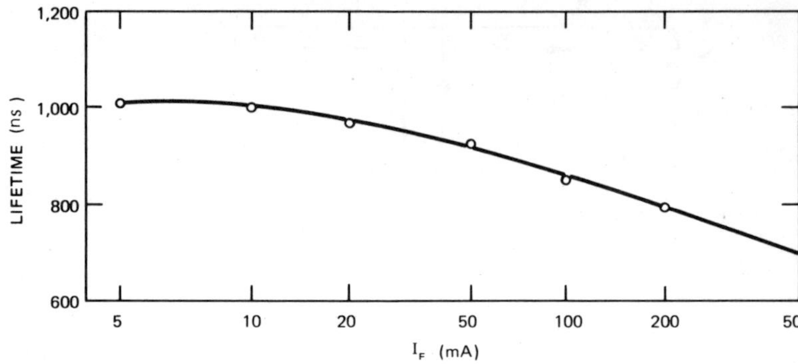

Figure II-12 **Typical Variation of PIN Lifetime with Forward Current (After Ciccolella, Johnston, and DeLoach [6 (p. 289)])**

made whereby the switching time can be adjusted. In the circuit of Figure II-11, R_R is made variable. The ratio I_R/I_F is adjusted so that $\tau_S = \tau$; the applicable condition is determined by analysis as follows.

If, at t = 0, I_F were turned off, the initial charge, Q_0 would decay at a rate proportional to the product of the instantaneous charge magnitude in the I region and the recombinate rate, $1/\tau$. Actually lifetime is somewhat dependent on the bias current level; a typical variation of PIN lifetime is shown in Figure II-12 [6] for a range of bias currents commonly used. Conventionally this variation of lifetime is ignored, not because it is insignificant, but because its inclusion would not permit the simple analysis which follows. Which is not to say that the analysis isn't useful. Common practice is to apply it, but one should be aware of its limitations.

With the constant lifetime assumption, suppose that the "driver" described in Figure II-11 provides no charge extraction current. Then for $t > 0$, the expression describing the instantaneous rate of charge (dq/dt) of charge in the I region, q, is

$$\frac{dq}{dt} = -\frac{q}{\tau} \quad (t > 0) \tag{II-18}$$

The solution of this differential equation is

$$q = Q_0 e^{-t/\tau} \qquad \text{(II-19)}$$

This equation shows the "natural recovery" curve in Figure II-11(b) and demonstrates the definition of lifetime as τ, the time constant of charge decay. In time $t = \tau$, q decays to 1/e or about 40% of its initial value. However, to make a practical measurement it is necessary to have a measurable quantity; this requirement is most easily fulfilled by providing a reverse current during recovery. Since the reverse current also removes charge from the I region, its effect must be included in the charge defining equation. Equation (II-18) then becomes

$$\frac{dq}{dt} = \frac{q}{\tau} + I_D \qquad \text{(II-20)}$$

This expression is called the *continuity equation* for stored charge; the name underlies the fact that stored charge is neither created nor destroyed instantaneously, but rather has time continuity. This equation is general and applies for charge building with I_D positive, as well as for recovery when the diode bias current direction is reversed and I_D is negative. The solution, which can be verified by substitution into Equation (II-20), is

$$q = Q_0 e^{-t/\tau} + I_D \tau \qquad \text{(II-21)}$$

Imposing the condition that the stored charge be depleted in time $t = \tau$, as shown graphically in Figure II-11(b), and noting that $Q_0 = I_F \tau$ and that $I_D = -(I_R - I_F)$, Equation (II-21) gives

$$\left| \frac{I_R}{I_F} \right| = e - 1 \approx 1.7 \qquad \text{(II-22)}$$

as the test condition under which $q = 0$ at $t = \tau_S = \tau$, permitting a convenient direct measurement. In practice, I_R and I_F can be monitored by connecting an oscilloscope across a small resistance in series with the diode under test. Switch S is realized using a pulse generator with repetition rate adjusted to permit the application of forward current, I_F, for a time which is long compared to the lifetime, τ, in order that the steady state charge, Q_0 equal to $I_F \tau$, is established before the recovery process is measured. A practical description of lifetime and switching speed measurements is given by McDade and Schiavone [7].

The Charge Control Model and Microwave Currents

We have just seen that, from transit time considerations, the PIN I region conductivity cannot follow a microwave signal because the diffusion of charge carriers isn't rapid enough to traverse the I region within the half period of an RF cycle. Moreover, from the preceding discussion of carrier lifetime it is clear that once charge is injected into the I region it resides there for 0.1-10 μs — the lower limit of which is even long compared to the 0.005 μs half period of, say, a 1 GHz signal. These two facts taken together indicate that the resistance behavior of the PIN at microwave frequencies can be described in terms of the charge present in the I region, q.

To illustrate this point, consider the diode I-V characteristic with superimposed RF excitations, as shown in Figure II-13. The I-V law shown is typical for a high voltage PIN. Under a forward bias current of 100 mA the I region becomes sufficiently conductive that its microwave impedance drops below 1 ohm of resistance (as we describe in the next section). If a microwave current having, say, 50 A peak amplitude (500 times the bias current) is then passed through the diode, the diode is found to remain in the low impedance condition despite the large "negative-going" half cycle of the RF waveform. The reason for this linear operation even under high RF current magnitudes is clear when the total charge movement produced by the RF signal is considered. Assuming a lifetime of 5 μs, typical of a high voltage PIN, the 100 mA bias results in a stored charge of 0.5 μC. However, during the negative-going portion of a 1 GHz sinusoid, the total charge movement is less than 0.025 μC, not even a tenth of the stored charge. This example epitomizes the charge control viewpoint that *it is the total stored charge produced by a bias which determines I region resistance rather than the instantaneous magnitude of an RF current.*

*Ryder** has likened the bias level on a PIN diode to "large signal" and the RF as the "small ac component," with respect to the amount of charge stored or removed from the I region. From the above example, the value of the charge control viewpoint is evident.

*R. Ryder (Bell Telephone Laboratories; Murray Hill, New Jersey) in a talk given at the NEREM Conference in Boston, circa 1970.

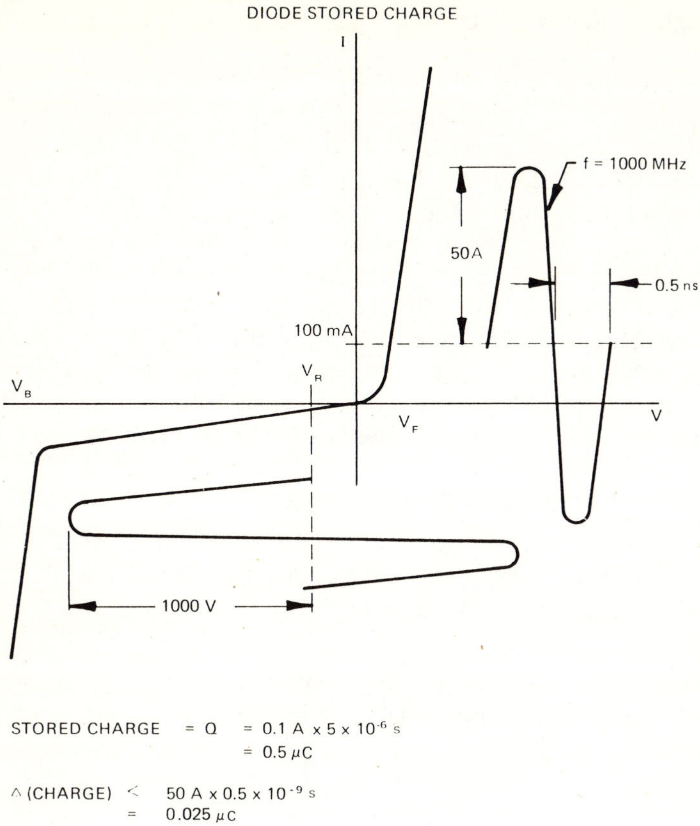

Figure II-13 **Example Comparing Charge Stored by Bias to Charge Movement Due to High Level Microwave Signal**

Under reverse bias, a relatively small voltage, about -100 V, is sufficient to hold off conduction of the diode under the application of an RF voltage whose peak voltage amplitude is as large as 1,000 V. Again, the brief duration of the half-period of the RF cycle is not sufficient to cause appreciable modulation of the I region of the diode, and the diode appears as a high impedance even with this large voltage magnitude applied.

One might ask why any reverse bias is necessary at all if the diode is nearly non-conducting at zero bias. First, reverse bias fully depletes the I region and its boundaries of charge. Thus, the diode has a higher microwave Q with reverse bias. Second, the role of a reverse bias is to maintain an average field which tends to prevent the accumulation of significant amounts of charge in the I region. The presence of excessive charge in the space, under high RF fields, can produce impact ionization, with a "runaway" current rise and resultant diode destruction. Nevertheless, under large RF excitation, impact ionization effects are often observed, resulting in a *pulse leakage current,* since it occurs only under the combined action of RF and reverse bias excitation. It is necessary that the driver circuit have *sufficiently low impedance* to be capable of providing this pulse leakage current (usually 1-5 mA) in a high power control device without causing an appreciable drop in the bias voltage supplied, if destructive diode conduction in the reverse bias state with high RF applied voltage is to be avoided.

2. Forward Biased I Region Resistance

Having demonstrated the suitability of the charge control approach for determining microwave properties, let us use it to calculate the conductivity and resistance of the I region under forward bias.

Conductivity, σ, is a bulk property equal to the ratio of current density, J, to applied electric field strength, E

$$\sigma = \frac{J}{E} \tag{II-23}$$

But J is the directed average rate of flow of electric charge. In terms of I region holes and electrons

$$\sigma = \frac{J}{E} = e\left(\frac{v_P \cdot p}{E} + \frac{v_N \cdot n}{E}\right) \tag{II-24}$$

Also, by definition, *mobility,* μ, is the average carrier velocity per unit of applied electric field, thus

$$\sigma = e(\mu_P p + \mu_N n) \tag{II-25}$$

where

$e = +1.6 \times 10^{-19}$ coulomb = magnitude of electron's charge

$\mu_{P,N}$ = mobility of holes and electrons respectively

p, n = respective injected hole and electron densities in I region

The formula for the resistance of a cylindrical conductor of electrical conductivity, σ, length W along the current path, and cross sectional area A is [9]

$$R = \frac{W}{\sigma A} \tag{II-26}$$

Using the dimensional notation of Figure II-10, the I region resistance is then

$$R_I = \frac{W}{eA(\mu_P p + \mu_N n)} \tag{II-27}$$

Three main assumptions* have been made in this derivation of R_I:

1) The I region as a whole is electrically neutral.
2) The bias current, I_0, injects holes and electrons which *recombine with each other in the I region;* the limitations of this assumption are discussed later.
3) The carrier lifetime is sufficiently long that both the holes and electrons are uniformly distributed within the I region. Another way of stating this point is that the average hole and electron diffusion lengths, L_P and L_N, are much longer than the I region width, W. This condition is usually valid for well designed PIN diodes and can always be verified by using the relation for diffusion length given below

$$L = \sqrt{D_{AP}\tau} \tag{II-28}$$

where

D_{AP} = ambipolar diffusion constant = $2D_PD_N/(D_P + D_N)$

τ = lifetime within the I region

In silicon, D_{AP} has an effective average value for holes and electrons, the *ambipolar diffusion constant,* of 15.6 cm^2/s [8]. Thus

*Fletcher, Neville H.: "The High Current Limit for Semiconductor Junction Devices," *Proceedings of the IRE, Vol. 45,* pp. 862-872, June 1957.

$$L = 40\sqrt{\tau\text{ (microseconds)}}\quad\text{(microns)}$$
$$L = 1.7\sqrt{\tau\text{ (microseconds)}}\quad\text{(mils)} \qquad \text{(II-29)}$$

For example, if the bulk lifetime is 10 μs the diffusion length is about 133 μ (5 mils).*

Under these combined assumptions, it follows that the injected hole and electron densities are equal and uniform

$$p = n \qquad \text{(II-30)}$$

and, furthermore, since they recombine with one another directly

$$\tau_P = \tau_N \qquad \text{(II-31)}$$

Then

$$R_I = \frac{W}{2e\,A\,\mu_{AP}\,p} \qquad \text{(II-32)}$$

where $\mu_{AP} = 2\mu_P\mu_N/(\mu_P + \mu_N)$, 610 cm^2/V-s in silicon [8], is the *ambipolar mobility*, i.e., the effective average of the hole and electron mobilities. But the injected charge is directly proportional to the bias current.

$$Q_P = epAW = I_0\tau \qquad \text{(II-33)}$$

Combining the last two equations gives

$$R_I = \frac{W^2}{2\mu_{AP}\,\tau I_0} \qquad \text{(II-34)}$$

This expression is applied frequently. We note from it that R_I is theoretically independent of I region area, being proportional to the square of I region width and varying inversely with mobility, lifetime, and bias current. However, care must be taken in the application of Equation (II-34) to practical situations. In particular, the following generalizations should be qualified:

1) *Holding all process steps the same except for varying A produces a selection of diodes with different capacitances but the same R_I for a given bias current.* This situation is true only if

*For an analysis of the case where this assumption is not made, see Leenov's paper, Reference 8.

τ remains constant; but generally, τ decreases with a decrease in A, since I region carriers are then nearer to the periphery where recombination can occur more rapidly.

2) *R_I decreases as $(1/I_0)$.* Again, this statement holds true only so long as τ remains constant. However, as I_0 increases, carrier density increases, and the recombination probability increases, decreasing τ. Furthermore, a saturation is reached when p and n increase sufficiently that substantial injection (holes into the N+ region and electrons into the P+ region) becomes significant, in violation of the second assumption used to derive Equation (II-34). Put simply, if there are high densities of electrons and holes in the I region, their chance for recombining increases, decreasing the average lifetime, τ.

3) *Above the transit time frequency, R_I is essentially independent of frequency.* This stipulation is only approximately true for most microwave PIN applications. Skin effect causes both the contact and I region resistances to increase somewhat with frequency.

Despite these limitations, Equation (II-34) is very useful and is typically invoked to estimate I region resistance at microwave frequencies. For example, consider a PIN with a 100 μ (4 mil) I region and a 5 μs lifetime operated with 100 mA bias current.

Using $\mu \approx 610$ cm^2/V-s

$$R_I = \frac{10^{-4}\ \text{centimeter}^2}{(2)(0.1\ \text{ampere})(5 \times 10^{-6}\ \text{second})(610\ \text{centimeters}^2/\text{volt-second})}$$

$$= 0.16\ \text{ohm} \tag{II-35}$$

This result is in reasonable agreement with the measured value of 0.3 Ω for a 1.56 mm (61 mil) diameter*, when one considers that the measured value includes resistive contributions of the ohmic contacts as well as those of the P+ and N+ regions. Furthermore, the lifetime at 100 mA is likely to be less than the 5 μs value which is measured at 10 mA — an additional factor contributory to a higher measured resistance than that calculated.

Using this example let us examine the role of skin effect in the forward biased I region. Using the parameters of the above example and solving Equation (II-26) gives $\sigma = 3\ (\Omega\text{-cm})^{-1}$. The skin depth, δ, in a conductor is given by [9]

$$\delta = \frac{1}{\sqrt{\pi f \mu_0 \sigma}} \tag{II-36}$$

*See data for the MA-47891 PIN diode in Table III-1, pp. 94-95.

where f = operating frequency (hertz)

$\mu_0 = 4\pi \times 10^{-9}$ henry/centimeter = free space permeability

σ = conductivity (ohm-centimeters)$^{-1}$

From Equation (II-36), the skin depth for $\sigma = 3\ (\Omega\text{-cm})^{-1}$ at 1 GHz is 0.09 cm, about equal to the diode radius. This diode example has a junction capacitance of about 2 picofarads and would not usually be used at frequencies much above 1 GHz. At higher frequencies a lower capacitance, and hence reduced diameter, would be employed. Thus, it can be seen I region* skin effect usually has but a moderate effect in PIN control devices in the 0.1 to 10 GHz frequency range.

Before leaving the subject of I region conductivity it is interesting to note what level of carrier density, p, was injected into the I region of this sample diode to produce $R_I = 0.16\ \Omega$. An estimate can be made using Equation (II-32) and $\mu \approx 610\ \text{cm}^2/\text{V-s}$, thus

$$p = \frac{W}{2eA\mu R_I} = 1.7 \times 10^{16}/\text{cubic centimeter} \qquad \text{(II-37)}$$

Since there is an approximately equal electron density, n, in the I region, the total free carrier density required to produce $R_I = 0.16\ \Omega$ is $3.4 \times 10^{16}/\text{cm}^3$. Recalling that the atom density is about $10^{23}/\text{cm}^3$, this figure represents *less than one carrier per million atoms.* It is therefore easy to see why the skin depth, so significant with metallic conductors at microwave frequencies, has only a moderate effect even under "high injection" levels in the I region of the PIN diode.

3. R_R and C_J Reverse Biased Circuit Model

Under reverse bias the I region is depleted of carriers and the PIN appears as an essentially constant capacitance to a microwave signal. The presence of dissipative losses can be taken into account by either a series or parallel resistance element in the equivalent circuit. In a well-made PIN, the I region has sufficiently high resistivity that most of the dissipation under low RF power conditions occurs in the ohmic contacts made to the diode and in the resistances of the P+ and N+ regions. Accordingly, a fixed series resistance, R_R, used to represent these losses can be expected to offer an equivalent circuit model which is applicable over a broader

*Skin resistance may be more important in the P and N regions and in the leads attached to them because it affects how the currents enter the I region.

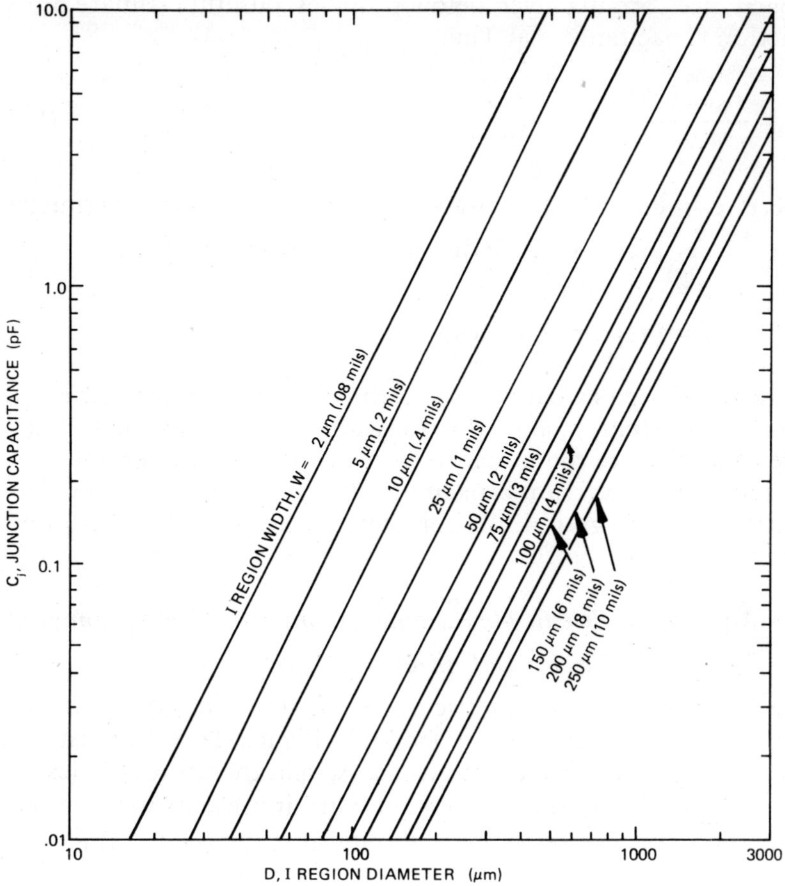

Figure II-14 **PIN Diode C_J vs. I Region Diameter and Thickness**

bandwidth than a parallel conductance. In any event, due to the ratio of diode capacitive reactance to practical RF circuit impedances, the dissipative losses of the PIN under reverse bias are usually much smaller than those under forward bias; thus, the choice of series or parallel R-C equivalent circuit under reverse bias usually can be made according to whichever offers greater computational convenience.

Because of the high relative dielectric constant for silicon (ϵ_R = 11.8), the fringing capacitance (in air) around the I region is relatively small and the capacitance calculated using the parallel plate capacitance formula given below provides a useful estimate of junction capacitance, C_J. Thus

$$C_J \approx \frac{\epsilon_0 \epsilon_R \pi D^2}{4W} \tag{II-38}$$

where $\epsilon_0 = 8.85 \times 10^{-14}$ farad/centimeter = free space permittivity

$\epsilon_R = 11.8$ = relative dielectric constant for silicon

D = junction diameter

W = I region thickness

For many design calculations — estimating thermal capacities, breakdown strength, and RF bandwidth — it is desirable to be able to interrelate the tradeoffs between I region dimensions (W and D) and junction capacitance (C_J). Figure II-14 shows Equation (II-38) graphically for typically available PIN I region widths.

4. Microwave Circuit Measurements and Cutoff Frequency, f_{cs}

Equivalent Circuit Definition and f_{cs}

The microwave equivalent circuit for the unpackaged PIN diode chip to be used in this text is shown in Figure II-15. In most applications the PIN diode is used as a switch; therefore, the less capacitance, the better an "open circuit" it presents with reverse bias. The lower the resistances, R_F and R_R, the smaller the dissipative losses, and, under forward bias, the more the diode resembles a "short circuit." A figure of merit has been defined [10] to relate the PIN's switching effectiveness, termed *switching cutoff frequency*, f_{CS}. The utility of this definition is apparent later in the discussion of performance limitations.

$$f_{CS} = \frac{1}{2\pi C_J \sqrt{R_F R_R}} \tag{II-39}$$

The equivalent circuit parameters are as defined in Figure II-15. Because the additional loss at high power is treated here by a separate equivalent circuit element, G_R, the *definition of f_{CS} as used in this text is limited to microwave power levels below the onset of nonlinear dissipation.* The effect of G_R is discussed in the next section.

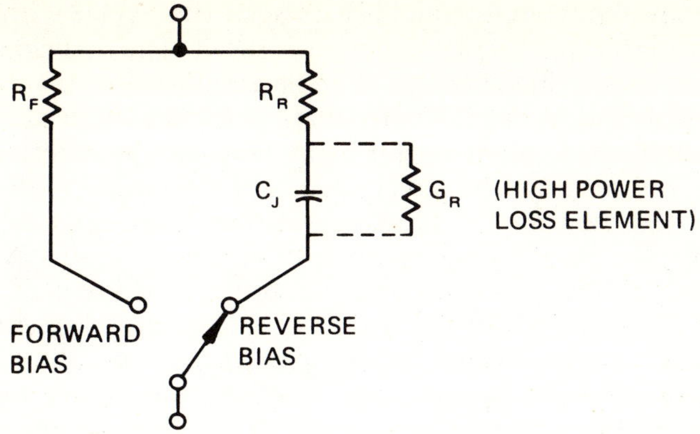

Figure II-15 **PIN Diode Chip Equivalent Circuit**

In principle, the values for R_F and R_R could be evaluated and f_{CS} could be specified under high power conditions. But it is not usually possible to obtain diodes characterized under high power as this task falls to the circuit designer; for this reason the separation of low and high power characterization is more consistent with actual practice.

Isolation Measurements

It was shown earlier, C_J measurements made at low frequency ($\approx$ 1 MHz) with sufficient reverse bias to deplete the I region provide a useable indication of the microwave capacitive reactance to be expected. However, the *resistances under forward, R_F, and reverse, R_R, bias conditions must always be determined by direct microwave measurements* since they include not only the inherent I region loss effects of the diode but P+ and N+ region as well as contact resistances, none of which is predictable with desirable analytic precision. Since, in most control device circuits, the greater microwave dissipation occurs under forward bias, the determination of R_F usually warrants the greater attention.

Many diode resistance measurement methods have been described. [11, 12] Ultimately the diode loss in the actual circuit of use is what is desired. For determining R_F, in either a test circuit or the actual circuit of use, the terminals where the diode is to be connected are short circuited and the loss of the circuit without diodes (i.e., the *cold circuit loss*) is measured. The additional circuit loss with diodes installed can then be attributed to the diodes themselves and, if the RF currents through the diodes can be estimated, the equivalent circuit parameters can be determined. Of course, by this time the insertion loss of the circuit under test is known and the value of knowledge of the diode equivalent circuit parameters is only of use in future design applications. Nevertheless, such direct evaluation in the circuit of end use is often required especially where diodes are circuit mounted in chip or beam lead configurations, requiring permanent bonding into the circuit to make the adequate ohmic contact necessary for accurate resistance determinations.

The two most common methods used to characterize PIN diodes, outside of the circuit of end use, are the *isolation* and *reflection* (or "slotted line") measurements.

To make an isolation measurement, the diode is used to interrupt a transmission line. When the diode is mounted in shunt with the line (Figure II-16), this method provides a sensitive measurement of the forward resistance, R_F. The isolation produced by a line shunting admittance Y is (the derivation is in Chapter V)

$$\text{Isolation} = \frac{P_A}{P} = \frac{V_A^2/Z_0}{V_L^2/Z_0} = |1 + YZ_0/2|^2 \qquad \text{(II-40)}$$

$$= 1 + GZ_0 + \frac{G^2Z_0^2}{4} + \frac{B^2Z_0^2}{4}$$

where $Y = Z^{-1} = G + jB$

To achieve the maximum test sensitivity, any series inductance introduced when mounting the diode across a transmission line is series resonated by a tunable capacitor. For this reason the measurement is most practical in the 0.5 to 1.0 GHz frequency range. Under these conditions the net series reactance, jX, of the mounted diode is zero and Equation (II-40) reduces to

$$\text{Isolation} = 1 + \frac{Z_0}{R} + \frac{Z_0^2}{4R^2} \qquad \text{(II-41)}$$

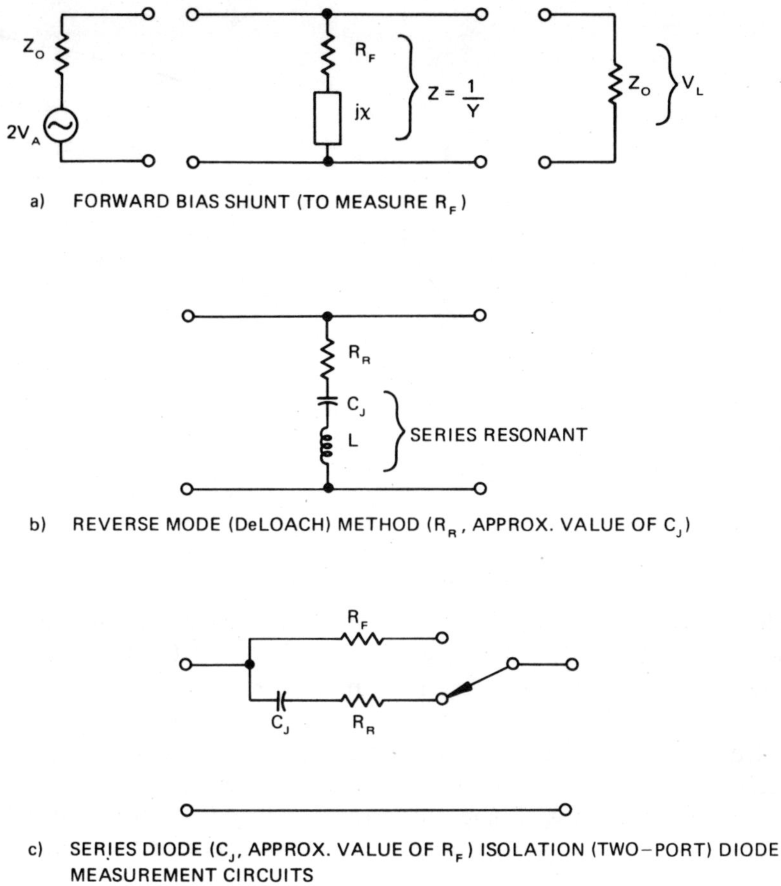

a) FORWARD BIAS SHUNT (TO MEASURE R_F)

b) REVERSE MODE (DeLOACH) METHOD (R_R, APPROX. VALUE OF C_J)

c) SERIES DIODE (C_J, APPROX. VALUE OF R_F) ISOLATION (TWO-PORT) DIODE MEASUREMENT CIRCUITS

Figure II-16 **Equivalent Circuits for Diode Measurements.**

Thus, for example, if a series resonated diode having forward resistance of 1 Ω shunts a 50 Ω line, the normalized conductance, GZ_0 equals $Z_0 R_F = 50$. The resulting isolation equals 676, or approximately 28 dB. The resistance limited isolation described by Equation (II-41) is encountered often in both diode measurement and SPST switch design. For convenient reference it is shown graphically in Figure II-17.

There are some fine points to be considered in performing this measurement. First, if the diode is mounted in a package, the

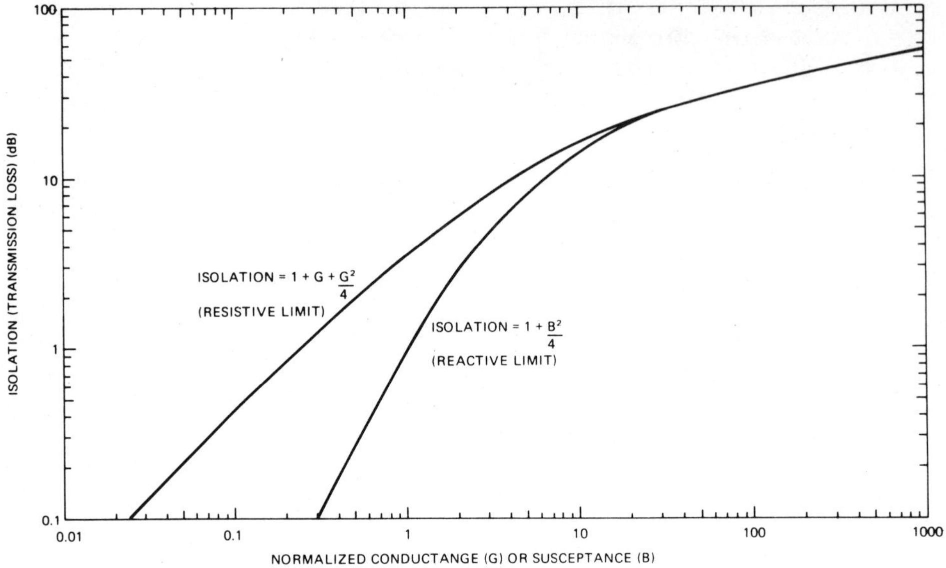

Figure II-17 **Isolation of Line Shunting G or B**

package capacitance transforms the effective resistance of the diode. This effect is usually negligible in the 0.5-1.0 GHz frequency band. Second, the circuitry used must not have significant leakage paths whereby power can reach the load from the generator by alternate paths such as higher order waveguide modes, fringing electric fields and so forth. This condition is readily tested by measuring isolation with the diode replaced by a short circuit of dimensions similar to the diode. Apart from ensuring that the leakage through the device is within acceptable limits, the isolation value so obtained gives an indication of the circuit *contact resistance. It is common practice to subtract contact resistance when quoting diode resistance.*

An interesting variation of the shunt mounted isolation measurement occurs when C_J series resonates with the mounted inductance of the diode (Figure II-16(b)). Then the diode shorts the transmission line under *reverse bias* and the isolation is a measure of R_R. From the isolation bandwidth an estimate of C_J is possible. This technique is usually used in waveguide at high frequencies, 5-15 GHz, to effect resonance with C_J. Switches built this way are often called *reverse mode*; the measurement technique is called the DeLoach method. [11] The reverse mode switching circuit is important for duplexer and radar receiver protector designs where isolation in the zero biased diode state is required, as well as in other *fail safe* applications where it is desirable that, should there be a failure of the driver to bias the diode, the high reflection state of the diode switch is obtained.

If the diode is mounted in series with the line (Figure II-16(c)) the high isolation condition gives a measurement of capacitive reactance, X_C, equal to $-(2\pi f C_J)^{-1}$, and hence, C_J. For an impedance $Z = R + jX$ in series with a line of characteristic impedance, $Z_0 = 1/Y_0$, the isolation, by duality, is given by the dual of Equation (II-41)

$$\text{Isolation} = \left|1 + \frac{ZY_0}{2}\right|^2 = 1 + RY_0 + \frac{(RY_0)^2}{4} + \frac{(XY_0)^2}{4} \qquad \text{(II-43)}$$

If $|X_C| > 15\, R_R$, as is almost always the case, the RY_0 terms in Equation (II-43) can be ignored with an error of less than 1%, and the reactance versus isolation can be read directly from the reactance dominated characteristic curve shown in Figure II-17. This method is especially useful for measuring the circuit mounted capacitance of low capacitance devices such as beam lead diodes.

Series mounted diodes require special equivalent circuit treatment. Figure II-18 shows schematically the electric field contours of a capacitor representing a reverse biased diode both within and without a series coaxial line mounting. Measured in free space, all E field lines terminate on the diode terminals directly, and a capacitance, C_0, is measured. When mounted in the coax line, however, some E field lines intercept the outer conductor. The effect is that the effective series capacitance, C, is less than C_0. An additional shunt capacitance, C_2, appears, but in most cases the effect of C_2 on the transmission line is negligble, since it serves to replace the distributed capacitance of the section of center conductor removed to install the diode. However, the fact that the mounted series capacitance, C_1, is less than the capacitance associated with the diode, C_0, means that a higher isolation is obtained in a switching circuit (generally a benefit). Moreover, in a phase shifter circuit a different phase shift than that anticipated will be obtained if this effect is overlooked.

The accuracy of the series measurement can be related to the loss and isolation measurement accuracies. Typically the series isolation

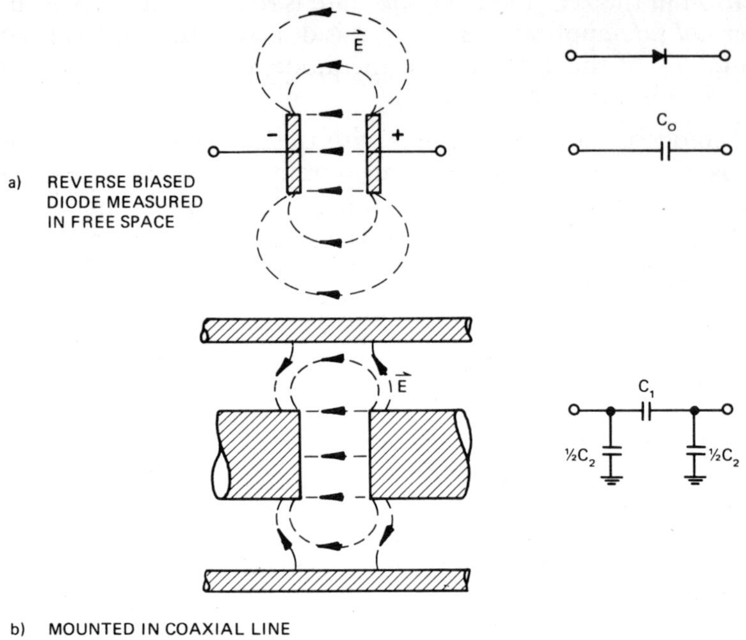

Figure II-18 **Change of Effective Series Capacitance with Circuit Mounting**

(or loss) measurement can be made to within an accuracy of ±0.1 dB for losses below 3 dB, and ±0.3-0.5 dB for isolation values from 10-40 dB. Thus, R_F can be determined to an accuracy of about ±2%; X_C, to about ±5% of the magnitude of Z_0. For most PIN diodes operated at sufficient forward bias to saturate the I region, $R_F < 1\ \Omega$; this measurement method would require impractically low Z_0 for meaningful measurements. However, for R_F measurements at low bias levels or with beam lead diodes wherein R_F = 2-10 Ω, the measurement is very practical using standard 50 Ω line. For example, a beam lead diode having $R_F = 5\ \Omega$ produces a 10% insertion loss of about 0.5 dB. The same diode, having C_J = 0.03 pF, has a reactance of -j795 Ω at 10 GHz, yielding, in the same test fixture, isolation under reverse bias at 10 GHz of 24 dB.

*Reflection Measurements**

Most of the principles described for isolation measurements are likewise applicable to reflection measurements wherein the diode is used to terminate the line (Figure II-19); the reflection coefficient (Γ equal to $\rho e^{j\phi}$) measurement is used to deduce diode parameters. Other things being equal, the sensitivity of this measurement method is about four times that of the matched load method described previously in the determination of R_F and R_R for a given line impedance and dissipation; therefore, it is used for most standard diode characterizations. Diode reactances under both forward and reverse bias can be determined from the reflection coefficient argument.

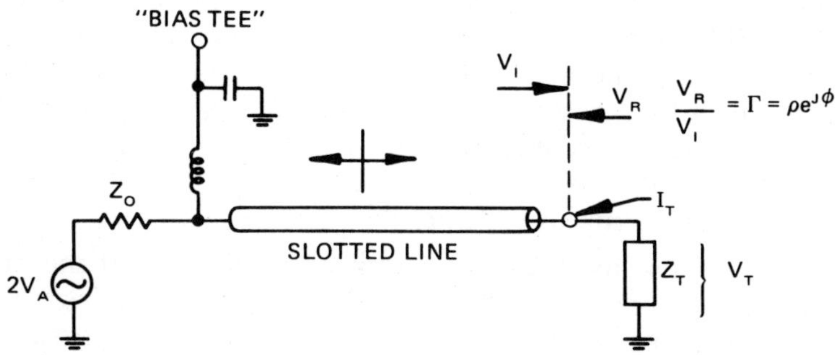

Figure II-19 **Reflection Measurement Equivalent Circuit**

The added sensitivity arises because, if the magnitude of Γ_L is either high or low compared to Z_0, the current (I_L) or voltage (V_L) at the end of the line is nearly double the value (V_A/Z_0 or V_A, respectively) experienced under matched load ($Z_L = Z_0$) conditions; the relative power absorbed in the diode consequently increases fourfold. For both a load impedance $Z_L = R_L + jX_L$ and a line with Z_0 characteristic impedance, the reflection coefficient at the load position is [13, 14]

$$\Gamma = \rho e^{j\phi} = \frac{Z_L - Z_0}{Z_L + Z_0} = \frac{(R_L - Z_0) + jX_L}{(R_L + Z_0) + jX_L} \qquad \text{(II-44(a))}$$

$$\phi = \tan^{-1}\left(\frac{X_L}{R_L - Z_0}\right) - \tan^{-1}\left(\frac{X_L}{R_L + Z_0}\right) \qquad \text{(II-44(b))}$$

*See Appendix J, The Smith Chart, for the development of reflection principles.

$$\rho = \frac{(R_L - Z_0)^2 + X_L{}^2}{(R_L + Z_0)^2 + X_L{}^2} \qquad \text{(II-44(c))}$$

The fractional dissipation in Z_L is

$$\text{Insertion Loss} = 1 - \rho^2 = \frac{4\,R_L Z_0}{(R_L + Z_0)^2 + X_L{}^2} \qquad \text{(II-45)}$$

Under forward bias, using $Z_0 = 50\ \Omega$, both R_L and X_L are usually much less than Z_0. The fractional power loss is approximately equal to $4R/Z_0$. Under the same approximation, the insertion loss ratio by the series isolation measurement is approximately equal to $1 + R/Z_0$ and the *fractional loss* is approximately R/Z_0, only one fourth that of the reflection measurement. Thus, for example, with $Z_0 = 50\ \Omega$ and $R_F = 1\ \Omega$, the measured loss is about 0.4 dB with the reflection method and 0.1 dB with the isolation method, giving (with ±0.1 dB accuracy) the determination of R_F with ±0.25 Ω and ±1.0 Ω accuracies, respectively. Accordingly, when the diode is mounted in series with the line, the reflection measurement is usually employed for determining R_F.

This method is also used for determining R_R and C_J, but the calculations are less convenient than for R_F because the series reactance, X, cannot be ignored. Furthermore, the impedance transformation effects of a diode package (the package being necessary if the diode is to be conveniently mounted at the end of a slotted line) are not negligible in the reverse biased condition. For routine diode evaluation the reflection coefficient magnitude, ρ, and phase, ϕ, are measured and the exact equations relating diode C_J and R_R are solved using a computer program.* It is common practice to use a coaxial line and obtain a zero impedance reference by short circuiting the line at the leading surface of the diode package (as shown in Figure II-20). Both a phase reference ($\phi = 180°$) and a loss reference ($\rho = 1.0$) are thereby established. The packaged diode impedance, Z_L, is then evaluated by solving Equation (II-46) for Z_L

$$Z_L = R_L + jX_L = Z_0\,\frac{(1+\Gamma)}{(1-\Gamma)} = Z_0\,\frac{(1+\rho e^{j\phi})}{(1-\rho e^{j\phi})} \qquad \text{(II-46)}$$

*Computer programming for circuit evaluation is discussed in Chapter VI.

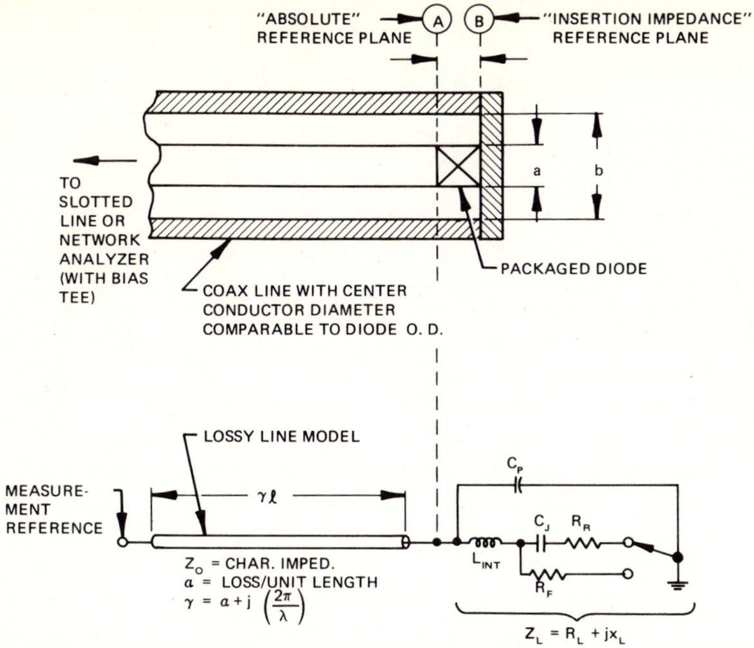

Figure II-20 **Schematic and Equivalent Circuit Detail for "Slotted Line" Measurement of Packaged PIN Diode**

It is illustrative of the measurement method to plot the reflection coefficients obtained under forward and reverse bias on the Smith Chart. For this example, consider a PIN with C_J = 2 pF, $R_R = R_F$ = 0.3 Ω mounted in a package having C_P = 1 pF and L_{INT} = 0.3 nH. At 3 GHz the *chip* impedances are 0.3 Ω and (0.3–j26.5) Ω under forward and reverse bias; they are transformed by the package to (0.37+j6.30) Ω and (0.15–j15) Ω, respectively. When normalized to Z_0 = 50 Ω the corresponding reflection coefficients are (from Equation (II-44)) Γ = 0.986/165.6° with forward bias and Γ = 0.976/-145.8° with reverse bias. These results are shown graphically in Figure II-21. The proximity of the points to the Smith Chart periphery (ρ = 1.0) underscores the need for careful measurements if R_F and R_R are to be evaluated accurately.

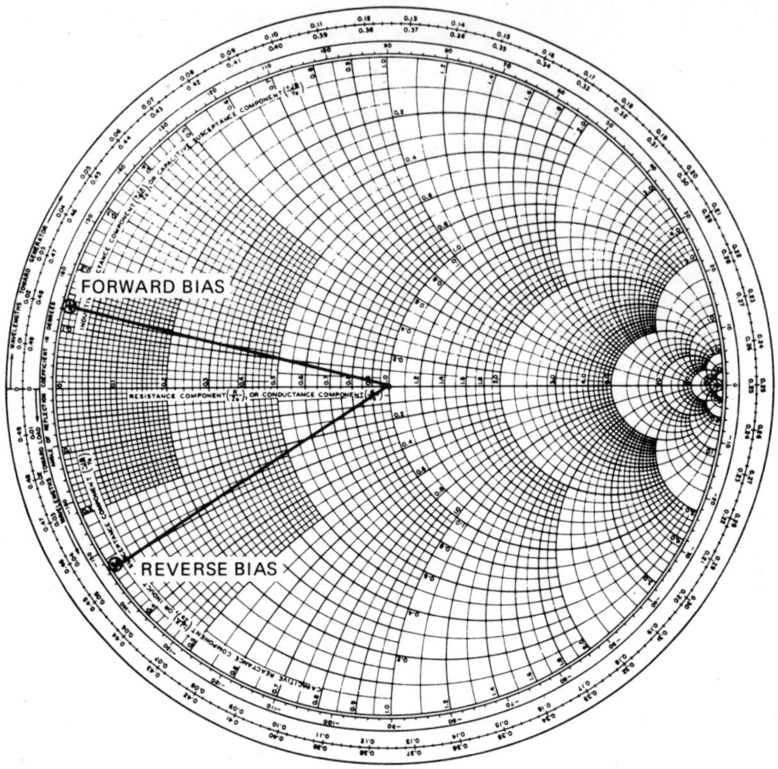

Figure II-21 **Reflection Coefficients for Measurement Example**

While the application of the principles of this method is straightforward, great care must be exercised if results with useful accuracy are to be obtained. Since a single frequency measurement of Γ produces only two bits of data, ρ and ϕ, it is necessary either to have foreknowledge of the values of package *parasitics* (internal inductance, L_{INT}, and package capacitance, C_P) or to perform the reflection measurement at more than one frequency in order to solve for C_P, C_J, R_R, and R_F. In practice the former method is usually followed. C_P is first determined using an empty diode package; this result for C_P is used with an internally shorted package having a wire or strap lead similar to that to be employed with the diodes to be measured. In this respect, the eventual accuracy of evaluation of C_J and R_R is dependent upon the reproducibility of

C_P and L_{INT} and their relative reactances compared to that of C_J. Furthermore, since the point of measurement and diode reference plane are separated by a line with finite loss, the resulting *lossy line transformation* (see Chapter VI, Equation (VI-2)) must be taken into account when R_F and/or R_R are small ($<$ 2% of Z_0) as is usually the case. This requirement necessitates a computer program to reduce the data if such measurements are to be made routinely.

Diode Inductance Measurements and Definitions

It should be noted that values measured for diode impedances depend to some extent on the test fixture — especially with inductive reactance, which can *only* be specified in terms of a return path. For example, the inductance per unit length of coaxial line having an outer conductor diameter, b, and an inner conductor diameter, a, is [9]

$$L = \frac{\mu_0}{2\pi} \ln \frac{b}{a} \tag{II-47}$$

and the characteristic impedance Z_0 is [9]

$$Z_0 = \frac{1}{2\pi} \sqrt{\frac{\mu_0}{\epsilon_0}} \ln \frac{b}{a} \tag{II-48}$$

For 50 Ω, the ratio b/a equals 2.3 for air dielectric coax. Suppose a packaged diode having 2.5 mm (0.1 in) length and effective diameter, a of 1.25 mm (0.05 in), is first measured under forward bias in a 50 Ω line having b equal to 15 mm (0.6 in); the inductance is 0.62 nH. If, however, the same measurement is performed using a smaller diameter 50 Ω coaxial line in which b equals 7.5 mm (0.3 in), the inductance is 0.45 nH.

Not only the absolute circuit dimensions but also the reference plane definition affects the inductance determination. For example, the above determination of inductance corresponds to reference plane A selection in Figure II-20. If, however, reference plane B were selected — by replacing with an equivalent length of center conductor to obtain a short circuit measurement reference — *insertion impedance* would be obtained. Insertion impedance is Z_L less the Z_S of the short circuit terminated length of the measurement line, ℓ, neglecting line loss

$$Z_S = jZ_0 \tan\left(\frac{2\pi\ell}{\lambda}\right) \tag{II-49}$$

where λ = wavelength at test frequency.

If, as is usually the case, $20\ell < \lambda$, the value of the tangent term can be replaced by its argument (within 3%). Furthermore, Z_S is an inductive reactance ($j2\pi fL_S$); for a coaxial line

$$\begin{aligned} L_S \text{ (nanohenries)} &\approx 0.0033 \cdot \ell \text{ (millimeters)} \cdot Z_0 \text{ (ohms)} \\ &\approx 0.084 \cdot \ell \text{ (inches)} \cdot Z_0 \text{ (ohms)} \end{aligned} \tag{II-50}$$

For the example cited, $\ell = 2.5$ mm, $L_S = 0.41$ nH, and the diode respective *insertion inductance** values determined from measurements in the two line sizes is 0.21 and 0.04 nH respectively.

These examples highlight the importance, especially for inductance measurements of specifying both the measurement fixture and reference plane selection. Similar reasoning indicates that if the actual circuit of use does not duplicate these conditions — and usually it doesn't — calculations of performance sensitive to inductance will be inaccurate unless the new conditions are taken into account.

C. High RF Power Limits

1. Forward Biased Limits

Under forward bias the PIN diode chip usually has an RF resistance of 1 Ω or less. Failure of the diode in this bias state will occur if the dissipative heating (I^2R_F) is sufficient to cause the diode temperature to rise sufficiently to induce metallurgical changes. For silicon and its dopants, this point is not reached until a temperature of about 1000°C. However, the metal contacts at the silicon boundaries introduce failure mechanisms in the vicinity of 300-400°C, at which temperatures common contact metals form eutectic alloys with silicon. For example, the gold-silicon eutectic occurs at 370°C [16]. Repeated or continuous exposure of silicon to the eutectic temperature in the presence of the corresponding metal can produce conducting filaments of metal-silicon alloy, which eventually "grow" across the I region of a PIN diode, short circuiting it. This structural change of the diode crys-

*Also sometimes called *excess inductance.*

tal is the most common diode failure mechanism with heat, even with reverse breakdown induced failures, described subsequently.

Failure of a diode does not occur instantaneously when an over stress is applied unless the resulting temperature greatly exceeds 300°C, as can occur with filamentry heating produced by avalanche breakdown in the reverse bias condition. This situation is also discussed subsequently. Except for the rapid failure induced by avalanche breakdown, thermally produced failures proceed over a time period related to the ratio of the operating temperature, T, to that which causes near instantaneous burnout. Rather extensive experiments carried out on computer diodes have shown that the mean time to failure can be described by the empirical relationship [15] given by Equation (II-51)

$$t_M(T) = Ae^{+Q/kT} \qquad \text{(II-51)}$$

where $t_M(T)$ = the mean time to failure at operating temperature T

A = a constant

Q = the "activation energy" constant

k = Boltzmann's Constant

T = the average device temperature in Kelvin (= °C + 273)

This expression is called the Arrhenius Law. It can be applied when the variation of operating life with temperature is determined by *only one* failure mechanism — for example, the formation of a particular alloy of the metallization system with the silicon.

To apply this relationship, the failure temperature, T_F, is first determined for the diode type; it depends on the semiconductor material (usually silicon for a PIN) and the metallization system. Next the device is operated at a lower temperature for a period until 50% of the samples under test fail, establishing a data point along the temperature-time graph. Additional data points at different temperatures are determined to allow for averaging of experimental data. This process, called *step-stress temperature testing,* is time-consuming because data points corresponding to hundreds and thousands of operating hours are required if the failure curve is to be established with sufficient accuracy to permit meaningful extrapolation to long life operation — on the order of years.

Care must be exercised that only the common *thermal* failure mode applies throughout the step stress tests. Careful analysis, usually including sectioning of failed diodes, is required to confirm the failure mode of each diode specimen used to establish the failure curve. The resulting temperature-time data plotted on semi-log paper form a straight line, permitting extrapolation for longer periods. Figure II-22 shows a typical plot for a surface-glass passivated, mesa type, high voltage PIN diode used in a phased array application. Notice that with a 200°C junction temperature (often cited as a safe operating limit for semiconductor devices) the anticipated mean life is 1000 hours (or 0.1 years), while for 140°C the anticipated mean life is extended to 1,000,000 hours (or 114 years). Accordingly the role of operating temperature must be given careful consideration if the estimate of anticipated life is to be meaningful.

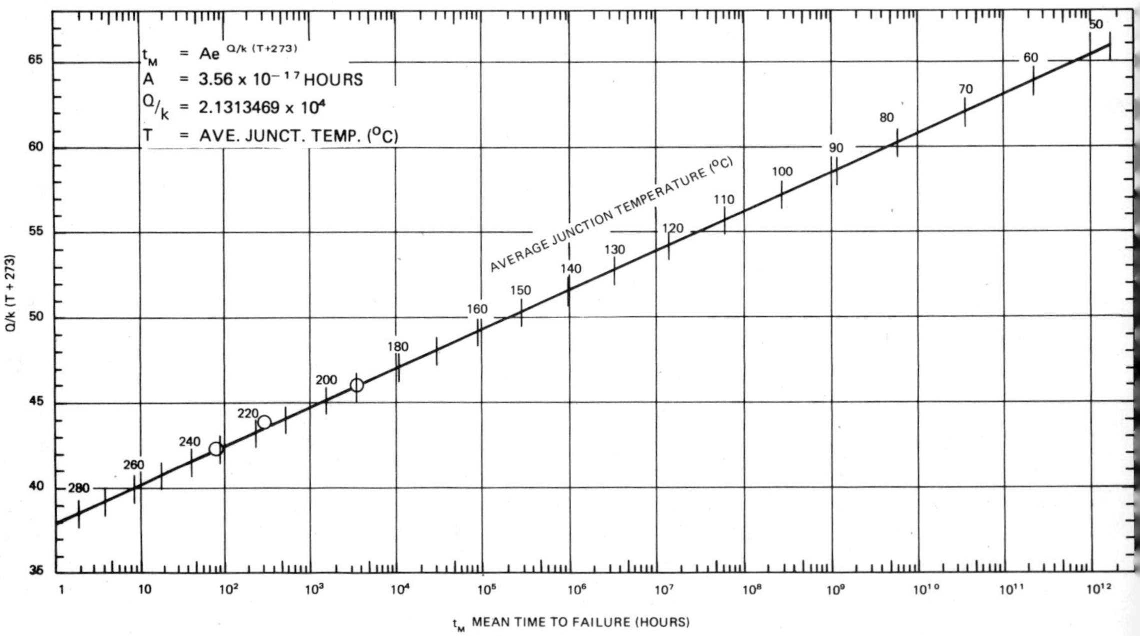

Figure II-22 PIN Life Expectancy vs. Temperature (Courtesy P. Ledger, Microwave Associates, Inc.)

2. Reverse Biased Limits

The reliability criteria apply for reverse biased operation as just discussed for forward bias. The junction temperature is again the result of ambient and RF heating. Unlike the forward biased condition, however, the fractional RF insertion loss does not remain nearly constant once the applied RF voltage has a magnitude which is comparable to either the reverse bias and/or the diode's reverse breakdown voltage. Under these conditions diode dissipation (is nonlinear and increases more rapidly than RF power, producing at times a runaway insertion loss). The onset of this rapidly increasing insertion loss nonlinearity can be used as a practical measurement that the destructive temperature has been reached in the reverse biased state, since diode failure usually occurs if the incident RF power level is increased much beyond this level. Figure II-23 shows a typical insertion loss versus RF power characteristic obtained with a reverse biased diode phase shifter. The mechanisms of reverse biased diode failure under RF voltage stress are

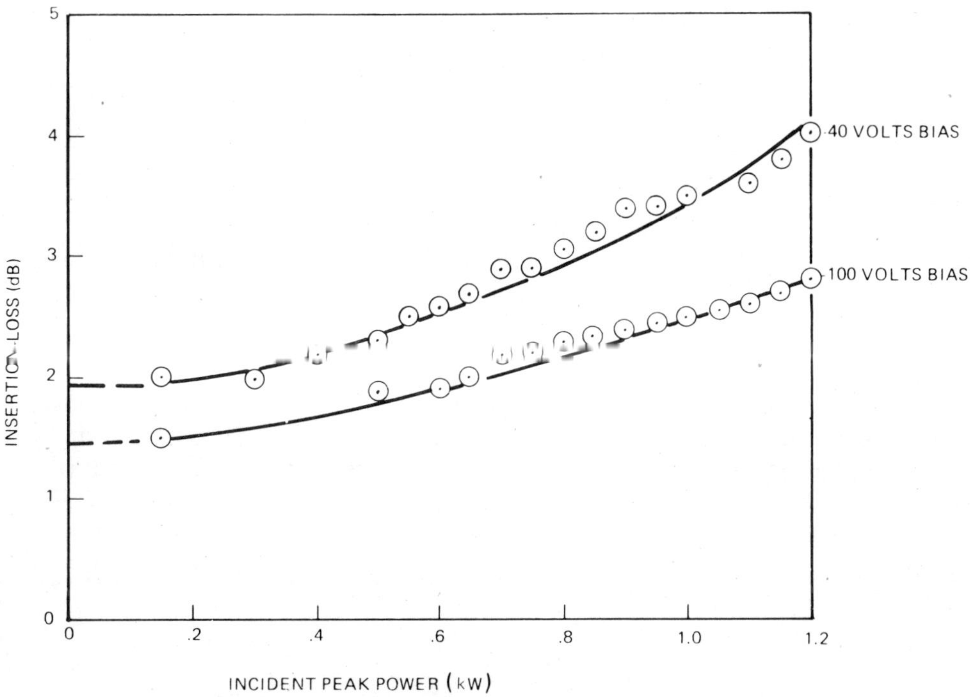

Figure II-23 **Insertion Loss vs. Peak Power for a 4 Bit X-Band Phase Shifter Under Reverse Bias in All Bits (Phase Shifter Described in Chapter IX, pp. 470-475)**

not sufficiently evaluated for a definitive theory to be developed, largely because of the difficulty of performing such measurements, with enough samples to have adequate statistical data. Qualitively, two conditions in which I region charge is generated occur and which one predominates depends, as described in Figure II-24, upon the relative magnitudes of the peak RF voltage, V_P; the bias voltage, V_{BIAS}; and the diode breakdown voltage, V_{BD}, as described in Figure II-24.

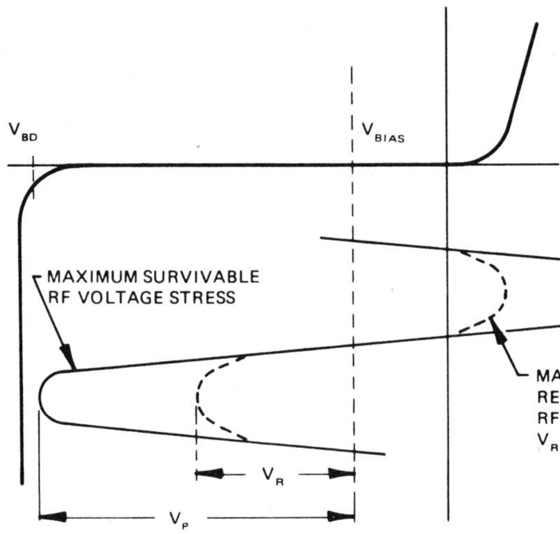

Figure II-24 **Operation at High RF Voltage and Reverse Bias**

The condition shown in Figure II-24 with voltage, V_P, is representative of typical operation near the failure limit. The RF voltage has a large excursion into the forward direction. Although, as has been shown earlier, the duration of this half cycle is insufficient to result in conduction by the diffusion transit (injection) of P region holes and N region electrons across the I region, *some charge* is introduced into the I region from these boundaries, and not all of it is extracted by the combined action of the reverse bias and the negative-going half of the RF cycle before the next forward-going RF voltage excursion. However small the magnitude of the injected charge may be, it can increase multiplicatively with each succeeding RF cycle through *impact ionization*; electron-hole pair production results when mobile carriers accelerated by the high RF field strike silicon atoms in the I region lattice with sufficient energy to promote valence band electrons to the conduction band. This cause of increased insertion loss can be identified experimentally by its bias voltage dependence. Increasing the magnitude of reverse bias voltage sweeps such injected charge out of the I region more effectively and thereby extends to a higher applied RF voltage the onset of rapid insertion loss increase, which precedes what, for present purposes, is called the *injection mode* failure mechanism.

The second mechanism causing nonlinear insertion loss is the *direct impact-ionization mode,* occurring when the combined RF + bias voltage exceeds the diode bulk breakdown, i.e, $V_P + V_{BIAS} > V_{BD}$. In this case no partial injection is needed to initiate impact ionization; the requisite electron-hole pairs are obtained directly by high electric field ionization of I region silicon atoms. One might think this mechanism would be eliminated by reducing the bias voltage, since this action would reduce the combined magnitude $(V_{BIAS} + V_P)$; but, in most practical cases, where the bias is 10-20% of V_{BD}, reduction of V_{BIAS} would only precipitate the injection-mode failure. An exception, of course, is when the bias is kept at half the breakdown (i.e., $V_{BIAS} = V_{BD}/2$). Then the RF waveform makes no injecting excursion into the forward direction. But for high power switching applications, a driver circuit to accomplish this end requires prohibitively high voltage transistors (300 V or more); the overall expense of the RF control circuit with driver could be more readily reduced by using a larger number of PIN diodes operated with less RF voltage stress.

Practically, the maximum sustainable RF voltage, V_P must be determined by measurement. Taking high power loss data for the diode — RF frequency, pulse length, and duty cycle of intended use — is the most direct and effective technique. In no case can $V_P + V_{BIAS}$ exceed V_{BD}, where V_{BD} is the bulk breakdown voltage of the diode.* The bias voltage is selected to be as large as is possible in a practical driver circuit (usually 10-20% of the diode's reverse bulk breakdown voltage) and the RF power level (from

*See Chapter III for bulk breakdown voltage evaluation of practical diodes.

which the corresponding voltage, V_P, can be calculated) is set at the point at which a statistical sample of diodes have been found to undergo rapid loss increase and/or failure. Failures due to either of the two modes described are usually evidenced by a permanent short formed by a conducting filament across the I region.

Most high power switching applications use the PIN diodes in a transmissive circuit with a matched load. Accordingly, failure or removal of the load, transmission line arcing, or any mechanism which affects the load match can result in a voltage reflection and possible RF voltage enhancement at the diode. Neglecting losses in the switching circuit and diodes, this reflection voltage enhancement can double the stress on the diodes. Such a condition, even if encountered only briefly, usually precipitates diode failure. Therefore it is good practice to rate the diode at a stress level $V_R \leqslant V_P/2$ (see Figure II-23) in order for the device to be able to survive such a total reflection. Since power is proportional to the square of voltage, *PIN diode devices should be rated at one-fourth or less of the power level at which, with matched load, they would be expected to undergo near instantaneous failure.* Even if provision has been made to minimize the likelihood of a totally reflecting load, consideration should be given to the following factors before opting a power safety factor of less than 4 to 1:

1) Diode failure is not an exactly reproducible event even with PIN's made by the same process within the same lot. A 2 to 1 variation in burnout is typical for a given process. Thus, a production run of diodes may have a considerably lower (or higher) burnout than experienced with a prototype test lot.
2) Most high power tests are conducted at room temperature while practical devices usually must perform at considerably higher temperature, reducing the power safety margin that is inferred from a room temperature test.
3) High power RF testing is often of short duration, an hour or less, due to the generally limited availability of high power testing facilities. However, semiconductor devices are usually expected to have useful lifetimes of years. Derating, according to Figure II-22, is necessary to accomplish long life.
4) PIN devices operated at or below one-fourth of their burnout power are typically found to be able to survive temporary driver failures wherein the high power RF signals are applied with the diodes at zero bias.

Generally PIN devices to control pulsed high RF power are limited in the reverse biased state by the maximum safe rated RF voltage stress, V_R, as is seen in the circuit discussions to follow. While a device which fails to meet circuit performance expectations may cause some user disappointment, it has been the author's observation that nothing quite equals the state of dissatisfaction resulting when solid-state control devices fail catastrophically due to overrating. No doubt it is for reasons such as this one that it has been industry practice in large phased array systems to design PIN phase shifters to survive operation into a short circuit load of any phase.

Only by carefully rating these devices can the good reliability which has come to be expected, indeed often assumed without question, of solid-state control be sustained. Accordingly, the designer should adopt as a minimum a policy of both designing diode control devices to sustain operation into a short circuit of any phase and testing throughout production to insure that this level, at least statistically, is maintained for the complete population of devices built.

References

[1] Phillips, A.B.: *Transistor Engineering and Introduction to Integrated Semiconductor Circuits;* McGraw-Hill, Inc., New York, 1962.

[2] Shockley, W.: *Electrons and Holes in Semiconductors;* D. VanNostrand Co. Inc., Princeton, N.J., 1953; pp. 318-321.

[3] Blakemore, J.S.: *Semiconductor Statistics* Chapters 5-9); Pergamon Press, New York, 1962.

[4] *Hewlett-Packard Application Note,* "The Step Recovery Diode;" Hewlett-Packard Inc., Palo Alto, California, December 1963.

[5] Moll, J.L.; Krakauer, S.; and Shen, R.: "PN Junction Charge Storage Diodes," *Proceedings of the IRE, Vol. 50,* pp. 43-53, January 1962.

[6] Watson, H.A., (ed.): *Microwave Semiconductor Devices and Their Circuit Applications;* McGraw-Hill, Inc., New York, 1969.

[7] McDade, J.C. and Schiavone, F.: "Switching Time Performance of Microwave PIN Diodes," *Microwave Journal,* pp. 65-68, December 1974.

[8] Leenov, D.: "The Silicon PIN Diode as a Microwave Radar Protector at Megawatt Levels," *IEEE Transactions on Electron Devices, Vol. ED-11, No. 2,* pp. 53-61, February 1964.

[9] Ramo, S. and Whinnery, J.R.: *Fields and Waves in Modern Radio;* John Wiley and Sons, New York, 1944 and 1953. Also revised by Ramo, S., Whinnery, J.R., and Van Duzer, Theodore, as *Fields and Waves in Communication Electronics;* John Wiley and Sons, New York, 1965.

[10] Hines, M.E.: "Fundamental Limitations in RF Switching and Phase Shifting Using Semiconductor Diodes," *Proceedings of the IEEE, Vol. 52,* pp. 697-708, June 1964.

[11] Deloach, B.C. Jr.: "A New Microwave Measurement Technique to Characterize Diodes and an 800 GHz Cutoff Frequency Varactor at Zero Volts Bias," *1963 IEEE MTT Symposium Digest,* pp. 85-91.

[12] Getsinger, W.J.: "The Packaged and Mounted Diode as a Microwave Circuit," *IEEE Transactions on Microwave Theory and Techniques, Vol. MTT-14, No. 2,* pp. 58-69, February 1966. Also see by the same author, "Mounted Diode Equivalent Circuits," *IEEE Transactions on Microwave Theory and Techniques, Vol. MTT-15, No. 11,* pp. 650-651, November 1967.

[13] Altman, Jerome L.: *Microwave Circuits;* D. VanNostrand Co., Inc., New York, 1964.

[14] Collin, R.E.: *Foundations for Microwave Engineering;* McGraw-Hill, Inc., New York, 1966.

[15] Peck, D.S.; and Zierdt, C.H. Jr.: "The Reliability of Semiconductor Devices in the Bell System," *Proceedings of the IEEE, Vol. 62, No. 2,* pp. 185-211, February 1974.

[16] Hansen, Max: *Constitution of Binary Alloys,* McGraw-Hill, Inc., New York, 1958.

Questions

1. What is the punchthrough voltage of a PIN diode having a 50μ (2 mil) I region width and I region resistivity of 300 Ω-cm?
2. What is the dielectric relaxation frequency of the PIN diode in Question 1?
3. What is the transit time of the PIN diode in Question 1?
4. If the PIN diode in Question 1 has an average carrier lifetime of $\tau = 2\ \mu s$, what is the microwave resistance of the I region, R_I, when a forward bias of 50 mA is applied?
5. For the conditions in Question 4 what is the value of R_I 1 μs, 10 μs, and 15 μs following the turn off of the forward bias? Neglect reactive effects and assume a constant lifetime of 2 μs, and that the forward bias had been on long enough to establish a steady state charge in the I region before turnoff.
6. If the diode in Question 5 has contact resistance of 0.2 Ω what isolation does it produce when it shunts a 50 Ω line with negligible inductance (SPST switch) with 50 mA bias? What is the isolation 1 μs, 10 μs, and 15 μs after the bias is turned off?

Chapter III

Practical PIN Diodes

A. Basic Parameters – I Region Thickness and Area

The selection of a PIN to perform a particular microwave switching function begins with the choice of the size of the I region – specifically, its thickness, W, and cross sectional area, A. In practice, however, this characterization is usually accomplished using the related variables junction capacitance, C_J, which depends upon both A and W, and bulk breakdown voltage, V_{BB}, which is proportional to W. What are the basic considerations in making these selections?

Consider first the I region thickness. A wide I region gives high bulk breakdown voltage, an advantage for high power switching applications. It also gives a low transit time frequency (see Figure II-9), which has two principal effects. First, the diode has less tendency to rectify at microwave frequencies, a benefit for high power switching but a drawback if the diode is to be a self actuated microwave limiter. The second effect is that the diode is slower to change its impedance when the bias is switched. We discuss switching speed further in Chapter IV; here, it is noted that wider I region means longer switching time

A wide I region also usually requires large bias current and voltage magnitudes. We saw in Equation (II-34) that the microwave resistance of the I region, R_I, under forward bias varies inversely with I_0; to achieve a low resistance, I_0 must be increased along with W if the same R_I is to be obtained. We avoid saying I_0 must increase as W^2 for given R_I since, as discussed in Chapter II, carrier lifetime also increases somewhat with W. The net effect, however, is an increase in R_I as W is widened. Similarly the wide I region diodes used for high RF power switching also require high reverse bias voltage if the I region is to be kept fully depleted of charge – and to remain non-conducting – during the application of large RF voltages.

Next consider the effect of I region area on performance. The larger the area (having chosen W), the greater the junction capacitance, with the particular value of C_J being estimated from Figure II-14. The most direct result of large capacitance is an inherent limitation on RF circuit bandwidth. In switching applications the capacitive reactance of the diode usually can be neutralized resonately using other circuit elements such as inductors and line lengths. However, such tuning always has frequency sensitivity which is more pronounced as the capacitance is increased. Thus, large area limits bandwidth. Not only does increased diode area reduce performance through increased capacitance but, other things being equal, it also increases manufacturing cost since fewer diode chips are obtained from a slice of silicon material.

Still another disadvantage of large area is that the diode cutoff frequency is lessened because, while the resistances under forward and reverse bias do decrease with area, they do not drop fast enough to compensate the increased capacitance, and so cutoff frequency (Equation (II-39)) decreases with A. We see later that a decrease in cutoff frequency increases the insertion loss that theoretically can be obtained in a loss equalized (same loss under forward and reverse bias) switch. However, before concluding that large area diodes always give higher loss, it should be noted that in many practical device designs considerably greater diode loss occurs under forward bias than under reverse; in such cases the reduction of R_F which occurs when junction area is increased provides an overall loss reduction.

Finally, larger area increases both pulsed and average high power handling capability. Pulsed heat sinking capacity (i.e., the amount of heat which the I region can absorb for a given temperature rise) increases with A since the I region volume increases accordingly. Average power dissipating capability likewise increases with junction area as, generally, a more efficient path is provided for heat flow from the I region to the metallic base to which the diode chip is mounted.

B. Table of Typically Available PIN Diodes

Once the I region width and area (or equally, the bulk voltage breakdown and capacitance) have been specified, the remaining characteristics are, within manufacturing limits, also established. Conversely, from a circuit design viewpoint, if alternate characteristics such as, say, switching speed and forward biased resistance are specified, then maximum I region thickness and minimum area effectively are established – although empirical measurements may be required to determine the relationships between these parameters and the junction dimensions. To facilitate the selection of PIN control diodes, Table III-1 lists a number of typical PIN diode chips available from Microwave Associates, Inc. along with their approximate electrical parameters. Table III-2 lists the chip dimensions and some of the package styles available. In using these Tables the reader should keep in mind that guaranteed specifications can be expected to be more conservative and also may be limited in scope. To some extent even the interpretation and measurement techniques vary among different manufacturers. Insofar as is possible, the commonly accepted definitions have been applied here; they are defined in the paragraphs below.

C. Definition of Characteristics

1. *W, I Region Width* – The distance separating the P+ and N+ regions, as shown in Figure III-1(a). Generally, for well-made diodes having 25 μm (1 mil) or larger I region width, a fairly sharp resistivity gradient occurs at the P+/I and I/N+ interfaces which can be viewed directly by chemical staining of sample diodes which have been cross-sectioned. With smaller W, a less pronounced resistivity gradient is obtained and the precision with which W can be determined is lessened.

All diodes obtained from a wafer are presumed to have approximately the same W. However, for power switching, considerable variations in high microwave voltage sustaining strength can be experienced, even among diodes obtained from the same slice. Figure III-1(b) shows how this variation can happen. All semiconductor diodes require metallic contacts in order that the chip can be attached with good thermal and electrical conductivity to a circuit. Such metals form alloys with the silicon, resulting in the desired low thermal and electrical resistance joint. However, the temperatures required to form these alloys (typically 400-700°C) are sufficient to drive alloy *stalactites* (from the top contact) or *stalagmites* (from the lower contact) through the heavily doped regions and into the I layer should there be crystalline faults in the starting crystal. To some extent, any starting material has some faults; therefore, some number of the diodes produced from each slice have these imperfections to varying degrees. Their consequence is to reduce the effective I region width, as shown diagrammatically in Figure III-1(b).

SYMBOL		ELECTRICAL & PHYSICAL PARAMETERS	UNITS	MA-47890
(1)	W	I REGION WIDTH	μm (mils)	150 (6)
(2)	V_{BB}	BULK BREAKDOWN VOLTAGE	kV	1.8
(3)	D	I REGION EFFECTIVE DIAMETER	mm (mils)	2.3 (92)
(4)	C_j	JUNCTION CAPACITANCE	pF	3
(5)	T_w	AVERAGE CARRIER TRANSIT TIME	ns	8500
(6)	I_p	TYPICAL BIAS: FORWARD	mA	250
(7)	V_R	REVERSE	V	200
(8)	R_F	FORWARD RESISTANCE @ 1 GHz @ I_F	Ω	0.2
(9)	R_R	REVERSE RESISTANCE @ 1 GHz @ V_R	Ω	0.2
(10)	f_c	CUTOFF FREQUENCY (REVERSE BIAS)	GHz	250
(11)	f_{cs}	SWITCHING CUTOFF FREQUENCY	GHz	250
(12)	τ	CARRIER LIFETIME @ 10 mA	μs	15
(13)	θ	THERMAL RESISTANCE	°C/W	1.5
(14)	HC	I REGION HEAT CAPACITY	μJ/°C	1100
(15)	τ_T	MINIMUM THERMAL TIME CONSTANT	μs	1650

Table III-1 Typical Parameters

A diode with these imperfections, of course, has a lower bulk breakdown (and RF voltage breakdown) than one which is free of such alloy stalactites and or stalagmites; therefore, all diodes having the same starting I region thickness cannot be assumed to have the same RF voltage sustaining capability. Such faults reduce the diode's bulk breakdown and presumably should be measureable at low frequency, such as with an I-V curve tracer. However, surface leakage paths usually mask the bulk breakdown behavior, as described below for V_{BB}. For this reason, individual high power RF testing is required to assure that each diode of a lot is adequate if the RF circuit design and power rating is to be extended to the limits described in Chapter II.

MA-47891	MA-47892	MA-47893	MA-47894	MA-47895	MA-47896	MA-47897	MA-47898	MA-47899	MA-47152	MA-47154	MA-47156
100 (4)				50 (2)			25 (1)		12 (0.5)	6 (0.25)	2 (0.1)
1.2				0.6			0.3		0.15	0.07	0.03
1.56 (61)	1.10 (43)	0.49 (19)	0.35 (14)	0.65 (26)	0.35 (14)	0.25 (10)	0.25 (10)	0.18 (7)	0.12 (5)	0.09 (3)	0.05 (2)
2	1	0.2	0.1	0.7	0.2	0.1	0.2	0.1	0.1	0.1	0.1
3800				950			250		55	14	1.5
150	100			50			25		25	25	25
100				50			25		10	10	10
0.3	0.4	0.8	1	0.4	0.7	1	0.9	1	1	1	1
0.3	0.5	3	6	0.6	3	4	2	4	4	4	3
250	300	250	250	350	250	400	400	400	400	400	550
250	350	500	600	350	550	700	600	800	800	800	800
8	5	4	3	2	1.5	1	0.8	0.5	0.2	0.1	0.02
3	4	15	25	7	12	15	15	25	30	35	40
340	170	34	17	30	8	4	2	1	0.2	0.06	0.007
1000	680	500	425	210	96	60	30	25	6	2	0.3

Of Available Diodes

2. V_{BB}, *Bulk Breakdown Voltage* – The *bulk breakdown voltage* of a diode is that reverse bias voltage magnitude which induces conduction through *impact ionization.* It is characterized by a "sharp" knee in the reverse part of the I-V characteristic, as shown in Figure III-2. It is also called *avalanche breakdown* because the current increases so rapidly when the avalanche breakdown voltage is exceeded that, unless the voltage supply is limited to a small current level (generally, a few microamperes because of the localized (filamentary) heating of the avalanche), the diode can be destroyed before the voltage can be turned off.

INCH / MM

131 (CHIP)

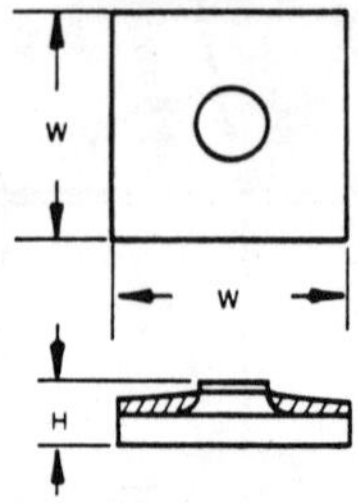

30

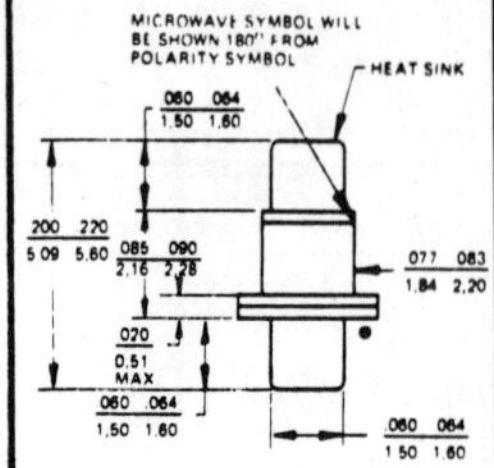

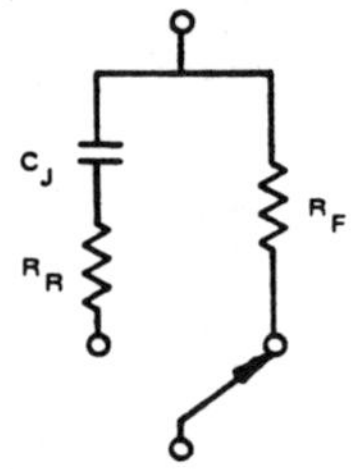

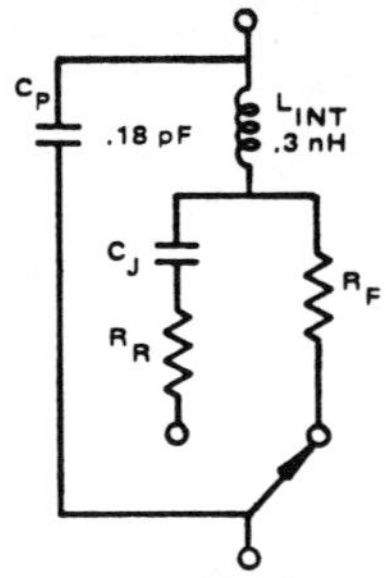

DIODE CHIP NO.		131 (CHIP) W	H	30
MA-47890-(*)	(1)	.110	.012	–
MA-47891	(2)	.085	.009	–
MA-47892	(3)	.065	.010	–
MA-47893	(4)	.025	.009	YES
MA-47894	(5)	.025	.009	YES
MA-47895	(6)	.033	.007	YES
MA-47896	(7)	.025	.007	YES
MA-47897	(8)	.015	.006	YES
MA-47898	(9)	.015	.006	YES
MA-47899	(10)	.015	.006	YES
MA-47152	(11)	.015	.006	YES
MA-47154	(12)	.015	.006	YES
MA-47156	(13)	.015	.006	YES

*The complete model number for the diode includes a suffix which describes the furnished in chip form and the MA-47890-150 is the same diode furnished with

Table III-2 Typical PIN Packages

54	109	150
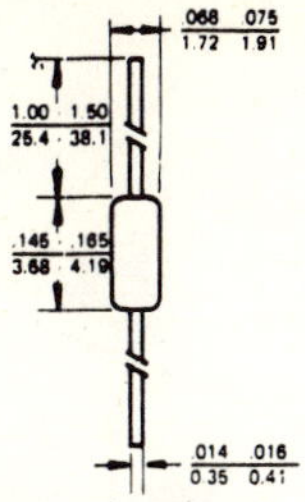	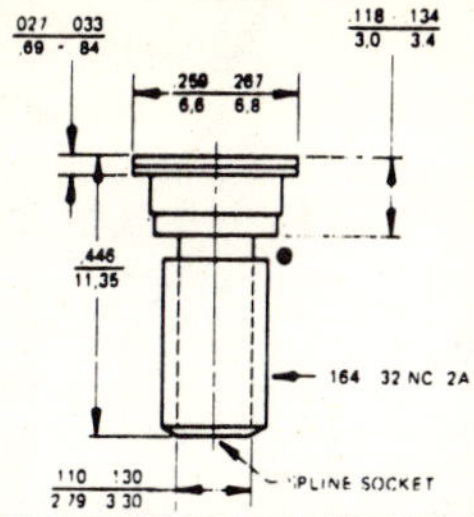	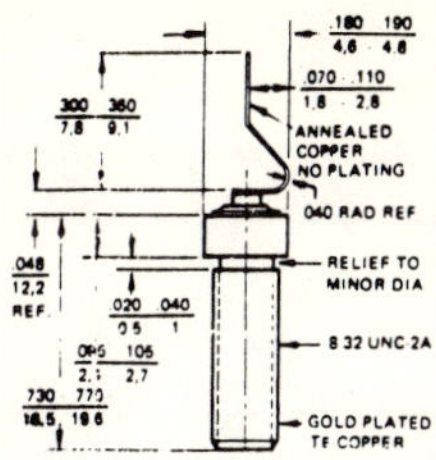
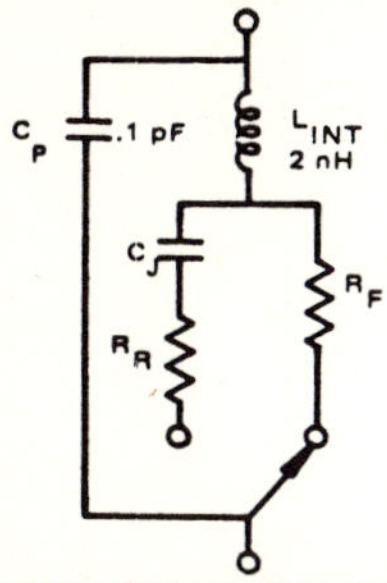	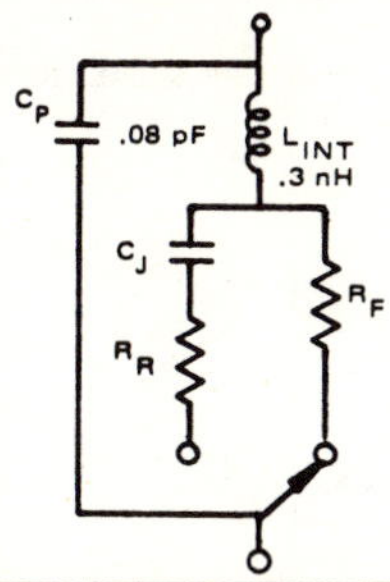	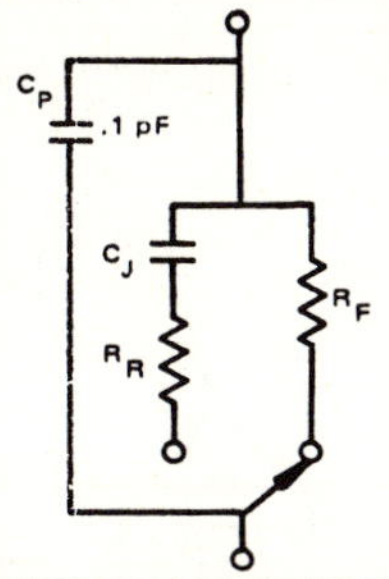
—	—	YES
—	YES	YES
—	YES	YES
—	YES	—
—	YES	—
YES	YES	—
YES	—	—
YES	—	—
YES	—	—
YES	—	—
—	—	—
—	—	—
—	—	—

package style. Thus, for example, the MA-47890-131 is a type (1) diode heat sink and top contacter strap.

and Their Equivalent Circuits

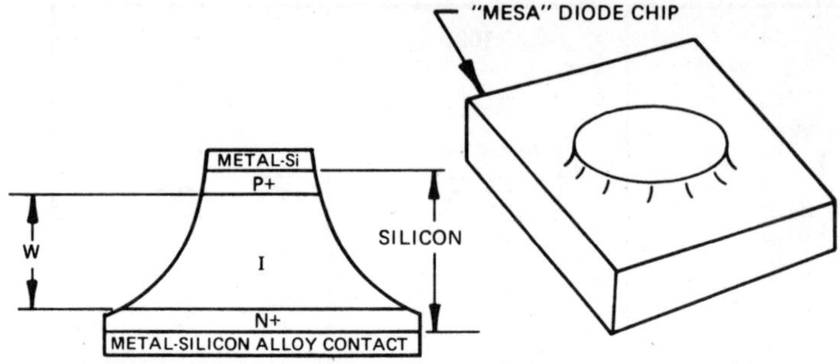

a) IDEAL DIODE WITH UNIFORM I REGION GEOMETRY OF WIDTH, W.

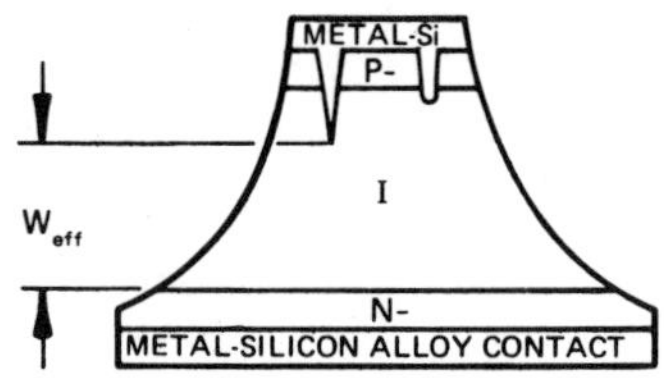

b) DEFFECTIVE DIODE HAVING REDUCED EFFECTIVE I REGION WIDTH, $W_{EFFECTIVE}$, CAUSED BY ALLOY "STALACTITE" FAULT.

Figure III-1 **Effective I Region Width of Practical PIN Diodes**

As a rule of thumb, it is found that the bulk breakdown voltage is estimable from the width of the I region of a PIN. In Figure I-13 the breakdown field in silicon as a function of impurity concentration is shown. PINs have a high resistivity I region, commensurate with the lowest impurity concentration (10^{15} atoms/cm^3) shown. It thus would be concluded that the breakdown voltage should be that which produces a field in excess of 300 kV/cm, or about 0.8 kV per mil of I region width. In practice, however, only about half this breakdown strength is realized, probably because, in a real diode, the applied voltage does not produce a uniform field between the P+ and N+ areas due to the presence of varying penetration of the contact metallization (stalactites and stalagmites) into the I region. Even if the I region upper and lower boundaries were perfectly planar, geometry factors at the periphery of these circular boundaries would produce higher local electric field stresses than would be estimated from the (V_{BB}/W) approximation. The V_{BB} listings in Table III-1 are based on an estimate of 0.3 kV per mil of I region.

The measurement of bulk breakdown can be, and usually is, hampered by the presence of surface contaminants on the diode which provide a weakly conducting path in shunt with the I region. The resulting *reverse leakage current* characteristic then differs from the ideal sharp knee shown in Figure III-2. Typical surface breakdown profiles are shown sketched in this figure. Since these currents are small (a few microamperes or less) in good diodes and their paths could not even support high RF currents; they cause no material change in the RF voltage sustaining capability or the RF impedance of the diode. As a result, their effect is customarily ignored in the estimate of RF power ratings. However, as can be seen in Figure III-2, the presence of such surface break-

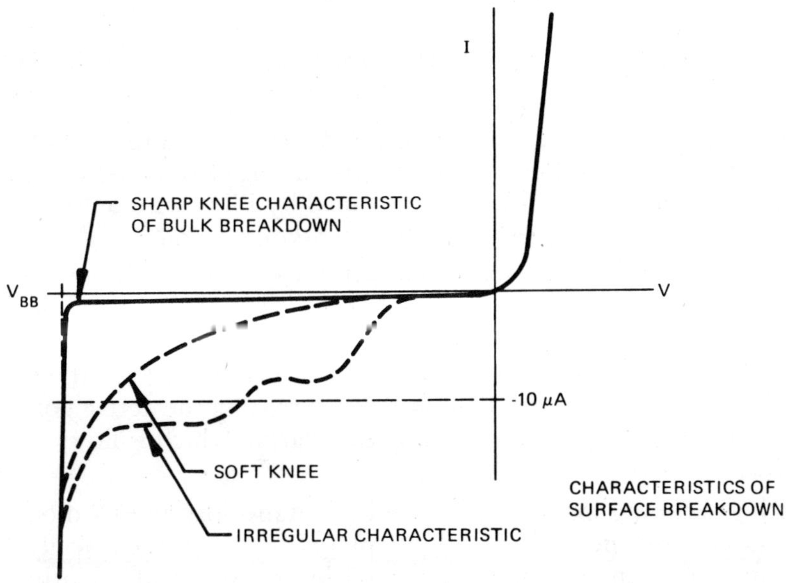

Figure III-2 **Ideal and Practical Low Frequency Breakdown Characteristics**

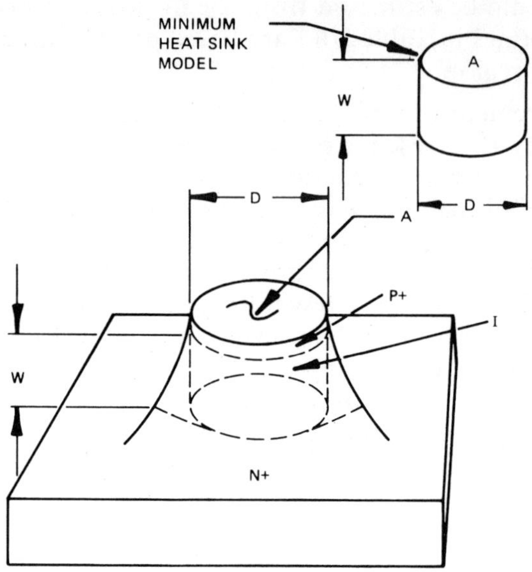

Figure III-3 **PIN Module Used for Heat Sinking Calculation**

down can result in a wide variation in dc breakdown readings for a given run of diodes. Because of the ease of measurement, *dc breakdown* at 10 μA, V_B, is usually performed by PIN manufacturers. As such, V_B is usually less than *bulk breakdown,* V_{BB}.

3. *D, I Region Effective Diameter* – The diameter of a right circular silicon cylinder (see Figure III-3) having height W which, using the parallel plate capacitance formula (Equation (II-38)), is calculated to have the same capacitance, C_J, as is measured for the actual diode. This value for D is used in estimating the active volume of the I region for the heat sinking calculation which follows below.

4. C_J, *Junction Capacitance* – The capacitance of the PIN diode measured with the I region fully depleted of charge. Practically, this measurement is usually made at 1 MHz using a sufficiently large reverse bias to deplete the I region. I region depletion is indicated by a knee in the C(V) characteristic as described in the punchthrough description (pp. 41-43).

5. T_W, *Average Carrier Transit Time* – The average time required for holes and electrons to cross the I region by diffusion. Using

average mobility values of 500 and 1500 cm^2/V for holes and electrons, respectively, and noting that at room temperature $kT/e = .026$ V, the average transit time (from Equations (II-11) and (II-12) is given by

$$T_W = \frac{W^2}{(kTe)\,\dfrac{\mu_P + \mu_N}{2}} \qquad \text{(III-1(a))}$$

$$T_W = \frac{0.038\ W^2}{\text{centimeter}^2}\ \text{(seconds)} \qquad \text{(III-1(b))}$$

The diffusion transit time provides a measure of the switching time that can be obtained easily with a PIN diode with I region width W. This point is discussed further in Chapter IV.

6. *I_F, Typical Forward Bias Current* — The steady state forward bias required to achieve the forward resistance, R_F.

7. *V_R, Typical Reverse Bias Voltage* — The steady state reverse bias voltage required to achieve the reverse resistance, R_R.

8. *R_F, Forward Resistance at 1 GHz at I_F* — The series resistance measured at low power at 1 GHz under forward bias current, I_F, defined according to the equivalent circuit in Figure III-4. See Chapter II for resistance measurement techniques. The resistance of the I region decreases with increasing bias current. The practical limit on the lowest value of forward resistance obtainable occurs when the I region is so saturated with charge that increased recombination (decreasing lifetime) prevents further decreasing of R_F by increasing I_F. Figure III-5 shows how forward resistance varies with the reciprocal of bias current for a typical PIN diode. The intercept of this curve with the resistance axis can be interpreted as the total combined contact and bulk resistances of the P+ and N+ regions. This resistance is that which would be experienced if the I region could be filled with an infinite charge density and its resistance could be reduced to zero.

9. *R_R, Reverse Resistance at 1 GHz at V_R* — The series resistance measured at low power at 1 GHz, under reverse bias voltage, V_R, defined according to the equivalent circuit in Figure III-4. See Chapter II for resistance measurement techniques.

It would at first appear that R_R should always be less than R_F because, under reverse bias, the I region has been depleted of charge

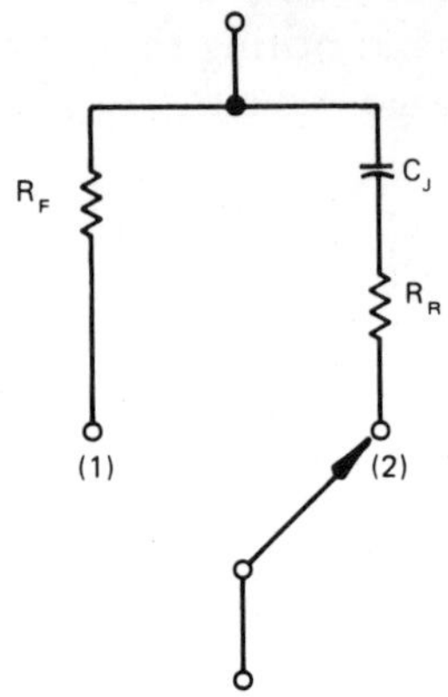

Figure III-4 **Series Equivalent Circuit for PIN Chip Under (1) Forward Bias and (2) Reverse Bias**

and all that remains are the ohmic contributions of the contacts and of the bulk resistivity of the P+ and N+ regions. But the real situation does not follow along these lines; generally, R_R is equal to or greater than R_F, because R_R includes a component resulting from the gradient in resistivity at the P+/I and I/N+ boundaries. Ideally, an abrupt change from high conductivity P+ or N+ material would occur at these I region interfaces; in practice, though, since this interface is formed at high temperature, some diffusion of impurities into the I region occurs and the resultant resistivity gradient produces a series resistance component. In addition, dissipation in the I region due to its finite Q is identified with R_R since, in the simple series R-C model of Figure III-4, no shunt conductance has been included. Practically, R_R is not predictable and resort must be made to microwave measurement for this value. For the diodes listed in Table III-1 a wide range of measurements and measurement frequencies are typically used. Values given in the table are estimates of what would be obtained at 1 GHz, in order that a comparable estimate of the relative performance of a large selection of diodes can be made.

10. f_C, *Cutoff Frequency* (Reverse Bias) — The frequency at which the reactance of C_J is numerically equal to the reverse resistance

$$f_C = \frac{1}{2\pi C_J R_R} \qquad \text{(III-2(a))}$$

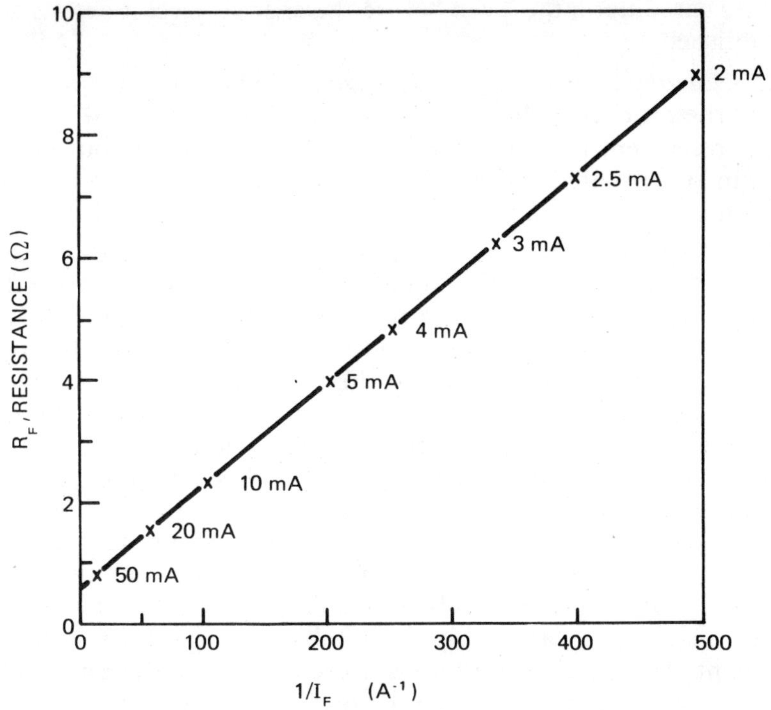

Figure III-5 **Typical Forward Resistance vs. Bias Current (at 100 MHz) (Data for Microwave Associates MA-47047 PIN Diode)**

The reverse bias cutoff frequency originated with varactor diodes. It is useful in some switching circuit analyses because the Q of the diode under reverse bias at any frequency is easily expressed as f_C/f.

11. f_{CS}, *Switching Cutoff Frequency* – The frequency at which the reactance of C_J is numerically equal to the geometric mean of the forward and reverse resistances. Using the equivalent circuit of Figure III-4, the defining relationship for f_{CS} is [1]

$$f_{CS} = \frac{1}{2\pi C_J \sqrt{R_F R_R}} \qquad \text{(III-2(b))}$$

This cutoff frequency definition is useful as an insertion loss figure of merit for PIN diodes used in switches. It is shown in Chapter V that an estimate of the best insertion loss performance can be made, based upon f_{CS}, even before the specific circuit details are established.

12. τ, *Carrier Lifetime* — The average carrier lifetime measured using the procedure shown in Figure II-11. Conventionally, the forward current is set at 10 mA for this measurement, but lower lifetime results at the higher forward bias levels used typically as shown in Figure II-12.

13. θ, *Thermal Resistance* — The ratio of the steady state temperature (°C) rise of the junction per watt of steady state power dissipated within it. In chip form the PIN diode has no means of dissipating heat. For the diodes listed a mesa diode geometry is typical, as shown in Figure III-6, and the thermal resistance values indicated are those which can be obtained when the N+ side of the chip is soldered to a copper heat sink. This value of thermal resistance is usually measured by mounting the diode chip on one of the metal bases described in the following diode package section. The base can then be attached to a larger (circuit) heat sink. Thermal resistance measurements for diodes which are ultimately to be used in chip form can be evaluated only on a sample basis, mounting representative chips in suitable packages for the measurement. The steady state temperature rise, ΔT, between the diode junction and the heat sink due to continuous power dissipation, P_D, in a diode with thermal resistance θ is

$$\Delta T = P_D \cdot \theta \tag{III-3}$$

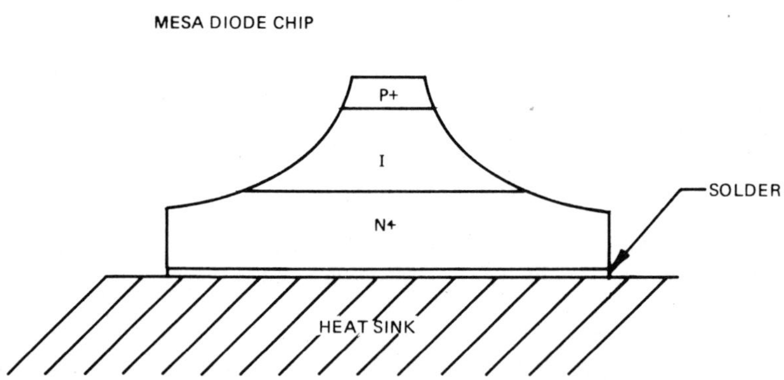

Figure III-6 Mesa PIN Diode Form

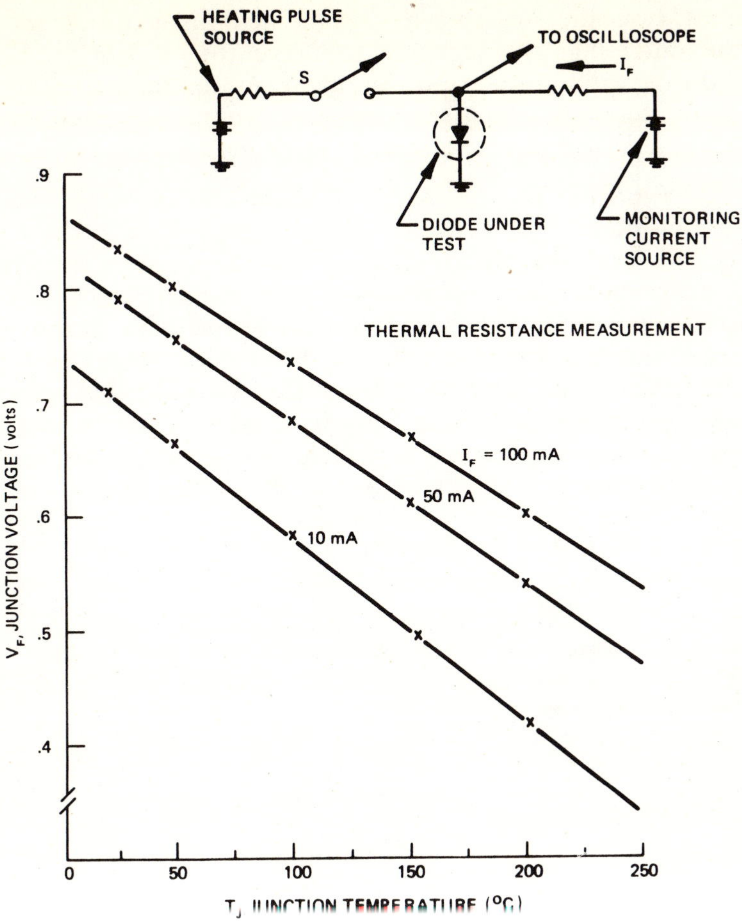

Figure III-7 **Typical PIN Forward Voltage Drop vs. Junction Temperature for Various Bias Currents**

The thermal resistance measurement is performed by dissipating a known amount of dc power in the diode and using the diode's junction voltage as a thermometer to measure the resulting temperature rise. Figure III-7 shows a typical plot of the forward voltage drop for a given applied current as a function of junction temperature. Such a characteristic is obtained by first mounting the diode in a temperature calibrated oven and sampling the dc forward voltage drop at various temperatures for a fixed value of forward current. Care must be taken to use a small enough forward current, typically 1 to 10 mA, that the heat dissipation it produces in the junction gives a negligible change in the junction temperature during calibration or measurement. Actual thermal resistance as encountered in a practical control device can be somewhat higher, since less than ideal thermal paths are available.

14. *HC, I Region Heat Capacity* – The amount of energy required for a unit increase (1°C) in the I region temperature in the absence of heat flow from the diode. HC is a measure of the amount of heat which can be absorbed by the diode during operation at pulses so short that little heat flows out of the diode into the heat sink during the pulse. In such cases, the average temperature rise of the I region calculated using the steady state thermal resistance would be unrealistically high. The HC is calculated by multiplying the specific heat and density* of silicon by the volume of the equivalent I region cylinder; thus, for silicon PIN diodes

$$HC = (\text{specific heat x density})_{\text{silicon}} \cdot V_{\text{I region}} \qquad \text{(III-4(a))}$$

$$HC = 0.176 \frac{\text{calorie}}{\text{gram-°Celsius}} \cdot \frac{2.43 \text{ grams}}{\text{cubic centimeter}} \cdot \frac{4.186 \text{ joules}}{\text{calorie}} \cdot \frac{\pi D^2 W}{4}$$

and simplifying we get

$$HC = \frac{1.4\ D^2W}{\text{cubic centimeter}} \ (\text{joules/°Celsius}) \qquad \text{(III-4(b))}$$

$$= \frac{0.023\ D^2W}{\text{cubic mil}} \ (\text{microjoules/°Celsius})$$

For example, the 3 pF, 150 μm (6 mils) I region width diode listed in Table III-1 has a minimum heat capacity of 1.4 x $(0.23)^2$ x (0.015) = 0.00111 J/°C or 1,110 μJ/°C. Since a joule is an amount of energy equal to one watt for one second, this PIN could absorb 1110 W of heat dissipation for 1 μs with a junction temperature rise of no more than 1°C, neglecting heat flow out of the diode. Suppose a 50°C pulsed heat rise is acceptable, as is usually the case. If the average power is low, this diode could safely be used in a pulsed power application in which the peak dissipation during the pulse under worst conditions of load mismatch is 55,500 W

*Values used for the physical and thermal parameters of silicon are from Reference 2.

for 1 μs (or 55 W for 1 ms and so forth). Using this heat sinking (zero heat flow) analysis method, the maximum I region temperature rise ΔT_M above the heat sink is limited to the value

$$\Delta T_M < P_D \cdot t/HC \tag{III-5}$$

where P_D is the pulse power dissipation within the I region and t is the pulse length. This refers to the temperature rise during each pulse, occurring during the short pulse length, assuming that the heat from previous pulses has been dissipated.

Strictly speaking, the heat dissipated in the PIN diode is not all developed within the I region. Nor is that which is dissipated in the I region uniformly distributed. Mortenson has shown [3] that within the I region somewhat higher temperatures can be reached than are monitored at the junction by using the diode's forward voltage drop as a thermometer. But as a first approximation, we use junction temperature, T_J, as the measure of peak diode temperature, referring to Reference 3 for those applications when a more precise thermal modeling is desired.

15. τ_T, *Minimum Thermal Time Constant* – The product of the steady state thermal resistance, θ, with the I region heat capacity, HC

$$\tau_T = \theta \cdot HC \tag{III-6}$$

The need for a thermal time constant can best be appreciated by considering a sample calculation showing how peak junction temperature rise is estimated (actually, bounded) using only θ and HC. For our 3 pF diode example, Table III-1 gives θ = 1.5°C/W and HC = 1110 μJ/°C. If, at time t = 0, a power, P_D, is dissipated in the I region, the junction temperature will increase by an amount, ΔT_M, over that of the heat sink found using Equation (III-5). For this example, let P_D = 66-2/3 W, then

$$\Delta T_M < \frac{0.06°\text{Celsius}}{\text{microsecond}} \cdot t \tag{III-7}$$

Since Equation (III-7) neglects cooling, the bound it gives increases without limit as t increases. However, we know that a steady state condition is reached when ΔT is large enough to permit the dissipation of P_D through the thermal resistance θ to the heat sink.

For this example, this point is found to occur, using Equation (III-3), when

$$\Delta T = 66\text{-}2/3 \text{ watts} \cdot 1.5 \frac{°\text{Celsius}}{\text{watt}} = 100°\text{Celsius} \tag{III-8}$$

It follows that, for this diode, Equation (III-7) is useful in estimating maximum temperature rises corresponding to $0 \leqslant t \leqslant 1667\ \mu s$. This situation is shown graphically in Figure III-8, with the graph representing the ΔT_M boundary using the heat sinking thermal and the thermal resistance model for longer pulses.

We know that the junction temperature does not have the abrupt rate of time change suggested by Figure III-8. In order to arrive at a more realistic temperature contour, let us consider the simplified diode thermal model shown in Figure III-9(a). The I region power dissipation, P_D, produces heat, $P_D \cdot t$, *some* of which serves to increase the I region temperature by, at most, ΔT_M. The resulting temperature rise, however causes heat to flow out of the I region through the resistance path, θ. At this "temperature node," the

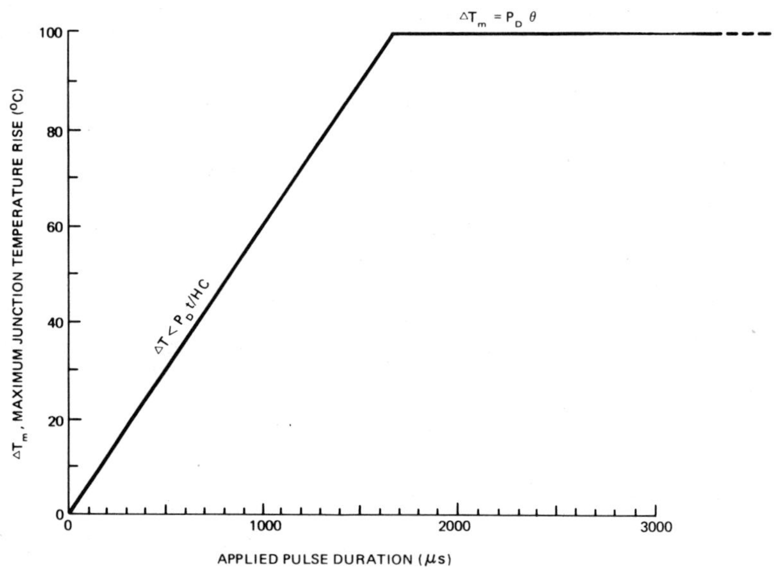

Figure III-8 **Example of Maximum Diode Temperature Rise Estimates Using θ and Heat Capacity**

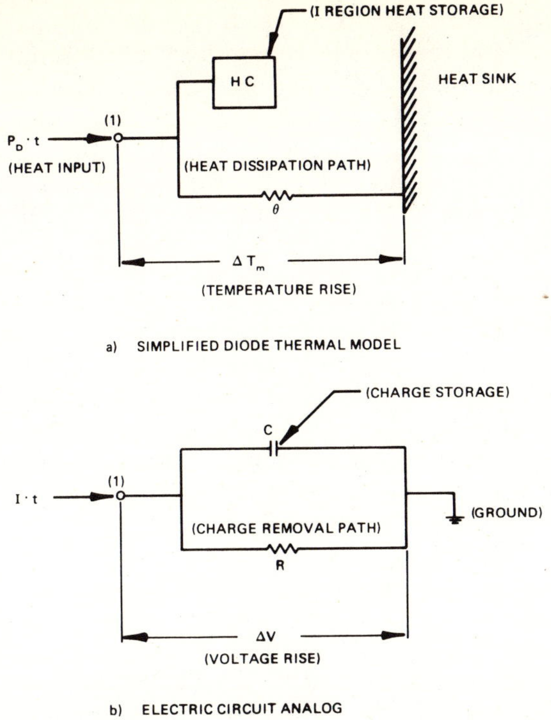

Figure III-9 **Simplified Thermal Models for Transient Diode Heating**

instantaneous total heat flow input, P_D, must equal the change in heat storage, $HC \cdot d(\Delta T_M)/dt$, plus the heat $\Delta T/\theta$ lost by the thermal path to the heat sink. This equality is described by the differential equation

$$P_D = HC \cdot \frac{d(\Delta T_M)}{dt} + \frac{\Delta T_M}{\theta} \qquad \text{(III-9)}$$

This equation is directly solvable for ΔT_M; however, for electrical engineers it is perhaps more readily visualized by considering the analog electric circuit shown in Figure III-9(b) for which

$$I = C \frac{d(\Delta V)}{dt} + \frac{\Delta V}{R} \qquad \text{(III-10)}$$

A charge input, I•t, at voltage anode (1) in the electric analog circuit corresponds to $P_D \cdot t$. Charge storage and dissipation by the capacitor, C, and resistor, R, respectively correspond to heat storage and dissipation by HC and θ in the thermal circuit. The voltage rise, ΔV, at node (1) in the electric analog corresponds to the temperature rise, ΔT_M, at node (1) which, in turn, corresponds to the I region temperature rise in the thermal model. The solution for the voltage rise in the electric analog has the well known exponential form,

$$\Delta V = IR \left[1 - e^{-t/\tau_E}\right] \quad \text{(III-11)}$$

where τ_E = RC is called the electrical time constant of the R-C network. Since the heat flow conditions in the thermal model can be described by the same mathematical expressions as the charge flow in the electric current, the solution for the temperature rise, ΔT_M, has the same form as that for ΔV, or

$$\Delta T_M = P_D \cdot \theta \left[1 - e^{-t/\tau_T}\right] \quad \text{(III-12)}$$

where $\tau_T = \theta \cdot HC$ is the thermal time constant.

The electric analog model should be kept in mind, not only for the physical insight it imparts, but also because the electrical model can be built more readily and its voltage response to various pulsed current excitations measured. Actual PIN diodes have a more complicated thermal model; a means for "measuring" the thermal response using an electric circuit analog may often be more practical than a direct mathematical solution of the thermal problem.

Using Equation (III-12) we can now make a more realistic approximation to the maximum junction temperature profile given in Figure III-8. In so doing, the transient temperature rise has been normalized as the ratio of the transient to the steady state temperature rise, $P_D \cdot \theta$. Similarly, the pulse length is normalized to the thermal time constant, τ_T. In this way, the resulting diode junction temperature rise versus time profile shown in Figure III-10 is general and can be applied to any diode, simply by converting to these normalized coordinates.

From this graph it can be seen that the I region heat storage accounts for 85% or more of the temperature rise for $t < 0.25\ \tau_T$. However, when $t = \tau_T$, the heat sinking model alone from

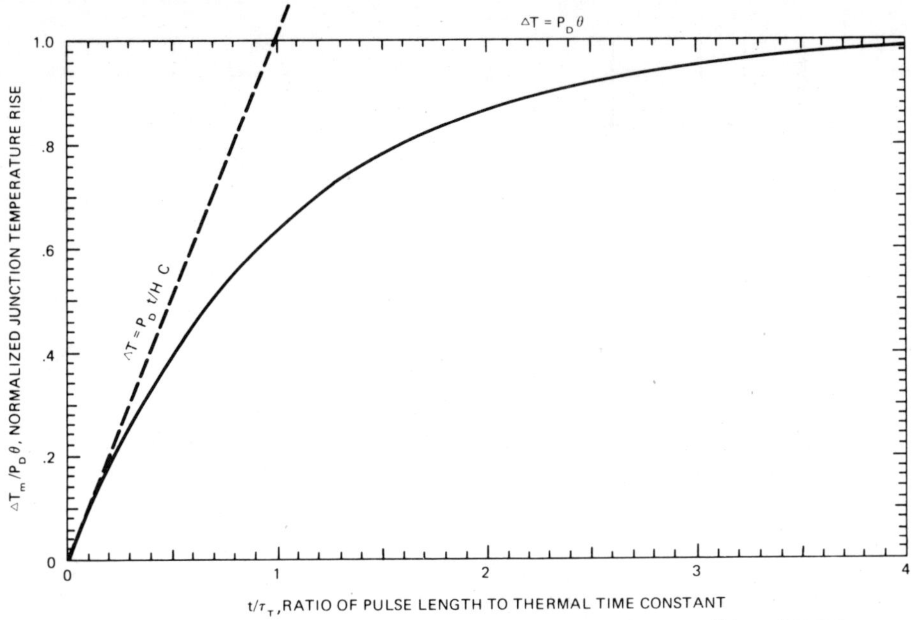

Figure III-10 **General Pulsed Temperature Rise Profile of PIN Diode Using Minimum Time Constant Model**

Equation (III-5) gives a ΔT_M which is 58% greater than that obtained with the time constant model from Equation (III-10). Finally, when $t > 2\ \tau_T$, ΔT_M is within 85% of the steady state value, $\Delta T = P_D\theta$. Thus with an accuracy of 15% or better

$$\Delta T_M < P_D \cdot t/HC \qquad 0 \leqslant t \leqslant 0.25\ \tau_T \qquad \text{(III-13(a))}$$

$$\Delta T_M \approx P_D \cdot \theta \qquad t > 2\ \tau_T \qquad \text{(III-13(b))}$$

and for the pulse length interval, $0.25\ \tau_T < t < 2\ \tau_T$, Equation (III-12) or Figure III-10 should be used for estimating ΔT_M.

In practice, a train of heating pulses, of the sort which occurs in a pulsed radar, is incident upon the diode. If the pulse width is approximately equal to or less than the diode's thermal time constant, and if the interpulse period is more than $5\ \tau_T$ (as is usually the case), the junction temperature, T_J, will cool between pulses to the temperature of the heat sink, T_S, (within 1% of ΔT_M). The diode temperature following the heating pulse has an exponential decay with time constant, τ_T, and is related to ΔT_M (reached at the end of the heating pulse) and T_S by

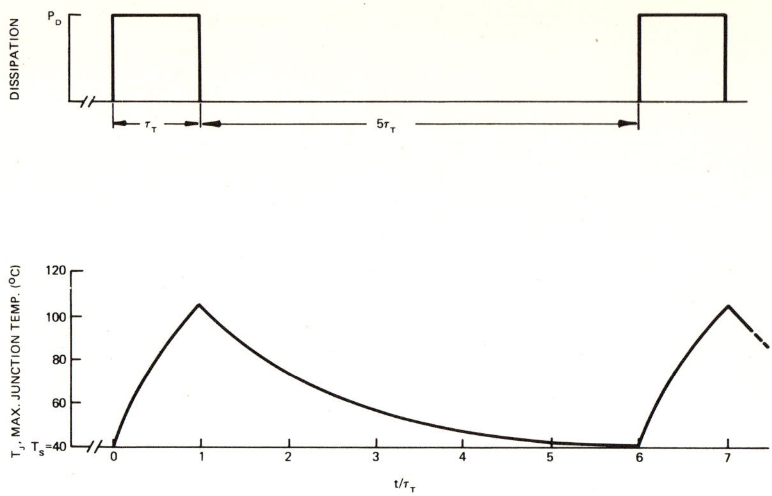

Figure III-11 **Sample Estimate of PIN Junction Temperature During Pulse Train**

$$T_J = T_S + \Delta T_M e^{-t/\tau_T} \qquad \text{(III-14)}$$

where t is measured from the end of the heating pulse. Figure III-11 shows a sample temperature characteristic with heating pulse length, τ_T: interpulse period, 5 τ_T: ΔT_M = 63°C; and heat sink at 40°C.

Throughout the discussion of pulsed heating we have emphasized that τ_T is the *minimum* thermal time constant and that ΔT_M is the *maximum* pulsed temperature rise. These delineations arise because most PIN device designs require estimates of the survivable device stresses; thus, *we are usually content with knowing the maximum, rather than the actual, T_J profile.* The estimates discussed, therefore, represent upper temperature bounds (subject to the assumption that the I region temperature is uniform throughout). The *actual T_J is likely to be lower.* To understand why, consider again the simplified thermal model in Figure III-9(a). Only the thermal capacitance, HC, of the I region was used, but the complete thermal path resistance, θ, to the heat sink was included in the minimum thermal time constant definition, $\tau_T = \theta \cdot HC$. In actuality, the heat sinking capacity of a real PIN diode includes additional silicon surrounding the active I region, the top weight, solder masses, and a portion of the metal base electrode, as shown

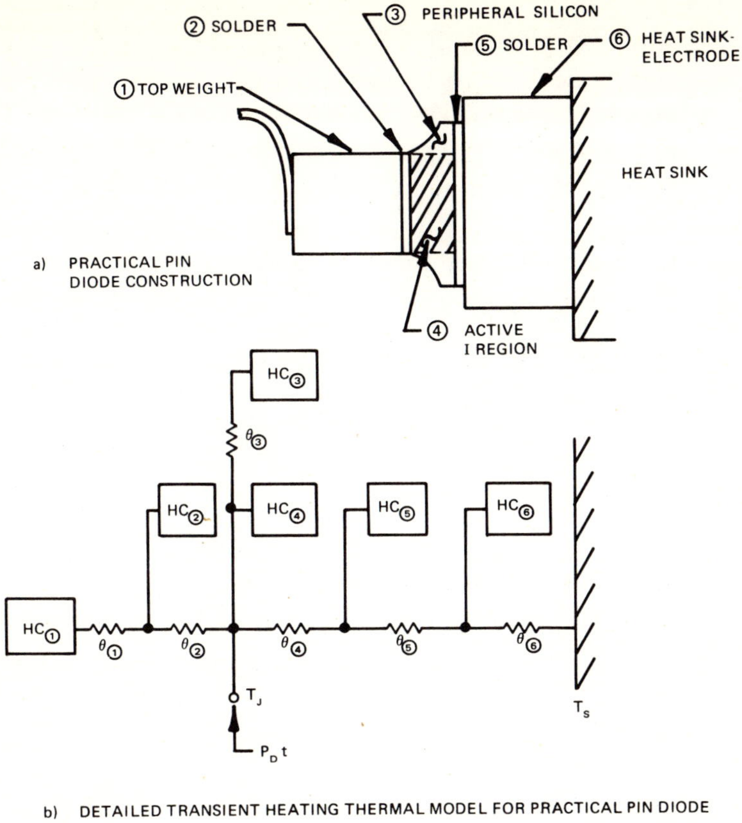

Figure III-12 **Constructional PIN Detail and its Detailed Thermal Model**

in Figure III-12(a). A corresponding thermal model, Figure III-12(b), has not one, but several temperature nodes; the final solution for ΔT_J involves a number of time constants. However, the difference to be obtained with the more detailed model is such as to predict a longer thermal time constant but no greater a peak temperature, because we have used $\theta = \sum_{n=4}^{6} \theta_n$ in our estimate for the thermal time constant, τ_T. The result then obtained with the more precise model is a *lower* value for ΔT than that obtained using the approximate method, thus Equation (III-12) gives a conservative result with respect to device survivability.

References

[1] Hines, M.E.: "Fundamental Limitations in RF Switching and Phase Shifting Using Semiconductor Diodes," *Proceedings of the IEEE, Vol. 52, No. 6,* pp. 697-708, June 1964.

[2] *ITT Reference Data for Radio Engineers – Fifth Edition;* Howard W. Sams and Co. Inc., New York, 1974.

[3] Mortenson, K.E.: "Analysis of the Temperature Rise in PIN Diodes Caused by Microwave Pulse Dissipation," *IEEE Transactions on Electron Devices, Vol. ED-13, No. 3,* pp. 305-314, March 1966. Also Mortenson's *Variable Capacitance Diodes,* Artech House, Dedham, Massachusetts, 1975.

Questions

1. Show that when pulsed power P_D is dissipated in a PIN diode for pulse lengths, t, small compared with the thermal time constant, τ_T, the maximum temperature rise, ΔT, is given approximately by

$$\Delta T \approx P_D \theta \left(\frac{t}{\tau_T}\right)$$

with an error bounded by

$$E < \frac{P_D \theta t}{\tau_T} \quad (50\%)$$

Hint: Use the series expansion

$$e^x = 1 - x + \frac{x^2}{2!} - \frac{x^3}{3!} + \ldots$$

2. Suppose a PIN diode has an effective I region diameter of 47 mils (1.2 mm) and thickness of 4 mils (0.10 mm). What is its minimum heat capacity, HC? If the thermal resistance, θ, is 5°C/W, what is its thermal time constant, τ_T?

3. If the diode in Question 1 is mounted on a surface held at 50°C, how much power can be dissipated within it if the maximum junction temperature, T_J, is not to exceed 150°C under a) CW dissipation; and b) widely spaced pulses of 2 ms, c) 1 ms d) 500 μs, e) 200 μs, f) 100 μs, g) 10 μs, h) 1 μs?
4. If, because of a poor thermal path in the circuit mounting, the diode in Question 3 has a total thermal resistance of 25°C/W, what is τ_T? How are the answers to Question 3 affected?
5. What pulse length ratings are affected the most by the increased thermal resistance in Question 4?

   Is the dissipation rating ever increased by increasing θ? Why?

Chapter IV

Binary State Transistor Drivers

A. What the Driver Must Do

The subject of transistor drivers for diode control devices is extensive enough to demand a separate volume. We do not attempt to make this chapter a complete working notebook for the driver designer, but rather a guide for the microwave circuit engineer who may be in the position of having either to fabricate an elemental driver to prove out a given device or to work with a driver circuit engineer to develop a circuit no more complex than necessary to satisfy the design objectives. As such, the sample drivers described show operating fundamentals and do not necessarily represent final production designs. This chapter treats only two-state, or *binary,* driver circuits in which either a forward bias current or a reverse bias voltage is to be applied to a PIN diode. Not covered are analog drivers as used in such circuits as linear attenuators and continuous phase shifters, where a precise bias current or voltage profile might be required. The binary circuits to be discussed are useful for discrete control devices including switches, phase shifters, and high power duplexers, and, as such, cover most of the driver requirements for PIN diode devices. We divide this treatment into the following performance categories:

1) *Input Signal Power Supply Buffer and TTL Compatibility* – The first function of a driver is to serve as a voltage and current buffer between a command control signal source and a set of power supplies. The power supplies must deliver to each PIN diode either a forward bias current (tens to hundreds of milliamperes at about IV) or a reverse bias voltage (tens to hundreds of volts, usually at less than 1 mA). The type of input control signal usually falls into one of two categories, unbal-

anced TTL or balanced TTL, depending on whether the signal need be transmitted over short or long transmission lines, respectively.

2) *Switching Speed* – Two main topics need to be considered with respect to the switching speed of a driver. First, how is it to be defined and measured? Second, how can switching speed be improved if the driver circuit under consideration in conjunction with the PIN diodes to be switched is not fast enough?

3) *High Power, Pulse Leakage Current Supply* – In PIN control devices operated under reverse bias with high RF peak power, the driver is required to supply some reverse bias current during that period when the RF pulse is applied to the diodes. The driver without this capability could precipitate diode failures due to thermal runaway in the reverse bias state.

4) *Fault Detection* – In many cases it is desirable to build circuitry into the driver which provides an indication of whether or not the driver or the PIN diodes have failed. We shall examine methods and assumptions concerning this function requirement, referred to as BITE, for "Built In Test Equipment."

5) *Complementary Drivers* – The simple two transistor driver is easily ganged to provide complementary bias states to two PIN diodes – forward bias to the first and reverse bias to the second or vice versa – as needed for SPDT switches.

B. The Driver as a Logic Signal – Power Supply Buffer (TTL Compatibility)

1. Unbalanced TTL Signal Input

Figure IV-1 shows schematically the output portion of a typical unbalanced TTL logic signal source. In a logic state *zero* (low), T_2 is driven into conduction with the result that any device connected to the logic signal source at terminals XY sees a path to ground through the saturated collector emitter of T_2. By convention, circuits which qualify as TTL signal sources must be capable of holding the voltage at the terminals XY to not more than 0.8 V while sinking up to 16 mA of current, I_X. Said simply, a circuit which provides a logic *zero* provides a "short circuit" for currents up to 16 mA.

In the alternative state *one* (high), T_2 is opened and T_1 is closed to connect terminal X to the +5 V power supply; by convention, this

LOGIC	V_{XY}	I_X
"ZERO"	0 – 0.8 V	0 – 16 mA
"ONE"	2 – 5 V	-100 μA MAX.

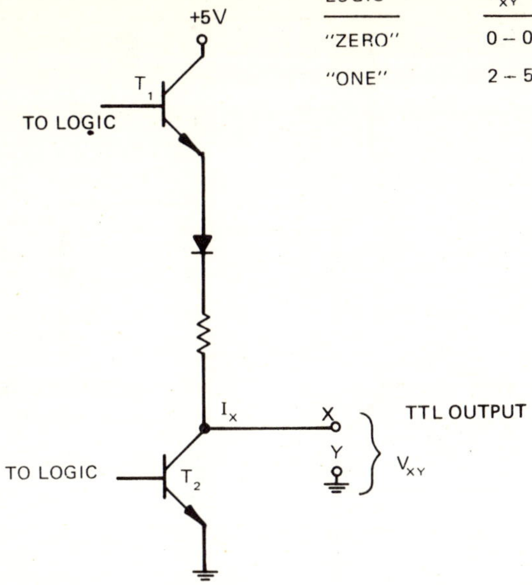

Figure IV-1 **Typical Source of an Unbalanced TTL Signal**

voltage is the standard used by TTL logic circuitry. In this case, however, the connection is made through a high impedance; *any load placed on terminal X must not draw more than 100 μA of current from this logic signal source.* By convention the voltage at the terminals XY must be at least 2 V and not more than 5 V in the logic *one* state. Said simply, a logic *one* corresponds to a high impedance at terminals XY with a static voltage between 2 and 5 V.

Any other voltage-current conditions at terminals XY are considered to be undefined and there is no assurance that one or the other state will be assumed by the driven circuitry when levels at terminals XY do not satisfy the defining conditions of either the *zero* or *one* states.

2. Sample Circuits

Figure IV-2 shows an elemental, one transistor driver which is TTL compatible and switches a PIN diode load between -25 V of reverse bias and a forward bias current whose magnitude is ad-

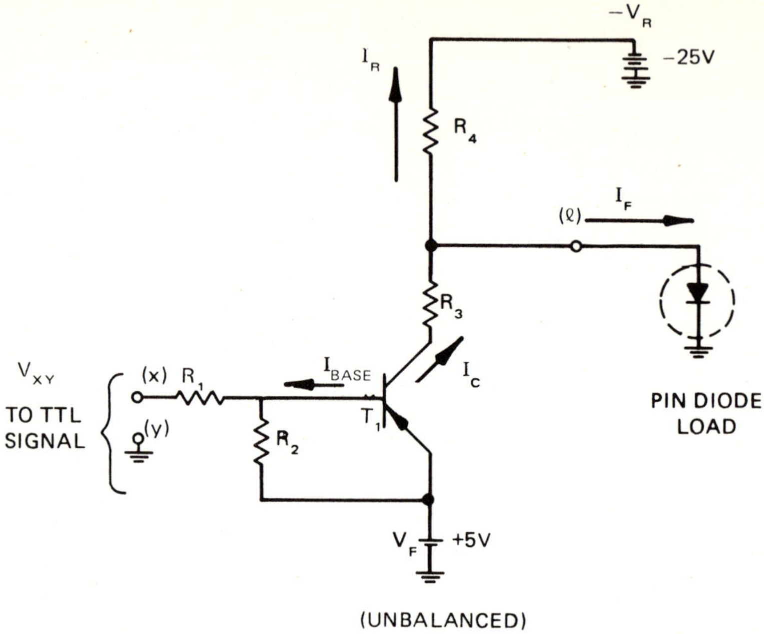

Figure IV-2 **A One Transistor (Unbalanced) TTL Compatible Driver**

justable by the choice of resistance value, R_3. The operation can be readily understood by considering the effect of the input signal at terminals XY. Suppose that a TTL *one* is applied. The voltage, V_{XY}, is then about 5 V; at most, only a fraction of a milliampere of current flows through R_1. The resistance of R_2 is chosen such that there will not be enough voltage at the emitter base of T_1 to turn it on. For example, for silicon transistors, approximately 0.5 V is required for a significant current to flow through the emitter-base, turning on the transistor. If R_2 is selected such that

$$R_2 \leqslant \frac{0.5 \text{ volt}}{0.0005 \text{ ampere}} = 1 \text{ kilohm} \qquad \text{(IV-1)}$$

then leakage current between the signal source and the 5 V supply of the driver will follow a path through R_2. No current will go through the emitter base of T_1, and T_1 therefore will be non-conducting. Under this condition, the PIN diode load is connected to the -25 V negative bias supply through resistor R_4, and the driver applies reverse bias to the PIN diode load when a TTL *one* is applied to its logic terminals XY.

Next, suppose that a TTL *zero* is applied at the logic input terminals XY. By convention, in the zero state, V_{XY} is between 0 and 0.8 V. Since this potential is much less than that of the +5 V at the emitter of T_1, a current will flow from the +5 V supply through the emitter-base of T_1, and through R_1 to the logic signal source. Thus, it is apparent why the logic signal source must be able to sink current. The resistance of R_1 is chosen so that no more than 16 mA of emitter-base current, I_{BASE}, is permitted to flow toward the logic signal source under this condition. Recognizing that V_{XY} might be as much as +0.8 V and that the voltage drop across the emitter base of T_1 is 0 to 0.8 V, a total of 3.5 to 4.3 V is dropped across R_1. The value of R_1 is chosen such that

$$R_1 \geqslant \frac{4.3 \text{ volts}}{.005 \text{ to } .008 \text{ ampere}} = 0.5 \text{ to } 0.8 \text{ kilohm} \tag{IV-2}$$

This current, I_{BASE}, switches on T_1, thereby creating a path from the +5 V supply to the PIN diode load through R_3.

Assume for the moment that the resistance of R_4 is sufficiently high to neglect its presence in the determination of the forward bias current reaching the PIN diode load. The magnitude of this current, I_F, is determined by the current gain factor ($\beta = I_C/I_{BASE}$) of the transistor in conjunction with the selection of the value of R_3. In this example, if it is desired that 25 mA of forward bias be applied to the load, then R_3 is chosen so that

$$R_3 \approx \frac{(5.0 - 0.7) \text{ volts}}{0.025 \text{ ampere}} \approx 0.2 \text{ kilohm} \tag{IV-3}$$

This estimate neglects the voltage drop across T_1, usually only a fraction of a volt in the saturated, turn-on condition. It is good practice to control I_F through R_3 rather than through the transistor gain, since the latter varies considerably both with temperature and from transistor to transistor.

We have neglected the current drawn from the -25 V supply when T_1 is turned on. In this case, the voltage at terminal ℓ (the load output of the driver) is equal to about +0.7 V; the result is a current drain from the -25 V reverse bias supply through the resistance, R_4 and R_3, the turned-on collector-to-emitter path of T_1, and finally through the +5 V power supply. With this simple one transistor driver the reverse bias is merely overridden by the forward bias when T_1 is turned on. To minimize the current drawn, R_4 must be made as large as possible. For example, if not more than 10 mA of current drain can be tolerated, then

$$R_4 \geqslant \frac{(25 + 0.7)\ \text{volts}}{.010\ \text{ampere}} = 2.5\ \text{kilohms} \tag{IV-4}$$

Clearly, the current drawn from the reverse bias supply can be made as small as desired, but only at the expense of the reverse bias current magnitude available for removing charge from the diode load. This current is needed to provide the pulse leakage current experienced by reverse biased PIN diodes which are subjected to high microwave voltage levels; it is also required for fast charge removal (switching) during the transition from forward to reverse bias.

In Chapter II we observed that the steady state charge in the I region of a forward biased PIN diode is equal to $I_F\tau$ where τ is the average carrier lifetime. For the present example suppose that the PIN diode has a carrier lifetime of 1 μs. Neglecting natural recombination within the I region, the time required for a reverse bias current, I_R, to remove this charge is

$$T_S = \frac{I_F \cdot \tau}{I_R} \tag{IV-5}$$

with I_F = 25 mA and I_R = 10 mA, the switching speed T_S = 2.5 μs. This period is long compared with τ = 1 μs; accordingly, the actual switching speed is, of course, shorter, since recombination accounts for much of the charge removal. To speed up this process, a second transistor, T_2, could be added to the driver circuit, as shown in Figure IV-3. The role of T_2 is to amplify the current through R_4, providing a more rapid removal of the I region charge in the PIN diode load. This function is frequently called *active pullup.* The addition of the active pullup transistor, T_2, requires an element in its emitter-base to insure that this transistor is turned off during the forward bias condition. In Figure IV-3 this function is accomplished with a diode, D_1. The forward voltage

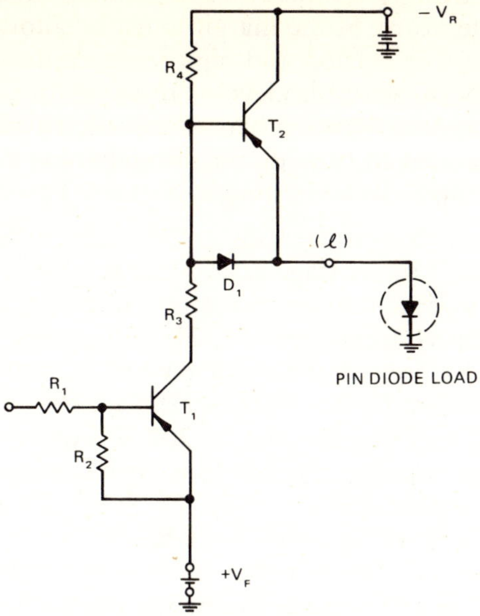

Figure IV-3 **Two Transistor (Unbalanced) TTL Compatible Driver**

drop across this diode is limited to about 0.7 V and has a polarity to reverse bias the emitter-base junction of T_2 so as to insure that it remains in a non-conducting state.

3. Diode Bias and Transistor Ratings

Because the size and power handling requirements of PIN diodes vary widely, so also do the reverse bias voltage and forward bias current requirements vary with both the diode type and application. Typical forward bias current and reverse bias voltage values are given in Table III-1. Forward bias currents typically are from 10 to 250 mA; reverse bias voltages, -10 to -250 V. Accordingly, transistors are required which are capable of switching these levels.

One of the first characteristics of interest in the selection of a driver transistor for a binary state PIN diode driver is the available current gain (β) of the transistor. In order to operate directly from the logic signal supply, the current gain factor must be sufficient to amplify the input current to the value required to supply the PIN diode(s). Attention should be given to the fact that transistor current gain is strongly temperature dependent, decreasing with increasing temperature. Some margin must be allowed between room temperature operation and high temperature operation if the required PIN forward bias level is to be assured. Generally, if more than 100 mA of forward bias is needed, an additional transistor may be needed to provide buffering between the logic signal source and the final bias switching transistor, T_1.

The most important consideration to be made in the selection of driver transistors is the voltage safety margin. The predicted operating lifetimes of switching transistors, resistors, and diodes have been given extensive treatment. [1, 2] As a general rule, transistors should have a breakdown voltage which is 150-200% of the bias voltage switched for long life operation. For example, PIN diodes operated with reverse bias of 200 V require a driver with transistors whose minimum breakdown voltages are in the 300-400 V range. It is this consideration which usually sets the maximum practical bias voltage which can be used with high power PIN control circuits. Often, higher RF power handling capability can be obtained per PIN diode if a higher reverse bias voltage is applied, but this action may result in greater overall expense for the phase shifter-driver combination, due to the cost of high voltage transistors, than that which would be incurred by designing an RF circuit which uses a greater number of PIN diodes operated at lower individual stress levels. Accordingly, the driver design is an important part of the PIN control circuit planning.

4. *Balanced (Input Impedance) Signal Input*

Frequently it is necessary to transmit the logic signal input to the driver over a long distance — tens to hundreds of feet. When a large number of such signal lines must be run, it is uneconomical to employ coaxial cables to provide the required immunity to the noise signals which can be induced on the long lines. The problem is similar to that encountered in local circuit telephone systems which employ twisted pair transmission lines. Such a transmission system can be highly resistant to stray signal noise pickup if both wires of the twisted pair are driven from and are terminated in balanced impedances. *An impedance is balanced when both terminals have the same impedance to ground.* A typical input circuit for a balanced driver is shown in Figure IV-4. The driver is activated by a *difference* in potential between the two wires of the twisted pair. If both wires are separated from ground by the same impedance, then equal amplitude and phase noise signals will be in-

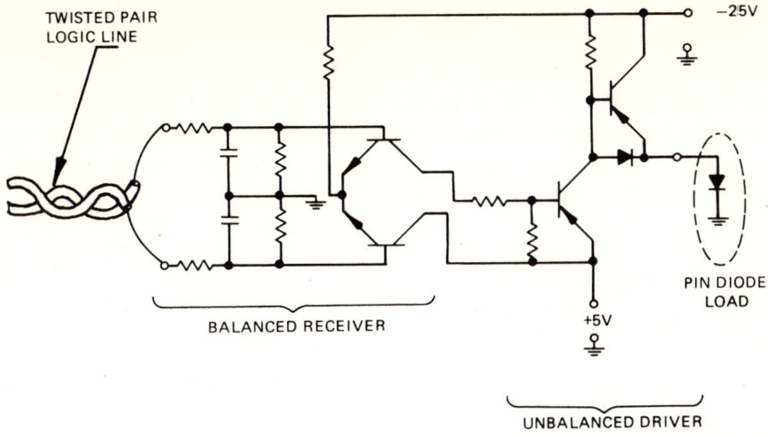

Figure IV-4 **Functional Schematic of Balanced Input Driver**

duced on both; no noise component will appear as a difference in potential between the two lines. With a circuit such as that shown in Figure IV-4 it is practical to build an input circuit which will switch when the difference in potential (called *differential mode voltage*) between the two input signal lines is 5 V, but which will not switch nor be damaged by an extraneous signal voltage (called *common mode voltage*) as high as 25 V with the same polarity applied simultaneously to both wires. While this balanced input circuitry requires more components than does the unbalanced driver circuit described in Figures IV-2 and IV-3, the added cost is less than the cost of shielded coaxial cables, and the inherent noise rejection capability is much greater. Balanced circuitry is very useful in large phased arrays where separate logic signals may have to be transmitted hundreds of feet to thousands of individual driver circuits that control PIN diode phase shifters in the antenna.

C. Switching Speed

1. Definitions and Measurements

The PIN diode presents a highly nonlinear and nonreciprocal load to the transistor driver circuit. Even if the driver could be terminated with a resistive load, the switching speed measured with this load would not be representative of the switching speed obtained with the PIN diode load. Thus, it is impossible to speak of a switching speed for the driver independent of its PIN diode load. Further-

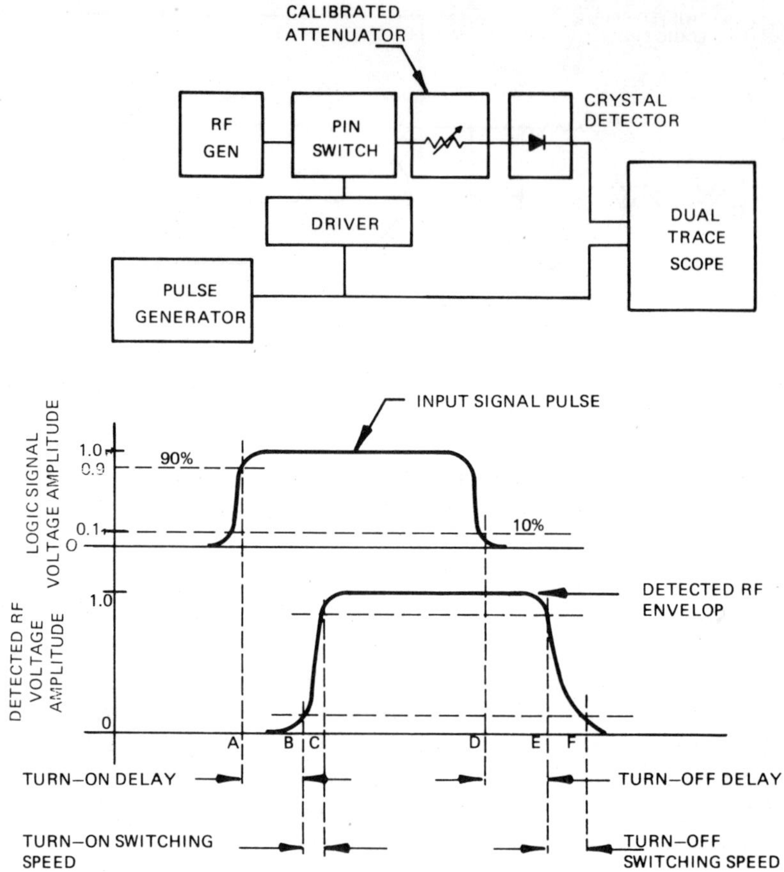

Figure IV-5 **Switching Speed Measurement and Definition for PIN Switch**

more, the speed with which the bias currents and voltages are switched into the PIN diode do not identify uniquely the speed with which the microwave control device in which the PIN diode is installed will perform its microwave switching function. Switching speed specifications and measurements relate to the combined performance of the PIN control device and its driver. Figure IV-5 shows both a typical circuit block diagram for switching speed measurement of a PIN diode switch and sample waveforms used to define the switching speed. A pulse generator with sufficiently fast switching speed is used as a "logic" signal source. The output of the microwave crystal detector on an oscilloscope is compared with the switching waveform from the pulse generator. The two resulting pulse shapes appear as shown in Figure IV-5. *There cannot be a standard definition for the switching speed of PIN control devices,* since the appropriate definition depends upon how the control device is to be used. For example, suppose the PIN switch shown has insertion loss of 1 dB in the transmit state and 30 dB of insertion loss (*isolation*) in the isolation state. The switching speed might be defined as the time required for the transition from 1.5 dB to 28 dB, if in the usage of the switch an additional 0.5 dB of loss and 2 dB less isolation can be tolerated during time periods near the switching transitions. The detected RF voltages corresponding to these loss values (points B and C in Figure IV-5) can be calibrated in the oscilloscope with the aid of the calibrated attenuator in the measurement circuit and the turn-on switching speed, which corresponds to the transition from the isolate to transmit condition of the switch measured, as indicated in Figure IV-5. Similarly, the turn-off transition time (points E to F) can be determined. Unless specific attention is given in the driver design and adjustment, these two switching transients generally are not equal.

In addition to the actual insertion loss transition measured for the PIN switch, a delay is encountered between when the signal pulse is applied and when the switching begins. Usually this time delay is less objectionable than the time required for switching, because it can be anticipated and the logic signal time of arrival and width can be adjusted to offset both the turn-on and turn-off switching delays (intervals AB and DE, respectively, in Figure IV-5). Compensation for such delays can, of course, be made only to the extent that these periods are reproducible. Where precise timing of the RF switching is important, variations in delay periods can be considered part of the corresponding switching transition times.

2. *Increasing Switching Speed*

Driver Considerations

The switching speed of a PIN control device is dependent upon both the driver and the PIN to be switched. Driver switching can be enhanced through the use of speed-up capacitors and coils, as shown in Figure IV-6 (the basic two transistor circuit shown in Figure IV-3 except for the addition of speed-up elements).

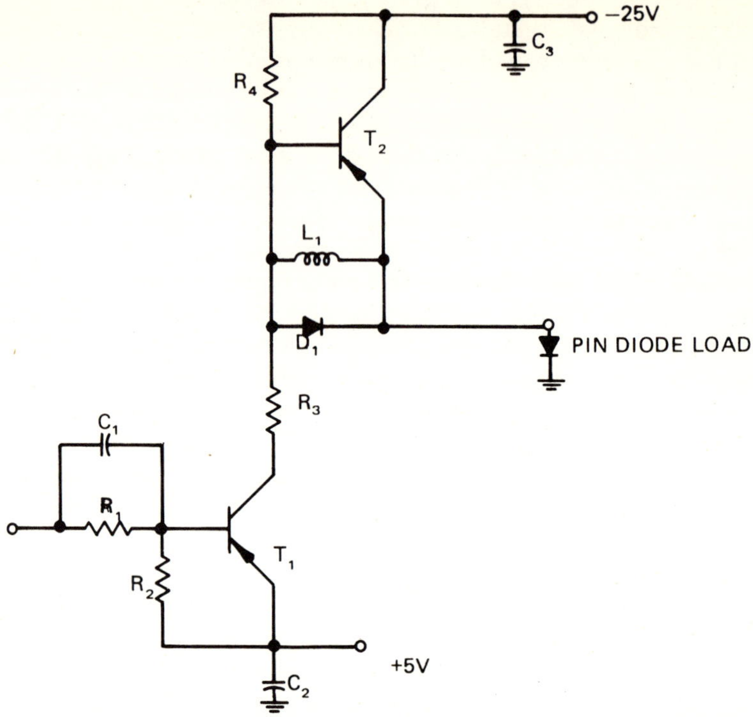

Figure IV-6 **Transistor Driver With Speed-Up Elements**

The driver turn-on can be speeded with a capacitor, C_1, across R_1, thereby providing a shunt path for the logic signal current to the leading edge of the applied logic signal through the emitter-base path of T_1. The value of C_1 should be made sufficient to pass the emitter-base current through T_1 for the duration of the period required for the turn-on of T_1. This value is readily estimated from the expression for the charging of a capacitor

$$\Delta V = \frac{1}{C} \int i \, dt \qquad \text{(IV-6)}$$

If, for example, we assume i, the maximum available current from an unbalanced TTL logic signal source, to be equal to $I_{EB} = 16$ mA; ΔT, the time allowed for the turn-on of T_1, to be 50 ns; and ΔV, the average signal level difference between a TTL *zero* and *one*, to be approximately 3 V, then

$$C = \frac{\Delta T}{\Delta V} \cdot I$$

$$C = \frac{50 \times 10^{-9} \text{ second} \times 15 \times 10^{-3} \text{ ampere}}{3 \text{ volts}} = 250 \text{ picofarads} \tag{IV-7}$$

Of course, the function of C_1 is effective only if the switching speed potential of T_1 is such as to realize the allowed turn-on time, ΔT; the transistor manufacturer's ratings must be consulted before an accurate assignment can be made for ΔT.

The driver turn-off is made more rapid by adding a shunting coil, L_1, to the diode, D_1, in the driver circuit of Figure IV-3, as shown in Figure IV-6. During the forward bias state the forward bias current, I_F, is established through L_1. When T_1 turns off, this current cannot change instantaneously through L_1; and so finds an alternate path through the emitter-base of T_2. The resulting current polarity and rapid increase with time through the emitter-base of T_2 quickly turns this transistor on, connecting the PIN diode(s) load to the reverse bias supply for a rapid removal of the stored charge obtained under forward bias and a fast transition to the reverse bias state.

The value of L_1 must be selected to be large enough that the induced voltage, ΔV, of the coil is sufficient to hold T_1 in conduction long enough to remove the stored charge in the PIN load. The size of the coil can be determined using the defining relationship for an inductance

$$V_i = L \frac{di}{dt} \tag{IV-8}$$

where i is the current through the coil of inductance L and V_i is the voltage induced with a polarity to oppose a change in the current. Assuming that an *average* induced voltage, $\overline{V}_i$, equal to 1 V is required for a time, ΔT, equal to 500 ns to switch off (i.e., remove all charge from) the PIN load, and that the forward bias current, I_1, is equal to 100 mA, the value for L is estimated from

$$L = \frac{\Delta T}{\Delta I} \cdot \overline{V}_i$$

$$L = \frac{500 \times 10^{-9} \text{ second}}{0.1 \text{ ampere}} \times 1 \text{ volt} = 5 \text{ microhenries} \qquad \text{(IV-9)}$$

The instantaneous induced voltage, $V_i(t)$, is, of course, not constant. The nature of an inductor is such as to induce whatever voltage magnitude is necessary to preserve continuity of the current through the inductor. The more rapid the turn-off of T_1, the greater the initial amplitude of $V_i(t)$ available to turn-on T_2.

The use of the speedup coil, L_1, *prevents T_2 from turning on at any time other than at the instant of the turn-off transient following a forward bias current, I_F.* Forward current through L charges the coil; interrupting this current produces an induced voltage to turn-on T_2. But without this initial condition the coil cannot turn-on T_2; indeed, the coil shorts out the emitter-base of T_2 at any other time.

This arrangement has both an advantage and a disadvantage. It is an advantage if a short circuit develops in the bias circuit or the diode itself; then the current supplied from the reverse bias supply, V_R, is limited to that which flows through R_4. This current is much less than the peak current supplied when T_2 turns on. Thus, the power supply and T_2 are protected by the coil from overheating due to the continuous high current drain of a short circuit. The disadvantage is that any charge injected into the reverse biased diodes by a high power RF pulse must be removed through R_4, without the current gain advantage of T_2. The resistance, R_4, must then be sized to permit ample reverse current leakage during any operation, as is discussed in Section D below.

Finally, the storage capacitors, C_2 and C_3, should be sized to provide the current transients needed for turn-on and turn-off of the PIN load with a small drop in supply voltage (usually 10% or less), both to prevent sagging of the power supply voltage during switching and to prevent the switching transients from propagating on the power supply lines. For the present example, with $I_F = 100$ mA, a forward supply voltage of 5 V, and an assumption that the PIN turn-on is less than 0.5 μs.

$$C_2 \geqslant \frac{0.5 \times 10^{-6} \text{ second}}{0.5 \text{ volt}} \times 0.1 \text{ ampere} = 0.1 \text{ microfarad} \qquad \text{(IV-10)}$$

The reverse bias switching current is not constant during the PIN switching transition from forward to reverse bias; however, the total charge removed is at most (neglecting carrier recombination within the PIN) equal to the magnitude of the forward bias current times the PIN carrier lifetime, $I_F \tau$. Assuming a carrier lifetime of 2 μs

$$C_3 \geqslant \frac{I_F \tau}{\Delta V} = \frac{0.1 \text{ ampere x } 2 \text{ x } 10^{-6} \text{ second}}{2.5 \text{ volts}} = 0.08 \text{ microfarad} \qquad \text{(IV-11)}$$

This capacitance is the minimum required; in practice, though, a larger capacitor, about 0.5 μF, typically may be used.

The sample driver in Figure IV-6 which uses 25 V reverse bias would be suitable for moderate applications with PIN diode I region widths up to 50 μm (2 mils); switching speed depends mainly on the particular I region width of the PIN diode used. In particular, with 50 μm (2 mil) I region diodes the carrier lifetime, τ, is typically 1 μs. The charge stored, Q_F, in the diode under forward bias, I_F, is

$$Q_F = I_F \tau \qquad \text{(IV-12)}$$

This charge is removed during the transition from forward to reverse bias by a current, I_R, through the collector of T_2. If we assume this current has a constant value for the PIN switching time, T_{SR}, and we equate the charge removed to the stored charge (i.e., $I_R \cdot T_{SR} = I_F \tau = Q_F$), then the switching speed of the forward to reverse bias transition is estimated from

$$T_{SR} = \frac{I_F \tau}{I_R} \qquad \text{(IV-13)}$$

The driver circuit limitation of the magnitude of I_R is set by the current gain (β) of T_2. In the circuit of Figure IV-6, the ratio of I_R / I_F is equal to β.

In the present example, suppose that the PIN has $\tau = 1$ μs at a forward bias of 100 mA. Then if T_2 has a $\beta = 10$, the reverse switching speed is

$$T_{SR} = \frac{\tau}{\beta} = \frac{1 \text{ microsecond}}{10} = 100 \text{ nanoseconds} \qquad \text{(IV-14)}$$

Use of Equation (IV-14) ignores recombination during the transition. This assumption is a good one when the switching speed is short compared with the carrier lifetime ($T_{SR} \ll \tau$). Furthermore this evaluation is conservative, to the extent that when carriers recombine during switching, the net charge to be removed is less and the switching time is faster.

Diode Construction

However, a more significant assumption made in the application of Equation (IV-14) to estimate switching speed is that the reverse current, I_R, is solely limited by the driver circuit. Effectively, this statement is true only when the carrier transit time across the I region is comparable to or less than the switching speed. We shall examine the limitations on switching speed posed by the PIN itself. However, beyond increasing the bias supply voltages to accelerate charges into and out of the PIN, there is little more that can be done practically to speed up switching.

Having designed a driver with the features described so far, the switching speed of the PIN generally is limited by its I region width rather than by the driver. The average diffusion limited carrier transit time at room temperature using average electron and hole mobility of 1000 cm²/V-s is (see pp. 48-52 for transit time discussion)

$$T_{DIFF} = 0.38\ W^2\ \frac{\text{nanoseconds}}{(\text{micron})^2} \qquad \text{(IV-15)}$$

Thus, if the driver applies only sufficient forward bias voltage to change the built-in junction potential (about +1 V) and only a slightly negative reverse bias voltage (about -1 V), the carriers within the I region will move to and from the P+ and N+ regions by diffusion and the switching speed will be that given by Equation (IV-15). The diffusion limited PIN transit time is shown graphically in Figure IV-7.

However, practical drivers, such as that shown in Figure IV-6, apply initial bias magnitudes more than the minimum ± 1 V for switching. In fact, the switching speeds practically obtainable with "fast" drivers are about five times faster than the diffusion limit. This increased speed is brought about by the introduction of an electric field in the I region due to the transient applied bias voltage. Carriers then move by *drift* into and out of the I region.

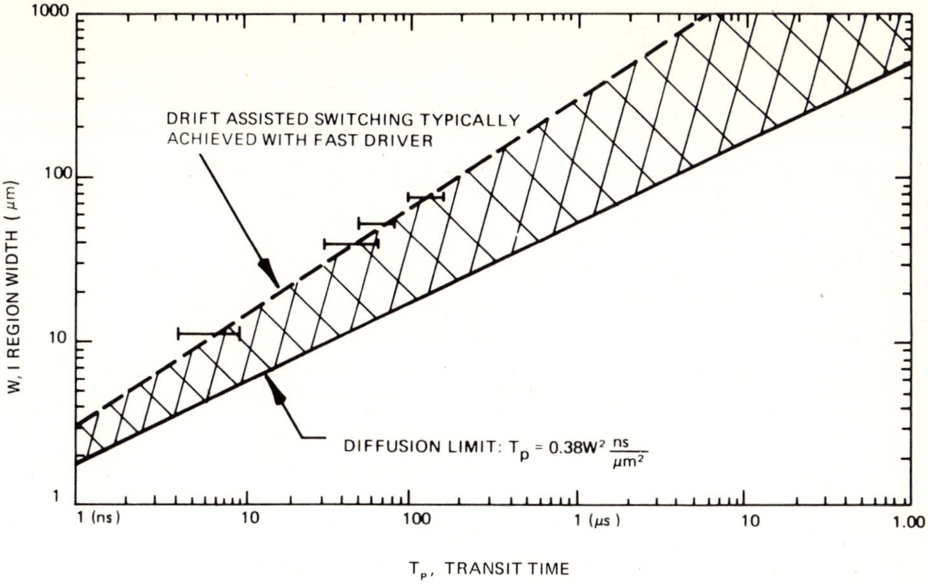

Figure IV-7 **Transit Time and RF Switching Speed of PIN Diodes of Various I Region Widths**

The drift velocity of carriers is related to the E field as shown in Figure IV-8. Above 10,000 V/cm, the velocity saturates; further increases of E field produce little increase in carrier velocity. If a field this large could be created across the I region of the PIN, switching speed would increase by orders of magnitude from that estimated by Equation (IV-15) for diffusion. The saturated carrier velocity, V_{DRIFT}, at $E = 10^4$ V/cm is about 6×10^6 cm/s. This figure could be achieved in the diode example with 50 μm I region with an applied voltage across the diode of only

$$V = E \cdot W = 10^4 \frac{\text{volts}}{\text{centimeter}} \times 50 \times 10^{-4} \text{ centimeter} = 50 \text{ volts}$$

The resulting drift transit time is

$$T_{DRIFT} = \frac{W}{V_{DRIFT}} = \frac{50 \times 10^{-4} \text{ centimeter}}{6 \times 10^6 \text{ centimeters/second}} = 0.8 \text{ nanosecond} \tag{IV-17}$$

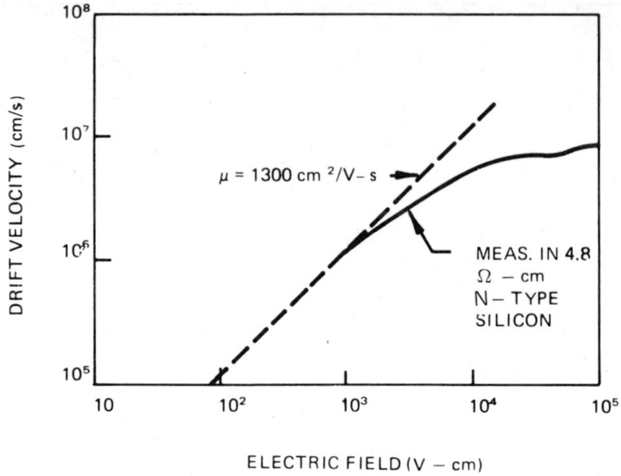

Figure IV-8 **Drift Velocity Saturation in Silicon (Curve Presented by Watson [3 (p. 91)])**

This figure is over three orders of magnitude faster than the diffusion transit time predicted by Equation (IV-15), which for this diode is 950 ns. Moreover, if drift saturated velocity can be achieved, switching speed should be linearly proportional to W, rather than W^2 as indicated for diffusion propelled carriers. Clearly, the application of excess bias voltage across the PIN diode terminals can be expected to improve switching speed, and it does. However, the enhanced switching speed is usually within an order of magnitude of that predicted using the diffusion transput mechanism. Some data points and an approximate Transit Time vs. W contour are sketched in Figure IV-7.

Analytically the drift assisted switching speed has not so far been found practical to treat because while the motion of a few carriers within an otherwise depleted I region can readily be estimated, the transient flow of the concentrated charge distribution necessary to make the I region a low resistance medium changes both the field magnitude and uniformity, inhibiting the prediction of the carrier velocity. Accordingly, the diffusion limited switching speed can be considered a worst case, and drivers capable of inducing transient electric fields within the I region produce considerably more rapid transitions between the forward and reverse bias states.

D. High Power Reverse Bias (Enhanced) Leakage Current Supply

Not only must a driver have sufficient reverse bias current capability to achieve the desired switching speed, it must also provide reverse current during high power RF pulses when the peak RF voltage exceeds the bias voltage, as shown in Figure IV-9. The *enhanced* reverse leakage current magnitude may not remain constant during the pulse. If the diode heats appreciably during the pulse, the enhanced current will increase as well. Likewise, if the average diode temperature rises, the leakage current will increase.

The magnitude of the leakage current is a function of the I region width, the magnitude of the peak RF voltage compared to the bias

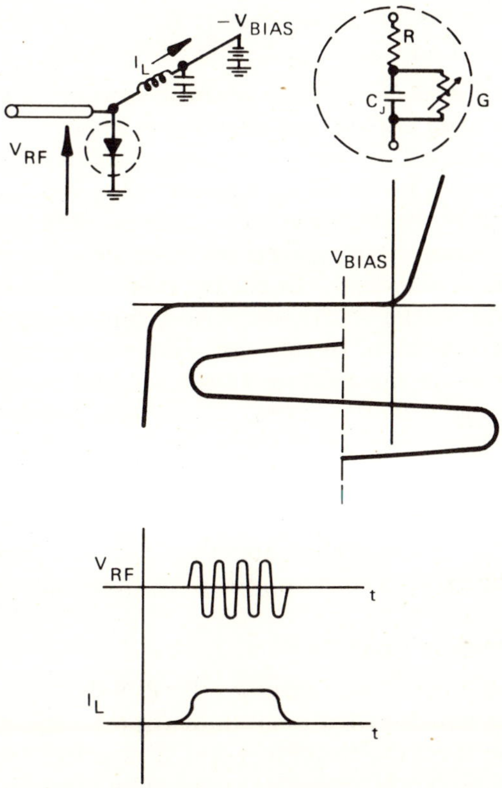

Figure IV-9 **Pulse Leakage Current Obtained with PIN Diode When Peak RF Voltage Magnitude Exceeds Reverse Bias Voltage**

voltage, and the duration of the time that the RF voltage causes the net field within the I region to be positive. This leakage measurement is not predictable and must be measured under conditions representative of actual use. Fortunately, in most applications the enhanced leakage is a few milliamperes or less, when the bias voltage is at least 1 volt per micron of I region width. However, should the reverse bias voltage "droop" during the pulse, the charge introduced into the I region may not be completely removed during the negative-going RF voltage sweep. The result is that the diode becomes more lossy (shunt conductance, G, shown in the equivalent circuit in Figure IV-9 increases), hotter, and the enhanced leakage current increases further. With a high impedance driver, a further drop in the bias voltage occurs. The process can become regenerative to the point that diode failure due to filamentary overheating occurs, creating an alloy fault which permanently shorts out the diode, as described in Chapter III.

Active pull-up drivers of the type shown in Figure IV-3 easily provide the enhanced current required because any charge in the PIN experienced while T_1 is non-conducting produces a current through R_4 which turns on T_2 and connects the diode to the reverse bias supply through a current path capable of tens to hundreds of milliamperes. However, use of a speed-up coil, L_1, as described in the driver circuit in Figure IV-6 inhibits the turn-on of T_2, except during the transition period immediately following a forward bias condition. In this case, R_4 must be made low enough in value to provide the enhanced leakage current without an appreciable change (less than 20%) of the supplied reverse bias voltage.

This action represents a trade-off among reverse bias power supply drain, switching speed, and RF power handling capabilities. The lower R_4 is made, the greater the drain from the reverse voltage supply when T_1 is turned on for the forward bias state.

E. Fault Detection Circuits

Frequently it is desirable to design the PIN driver with a built-in fault monitor which provides an indication whenever a malfunction occurs in the PIN, the driver, the bias power supplies, or simply the logic signal wiring network. The most complete fault sensing requires an RF detector mount which samples the microwave signal after it has been acted upon by the PIN, but this arrangement is usually too complex and costly. Furthermore, any

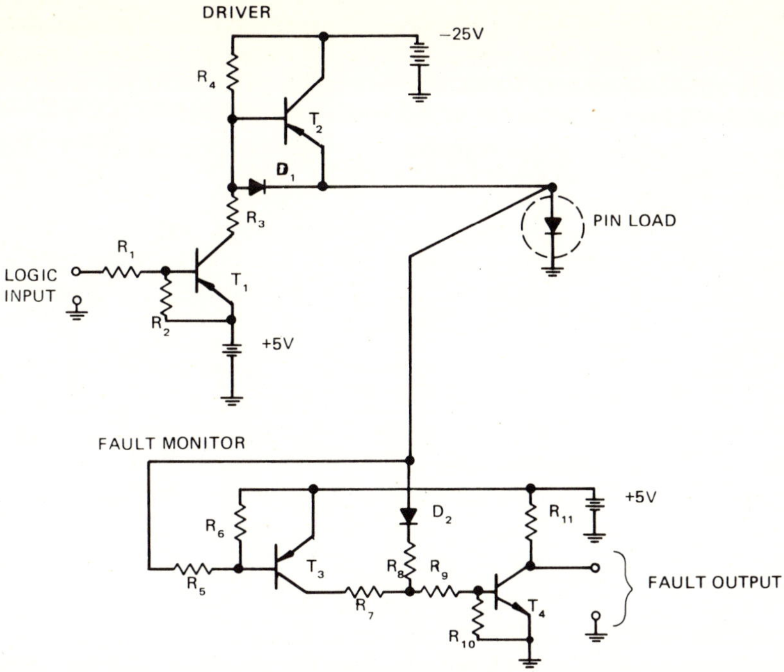

Figure IV-10 **Driver Circuit With Fault Monitor (BITE-Built In Test Equipment)**

malfunction associated with the PIN nearly always is evidenced by a change in its capacity to reflect a proper load on the driver in either or both the forward and reverse bias states.

Figure IV-10 shows the functional form of a fault detector used with the driver of Figure IV-3; it provides an unbalanced TTL output signal with the opposite sense of the logic signal input to the driver when there is no fault, and with the same sense when there is a failure. Thus, by comparing the fault output signal with the applied logic signal, an indication can be obtained of how well the PIN control circuit is functioning.

By comparing the fault and logic signals in both the forward and reverse bias states of the PIN, this circuit can reveal the following kinds of malfunctions:

1) *Shorted PIN Diode (or shorted diode socket)*
2) *Open Circuit PIN*

3) *Low Voltage Reverse Bias Supply*
4) *Loss of Forward Bias Supply*
5) *Loss of Logic Input Signal (or lack of driver circuit response to logic signal)*

The operation of the circuit can be understood by considering its function under each of these malfunctions.

1. *Shorted PIN Diode (or shorted diode socket)* — Under normal operation with a logic *one* applied to the driver input, -25 V reverse bias is applied to the PIN diode. This voltage is also fed through a voltage divider consisting of R_5 and R_6 to turn on T_3. When T_3 turns on, so also does T_4, producing a TTL *zero* at the fault output. This situation is opposite the applied logic *one* required for reverse bias; thus, proper operation is indicated.

Should either the PIN itself or the RF circuit in which it is embedded become shorted, the -25 V reverse bias will not be established at the PIN. In the example circuit being considered, unless a current limited -25 V supply is used, T_2 most likely will be burned out as well. With a lack of voltage at the PIN, T_4 is not turned on and the fault output is *one*, the same as the logic input, indicating a malfunction.

2. *Open Circuit PIN* — Under normal operation a logic *zero* at the driver input turns on T_1 and delivers a forward bias current to the PIN, across which the voltage drops to less than +1 V. The transistors T_3 and T_4 in the fault circuit are then non-conducting, a logic *one* appears at the fault output — opposite to the zero at the driver input — and normal operation is indicated.

However, should the diode fail to draw its normal bias current, there is little or no voltage drop across R_3, the voltage at the PIN driver terminal will rise above +2 V and turn on T_4 through D_2. This occurrence indicates a malfunction by producing a *zero* at the fault output, coinciding with the *zero* at the logic input.

3. *Low Voltage Reverse Bias Supply* — The voltage divider consisting of R_5 and R_6 can be adjusted to turn on T_3 at any preselected reverse bias voltage amplitude, thereby assuring that a satisfactory level of reverse bias is available to the PIN. This resistive voltage divider is shown for illustration; however, more precise voltage reference circuits can be used as required to establish this negative reference below which bias magnitude level a fault condition is indicated.

4. *Loss of Forward Bias Supply* – Loss of the +5 V forward bias supply causes the PIN to remain in the reverse bias state since forward bias current through D_1 is required to turn T_2 off and decouple the PIN from the -25 V supply. Negative PIN bias produces a *zero* output from the fault circuit – provided that the +5 V supply for the fault circuit is separate from the +5 V forward bias supply – which is the same as the *zero* at the logic input, thereby indicating a malfunction, when a *zero* input is present.

5. *Loss of Logic Input Signal (or lack of driver circuit response to logic signal)* – Should an interruption occur in the logic signal line to the driver, the PIN will remain in reverse bias and the logic output will always be a *zero.* A malfunction will be indicated when the system calls for PIN forward bias and a logic *zero* is presumed to be applied at the driver logic input. This type of malfunction indication, of course, can be obtained only if the comparison logic input and fault output levels is made ahead of the logic line break, such as at the system logic control network.

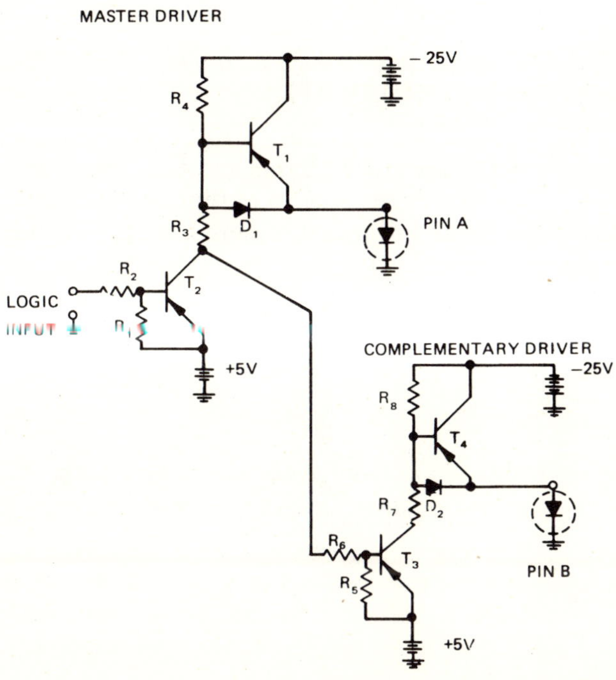

Figure IV-11 **Complementary Driver Pair**

F. Complementary Drivers

In some switching applications it is necessary to operate pairs of PIN's in opposite bias states, as in a SPDT switch which is used to switch a microwave signal from a common source to one of two loads. The basic driver circuit of Figure IV-3 is easily adapted to accommodate this function while requiring only a single input logic level, as shown in Figure IV-11.

PIN diode A receives reverse bias from the master driver with a logic *one* applied. At the same time, T_3 in the complementary driver is turned on by emitter-base current through the R_6, R_3, and R_4 path to the -25 V supply. This current is limited mainly by R_4 which must be selected to provide adequate forward bias to PIN diode B after amplification by the β of T_3.

A logic *zero* at the input puts forward bias on PIN diode A, turning off T_3 and thereby reverse biasing PIN diode B.

References

[1] *Reliability Stress and Failure Rate Data for Electronic Equipment,* MIL-HDBK-217.

[2] *ITT Reference Data for Radio Engineers – Fifth Edition* (Chapter 40); Howard W. Sams and Co., Inc., New York, 1974. (This reference gives an extensive treatment to reliability definitions and measurements; it also includes an extensive bibliography.)

[3] Watson, H.A., (ed.): *Microwave Semiconductor Devices and Their Circuit Applications,* McGraw-Hill, Inc., New York, 1969.

The author wishes to acknowledge the special assistance of R. Ziller at Microwave Associates, Inc. in the preparation of this chapter.

Questions

1. Show, using the charge continuity expression in Equation (II-21) that, if recombination during the switching period from forward to reverse bias is taken into account, the switching time is given by the transcendental relationship

$$T_S = \left(\frac{I_F \tau}{I_R}\right) e^{-T_S/\tau}$$

(which reduces to Equation (IV-5) when $T_S \ll \tau$)

2. Find the actual switching time, T_S, for the example given below Equation (IV-5) (in which $\tau = 1\ \mu s$, $I_F = 25$ mA, and $I_R = 10$ mA) using the result derived above in Question 1. Hint: find a solution to the transcendental function by trial and error – guessing.

3. Find the ratio T_S/τ when $I_R = -I_F$, using the result from Question 1.

Chapter V

Fundamental Limits of Control Networks

A. Introduction

This chapter contains a general treatment of the best switching, duplexing, and phase shifting performance obtainable from PIN diodes, recognizing that practical PIN's must necessarily have both finite capacitance and resistance and bounds on the RF voltages and currents that can be sustained. It is fortunate for the designers of PIN circuitry that such maximum RF performance capabilities can be defined and evaluated analytically — even though the diode parameters may have to be measured experimentally. It is possible to say in advance, given the listing of diode parameters, just what switching insertion loss and power handling capability can be obtained per diode. From this estimate a decision can be made about how many diodes are needed and whether the task is practical *even before the specific switching circuit has been selected.*

Furthermore, the general circuit treatment given in this chapter not only defines the absolute performance limits to be obtained but suggests circuit configurations in which this behavior might be realized. To understand the limits of PIN circuitry it is necessary to look at microwave networks in a very basic way. But the effort is more than rewarded by the understanding of the PIN switching limitations and, equally important, the conditions under which they can be applied.

Hines [1] showed that for practical microwave circuits (i.e., essentially lossless, linear, passive, and reciprocal) the insertion loss introduced by the diode resistance is determined by the switching function and the ratio of the operating frequency, f, to the switching cutoff frequency, f_{CS}, of the diode used in the circuit. Furthermore, the product of the RF voltage and current stresses on the diode in its two states is related directly to the incident power on the switch and the type of switching function (i.e., SPST, SPDT, phase shift, and so forth). Hines' work indicates that it is unlikely that there is some hitherto undiscovered circuit by which, say, a little glass packaged computer diode could switch the megawatt output of a complete radar. His conclusions are the more remarkable for the method by which they were derived. The analysis treats the ideal switch as a switchable reflection circuit; the fundamental equations, as he derived them, relate to the reflection coefficient of the circuit. Strictly speaking, the results were proved only for reflection type circuits. However, they appear to yield valid limits to multi-throw switching, where their validity has not been demonstrated formally; subsequent analysis of matched transmission phase shifter limits produce approximately the same results for small phase shift values as those Hines obtained for reflection phase shifters.

Rather than repeat the reflection based derivation, we examine the SPST ideal switch by considering an alternative derivation which treats the switch as a general lossless three-port network. Before doing so, however, let us examine the simple Single-Pole Single-Throw (SPST) switching circuit which consists of a single diode mounted either in shunt or in series between a matched generator and load.

B. The Simple Loss (or Isolation) Formula

There is a wide variety of PIN microwave circuits which accomplish multi-throw switching, multi-bit phase shifting, and routing of signals in waveguide. So broad have these functions become that it is easy to lose sight of the fact that a diode is only a two terminal device; as such, it yields only SPST switching. Achievement of multi-throw switching must, therefore, be accomplished by suitable switching circuitry containing several PIN diodes. By the same token, switching in waveguides in which the RF energy is not concentrated between two conductors, is effected only by designing transitions between the waveguide and a TEM (usually coax) transmission structure. These "transitions," of course, may be simply posts in the waveguide in which PIN diodes are embedded.

Either a series or shunt mounting of the diode in a "two wire" transmission line can be analyzed using the same equations.

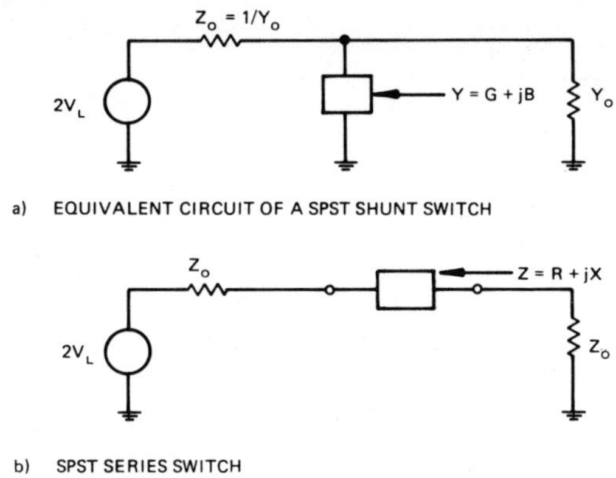

Figure V-1 **Equivalent Circuits of SPST Switches**

Consider first the shunt mounting. The simplest circuit (see Figure V-1(a)) for a microwave switch consists of an admittance (representing the diode) mounted across a transmission line between a generator and a load, both of which are matched to the characteristic impedance of the line. We define insertion loss (or isolation) according to

$$\text{Insertion Loss (or Isolation)} = P_A/P_{LD} \tag{V-1}$$

where P_{LD} is the actual power dissipated in the load and P_A (equal to $V_L{}^2/Z_0$) is the "available" power which would have been delivered by the generator to the load if the admittance, Y, had not been present. When the diode is biased to produce a large admittance (low impedance), the line is "shorted out" and the microwave power is reflected back to the generator. In the opposite diode state, the diode, with whatever tuning elements are used, is made to have a low admittance (high impedance) and the incident microwave power passes by it to the load. This process provides SPST reflective switching. A particularly useful and simple relationship exists for the insertion loss (or isolation) produced in a matched system with Z_0 impedance by the shunt circuit of admittance Y. The formula is readily derived by evaluating the load voltage, V_{LD}, with Y in place shunting the load and generator

$$V_{LD} = 2\ V_L \cdot \frac{\dfrac{1}{Y_0 + Y}}{Z_0 + \dfrac{1}{Y_0 + Y}}$$

Simplifying

$$V_{LD} = \frac{2\ V_L}{2 + YZ_0} \tag{V-2}$$

The *power insertion loss ratio* is just the ratio of the square of the magnitude of the available line voltage (under matched conditions), V_L, to the load voltage, V_{LD}, or

$$\text{Insertion Loss} = \left| \frac{V_L}{V_{LD}} \right|^2$$

$$\text{Insertion Loss} = \left| 1 + \frac{YZ_0}{2} \right|^2 = 1 + GZ_0 + \frac{(GZ_0)^2}{4} + \frac{(BZ_0)^2}{4} \tag{V-3}$$

The term YZ_0 is recognized as the admittance of Y normalized to the transmission line impedance, Z_0. The loss ratio is usually defined as a number greater than unity for finite loss; when expressed in decibels, the resulting quantity is positive. No confusion need arise since passive admittances always reduce the power flow between matched generator and load. It should also be noted that the insertion loss calculated with Equation (V-3) *includes both power reflected by and dissipation within the shunt admittance Y.* Equation (V-3) is shown graphically in Figure II-17 for two important conditions, when Y is composed entirely of a conductance, G, or entirely of a susceptance, B.

The series Z form of the SPST switch, shown in Figure V-1(b), is analyzed in the same fashion. The load voltage is

$$V_{LD} = \frac{2\ V_L}{2 + (Z/Z_0)} \tag{V-4}$$

which is identical to Equation (V-2) for the shunt switch with the normalized impedance, Z/Z_0, replacing the normalized admittance, YZ_0. The corresponding insertion loss formula for the series impedance is,

$$\text{Insertion Loss} = \left|1 + \frac{Z}{2Z_0}\right|^2 = 1 + R/Z_0 + \frac{(R/Z_0)^2}{4} + \frac{(X/Z_0)^2}{4} \tag{V-5}$$

Equations (V-3) and (V-5) can be used to estimate both insertion loss or isolation for a switch merely by inserting the appropriate value for the admittance, Y, which pertains to the control element bias state under consideration.

The equivalent circuits shown in Figure V-1 are so simple that everything about the switch performance — insertion loss, isolation, RF voltage, and current stresses on the diode — is easily calculated. It would be nice if all switching circuits could be analyzed so readily. The fact is that, for single frequency operation, all switching circuits can be reduced to these simple shunt and series models with a suitable choice of the characteristic impedance, Z_0. We prove this assertion in the next section because it underlies the fundamental switching limits which follow.

C. The General Three-Port SPST Equivalent Circuit

1. A Theorem about Lossless Three-Ports

It is always possible at any specified frequency to place a short-circuit across one arm of a lossless reciprocal three-port transmission-line junction in such a position so as to decouple the other two arms.*

This result is a remarkable one which taxes the imagination, since it would appear initially that an exception could be found. The theorem is very general and applies directly to a SPST diode switching circuit. At one port is the generator, at the second is the load, and at the third is the diode. The extension of the theorem to the diode switch follows shortly, but first consider the statement of the theorem. Figure V-2 shows a general con-

*The "three-port" is defined only if *finite mutual coupling actually exists* between the three ports, thus precluding the trivial case in which one or more ports are not connected.

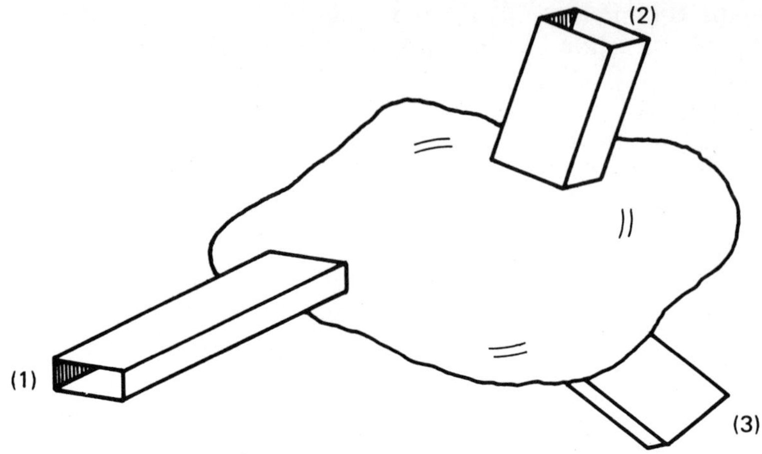

Figure V-2 **A General Lossless Three-Port Network**

ducting enclosure with three waveguide ports, although any other non-radiative lossless circuit with suitable transmission lines for the propagation of the frequency of interest is equally appropriate. The theorem states that a suitable location (reference plane selection) can always be found in any port such that a short circuit at that position completely blocks the transmission of any energy between the remaining two ports.

To help visualize the generality of this theorem suppose, for example, that a metal trash barrel has the lid and three waveguide terminals welded on, as shown in Figure V-3. Then, assuming that the structure is perfectly conducting (lossless) and that none of the waveguides is initially blocked (and, therefore, the device is a three-port), the theorem applies to this structure. A sliding short circuit in, say, Port 2 has a position at which a signal incident at Port 1 is blocked totally from exiting at Port 3. Of course the terminals can be interchanged, with the source, load, and short assigned separately to any permutation of the three ports. The decoupling situation can be visualized by imagining that the movement of the short circuit affects the electric field pattern within the structure so that at the frequency of interest a null occurs at the entrance to one or both of the other two ports, fully decoupling the power flow between them.

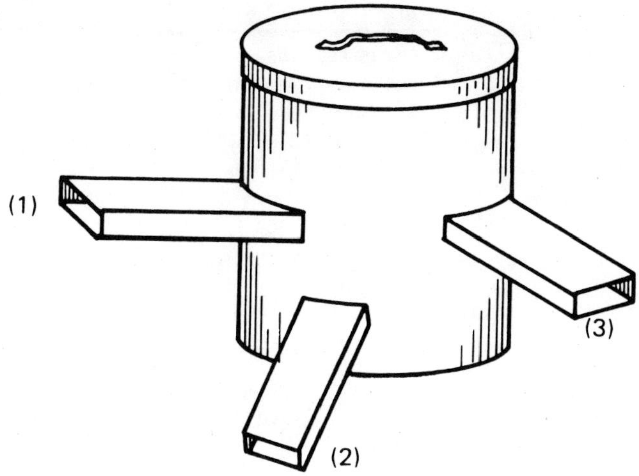

Figure V-3 "Lossless Trash Barrel" Three-Port Example

The proof of the theorem is carried out using an impedance matrix description for the three-port, as follows:*

Define the voltages and currents at the three ports using the convention shown in Figure V-4. Since waveguide terminals could be used, the voltages and currents are interpreted to be normalized to the characteristic impedance of the waveguide port.**

The normalized impedance matrix for the network is defined as,

$$v_1 = z_{11}i_1 + z_{12}i_2 + z_{13}i_3 \qquad \text{(V-6(a))}$$

$$v_2 = z_{21}i_1 + z_{22}i_2 + z_{23}i_3 \qquad \text{(V-6(b))}$$

$$v_3 = z_{31}i_1 + z_{32}i_2 + z_{33}i_3 \qquad \text{(V-6(c))}$$

A short placed in arm 3 can be transformed by the line length at that port to a pure imaginary impedance, Z_3, at reference cross section (3) and then

*This derivation follows that of the Microwave Circuits lectures by Prof. Herman A. Haus of the Massachusetts Institute of Technology, Cambridge, circa 1960.

**Actually, a three-mode single waveguide could be attached to a lossless cavity and the theorem would apply to the separate three modes in the waveguide – provided, of course, that reference planes and shorts could be independently established for the three modes.

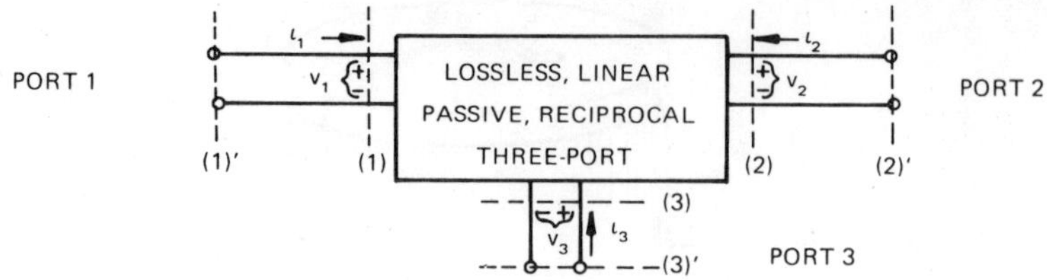

Figure V-4 **Reference Terminal Notation for the Three-Port Equivalent Circuit**

$$v_3 = -Z_3 i_3 \tag{V-7}$$

If i_3 is eliminated from Equations (V-6(a)) and (V-6(b)), then

$$v_1 = \left(z_{11} - \frac{z_{13}^2}{z_{33} + Z_3}\right) i_1 + \left(z_{12} - \frac{z_{13}z_{23}}{z_{33} + Z_3}\right) i_2 \tag{V-8(a)}$$

$$v_2 = \left(z_{12} - \frac{z_{13}z_{23}}{z_{33} + Z_3}\right) i_1 + \left(z_{22} + \frac{z_{23}^2}{z_{33} + Z_3}\right) i_2 \tag{V-8(b)}$$

Arms 1 and 2 are decoupled when

$$z_{12} = \frac{z_{13}z_{23}}{z_{33} + Z_3} \tag{V-9}$$

Since the three junction is assumed to be reciprocal and lossless, all of its elements, z_{ij}, are pure imaginary; also, diagonal symmetry exists, with $z_{ij} = z_{ji}$. But Z_3 is itself pure imaginary and may assume any value between $-j\infty$ and $+j\infty$; therefore, it is always possible to satisfy Equation (V-9), and the theorem is proved.

2. *Equivalent Circuit for a Three-Port*

If reference planes are chosen in each arm of the three-port at those positions at which a short decouples the remaining two arms, and the new impedance matrix terms denoted with prime superscripts, then from Equation (V-9) (and the two similar equations derived alternately by eliminating i_2 and i_1, from the matrix in Equation (V-6)) we obtain,

$$z'_{12}z'_{33} = z'_{13}z'_{23} \quad \text{(V-10(a))}$$

$$z'_{23}z'_{11} = z'_{21}z'_{31} \quad \text{(V-10(b))}$$

$$z'_{31}z'_{22} = z'_{32}z'_{12} \quad \text{(V-10(c))}$$

Multiplying Equations (V-10(a)) and (V-10(b)) gives

$$z'_{11}z'_{33} = (z'_{13})^2 \quad \text{(V-11)}$$

Since z'_{13} must be pure imaginary for a lossless reciprocal junction, it can be concluded that z'_{11} and z'_{33} are both pure reactances of the same sign. Similarly, z'_{11} and z'_{22} may be shown to have the same sign. Then,

$$z'_{11} = n_1^2 z \quad \text{(V-12(a))}$$

$$z'_{22} = n_2^2 z \quad \text{(V-12(b))}$$

$$z'_{33} = n_3^2 z \quad \text{(V-12(c))}$$

where n_1, n_2, n_3 are positive real numbers and z is pure imaginary. With no loss of generality z may be chosen equal to ±j, and the off diagonal impedance elements can be evaluated using Equation (V-11). The impedance matrix then becomes

$$z = \pm j \begin{bmatrix} n_1^2 & n_1 n_2 & n_1 n_3 \\ n_1 n_2 & n_2^2 & n_2 n_3 \\ n_1 n_3 & n_2 n_3 & n_3^2 \end{bmatrix} \quad \text{(V-13)}$$

An equivalent circuit which satisfies Equation (V-13) is shown in Figure V-5. Thus we conclude that, with suitable reference plane selection, the most general equivalent circuit for a three-port at a single frequency involves only a reactive element (which can be either inductive or capacitive depending upon the particular circuit) and real impedance transformations at each of the three ports (the magnitudes of which also depend upon the network to which this equivalent circuit is applied). The availability of such a simple equivalent circuit for all "brassware" networks from which SPST diode switches can be made is of great value. With it we can now define the limiting performance which can be obtained with diodes having finite resistive loss and limits on sustainable RF voltages and currents.

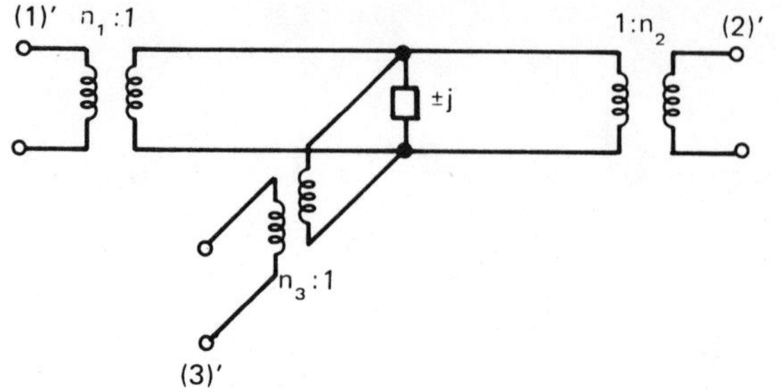

Figure V-5 **A General Equivalent Circuit for the Lossless, Reciprocal, Linear Three-Port with Reference Ports Defined at Short Circuit Decoupling Planes**

D. Switching Limits

1. Ideal SPST Switch

The PIN diode can be represented for switching purposes by an ideal knife switch with associated capacitance and resistances, as shown in Figure V-6. The capacitive reactance of C_J along with any package parasitic reactances are considered to be part of the switching three-port network. Since a lossless network model has been used in the three-port equivalent circuit, it is necessary

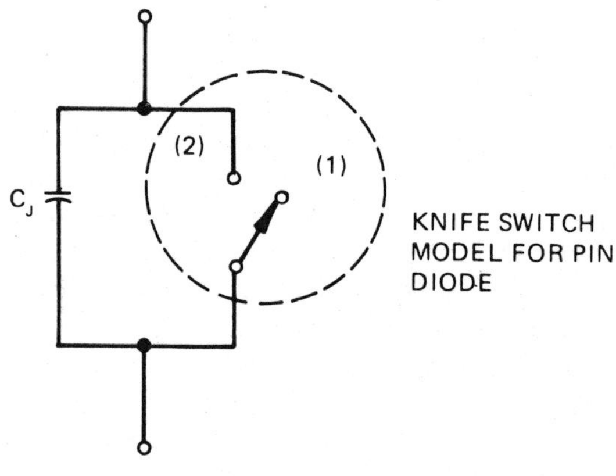

Figure V-6 **PIN Diode Equivalent Circuit Used for Switching Limit Derivation**

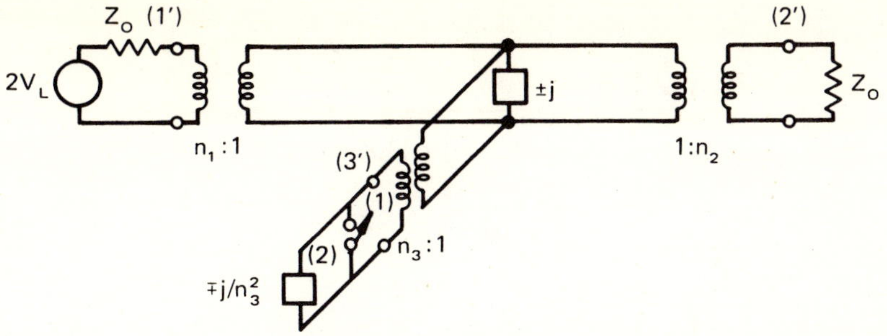

Figure V-7 **The Knife Switch Diode Model Used With a General Lossless Three-Port to Provide SPST Switching Between Matched Generator and Load**

to assume that the diode resistances, R_F and R_R, are small enough to be neglected; they are included later to determine insertion loss limits. At this point these resistances are treated as small perturbations to the network; i.e., their magnitudes are so small that the RF voltage and current magnitudes within the three-port do not change appreciably from the values calculated when $R_F = R_R = 0$.

Under these assumptions, the diode knife switch is used to provide SPST switching between matched generator and load of impedance Z_0 using the general three-port circuit as shown in Figure V-7. In the open condition (1) of the knife switch, we want all of the power available from the generator to be delivered to the load. This need requires that suitable reactive tuning be provided to neutralize resonately any inherent reactance in the three-port (which includes the reactances of the diode junction and its package). This tuning can be added to the equivalent circuit at any port; in Figure V-7, it is shown attached to the switch port (3′). In addition to the reactive tuning, any three-port which affords non-reflective transmission between matched generator and load must have symmetry about these ports, in particular

$$n_1 = n_2 \tag{V-14}$$

Under these conditions a new equivalent circuit having only a single impedance transformation which includes the n_1 to n_3 transformation separating the generator-load ports from the

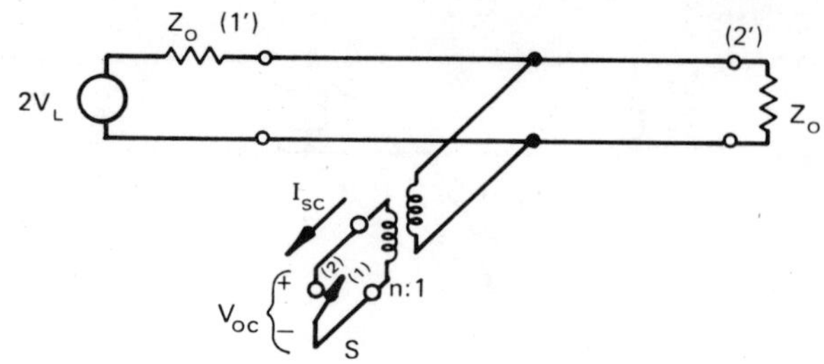

Figure V-8 **Reduced Equivalent Circuit of the Ideal SPST Switch Between Matched Generator and Load**

switching port can be defined, as shown in Figure V-8. This circuit affords matched, lossless transmission between generator and load in "open circuit" state (1) with the knife switch open; it supplies total isolation of generator and load in the "short circuit" state (2) when the knife switch closes.

2. Power Limit

In the open position maximum power travels to the load, equal to

$$P_{LD} = V_L I_L \tag{V-15}$$

and in this state the knife switch element has across its terminals a voltage

$$V_{OC} = nV_L \tag{V-16}$$

When the switch, S, closes all power is reflected to the source at (1′) and the switch element carries a current

$$I_{SC} = \frac{2I_L}{n} \tag{V-17}$$

By selection of n, the ratio of the magnitudes of V_{OC} and I_{SC} may be made to have any value. If the maximum tolerable values for V_{OC} and I_{SC} are defined as V_M and I_M respectively, n can, in principle, always be chosen so that, at some power $P = P_M$, $V_{OC} = V_M$ and $I_{SC} = I_M$.

Under these conditions the maximum potential of the switching element in a SPST switch is utilized and the "power switched" is

$$P_M = \frac{V_M I_M}{2} \tag{V-18}$$

Equation (V-18) indicates that *either* the voltage stress (under reverse bias) or the current stress (under forward bias) can be reduced by adjustment of the impedance level of the switching circuit; reduction of one, though, is done at the expense of a proportionate increase in the other, as the product of these two stress levels is a constant for a given power level of operation. Thus, by determining (either by calculation or experiment) the V_M and I_M ratings for the diode (at the RF frequency, pulse length, duty cycle, operating temperature, bias levels, etc.), *the maximum RF power level of operation for SPST switching can be calculated from Equation (V-18) before any effort is made to define a specific switching circuit.* Of course, it may develop that the required switching impedance level is impractical to realize or that the dissipative losses under forward and reverse bias conditions are too unbalanced, dictating operation at lesser power in an impedance level selected according to some criterion other than maximum power capability. The expression of Equation (V-18) is nonetheless useful as an upper bound on PIN diode power handling capability.

3. Minimum Insertion Loss

As already noted, the switch evaluation is made with the assumption that the diode resistances, R_F and R_R, are small enough as to have negligible effects on the magnitudes of the RF stresses, V_{OC} and I_{SC}, applied to the diode by the circuit. The specific dissipations in R_F and R_R depend on how the line impedance is transformed at the diode. For example, if the diode shunts a low impedance line, it will pass little current under reverse bias but a large current under forward bias. A switch so designed would have lower insertion loss and lower isolation than one designed with a higher impedance line.

There is no unique ratio of dissipative losses in the two diode states. However, if the line impedance is selected so that the losses under forward and reverse bias are equalized, the average loss value will be at a minimum. This situation applies because the dissipation values are proportional to the square of the voltage and current stresses; reducing either near the equal loss condition causes the other to increase more rapidly, resulting in a

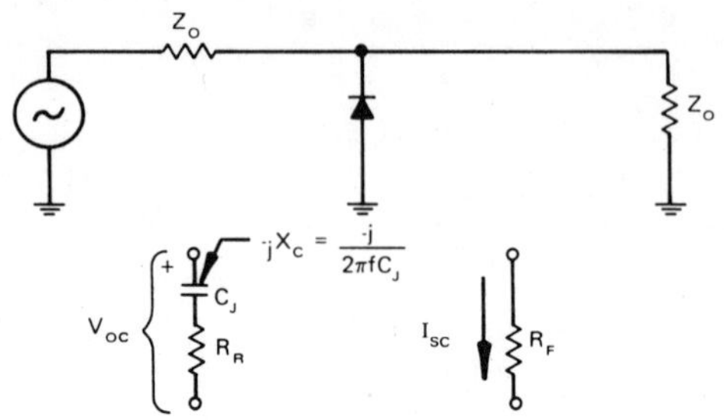

Figure V-9 **Equivalent Circuit Used to Determine Insertion Loss of Ideal Switch**

higher loss averaged over both states.

Under the condition of equal power dissipation in both states, the ratio of the power dissipated within the diode, P_D, to the power switched, $P_A = V_L{}^2/Z_0$, can be related to the diode parameters as follows. The open and short positions of the knife are replaced by the reverse and forward biased equivalent circuits of the PIN chip. These experience V_{OC} and I_{SC} excitations, respectively, as shown in Figure V-9. The power dissipated under reverse bias is

$$P_{DR} = \left| \frac{V_{OC}}{X_C} \right|^2 R_R \tag{V-19}$$

and under forward bias

$$P_{DF} = |I_{SC}|^2 R_F \tag{V-20}$$

Under the equalized loss condition $P_D = P_{DR} = P_{DF}$. Equating P_D to the product of the square roots of P_{DR} and P_{DF} gives

$$P_D = \frac{|V_{OC} I_{SC}| \sqrt{R_F R_R}}{|X_C|} \tag{V-21}$$

Multiplying Equations (V-16) and (V-17) together gives

$$|V_{OC}\, I_{SC}| = 2|V_L\, I_L| = 2P_A \tag{V-22}$$

Combining Equations (V-21) and (V-22)

$$\frac{P_D}{P_A} = \frac{2\sqrt{R_F R_R}}{|X_C|} \tag{V-23}$$

This expression is the minimum ratio of dissipated to available power achievable in a SPST switch with a diode having these X_C, R_R, and R_F characteristics; it can be cast in a useful frequency ratio form by defining a *switching cutoff frequency*, f_{CS}, or figure of merit for the PIN when used for switching.

$$f_{CS} = \frac{1}{2\pi C_J \sqrt{R_F R_R}} \tag{V-24}$$

Note that *only the diode junction capacitance* appears in the f_{CS} definition. Package inductance and capacitance, if present, usually have negligible ohmic losses; as such, they can be lumped into the lossless three-port which represents the SPST switch. Package "parasitic" capacitance and inductance reduce the operational bandwidth but do not affect the cutoff frequency and, hence, do not alter the minimum insertion loss theoretically obtainable with the diode.

If the right side of the expression for f_{CS} in Equation (V-24) is multiplied by f/f (unity) where f is the frequency of the SPST switch at which the minimum loss is to be estimated, then

$$f_{CS} = \frac{f}{2\pi C_J f \sqrt{R_F R_R}} = \frac{f\,|X_C|}{\sqrt{R_F R_R}} \tag{V-25}$$

This expression is readily substituted into the loss ratio of Equation (V-23), and

$$\frac{P_D}{P_A} \approx 2\left(\frac{f}{f_{CS}}\right) \qquad 10f < f_{CS} \tag{V-26}$$

This approximation relies on the small loss perturbation and gives a value accurate within a few percent when the operating frequency, f, is less than a tenth of the switching cutoff frequency.

For example, a diode having $C_J = 1$ pF and $R_F = R_R = 0.5\ \Omega$ has a switching cutoff frequency of 318 GHz. The minimum insertion loss ratio obtainable with the diode in a matched (VSWR = 1)

SPST switching circuit dissipating equal RF power in both bias states at 3 GHz is 2 x 3/318 = 0.019. This figure is about 2%, or less than 0.1 dB. Equation (V-26) is useful because diode cutoff frequencies can be tabulated for available diodes (see Table III-1) and the minimum SPST switching loss can be evaluated directly.

Strictly speaking, to use Equation (V-26), *f_{CS} must be evaluated using diode resistance values which apply at the switching frequency, f.* For example, if $R_F = R_R = 0.5\ \Omega$ is measured at 1 GHz, then, because of skin effect, somewhat different values for these resistances will be encountered where the switch is to be designed at, say, 3 GHz. The value for f_{CS} must be recalculated using the 3 GHz resistances to estimate minimum loss ratio in a 3 GHz switch. In practice, however, skin effect has only moderate influence in the 1-10 GHz frequency range (see Section II-B-2, pp. 62-66, for skin effect in forward biased I region resistance); useful approximations in this frequency range can be made using f_{CS} values evaluated directly from 1 GHz measurements of diode resistances.

It so happens that the same minimum insertion loss can be achieved for SPDT as for SPST switching, because the SPST switch (shown in Figure V-9) subjects the diode to twice the line current, $2I_L$, in the forward biased state and the line voltage, V_L, in the pass state. Effectively, the diode carries this double current because, in the isolation state, it must reverse the flow of RF power. (If a series diode SPST switch has been used the diode would see only I_L but $2V_L$.) Now consider the SPDT tee switch shown in Figure V-10. It subjects each diode to line voltage, V_L, and line current, $I_L = V_L/Z_0$. The net dissipation is unchanged, but it is divided between the two diodes (assuming Z_0 is adjusted to give equal loss in both impedance states of each diode). We gain an additional output port, with no increase in insertion loss. Extra isolation (nearly 6 dB) is obtained because only half voltage is applied to the isolating diode, the generator always being presented with a matched load. Thus SPDT switching performance is better than that for SPST, for diodes with the same f_{CS}.

4. *Linear Approximation to Insertion Loss (dB)*

In Equation (V-1) the insertion loss ratio was defined as the ratio of the available power to the power reaching the load in the presence of the switching circuit. For finite loss, this ratio

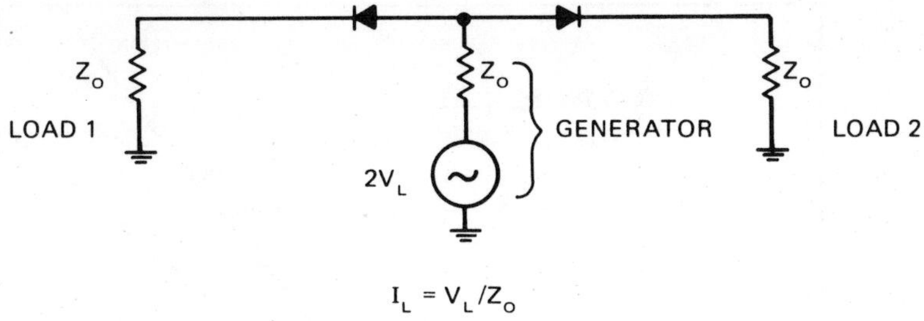

Figure V-10 **SPDT Switch Uses Two Diodes, but Each is Subjected to Only $V_L I_L$ Stress, Half That Experienced by the Single Diode in a SPST Switch**

is always greater than unity. For a switch design having good transmission match (VSWR $\approx$ 1), the reflection loss is negligible and the insertion loss ratio is approximately

$$\text{Insertion Loss (decibels)} \approx 10 \log_{10} \left(\frac{1}{1 - P_D/P_A} \right) \tag{V-27}$$

Frequently it is necessary to estimate loss in decibels when Math Tables or calculators are unavailable; for small loss situations, when $P_D \ll P_A$, Equation (V-27) can be further approximated as

$$\text{Insertion Loss} \approx 10 \log_{10} \left(1 + \frac{P_D}{P_A} \right), \ (P_D \ll P_A) \tag{V-28}$$

Then, noting that $\log_{10}(1+x)$ can be approximated for small values of x by its Taylor expansion [2] about x = 0 as

$$\log_{10}(1+x) = 0.43 \left(x - \frac{x^2}{2} + \frac{x^3}{3} - \frac{x^4}{4} + \ldots \right) \tag{V-29}$$

$$(\text{for } -1 < x < 1)$$

Using only the first term of the Taylor expansion of the logarithm expression and the approximation of Equation (V-28), Equation (V-26) for the minimum loss of a SPST or SPDT switch can be simplified to

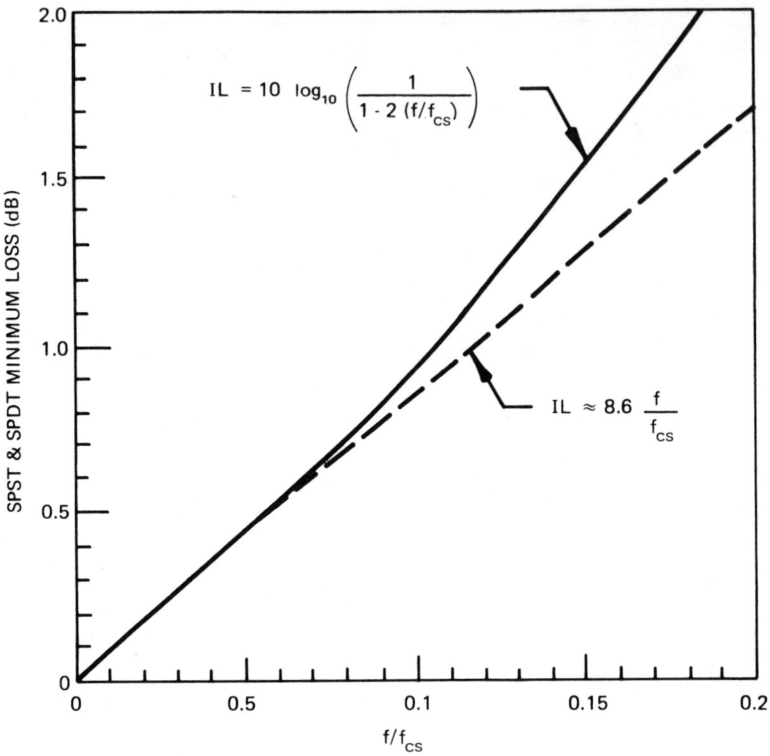

Figure V-11 **Minimum Insertion Loss Ratio (dB) for SPST and SPDT Switches vs. f/f_{CS}**

$$\text{Insertion Loss (decibels)} \approx 8.6 \, \frac{f}{f_{CS}} \qquad \text{(V-30)}$$

$$(f \ll f_{CS})$$

The loss per Equation (V-27) in decibels is plotted versus f/f_{CS} in Figure V-11; the linear approximation of Equation (V-30) is also shown. It can be seen that the linear approximation is accurate within ten percent for $f < 10\ f_{CS}$, a sufficient accuracy for most engineering purposes.

E. Duplexing Limits

The same relationships hold for a duplexer as for a SPDT switch; however, the ratings for the device reflect the fact that one state

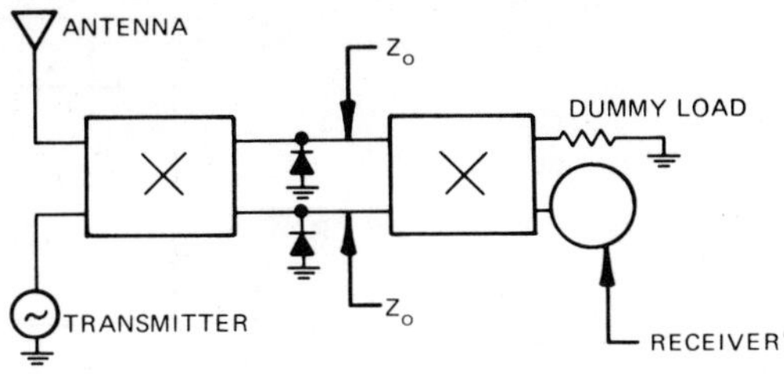

Figure V-12 **Balanced Duplexer Using 3 dB, 90° Hybrid Couplers**

must sustain the incident high power of a transmitter while the other incurs only the low power of the received signal. Figure V-12 shows the typical "balanced duplexer" SPDT switch configuration. Since diodes have only a thermal limit under forward bias (both thermal and voltage limits would apply for high power operation at reverse bias), the duplexer is designed to connect the transmitter to the antenna in the forward bias state. The balanced circuit shown has the added benefit that the couplers contribute to the isolation between the transmitter and receiver to the extent of their directivity, usually about 20 dB, most of the high power leaking past the diodes being absorbed in the dummy load (if the antenna is well matched).

To determine the maximum power handling capacity of a duplexer, assume that a maximum power, P_{DM}, may be dissipated in the forward biased diode switch element and that a maximum insertion loss, Insertion $Loss_M$, is allowable in the "receive" condition. The impedance level of the switch in principle, can then, be adjusted so that these two conditions always are met. The maximum power sustainable by the duplexer is then calculated as follows. If a single diode with forward resistance R_F shunts a Z_0 transmission line (Figure V-9) on which is incident a power $P_M = I_L{}^2 \cdot Z_0$, for $R_F \ll Z_0$, the short circuit current, $2I_L$, is experienced by the diode as it reflects the power. The power dissipation in R_F is

$$P_{DM} = (2I_L)^2 R_F$$

$$P_{DM} = \frac{4R_F P_M}{Z_0} \qquad \text{(V-31)}$$

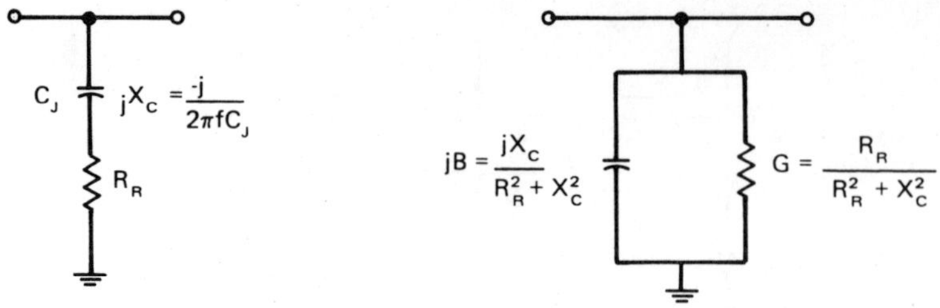

Figure V-13 **Series and Parallel Equivalent Circuits for Reverse (or Zero) Biased Diode at Frequency f**

In the reverse (or zero) biased state the diode has a series $R_R C_J$ equivalent (Figure V-9). Converting to admittance form (Figure V-13), the insertion loss ratio is given by Equation (V-3). However, if the capacitive susceptance is matched (resonated at f by parallel circuit inductance) the total susceptance at the diode is zero. Furthermore, if $R_R \ll X_C$ as is usually the case, then

$$GZ_0 \approx \frac{Z_0 R_R}{X_C{}^2} = Z_0 R_R (2\pi f C_J)^2 \qquad \text{(V-32)}$$

and the insertion loss ratio, from Equation (V-27) can be approximated as

$$\text{Insertion Loss} \approx 1 + GZ_0 \approx 1 + Z_0 R_R (2\pi f C_J)^2 \qquad \text{(V-33)}$$

The error in making this approximation approaches zero as R_R and the loss approach zero, and is less than ten percent if $10R_R < |X_C|$ and $GZ_0 < 0.1$. The *fractional loss,* as a fraction of the power incident on the duplexer in the low power receive mode is approximately equal to

$$\text{Loss} = \text{Insertion Loss} - 1 \qquad \text{(V-34)}$$

which we designate as "Loss" to avoid confusion with the insertion loss ratio.

If the duplexer is designed with as high a Z_0 as possible, then power handling will be maximized since, from Equation (V-31), the maximum power sustainable by one diode is

$$P_M = \frac{Z_0 P_{DM}}{4R_F} \tag{V-35}$$

and if N diodes are mounted in parallel and share the RF current equally the maximum power sustainable is

$$P_M = \frac{Z_0 P_{DM} N^2}{4R_F} \tag{V-36}$$

The N^2 rather than the N factor arises because the net dissipation is reduced due to the parallel combination of N resistances. However, the higher a value used for Z_0, the greater the loss, which is, from Equations (V-32), (V-33), and (V-34), for a single diode

$$\text{Loss} = Z_0 R_R (2\pi f C_J)^2 \tag{V-37}$$

and for N diodes mounted in parallel (provided, of course, that their net normalized conductance, NGZ_0, is much smaller than 1)

$$\text{Loss} = NZ_0 R_R (2\pi f C_J)^2 \tag{V-38}$$

Accordingly, a compromise must be made between the high power handling to be achieved under transmit conditions from a given number, N, of diodes and the fractional loss, Loss, tolerable under receive conditions. This compromise is effected by the choice of Z_0. Combining Equations (V-36) and (V-38) to eliminating Z_0 gives the fundamental equation relating P_M and the maximum tolerable loss in the receive condition, Loss_M

$$P_M = \frac{(\text{Loss}_M) P_{DM} N}{4R_R R_F (2\pi f C_J)^2} \tag{V-39}$$

This equation is more usefully expressed in terms of frequency normalized to f_{CS} by substituting Equation (V-25) into Equation (V-39), which yields

$$P_M = \frac{(\text{Loss}_M) P_{DM} N (f_{CS}/f)^2}{4} \tag{V-40}$$

For example, if a duplexer is to be designed using 12 diodes, each of which can dissipate a peak power of 100 W under the maximum radar system pulse length to be used, the ratio $(f_C/f) = 100$, and if a receive loss of 0.5 dB is tolerable

$$P_M = \frac{(0.1)\,(100 \text{ watts})\,(12)\,(100)^2}{4}$$

$$P_M = 300 \text{ kilowatts} \tag{V-41}$$

The impedance level at which this switching should be performed can be calculated directly by solving Equation (V-38) for Z_0, and recognizing that in the balanced duplexer (Figure V-12) the Z_0 is that of the SPST switches, each of which has (N/2) diodes in parallel. Thus

$$Z_0 = \frac{(\text{Loss}_M)}{(N/2)\,R_R\,(2\pi f C_J)^2} \tag{V-42}$$

If, in the present example, $C_J = 1$ pF, $R_R = 1\ \Omega$ and $f = 3$ GHz, then the proper value for the line impedance of the SPST switches between the hybrids is,

$$Z_0 = \frac{0.1}{6(1 \text{ ohm})(2\pi \times 3 \times 10^9 \text{ hertz} \times 1 \times 10^{-12} \text{ farad})^2}$$

$$Z_0 = 47 \text{ ohms} \tag{V-43}$$

Thus, in this example, diodes mounted in a standard 50 Ω line would give nearly the desired performance.

The optimization discussed so far has been with respect to peak power sustaining capability under forward bias and receive condition insertion loss. The balanced duplexer also has an insertion loss on transmit called *Arc Loss*, a term which originated when the diode positions shown in Figure V-12 were occupied by gas tubes which ionized under high transmitter power to reflect power to the antenna terminal. The *arc loss is the power dissipated in the switching elements in the transmit state of the duplexer.* This figure is given by Equation (V-31), modified for the balanced duplexer to account for the parallel combination of N/2 diodes in two SPST switches, each of which has $P_M/2$ incident power.

$$\text{Arc Loss} = \frac{(N/2)P_{DM}}{(P_M/2)} = \frac{4R_F/(N/2)}{Z_0}$$

$$\text{Arc Loss}^* = \frac{8R_F}{NZ_0} \qquad \text{(V-44)}$$

If in the example given $R_F = 1\ \Omega$, the arc loss is 0.67% or only 0.06 dB.

One can see that having chosen N and Z_0 to satisfy the requirements of P_M and $Loss_M$, there is no further design latitude to control the magnitude of arc loss. In the present example the arc loss is small because R_F is low and N is large. However, R_F is sometimes larger, especially when self forward biasing with high RF power is used and design constraints dictate fewer diodes. In such cases, Z_0 might have to be selected to afford a particular arc loss with forward bias and a particular receive loss with reverse bias*. The design then reduces to the SPDT switching circuit parameter selection. In short, design of switches, duplexers, and phase shifters generally cannot be simultaneously optimized with respect to insertion loss and power handling capability in both impedance states of the diode.

F. Phase Shifting Limits

1. Power Limit

The general SPST model can be used to form a reflection phase shifter by using it to switch a reactance at one port of a circulator,** as shown in Figure V-14. This reactance can be realized practically as a length of short- or open-circuited transmission line, as an inductor, or even as the junction capacitance of the real diode (represented as the knife switch in Figure V-14). The derivation of the power and insertion loss limits follows the same procedure as that used for switches and duplexers, except that the useful function provided is the change in angle of the reflection coefficient — *phase shift.* *** The first fundamental question is — "What voltage, V_{OC}, and current, I_{SC} does the switch experience in its two respective states for a given incident power, P, and phase shift, $\Delta\phi$?"

*A graph relating arc loss and isolation is given in Figure VII-46.

**Actually, the reflection phase shifter is realized practically using a pair of diodes to terminate a hybrid coupler, as described in Chapters VI and IX; but the model used here is simpler for deriving the phase shifting limits of PIN diodes.

***Phase shift is defined in Chapter X, pp. 389-391; reflection coefficient is developed in connection with the Smith Chart, Appendix J, pp. 537-549.

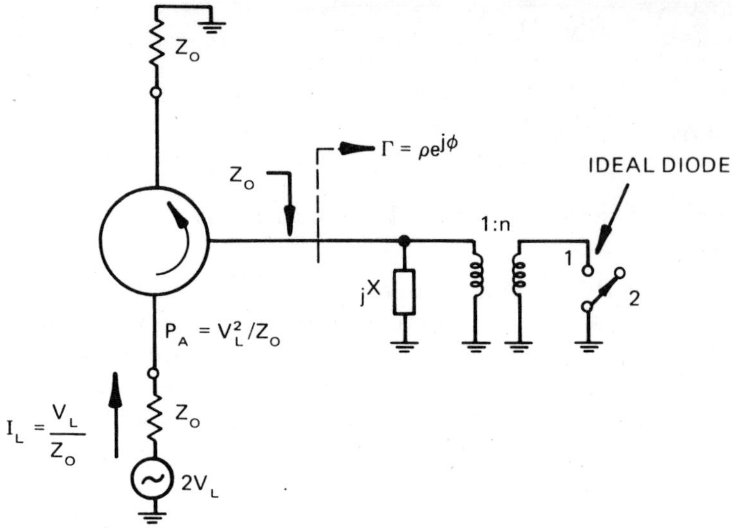

a) IDEAL PHASE SHIFTER MODEL USED TO CALCULATE PERFORMANCE LIMITS

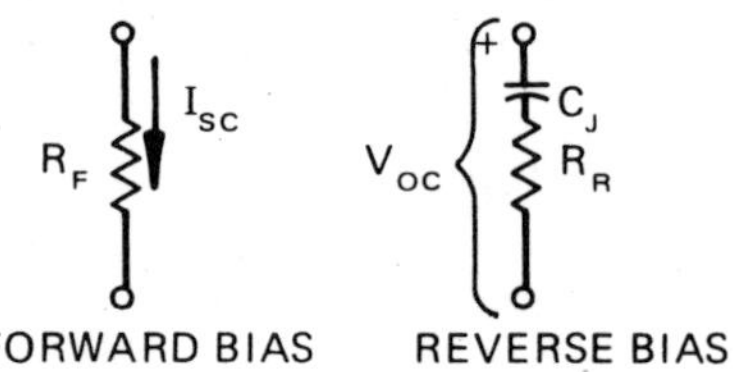

b) EQUIVALENT CIRCUIT FOR JUNCTION USED IN LOSS CALCULATION

Figure V-14 **Circuit Models Used For Phase Shift Limit Derivation**

Consider first state (1) when the switch closes and carries the short circuit line current $2I_L$ transformed through the turns' ratio, n.

$$I_{SC} = 2I_L/n \qquad \text{(V-45)}$$

The reflection coefficient is

$$\Gamma_{SC} = -1 \qquad \text{(V-46)}$$

If the switch is opened, state (2), we observe that jX and Z_0 form a voltage divider which the knife switch also shunts through the turns' ratio, n. Thus

$$V_{OC} = 2V_L n \left(\frac{jX}{Z_0 + jX}\right) \tag{V-47}$$

The reflection coefficient in the open state is given by

$$\Gamma_{OC} = \frac{Z_0 - jX}{Z_0 + jX} \tag{V-48}$$

The change in reflection coefficient is the complex difference

$$\Gamma_{OC} - \Gamma_{SC} = \frac{Z_0 - jX}{Z_0 + jX} - \frac{Z_0 + jX}{Z_0 + jX} = \frac{-2jX}{Z_0 + jX} \tag{V-49}$$

The angle difference described by $\Gamma_{OC} - \Gamma_{SC}$ is the phase shift obtained from the network. Notice that the quantity $2jX/(Z_0 + jX)$ also appeared in the expression for V_{OC} (Equation (V-47)). Hence, by substitution, we can write

$$\Gamma_{OC} - \Gamma_{SC} = -\frac{V_{OC}}{nV_L} \tag{V-50}$$

Next eliminate n using Equation (V-45)

$$\Gamma_{OC} - \Gamma_{SC} = -\frac{V_{OC}}{V_L} \cdot \frac{I_{SC}}{2I_L} \tag{V-51}$$

But the power incident $P_A = V_L I_L$ and so

$$\Gamma_{OC} - \Gamma_{SC} = -\frac{V_{OC} I_{SC}}{2P_A} \tag{V-52}$$

This expression is more useful when phase shift $\Delta\phi$, appears explicitly instead of as $(\Gamma_{OC} - \Gamma_{SC})$. To express it in this form, note the diagram representing the subtraction of the unit vectors, Γ_{OC} and Γ_{SC}, shown in Figure V-15. The magnitude of the resulting vector corresponding to this difference in complex reflection coefficients is shown in the figure. When the phase shift, $\Delta\phi$, is zero,

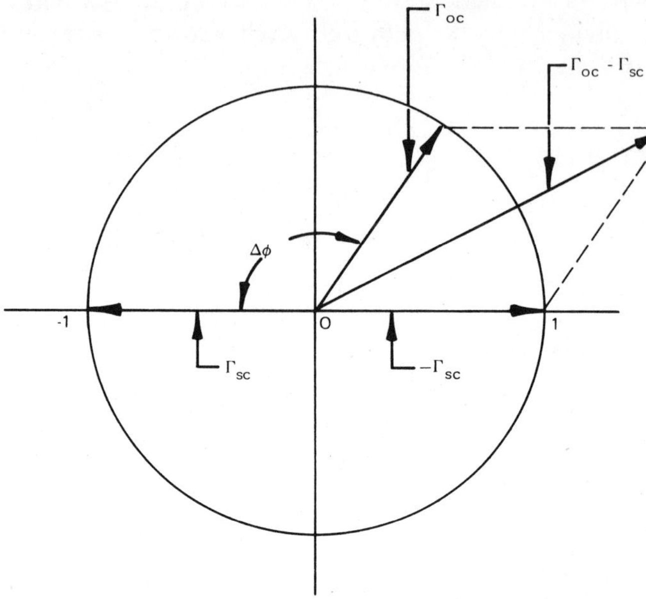

Figure V-15 **Diagram Used to Relate Phase Shift Angle $\Delta\phi$ to Change in Vector Magnitude of Reflection Coefficients of Phase Shifter**

$\Gamma_{SC} = \Gamma_{OC}$ and the difference vector has a magnitude of 0. For all other phase shift values from 0° to 360°, the reflection coefficient difference has a magnitude between 0 and 2. Equation (V-52) gives the vector change in reflection coefficient of a single step phase shifter. This change can be related directly to phase shift in degrees for more convenient design estimation. Before making this transition, however, an important characteristic of multi-step phase shifters can be observed from the present reflection change formula of Equation (V-52).

Consider Figure V-16 which shows two diagrams depicting a method for obtaining 180° of phase shift. In Figure V-16(a), Γ_{OC} and Γ_{SC} are 180° opposed and represent two impedance points diametrically opposite one another when plotted on the Smith Chart. The reflection coefficient diagrams shown in Figure V-16 can be overlayed on a Smith Chart with arbitrary initial angle reference, posing an infinite set of switched reactances which provide any particular amount of phase shift. The phase shift for the single step of Figure V-16(a) is 180° and the magnitude of $(\Gamma_{OC} - \Gamma_{SC})$

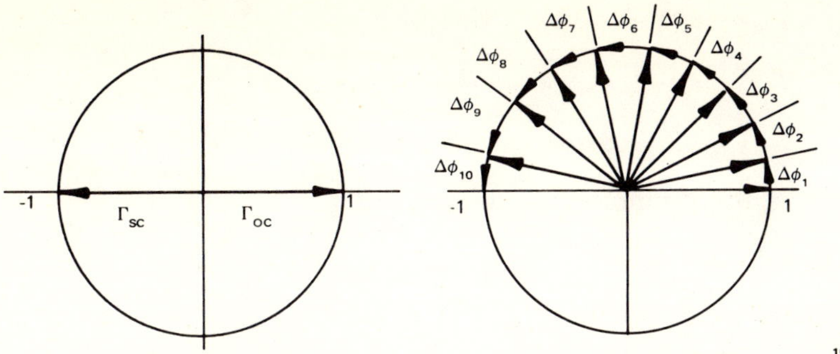

a) SINGLE STEP 180° PHASE SHIFTER $|\Gamma_{oc} - \Gamma_{sc}| = 2$ b) TEN STEP 180° PHASE SHIFTER, $\sum_{N=1}^{10} |\Gamma_{oc} - \Gamma_{sc}| \approx \pi$

Figure V-16 **Comparison Showing That Single Step 180° Phase Shifter Requires ($2/\pi$) Less Reflection Coefficient Magnitude Change (and Proportionately Lower Diode Stress) Than Multistep 180° Phase Shifter**

is 2. In contrast, the 10 step phase shift contour shown in Figure V-16(b) yields a total phase shift of 180° but the sum of the magnitudes of the reflection coefficient changes approaches the ratio of the half circumference of the unity radius circle or π. The corresponding net stress products ($I_{SC} V_{OC}$) experienced by the diodes in the multistep phase shifter are therefore about 1.4 times as large as that for the single 180° step diodes. This situation is the worst for the buildup of the reflection magnitude sum. Subdivision of any other phase change produces a lower buildup factor. Although the loss and stress can be up to 40 percent greater when the phase shifting task is distributed, this penalty is often a small one to pay for a circuit which conveniently divides the total phase shifting load among several diodes. From a theoretical viewpoint, of course, it is important to note that the power handling and diode insertion loss are optimized when the required phase shift is effected in a single step.

Conversion of Equation (V-52) to read in phase shift directly is accomplished as follows. Referring to Figure V-15 in which the vector difference corresponding to ($\Gamma_{OC} - \Gamma_{SC}$) is shown graphically

$$\Gamma_{SC} = -1$$

$$\Gamma_{OC} = -\cos\Delta\phi + j\sin\Delta\phi$$

$$|\Gamma_{OC} - \Gamma_{SC}|^2 = (1-\cos\Delta\phi)^2 + \sin^2\Delta\phi$$

$$= 1-2\cos\Delta\phi + \cos^2\Delta\phi + \sin^2\Delta\phi$$

$$= 2-2\cos\Delta\phi$$

However, by trigonometric identity

$$\cos\Delta\phi = 1-2\sin^2(\Delta\phi/2)$$

and hence we can write

$$|\Gamma_{OC} - \Gamma_{SC}|^2 = 2-2+4\sin^2(\Delta\phi/2)$$

$$|\Gamma_{OC} - \Gamma_{SC}| = 2\sin(\Delta\phi/2) \tag{V-53}$$

Substituting Equation (V-53) into Equation (V-52), noting that the algebraic signs are related only with which state is defined as the zero phase reference, gives

$$V_{OC}I_{SC} = 4P \sin(\Delta\phi/2) \tag{V-54}$$

We have now related the ideal diode stresses, V_{OC} and I_{SC}, to the incident power, P, and the phase shift performance, $\Delta\phi$. This expression can be used directly to relate the maximum power for a given phase shift which a diode capable of sustaining V_M voltage at reverse bias and I_M current at forward bias can sustain.

$$P_M = \frac{V_M I_M}{4\sin\left(\frac{\Delta\phi}{2}\right)} \tag{V-55}$$

where P is in watts when V_M and I_M are in volts and amperes rms.

This expression gives the Hines power limit for a single PIN diode used in a reflection phase shifter.

2. Minimum Insertion Loss

The preceding analysis can also be used to determine the insertion loss obtainable. Notice in Figure V-14 that reducing turns' ratio, n, reduces the V_{OC} stress on the diode (and its dissipation if it were to have had some finite conductance in the reverse bias, open, state). This reduction of V_{OC} is accompanied by a proportionate increase in I_{SC}, however. The minimum average loss is obtained when n is selected to produce equal diode dissipation in both states. Of course, this action may not necessarily coincide with the choice for n which gives maximum power handling (which would be when $I_{SC} = I_M$ and $V_{OC} = V_M$).

To determine the magnitude of insertion loss under the loss equalized condition, the chip equivalent circuit for the diode shown in Figure V-14(b) is used. A perturbation analysis is again employed wherein the diode resistances are assumed to be small enough for V_{OC} and I_{SC} to have substantially the respective values calculated for them in the lossless case. The junction capacitance (if not negligibly small) is assumed to be parallel resonated with a lossless inductor.

With forward bias the dissipation is then

$$P_{DF} = I_{SC}^2 \cdot R_F \tag{V-56}$$

and under reverse bias, assuming the diode resistance is small compared to the reactance of C_J

$$P_{DR} = (2\pi f C_J)^2 \cdot V_{OC}^2 \cdot R_R \tag{V-57}$$

Under the condition that these dissipation values are equal to P_D

$$P_D = I_{SC} \cdot V_{OC} \sqrt{R_F R_R}\; 2\pi f C_J \tag{V-58}$$

Using the definition for f_{CS} in Equation (V-24) and the value for the product $I_{SC} V_{OC}$ in Equation (V-54), the fractional insertion loss under the loss equalized condition, which is the ratio P_D/P_A, can be written

$$\text{Loss}_E \approx 4 \left(\frac{f}{f_{CS}}\right) \cdot \sin\left(\frac{\Delta\phi}{2}\right) \tag{V-59}$$

This expression is an approximation accurate within ten percent when the loss in both states is below about 0.5 dB (for a single phase shift step). The accuracy usually holds in a reasonably loss balanced circuit with $10\,f < f_{CS}$.

Similar limit expressions for P_M and Loss_E are obtained for the transmission phase shifter in Chapter IX; they are listed in Table V-1.

This result is what might be expected. Loss increases in proportion to the amount of phase shift and to the ratio of (f/f_{CS}). This is the Hines insertion loss expression for PIN phase shifters. For values of loss of $(P_D/P_A) < 0.1$, the loss in decibels can be approximated by multiplying the numeric value of the fractional loss by the factor 4.6 (using the first term of the Taylor expansion) as described in the previous section for duplexers. Thus, for a single step phase shifter

$$\text{Loss}_E \approx 18.4 \frac{f}{f_{CS}} \sin\left(\frac{\Delta\phi}{2}\right) \quad \text{(decibels)} \tag{V-60}$$

It is generally desirable to cover the 0-360° range with even increments which are as small as is practical. This division for diodes is done most efficiently by cascading phase shifter sections whose step sizes are a binary division of 360°. Thus, for example, a *four-bit* phase shifter consists of four cascaded phase shifter circuits having 180°, 90°, 45°, and 22½° of phase shift, respectively. With this combination any value of phase shift between 0° and 360° can be realized with an error no greater than 11¼°, or half the smallest bit. (360° is considered equivalent to 0°.)

For 180°, 90°, 45°, and 22½°, the respective values for $\sin(\Delta\phi/2)$ are 1.000, 0.707, 0.383, and 0.195. Thus, the minimum insertion loss for a multi-bit phase shifter can be written

$$\text{Loss}_E = K \frac{f}{f_{CS}} \quad \text{(decibels)} \tag{V-61}$$

where K is related to the number of bits as follows:

Number of Bits	*1*	*2*	*3*	*4*
K	18	31	38	42

For example, to achieve a maximum diode loss of 1.0 dB with a four-bit phase shifter at 10 GHz, a minimum diode switching cutoff frequency of 420 GHz is required. PIN diode cutoff frequencies lie in the range 300-1000 GHz; therefore, low loss phasers are indeed practical. The minimum diode insertion loss for four bit phase shifters using various cutoff frequency diodes is shown graphically as a function of frequency in Figure V-17.

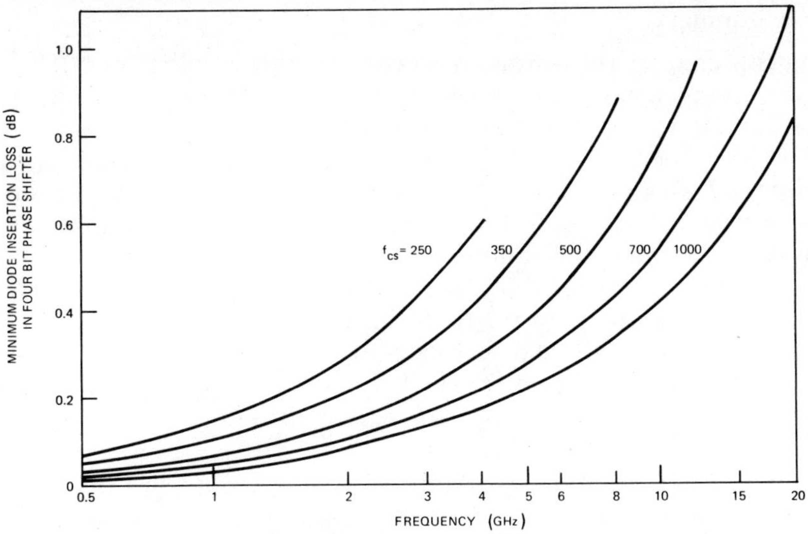

Figure V-17 **Minimum Diode Insertion Loss of Four-Bit Phase Shifter vs. f and f_{CS}**

DEVICE	POWER	LOSS
SWITCHES (SPST & SPDT)	$P_M = \dfrac{V_M I_M}{2}$	$\dfrac{P_D}{P_A} \approx 2f/f_{CS}$ $(10\,f < f_{CS})$
DUPLEXERS	$P_M \approx \dfrac{(LOSS_M)\ P_{DM}\ N\ (f_{CS}/f)^2}{4}$ $\approx \dfrac{(LOSS_M)\ P_{DM}\ N}{4\ R_F\ R_R\ (2\pi f)^2}$	(BALANCED SHUNT DUPLEXER) IF $R_F << Z_O$, ARC LOSS $\approx \dfrac{4R_F/(N/2)}{Z_O}$ IF $R_R << (2\pi f C_J)^{-1}$ RECEIVE LOSS $\approx Z_O R_R\ (2\pi\ f\ C_J)^2$
PHASE SHIFTERS (SINGLE STEP)	(REFLECTION CIRCUIT) $P_M = \dfrac{V_M I_M}{4\sin(\Delta\phi/2)}$ (TRANSMISSION* CIRCUIT) $P_M = \dfrac{V_M I_M}{2\tan(\Delta\phi)}$	(REFLECTION CIRCUIT) $\dfrac{P_D}{P_A} \approx 4\left(\dfrac{f}{f_{CS}}\right)\sin\left(\dfrac{\Delta\phi}{2}\right)$ (TRANSMISSION* CIRCUIT) $\dfrac{P_D}{P_A} \approx 2\left(\dfrac{f}{f_{CS}}\right)\tan(\Delta\phi)$ $10\,f < f_{CS}$

$$f_{CS} = \frac{1}{2\pi C_J \sqrt{R_F R_R}}$$

*SEE EQUATIONS (IX-13) AND (IX-14)

Table V-1 **Limit Equation Summary for PIN Control Circuits**

G. Summary

In this chapter the optimum power handling and insertion loss performance has been derived for switches, duplexers, and phase shifters in terms of the microwave voltage and current sustainable by the PIN and its resistance and capacitance parameters, neglecting the losses introduced by the circuitry. Simultaneous optimization with respect to both power handling and insertion loss is not possible without the ability to optimize both diode and circuit parameters, but such diode variability is generally not practical. Nevertheless, the limit relationships of power handling and insertion loss for the three device categories, summarized in Table V-1, form an important base for the PIN control field from which practical designs can be started.

References

[1] Hines, M.E.: "Fundamental Limitations in RF Switching and Phase Shifting Using Semiconductor Diodes," *Proceedings of the IEEE, Vol. 52, No. 6,* pp. 697-708, June 1964.

[2] *CRC Standard Mathematical Tables – Tenth Edition;* Chemical Rubber Publishing Co., Cleveland, Ohio, 1955; p. 14 (change of logarithm base) and p. 320 (Taylor series expansion for natural logarithm).

Questions

1. Suppose the diode described in Question 2, Chapter III has $C_J = 1.2$ pF, $R_F = 0.4\ \Omega$, and $R_R = 0.5\ \Omega$. What is f_{CS}?
2. Using the P_D values calculated in Question 3, Chapter III for the above diode, calculate V_M and I_M, for RF power dissipated at 1 GHz in the diode under a) CW conditions; and b) widely spaced pulses of 100 μs, c) 10 μs, and d) 1 μs. Neglect reverse voltage breakdown effects in these calculations.
3. If the bulk breakdown voltage of this diode is 1200 V, and -100 V of reverse bias is to be used, what is the maximum V_M rating for the diode when used in a SPST switch which is to operate into a fully reflective load of any phase mismatch?
4. With a V_M limit of 389 V rms, how much RF power, P_M, can the diode in Question 2 switch when mounted across a transmission

line of appropriate impedance, Z_0, with a) CW power; and b) widely spaced pulses of 100 μs, c) 10 μs, and d) 1 μs? What is the required Z_0 for each case? Assume lossless parallel resonant tuning is used so that the diode does not introduce a reflection on the line when reverse biased.

5. If one wishes to avoid the use of switching impedance levels lower than 20 Ω, what is P_M for the diode in Question 4?

6. What power can be switched at 1 GHz by a balanced duplexer designed using four diodes and operated into a totally reflecting load using the diode described above (for which $C_J = 1.2$ pF, $R_F = 0.4\ \Omega$, $R_R = 0.5\ \Omega$, and $P_D = 20{,}010$ W for 1 μs pulses) assuming a diode loss in the receive condition of 0.2 dB (5%) is tolerable? What is the required switching impedance, Z_0, at the diodes?

7. Suppose, instead of applying a reverse bias of -100 V to obtain $R_R = 0.5\ \Omega$ for the duplexer in Question 6, zero bias is used for the receive state and that the resulting $R_R = 8\ \Omega$. What are the corresponding values for f_{CS}, P_M, and Z_0?

Chapter VI

Mathematical Techniques and Computer Aided Design (CAD)

A. Introduction

1. Synthesis, Analysis, and Cut-and-Try

The methods taught and used until the mid-sixties for analysis of microwave networks were based primarily on hand calculations; at that time, emphasis was on techniques for reducing mathematical expressions to easily visualized forms. Such an approach might be termed the *analytic* method — the process of developing the understanding of the mathematics that pertain to a particular microwave circuit model. Once a model is visualized, such a procedure makes it nearly always possible to perceive what variable changes in the circuit or the diodes are needed for optimization with respect to some desired design performance characteristic. The result is *synthesis,* wherein approaches are envisioned which are known to have the desired properties before they are tried in the specific situation. There is no question that the analytic understanding which leads to synthesis should be sought whenever practical. In the past, lacking an analytic model for the microwave circuit, a designer was committed to a tedious cut-and-try, *empiric* approach. Prior to the current ready availability of the digital computer, the empiric approach had to be carried out in the laboratory with actual models; its range was limited because of both the labor required and the difficulty in maintaining sufficiently controlled experiments to isolate the effect of each parameter variation.

Recently, however, there has been a proliferation in number and kind of computer aided design tools available to the engineer. Minicomputers, programmable calculators, simplified language, time-shared computers, and so forth, have made it possible to perform precisely controlled experiments by cut-and-try techniques very quickly on the computer. Methods of analysis which previously were impractical for any calculation are now preferred because of their programming simplicity; enormous capacity computers can perform easily such repetitous calculations with high accuracy.

The result has been such a pendular swing in the relative effectiveness of the empiric over the analytic approach that those engineers having a mastery of even basic computer programming techniques enjoy an enormous advantage in predicting circuit performance. Even with a very limited understanding of switching principles, they can explore and obtain partially optimized design parameters for most of the typical switch, duplexer, and phase shifter circuit configurations much more rapidly and accurately than can one who, though versed in circuit theory, cannot "try his designs on the computer."

Of course, neither pure synthesis nor thoughtless cut-and-try analysis of switch design are recommended here. The truly well-versed designer uses these two methods in concert, selecting a circuit approach and basic parameters known to most likely meet his needs and then optimizing the model using computer aided design (CAD). For some readers, this notion is self-evident; for others, computer programming is viewed as a secretarial skill best assigned to someone (else) adept at programming. Such an impression is, of course, distorted. The PIN diode circuit designer, and indeed the modern engineer in any field, needs to attain the same skill today with a computer as had once been necessary with a slide rule. Only through fluency in a computer language can one be in a position to blend knowledge of the fundamentals of semiconductor devices and microwave circuits with the need to evaluate proposed new designs against the ever growing myriad of possible parameter variations which must be considered.

It is outside the proper scope of this book to present a thorough introduction to programming techniques, *per se.* Furthermore, most engineers, including those who do not regularly use CAD, have already been exposed to one or more languages. In this chapter we assume that the reader has already learned, or can readily learn, a computer language, although a FORTRAN IV summary for review or brief introduction is presented to stress the programming steps most useful for PIN circuit evaluation. We examine closely those mathematical and programming techniques by which the designer can develop a personal library of circuit design programs.

2. Choosing The Computer Aided Design Medium

There are three basic levels of computer aided designs available from current equipment:

1) The programmable calculator
2) The minicomputer
3) The time-shared or batch process large scale computer.

Unfortunately, the programming languages for these three options are generally not interchangeable. The first option, the programmable calculator is relatively inexpensive. At this time, however, the hand calculators require programming in their own machine language and are limited to programs of about 100 rather basic steps. The main advantages of the programmable calculator are its small size and consequent portability; the chief drawbacks are that each programmable calculator presently requires that the user learn a language unique to that machine and that the range of programs is limited to fairly simple design situations. In addition, a printed output summarizing the parameters examined and the resultant circuit evaluation is not always obtained.

The second option, the minicomputer, offers a wider range of programming options and a printout, but the portability is sacrificed. However, for small engineering groups it affords a method of outright purchase of the CAD facilities and their economical operation.

The third option is the use of a full scale computer. With standard programming, languages such as BASIC, FORTRAN IV, and COBOL (generally employed with business rather than scientific functions) can be used. A large computer implies multiple users for economic operation. If an engineering firm is large enough, the facility might be purchased outright; an alternative is computer rental using a "time-sharing" terminal with telephone link. The latter method is often less costly in the long run than use of an in-house large scale computer because, with time-sharing, programs can be written and compiled in real time, shortening the engineering time required for programming and de-bugging. In addition, most time-sharing services offer their users programming help at little or no additional cost.

3. Choosing the Language

If your decision involves only the hand programmable calculator and the minicomputer, there may be no choice of language. If, however, you elect to use a large scale computer or time-sharing facility, a variety of computer languages is usually available. There are "interpretive" languages, such as APL, and some proprietary languages which compile directly, the conversion to machine language and any appropriate debugging diagnostics being furnished as you generate the program. Many engineers gain their initial computer programming experience through such interpretive languages – program debugging is generally easier; for those who have had no previous programming experience, this feature presents a desirable first step because it permits solution of real problems in a short period of time, helping to make acquisition of the computer language a "pay-as-you-go proposition." A drawback of most commonly available interpretive languages is that directly writing programs involving complex numbers is not possible – complex expressions must be rewritten in terms of real and imaginary parts. To the electrical engineer this disadvantage is a serious one; FORTRAN IV is the logical consideration as a resolution for this problem.

With the FORTRAN languages (as well as with COBOL and BASIC), programs are compiled (that is, converted from standard statements intelligible to the programmer to statements acceptable to the particular computing machine being used) in batch form. In other words, the entire program is written first, then put into the machine, at which time one finds out which statements contain errors. For this reason the FORTRAN language requires more programming discipline; even so, programs written in FORTRAN (or COBOL or BASIC) can be expected to take longer to debug than if they were in an interpretive language, because relocating improper programming statements after the program is completed is generally more difficult. Nonetheless, the advantage to the microwave engineer of complex notation more than offsets such programming inconvenience in programs of any appreciable size. It is with FORTRAN IV that the author generates most CAD programs and in which the various programming examples given in this chapter are written.

There are other languages in addition to FORTRAN specifically designed for scientific programming, BASIC and APL being widely available and reasonably standardized. No doubt, in the future, new programming languages, as well as improved versions of existing languages, will become available and therefore will need to be considered. At this time, however, FORTRAN IV appears to offer both the flexibility needed by a microwave circuit designer and the ready portability from one computer system to another.

4. Proprietary Network Analysis Programs

Some time-shared services offer programs written especially for network analysis. Generally these programs are proprietary, and the actual programming steps are confidential. The user has access to the compiled version of the program; with the help of a brief instruction manual, one is able to evaluate performance of a *general network topology* directly without any need of programming experience. Such general topology programs are unquestionably valuable; at the end of this chapter we generate one for two-port networks. There will always be a use for the general purpose program, since many temporary analysis problems are too small to warrant the expenditure of time and funds necessary to write a *dedicated* program, useful only for that particular application.

However, total reliance on proprietary general purpose programs has two pitfalls for the microwave designer. First, the separation of the user from the direct experience of how the computer is solving problems may cause the stimulus to extend the CAD functions beyond those directly available with general programs to be lost. Second, much better appreciation of the limitations of CAD comes from the intimate knowledge gained in writing the actual steps taken in the program to solve the problem. Certain treatments of round-off, inverse trigonometric functions, cumulative error, available computational assumptions, and so forth, can produce misleading, even erroneous results. Access to most of the available general analysis programs is impractical because they are proprietary and the programs themselves are very long and too complicated for practical user evaluation. Therefore, it is advantageous for the designer to base everyday CAD work on a library of programs that is intimately familiar, because of either personal generation or access to their source program steps.

5. FORTRAN IV Language Reference

It is desirable to select a reference text for a standard computer language which is written independently of a particular equipment manufacturer or computer service. Even "standard languages" undergo minor changes when made available on particular systems;

these alterations usually take the form of "special options" which may be used only on the particular system, not in general on all those systems using the standard language. Over a period of years the circuit designer can develop a rather extensive library of computer programs; to rewrite this collection for a specific system would be extremely tedious. Therefore, it is desirable to write such programs in as "pure" a version of the standard language as possible to minimize the need for eventual rewriting should the computer equipment or computer service have to be changed. The author has found McCracken's book [1] on FORTRAN IV programming particularly useful; references made within this chapter to standard FORTRAN IV programming can be found treated in that volume. FORTRAN IV, as the title implies, represents an evolved version of the original FORTRAN (an anachronym for FORmula TRANslation) language. All the original versions of FORTRAN are compatible (in an upward fashion) with the current standard FORTRAN IV; that is, programs written in the earlier FORTRAN languages can be compiled and run on equipment designed for FORTRAN IV, though the reverse is not necessarily true. It should be noted that only in the FORTRAN IV version of FORTRAN is the use of complex number notation assured; microwave designers should check the level of FORTRAN available before committing to a particular computer service or equipment.

B. CAD Mathematical Analysis Approaches

1. Step-by-Step, No Approximations

Frequently in engineering, one becomes so accustomed to making approximations that the exact specifics are forgotten. Consider, for example, the transformation of an impedance along a length of uniform transmission line. This problem is solved typically by using the Smith Chart*; almost always, the line loss effects on the incident and reflected waves are handled in an approximate manner or ignored altogether, because an exact treatment requires spiral paths on the Smith Chart or the use of hyperbolic functions for the analytic expression (see Figure VI-1) for the transformation of an impedance by a lossy line. We refer to this basic transformation as a *Smith Chart transformation,* defined for both lossless and lossy lines, respectively, as Equations (VI-1) and (VI-2) in Figure VI-1. The same equations apply for the transformation of a load admittance, Y_L, transformed along a line of characteristic admittance Y_C to an input admittance, Y_{IN}, if Y_L, Y_C, and Y_{IN} re-

*The Smith Chart is derived and explained in Appendix J.

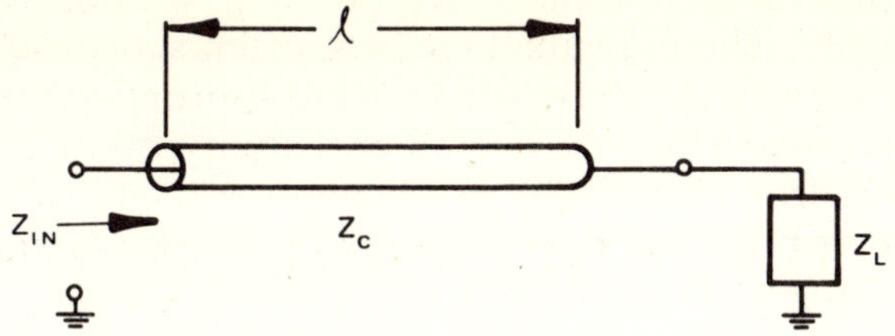

ℓ	=	PHYSICAL LENGTH	(METERS)
γ	=	$\alpha + jB$ = PROPAGATION CONST.	________
θ	=	$\beta\ell = \frac{2\pi\ell}{\lambda}$ = ELECTRICAL LENGTH	(RADIANS)
α	=	$\frac{\text{LINE LOSS (dB}/\lambda)}{8.685889638}$ = ATTENUATION CONST.	(NEPERS/METER)

LOSSLESS LINE ($\alpha = o$)

$$Z_{IN} = Z_C \left[\frac{(Z_L/Z_C) + j\tan\theta}{1 + j(Z_L/Z_C)\tan\theta} \right] \quad \text{(VI-1)}$$

LOSSY LINE ($\alpha \neq o$)

$$Z_{IN} = Z_C \left[\frac{(Z_L/Z_C) + \tanh(\gamma\ell)}{1 + (Z_L/Z_C)\tanh(\gamma\ell)} \right] \quad \text{(VI-2)}$$

Figure VI-1 Smith Chart Transformation (See Appendix J)

place Z_L, Z_C, and Z_{IN}, respectively. This applicability can be demonstrated readily by making these substitutions on the right sides of the equations and solving for $Y_{IN} = 1/Z_{IN}$.

With the computer, of course, once the program is written, the handling of complex arguments and hyperbolic tangents is no more difficult for the programmer, and usually little more expensive in terms of computer time, than are the lossless line calculations. Later in this chapter we show how the lossy Smith Chart transformation, defined by Equation (VI-2), is addressed with a FORTRAN *subroutine* and thus can be included in any program without repeated regeneration. This convenience in terms of handling complicated but exact formulations on a recurring basis is what makes CAD such a powerful tool.

Frequently, those who have been involved for many years with engineering that has not employed computer aided design have formed habits whereby multi-step problem solutions are avoided automatically. The individual realizes, often subconsciously, that each separate calculation made by hand is prone to error; in a multi-step calculation, the overall probability of arriving at a correct answer drops drastically as the number of steps increases. For this reason it takes some re-learning of problem solving to orient one's thinking so as to make effective use of CAD.

Once a program has been written, the same procedure is followed by the computer each time the program is executed. The implication, though, is not that if the program is correct, the result is always correct. While the execution is always consistent, there can be unanticipated combinations of inputs for which the program's operation is not valid. Nevertheless, the emphasis is on generating as flawless a program as is possible. In this respect *it is generally easier to debug a program which consists of a large number of simple steps than it is to correct one having a relatively few number of complicated steps.*

As an example of the *small step approach,* consider the equivalent circuit for the single pole, double throw (SPDT) switch shown in Figure VI-2. In its simplest form the SPDT switch consists of a T-junction with two diodes interposed between alternate matched loads fed by a common generator. The "on" diode is represented by the forward bias impedance, Z_F; the "off" diode, by the reverse diode impedance, Z_R. The operation of the switch as shown is not dependent on line lengths; consequently, the analysis is reduced to that of a simple voltage divider network wherein the fraction of the generator voltage appearing across the switched-on load, V_{LOADF}, or the blocked load, V_{LOADR}, is used to estimate insertion loss or isolation, respectively. Using 2V as the generator voltage, relative powers delivered to the direct and blocked loads are simply proportional to the square of the respective load voltages.

Were the solution for the voltages at these loads to be carried out by hand for a number of different frequencies, it would make sense to find single analytic expressions for the voltage transfer function between the generator voltage and each load. In this way a minimum number of hand calculations involving complex numbers would have to be carried out, thereby resulting in the greatest accuracy and the smallest probability of a computational error. However, in setting up the analysis for a computer program, such brevity is exchanged beneficially for *step-by-step simplicity.* A sample procedure might be as follows. Initially, calculate the im-

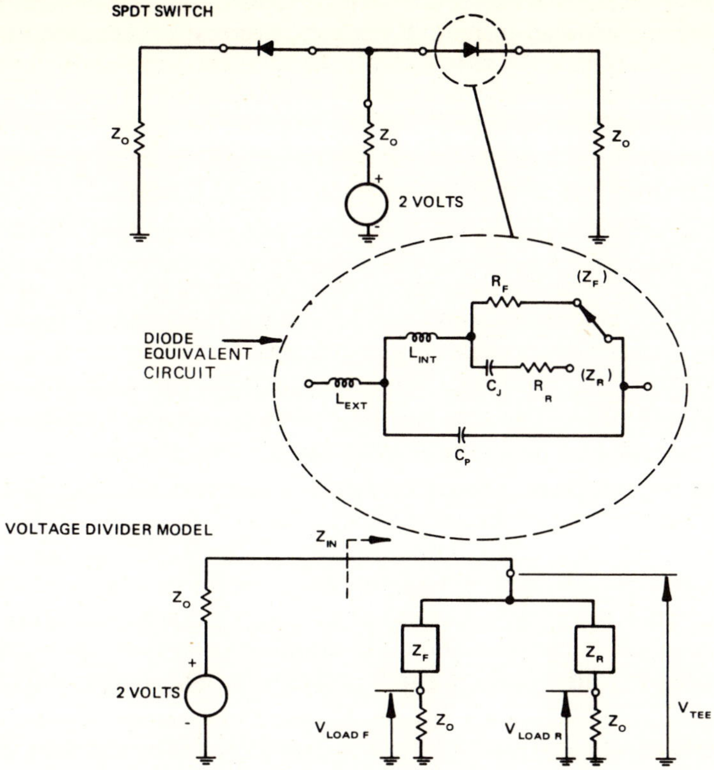

Figure VI-2 **Example of Step-By-Step Circuit Analysis for SPDT Switch (see FORTRAN IV analysis program in Figure VI-19 and sample program execution in Figure VIII-43)**

pedance, Z_{IN}, at the input of the SPDT switch. Secondly, calculate the input voltage at the switch, V_{TEE}; thirdly, calculate the voltages across each of the output loads. This process is summarized in Equations (VI-3) through (VI-6).

$$Z_{IN} = (Z_F + Z_0)(Z_R + Z_0)/(Z_F + Z_R + 2Z_0) \qquad \text{(VI-3)}$$

$$V_{TEE} = (2 \text{ volts})\, Z_{IN}/(Z_{IN} + Z_0) \qquad \text{(VI-4)}$$

$$V_{LOADF} = V_{TEE} Z_0/(Z_F + Z_0) \qquad \text{(VI-5)}$$

$$V_{LOADR} = V_{TEE} Z_0/(Z_R + Z_0) \qquad \text{(VI-6)}$$

The resulting switch insertion loss, isolation, and input VSWR can be determined readily from these expressions, as is shown in the program example for this switch which follows in this chapter.*

2. *Transmission (ABCD) Matrix*

Even with the relatively simple SPDT switch circuit just considered, it can be seen that the explicit equations needed to determine the transfer function of the network require care in formulation in order to avoid errors. This point is particularly applicable when the network has a large number of elements which are separated by transmission lines of various lengths, characteristic impedances, and attenuation constants. In short order such networks render not only hand calculation impractical, but even the explicit expression of closed form solutions for relative voltages and impedances (as in the SPDT switch example just given).

For minimum phase networks, (that is, networks having only a single path between input and output ports) the analysis can be handled in terse mathematical form with the ABCD matrix defined in Figure VI-3. The programming usefulness of the ABCD matrix analysis, which consists of a series of cascaded circuit elements, arises because of the definition between the dependent variables, v_1 and i_1, and the independent variables, v_2 and i_2. These variables are respectively the input voltage and current and the output voltage and current for the network. Thus, the output of one network element is the input for the following element, and so on. Figure VI-4 contains a brief library of ABCD matrices representing a series impedance, a shunt admittance, a length of transmission line, and an ideal transformer. *Although brief, this library is large enough to construct an overall cascade matrix for practically any two-port PIN diode control circuits likely to be encountered.*

The matrix multiplication operation is diagrammed in Figure VI-3. The rows of the matrix to the left are multiplied by the respective columns of the matrix on the right to produce the new ABCD matrix element terms as shown. Since the independent variables are usually the output voltage and current of a network, it is necessary to combine the cascade of elements by multiplying from the load toward the generator — multiplying the last element matrix by the matrix immediately preceding it, then multiplying the resulting matrix by the matrix to its immediate left, and so on toward the generator.

*See computer program SPDT, pp. 221-223.

ABCD MATRIX DEFINITION:

$V_1 = A V_2 + B I_2$

$I_1 = C V_2 + D I_2$

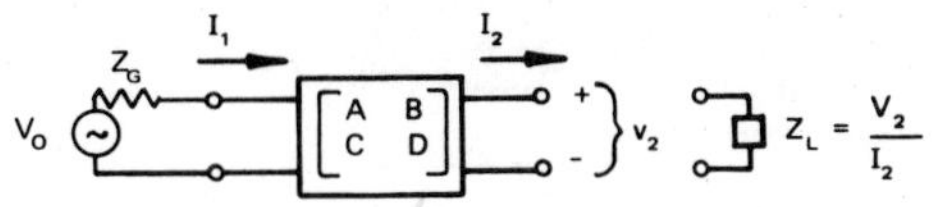

EVALUATION OF TERMS:

When $Z_L = 0, V_2 = 0$ When $Z_L = \infty$ $I_2 = 0$

$A = \left.\frac{V_1}{V_2}\right|_{I_2 = o}$ $B = \left.\frac{V_1}{I_2}\right|_{V_2 = o}$ $C = \left.\frac{I_1}{V_2}\right|_{I_2 = o}$ $D = \left.\frac{I_1}{I_2}\right|_{V_2 = o}$

FOR RECIPROCAL CIRCUITS: $AD - BC = 1$

INPUT IMPEDANCE: $Z_{IN} = \frac{V_1}{I_1} = \frac{A Z_L + B}{C Z_L + D}$

INSERTION LOSS $= \left|\frac{V_O}{2V_2}\right|^2 = \frac{1}{4}\left|A + \frac{B}{R_O} + CR_O + D\right|^2$ (When $Z_L = Z_G = R_O$)

CASCADE COMBINATION:

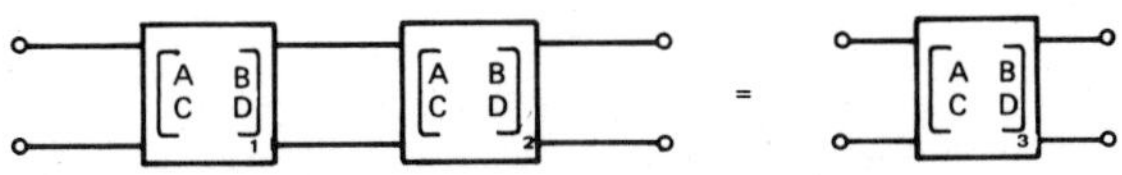

WHERE,

$$\begin{bmatrix} A_3 & B_3 \\ C_3 & D_3 \end{bmatrix} = \begin{bmatrix} A_1 & B_1 \\ C_1 & D_1 \end{bmatrix} \cdot \begin{bmatrix} A_2 & B_2 \\ C_2 & D_2 \end{bmatrix} = \begin{bmatrix} (A_1 A_2 + B_1 C_2) & (A_1 B_2 + B_1 D_2) \\ (C_1 A_2 + D_1 C_2) & (C_1 B_2 + D_1 D_2) \end{bmatrix}$$

Figure VI-3 **The ABCD Matrix Definition and Operations**

Since the matrix multiplication is not commutative, i.e., [A] x [B] is not necessarily equal to [B] x [A], care must be exercised in setting up this multiplication order. Multiplying generalized matrices (from Figure VI-4) of elements can be used as a method by which to derive general analytic expressions. The procedure of multiplying a cascade of ABCD matrices is demonstrated in Figure VI-5; the result is an important derivation, the equivalence between a section of transmission line and π circuit. At the frequency for which the transmission line is a quarter wavelength, it is seen that the reactances of the series and shunt elements in the equivalent π circuit are numerically equal to the characteristic impedance of the line section. This result is a valuable one since physical space limitations often favor substitution of a distributed line (over modest bandwidth) using "lumped" circuit elements. Con-

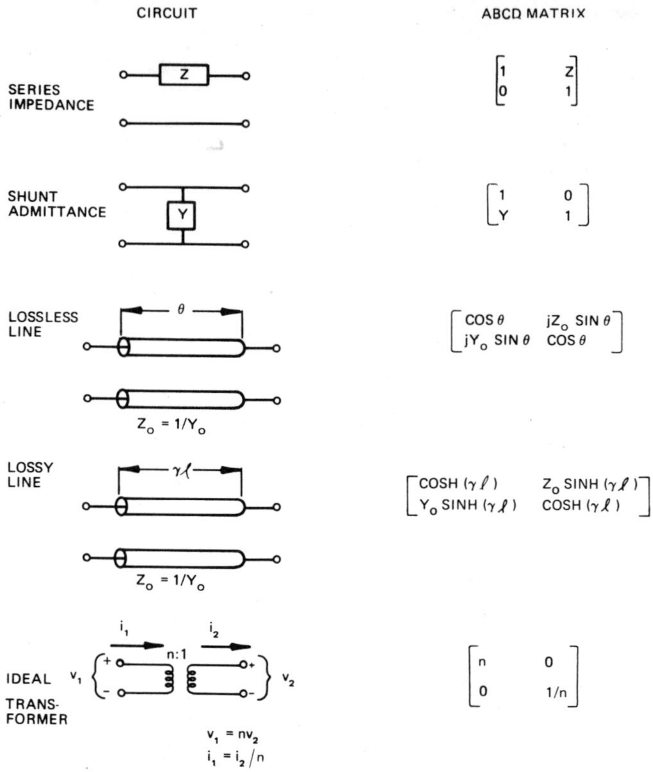

Figure VI-4 **Common ABCD Matrices Required for PIN Control Circuits**

sider how much more difficult this result would have been to derive without using the ABCD matrix. Later we test this equivalence numerically over a band of frequencies (see Figure VI-31).

3. Z and Y Matrices, Conversions and Branched Networks

Use of an ABCD matrix cascade analysis applies only for a *minimum phase* network for which there is only a single electrical path along which energy can propagate from the input of the network to the output. Networks having multiple paths between input and output are called *non-minimum phase*, or *branched*, networks; they cannot be handled solely with the use of the ABCD matrix. Two common interconnections found in branched networks are shown in Figure VI-6 – the *series-series* and the *parallel-parallel* combinations. The overall matrix representation for these net-

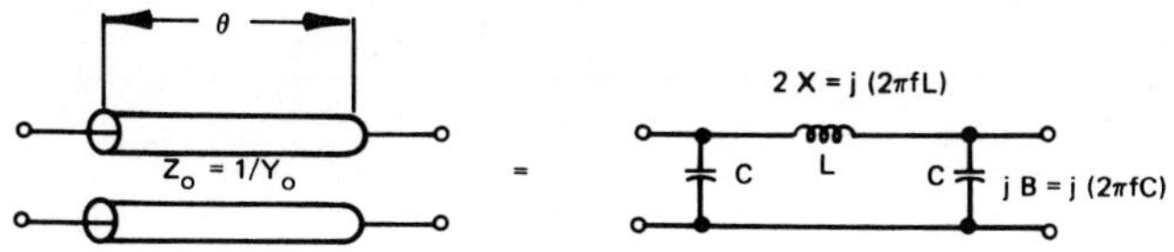

$$\begin{bmatrix} \cos\theta & jZ_o \sin\theta \\ jY_o \sin\theta & \cos\theta \end{bmatrix} = \begin{bmatrix} 1 & 0 \\ jB & 1 \end{bmatrix} \cdot \begin{bmatrix} 1 & jX \\ 0 & 1 \end{bmatrix} \cdot \begin{bmatrix} 1 & 0 \\ jB & 1 \end{bmatrix}$$

$$= \begin{bmatrix} 1 & 0 \\ jB & 1 \end{bmatrix} \cdot \begin{bmatrix} (1 - BX) & jX \\ jB & 1 \end{bmatrix}$$

$$= \begin{bmatrix} (1 - BX) & jX \\ jB\,(2 - BX) & (1 - BX) \end{bmatrix}$$

$$\frac{B}{C} = Z_o^2 = X/(2B - B^2X)$$

$$A = D = \cos\theta = 1 - BX$$

QUARTER WAVE SECTION OF LINE: $(\theta = 90^\circ)$

$$X = \frac{1}{B} = Z_o$$

Figure VI-5 **Derivation of π Equivalent to Quarter-Wave Transmission Line Using ABCD Matrix Multiplication***

works can be obtained by simple addition using the impedance and admittance matrices, respectively. These combinations can, in turn, be combined in an overall ABCD matrix cascade by converting the separate path ABCD matrix elements to the appropriate impedance or admittance parameters, performing the required addition, and then reconverting the results to ABCD parameters for inclusion in the remaining cascade network. Conversion formulas [4, 5] for making transformations among these matrix notations are given in Figure VI-7. Notice that the defining equations for the impedance and admittance matrices shown in Figure VI-6 are the same as the defining voltage and current directions used for the ABCD matrix defined in Figure VI-3 *except for the direction of the output current,* I_2. The matrix conversion formulas given in

*This "equivalence" of course, is valid only at the frequency for which the equations in Figure VI-5 are satisfied. For a demonstration of the practical bandwidth of this equivalence, refer to the computer evaluation in Figure VI-31 of both circuits.

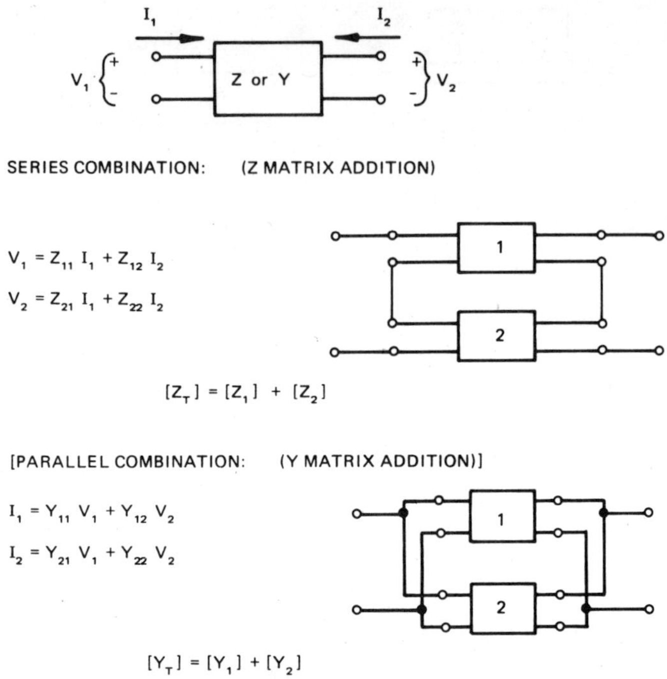

Figure VI-6 **Series and Parallel Network Combinations Respectively Represented by Impedance and Admittance Matrix Additions**

Figure VI-7 take this reversal in the direction of I_2 into account. While such matrix conversions would be far too unwieldy for hand calculations, they are easily employed using computer program subroutines.

4. Even and Odd Mode Analysis

Application to Symmetric Networks

The general matrix methods described in the previous section for analyzing networks with branches have general applicability; however, they have the disadvantage that physical insight is lost once one represents the network in matrix form and then converts those matrices to still another form to perform the required additions in a branched circuit analysis. An alternative method of analyzing branched circuits which often yields closed form analytic expressions can be applied when the network has symmetry

$$\begin{bmatrix} Z_{11} & Z_{12} \\ Z_{21} & Z_{22} \end{bmatrix} = \frac{1}{Y_{11}Y_{22} - Y_{12}Y_{21}} \begin{bmatrix} Y_{22} & -Y_{12} \\ -Y_{21} & Y_{11} \end{bmatrix}$$

$$\begin{bmatrix} Y_{11} & Y_{12} \\ Y_{21} & Y_{22} \end{bmatrix} = \frac{1}{Z_{11}Z_{22} - Z_{12}Z_{21}} \begin{bmatrix} Z_{22} & -Z_{12} \\ -Z_{21} & Z_{11} \end{bmatrix}$$

$$\begin{bmatrix} Z_{11} & Z_{12} \\ Z_{21} & Z_{22} \end{bmatrix} = \frac{1}{C} \begin{bmatrix} A & (AD-BC) \\ 1 & D \end{bmatrix}$$

$$\begin{bmatrix} Y_{11} & Y_{12} \\ Y_{21} & Y_{22} \end{bmatrix} = \frac{1}{B} \begin{bmatrix} D & -(AD-BC) \\ -1 & A \end{bmatrix}$$

$$\begin{bmatrix} A & B \\ C & D \end{bmatrix} = \frac{1}{Z_{21}} \begin{bmatrix} Z_{11} & (Z_{11}Z_{22} - Z_{12}Z_{21}) \\ 1 & Z_{22} \end{bmatrix}$$

$$\begin{bmatrix} A & B \\ C & D \end{bmatrix} = \frac{1}{-Y_{21}} \begin{bmatrix} Y_{22} & 1 \\ (Y_{11}Y_{22} - Y_{12}Y_{21}) & Y_{11} \end{bmatrix}$$

Figure VI-7 **Matrix Conversion Formulas**

with respect to the branches. This method is called *even and odd mode analysis*; it relies upon the principle of superposition of symmetric and anti-symmetric excitations onto a network to solve for the final circuit response.

Switched Path Analysis Example

The method is described most simply through an example. We choose as the illustrative case a switching circuit which finds application in phase shifter design, the *switched path,* or *delay line,* phase shifter. The circuit to be analyzed is shown in Figure VI-8(a) and consists of a generator connected to a load via two separate paths. Access to these paths is controlled by means of impedances Z_1 and Z_2, respectively for path 1 and path 2; these impedances would be realized with switchable control diodes mounted in series with the transmission line center conductor. In operation, if Z_1 is made small and Z_2 is made large, energy propagates from the generator to the load via path 1. Complementary switching of the two impedance magnitudes causes most of the energy to take the alternate path, path 2, to the load. The difference in path lengths gives a time delay, or phase shift, the value of which is equal to the net electrical length difference between the two paths, provided the impedances essentially switch between "open" and "short" cir-

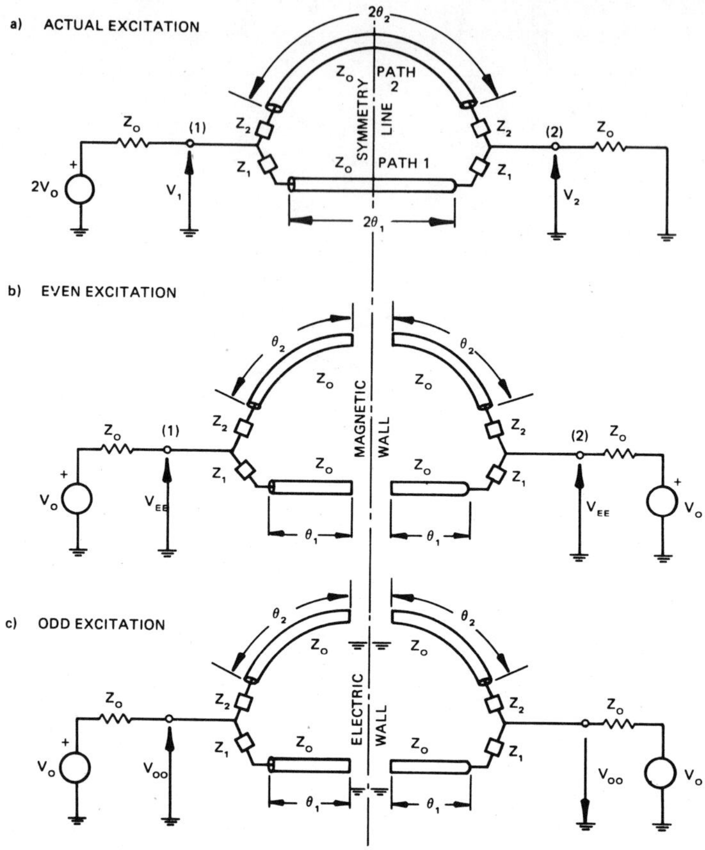

Figure VI-8 **Evaluation of Switched Path Phase Shifter Using Even and Odd Mode Analysis Technique**

cuits. Such ideal switching behavior is not, however, obtained in practice because the small capacitive coupling of the "off" diodes can produce absorptive resonances in the "off" path. Actual operation can be analyzed by writing the admittance matrices for the two paths, then adding them together to determine the complex voltages V_1 and V_2. From these voltages the actual transmission phase, insertion loss, and input VSWR to the network can be calculated, taking the finite practical impedances, Z_1 and Z_2, fully into account.

Alternatively the circuit can be analyzed using the even and odd mode analysis by noting the symmetry of the network about the symmetry line shown in Figure VI-8. In the actual circuit (Figure VI-8(a)), voltage is supplied at the generator port; in turn, two voltages are developed, V_1 at the input to the network, and V_2 at the output. From a voltage excitation point of view, however, this situation is indistinguishable from what would result with the superposition of the even and odd voltage excitations applied at both ends of the network, as shown respectively in Figure VI-8(b and c). The even excitation shown consists of in-phase voltage generators having magnitudes V_0; each separate mode has half the generator voltage magnitude of the actual excitation, $2V_0$. These generators, in turn, produce respective even and odd voltages, V_{EE} and V_{OO}, at the input (port 1) and the output (port 2) of the even and odd mode networks.

The voltages which we seek at ports 1 and 3, V_1 and V_2, are then related to the even and odd voltages, V_{EE} and V_{OO}, at the inputs to the even and odd networks according to

$$V_1 = V_{EE} + V_{OO} \tag{VI-7}$$

$$V_2 = V_{EE} - V_{OO} \tag{VI-8}$$

The computational advantage gained is that the *even and odd mode analysis reduces the number of network ports to be analyzed by a factor of two.* With the even excitation a current null exists on the symmetry line; therefore we may break the network at this point and install magnetic walls (open circuits). Analysis of both sides of the circuit is identical. A solution for V_{EE} is obtained in terms of the applied voltage V_O. In the case of the odd excitation a voltage null occurs on the symmetry line and we may install electric walls (short circuits) at the symmetry plane dividing the network. Again, the calculations performed on both sides of the network are identical — solving for V_{OO} in terms of the applied voltage, V_O. However, *opposed polarities* at ports 1 and 2 are applied, as shown in Figure VI-8(c). Accordingly, V_1 and V_2, as expressed in Equations (VI-7) and (VI-8), are both composed of identical voltage components, V_{EE} and V_{OO}; these components, though, combine with aiding and opposing polarities, as indicated in those two equations.

The specific equations required for solution of V_{EE} and V_{OO} are then

$$V_{EE(OO)} = V_O Z_{EE(OO)}/(Z_{EE(OO)} + Z_0) \tag{VI-9}$$

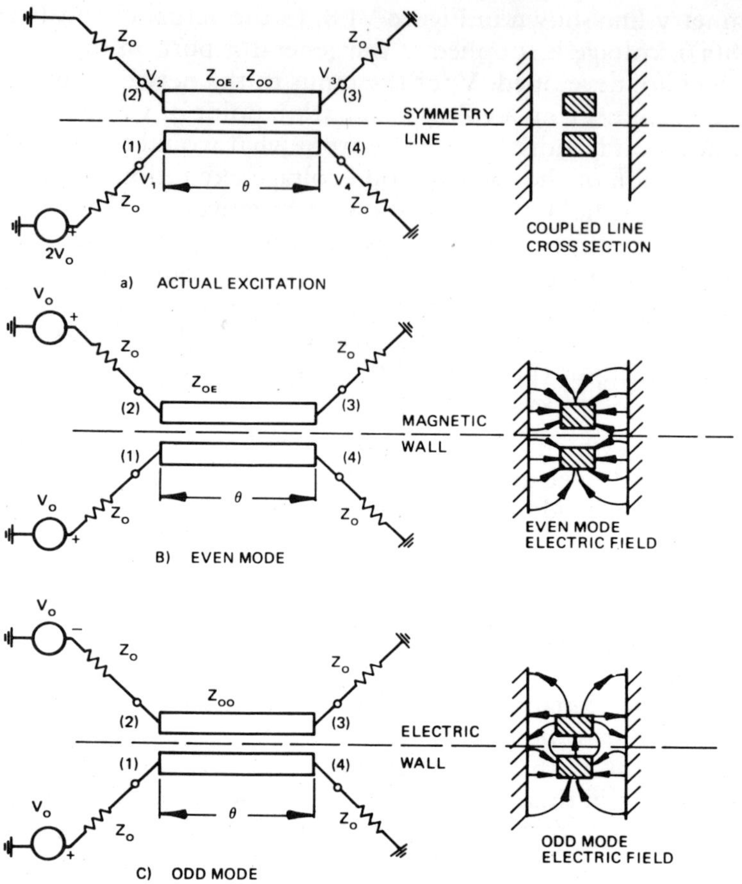

Figure VI-9 **Backward Wave Coupler Analysis Circuits**

where

$$Z_{EE(OO)} = Z_{E1(O1)} Z_{E2(O2)} / \left[Z_{E1(O1)} + Z_{E2(O2)} \right]$$

$$Z_{E1} = Z_1 + j\, Z_0 \cot \theta_1 \qquad Z_{O1} = Z_1 + j\, Z_0 \tan \theta_1$$

$$Z_{E2} = Z_2 + j\, Z_0 \cot \theta_2 \qquad Z_{O2} = Z_2 + j\, Z_0 \tan \theta_2$$

Backward Wave Coupler Example

The even and odd mode analysis is applicable and especially effective for analyzing networks which have transmission lines with distributed mutual coupling. One of the most useful and remarkable circuits in this category is the backward wave hybrid coupler;

it consists of two quarter wavelength sections of transmission line placed in close enough proximity so that energy propagating on one couples to the other. This set-up is shown schematically in Figure VI-9(a). The device is a four-port network with symmetry about all four ports.

We now examine the *homogeneous* case, in which the electrical length is the same for even and odd modes. If power is incident at port 1, a portion of that power, of course, exists at port 4. What is remarkable is that, with the right choice of impedances Z_{OE} and Z_{OO}, some energy is coupled backwards and exits at port 2, but *no energy exits at port 3!* Furthermore, the input port remains perfectly matched, and the direct and coupled waves are 90° out of phase *at all frequencies.* The even and odd mode analysis provides one of the best methods of obtaining closed form analytic expressions for this device. In solving for a closed form expression for the coupled energy at ports 2 and 4 in a backward wave coupler, we demonstrate the usefulness of this analysis. This particular example is chosen because of the wide use of this remarkable coupler in PIN control circuitry, particularly in diode phase shifters.

The analysis begins by dividing the network along the symmetry line which passes between the coupled lines, as shown in Figure VI-9(a). Operation with a signal generator attached at port 1 and matched loads at ports 2, 3, and 4 can then be represented by the respective superposition of even and odd excitations as shown in Figure VI-9(b and c). The concept of even and odd modes is more clearly seen in this example from the resulting field patterns which these excitations produce in the cross section of the coupler.

Even mode electric field lines begin on the ground planes and terminate symmetrically on the two center conductors. For the even mode there is no electrical potential difference between the two conductors at any cross section along their coupled length; therefore, a magnetic wall may be placed on the symmetry line without changing the field magnitudes. A magnetic wall represents a plane along which all electric field lines are parallel and is the field equivalent of an open circuit.

For the odd mode shown in Figure VI-9(c) the electric field lines, likewise, form paths from the ground planes to the center conductors. However, field polarities from the ground planes are opposite for each conductor. Furthermore, in the near zone of the two center conductors the electric field lines travel between the conductors since, for the odd mode, the conductors have the same volt-

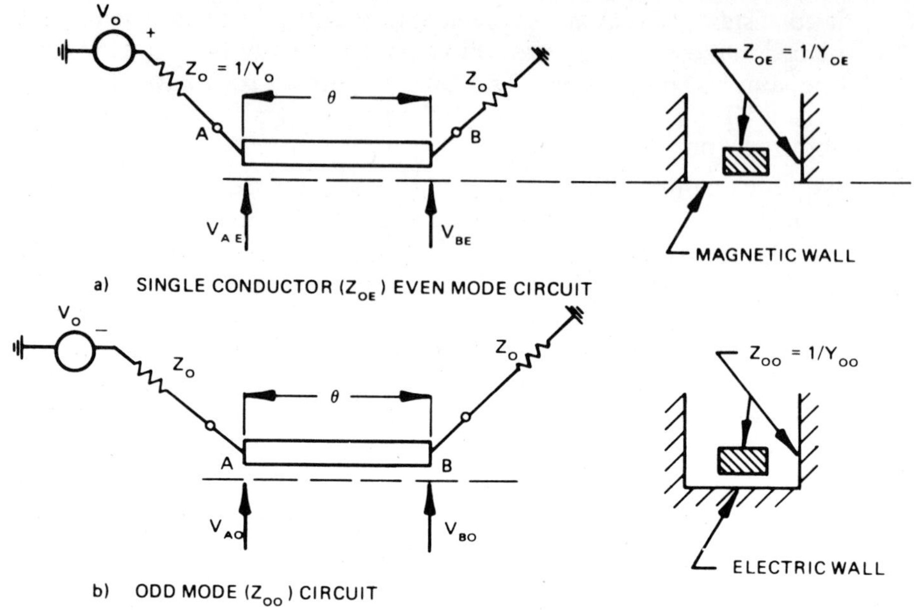

INPUT VOLTAGES: $(V_{AE,\,AO}/V_0) = Z_{AE,\,AO}/(Z_{AE,\,AO} + Z_0)$ (VI-11)

WHERE $Z_{AE,\,AO} = Z_{OE,\,OO}\,(Z_0 + TZ_{OE,\,OO})/(Z_{OE,\,OO} + TZ_0)$

OUTPUT VOLTAGES: $(V_{BE,\,BO}/V_0) = [2C + SY_0\,Z_{OE,OO} + Z_0\,Y_{OE,\,OO}]^{-1}$ (VI-12)

WHERE $T = j\tan\theta$; $C = \cos\theta$; $S = j\sin\theta$

Figure VI-10 **Single Uncoupled Transmission Line Models Used for Even and Odd Mode Analysis of the Backward Hybrid Coupler**

age magnitude but opposite polarities. As shown, electric field lines infringe on the symmetry plane at right angles; therefore, for this mode an electrical wall, or conducting plane, can be inserted without change in the field patterns or magnitudes. The electric field is normal to this plane, so the electric wall is more easily visualized as a conducting boundary or short circuit.

The analysis of the four-port hybrid coupler consists of solving for the voltages at the four ports, which shall be defined as V_1, V_2, V_3, and V_4, respectively. The analysis using the even and odd mode approach consists of determining the even and odd voltage magnitudes related to one-half of the coupler – shown diagrammatically in Figure VI-10. Consider the voltages on the upper conductor. In the even mode the network consists of a single transmission line of characteristic impedance Z_{OE} interconnecting generator and load of characteristic impedance Z_0. The even mode generator voltage, V_0, is one half the actual generator voltage for the complete excitation shown in Figure VI-9(a). The even mode impedance is defined as that characteristic impedance which a wave "sees" propagating on one of the center conductors of the coupler when the other center conductor is at the same potential.

The same analysis is carried out for the odd mode excitation shown in Figure VI-10(b). Again the analysis is that of a two-port network consisting of a length of line having the same electrical length as the coupling section and with characteristic impedance of the odd mode, Z_{OO}. The odd mode impedance is that characteristic impedance which a wave experiences propagating on one of the center conductors when the other center conductor has the same voltage magnitude but the opposite potential. The odd mode impedance is calculated from the geometry of one of the center conductors embedded between the ground planes with a short circuiting conducting plane installed at the symmetry plane of the coupler, as shown in Figure VI-10(b).

To avoid confusion with the reference port designation given for the four-port coupler in Figure VI-9, let the two-port terminals in Figure VI-10 be designated A for the input and B for the output, as shown. Then the even and odd mode analysis for the circuit in Figure VI-10 requires the solution for even and odd voltages at ports A and B; designate these voltages as V_{AE}, V_{BE}, V_{AO}, and V_{BO}. The solutions for the actual voltages on the four-port hybrid coupler are then as given below, with polarities as indicated in Figures VI-9 and VI-10.

$$V_1 = V_{AE} - V_{AO} \quad \text{(VI-10(a))}$$

$$V_2 = V_{AE} + V_{AO} \quad \text{(VI-10(b))}$$

$$V_3 = V_{BE} + V_{BO} \quad \text{(VI-10(c))}$$

$$V_4 = V_{BE} - V_{BO} \quad \text{(VI-10(d))}$$

Expressions for V_{AE}, V_{AO}, V_{BE}, and V_{BO} are given in Equations (VI-11) and (VI-12) in Figure VI-10. The input voltages are determined by finding the impedance at the input to the transmission line using the Smith Chart Transformation, Equation (VI-1). The output voltages are determined by writing the ABCD matrix for the cascade network which includes the generator impedance, the even or odd mode transmission line (whichever is appropriate), and the load. In this way, the ratio of generator and corresponding load voltage is simply the A term of the matrix, as shown in Figure VI-11, again demonstrating the broad usefulness of the ABCD matrix approach.

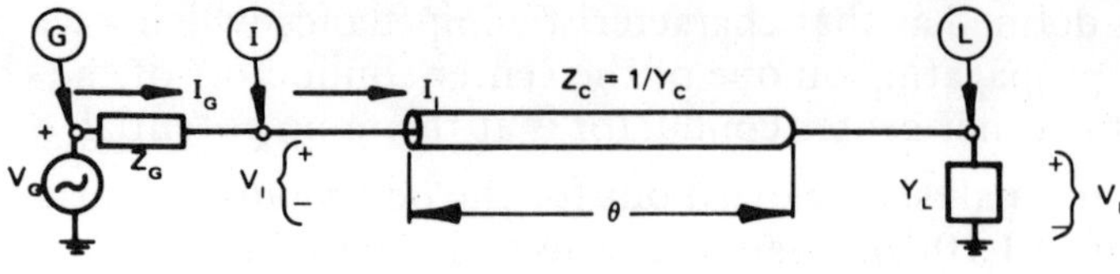

LET $K = \cos\theta$ $\quad S = j \sin\theta$

$$\begin{bmatrix} A & B \\ C & D \end{bmatrix}_{IL} = \begin{bmatrix} K & SZ_C \\ SY_C & K \end{bmatrix} \cdot \begin{bmatrix} 1 & 0 \\ Y_L & 1 \end{bmatrix}$$

$$= \begin{bmatrix} K + SY_L Z_C & SZ_C \\ SY_C + KY_L & K \end{bmatrix}$$

$$V_I = A_{IL} V_L$$
$$I_I = C_{IL} V_L$$

$$\begin{bmatrix} A & B \\ C & D \end{bmatrix}_{GL} = \begin{bmatrix} 1 & Z_G \\ 0 & 1 \end{bmatrix} \cdot \begin{bmatrix} K + SY_L Z_C & SZ_C \\ SY_C + KY_L & K \end{bmatrix}$$

$$= \begin{bmatrix} K(1 + Z_G Y_L) + S(Y_L Z_C + Y_C Z_G) & KZ_G + SZ_C \\ KY_L + SY_C & K \end{bmatrix}$$

$$V_G = A_{GL} V_L \qquad \text{(VI-13)}$$
$$I_G = C_{GL} V_L \qquad \text{(VI-14)}$$

Figure VI-11 **Solution for Input and Generator Voltages in Terms of Load Voltages and Currents Using ABCD Matrices**

Summary of Hybrid Properties

When the solutions for the four-port hybrid coupler are carried out for arbitrary Z_{OE} and Z_{OO}, the results (Equations (VI-15) and (VI-16) in Figure VI-12) do not suggest that the backward wave coupler has any particularly unusual properties. It is to the credit of the early researchers (see Reference 6 for a bibliography) in this field that the broadband nature of this device has been revealed. The special matched hybrid properties are obtained under the condition that the even and odd mode impedance product is equal to the square of the terminating impedance (Equation (VI-17) in Figure VI-13). Substitution of this relation into Equations (VI-15) and (VI-16) permits them to be simplified to Equations (VI-18), (VI-19), (VI-20), and (VI-21) for the voltages at the four ports of the hybrid (V_1, V_2, V_3, and V_4). Not only are the expressions simple; they are surprising for the hybrid's physical characteristics which they describe. Specifically, for the matched coupler with matched terminations, for *all values of coupling coefficient, k,* and *for all frequencies:*

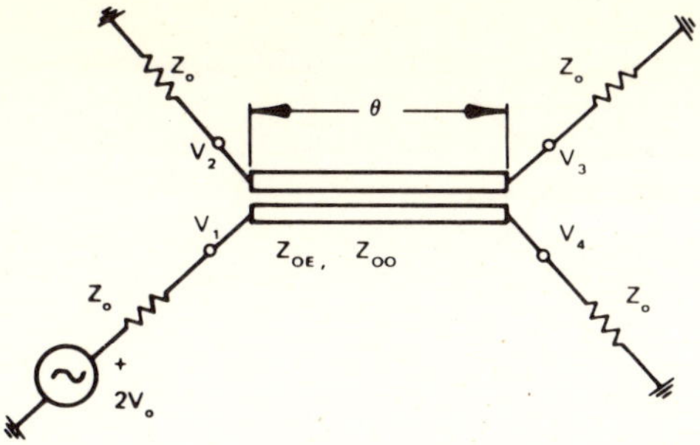

$$V_1 = V_{AE} + V_{AO}$$
$$V_2 = V_{AE} - V_{AO}$$
$$V_3 = V_{BE} - V_{BO}$$
$$V_4 = V_{BE} + V_{BO}$$

$$T = j \tan\theta$$
$$C = \cos\theta$$
$$S = j \sin\theta$$

$$\frac{V_{AE} \pm V_{AO}}{V_o} = \frac{[Z_{OE} Z_{OO} (2Z_o^2 + T^2 Z_{OE} Z_{OO})(1 \pm 1) + Z_o^2 T^2 (Z_{OE}^2 \pm Z_{OO}^2)] + T[Z_O Z_{OE} Z_{OO} (Z_{OE} + Z_{OO})(1 \pm 1) + Z_O (Z_{OE} Z_{OO} + Z_O^2)(Z_{OE} \pm Z_{OO})]}{[4Z_O^2 Z_{OE} Z_{OO} + T^2 (Z_O^4 + Z_O^2)(Z_{OE}^2 + Z_{OO}^2) + Z_{OE}^2 Z_{OO}^2] + T[2Z_O (Z_{OE} Z_{OO} + Z_O^2)(Z_{OE} + Z_{OO})]} \tag{VI-15}$$

$$\frac{V_{BE} \pm V_{BO}}{V_O} = \frac{[2C + S(Y_O Z_{OO} + Z_O Y_{OO})] \pm [2C + S(Y_O Z_{OE} + Z_O Y_{OE})]}{[4C^2 + S^2 (Y_O Z_{OE} + Z_O Y_{OE})(Y_O Z_{OO} + Z_O Y_{OO})] + S[2C(Y_O Z_{OE} + Z_O Y_{OE}) + 2C(Y_O Z_{OO} + Z_O Y_{OO})]} \tag{VI-16}$$

Figure VI-12 **General Solution for the Voltages at Four Ports of the Backward Wave Coupler Having Equal Electrical Lengths for the Even and Odd Modes (Homogeneous Coupler)**

1) The input is matched ($V_1 = V_0$).
2) The directivity is infinite ($V_3 = 0$).
3) The direct (V_4) and coupled (V_2) voltages are exactly 90° out of phase (phase quadrature).

Equations (VI-17) to (VI-22) strictly define the condition for a homogeneous backward wave coupler. In practice, it is common to assume that the match condition of Equation (VI-17) applies by proper design. It is then only necessary to specify the center frequency coupling, in dB. Thus, for example, a 20 dB coupler is one for which $k = 0.1$. Alternatively, for a $Z_0 = 50\ \Omega$ system, a 20 dB coupler has $Z_{OE} = Z_0\sqrt{(1+k)/(1-k)} = 55.28\ \Omega$.

5. Scattering Matrix and the Reflectivity Terminated Hybrid

s *Parameter Definitions*

We have seen how the choice of independent and dependent variables for a network imparts certain computational advantages to the resulting matrices that define the network. The *z* parameters are useful for series combination of circuits; *y* parameters, for parallel combination; and ABCD parameters, for cascade combination.

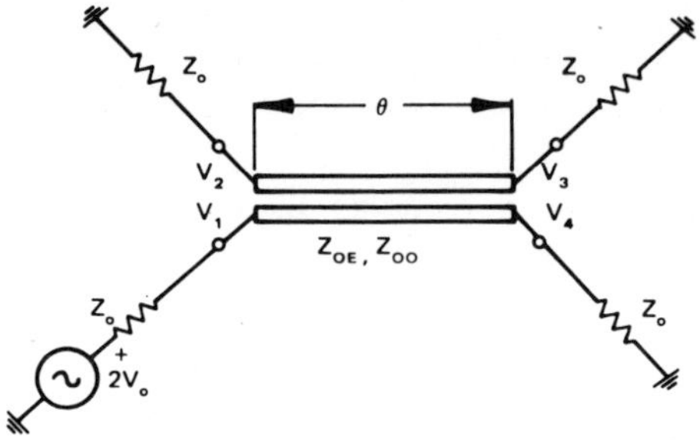

(VI-17)	$Z_{OE}\, Z_{OO}$	=	Z_o^2	MATCHING CONDITION
(VI-18)	V_1/V_o	=	1	PERFECT MATCH
(VI-19)	V_3	=	0	INFINITE DIRECTIVITY
(VI-20)	$\dfrac{V_2}{V_o}$	=	$\dfrac{j\,k \sin\theta}{(\cos\theta)\sqrt{1-k^2}+j\sin\theta}$	QUADRATURE (90°) COUPLED VOLTAGE
(VI-21)	$\dfrac{V_4}{V_o}$	=	$\dfrac{\sqrt{1-k^2}}{(\cos\theta)\sqrt{1-k^2}+j\sin\theta}$	DIRECT VOLTAGE
(VI-22)	k	=	$\dfrac{Z_{OE}-Z_{OO}}{Z_{OE}+Z_{OO}}$	VOLTAGE COUPLING COEFFICIENT

Figure VI-13 **Equations for the Matched Backward Coupler**

The *s* parameters [2, 7] defined in Figure VI-14, are useful for microwave networks when incident and reflected wave amplitudes are more readily measured and/or specified than are voltage and current variables. Incident waves are denoted by *a* variables, and can be thought of as excitations of the network; exiting reflected waves are denoted by *b* variables. These variables can be considered to represent reflections from the network, although the b_p wave exiting port *p* may be the result of *a* excitations at several ports, rather than at the *p* port exclusively. The amplitudes of *a* and *b* waves are normalized with respect to power, as indicated in Figure VI-14, so that characteristic impedance at the network ports need not be specified (an advantage for waveguide devices). We now analyze the hybrid for the case when symmetric reflections are placed on the direct and coupled ports, as shown in Figure VI-15. Before performing the matrix analysis, consider this operation of the hybrid from a physical viewpoint.

To describe the operation, suppose the coupling coefficient is adjusted for equal power (3 dB coupler) at the direct and coupled ports; then $k = 0.707$ and, at center frequency, f_0, $\theta = 90°$. The direct and coupled voltages, V_4 and V_2, respectively, would have equal amplitudes if matched loads were used to terminate ports 4

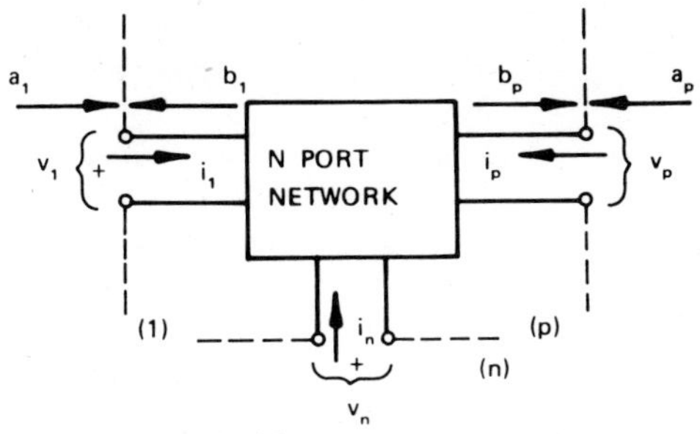

$[b] = [s] \cdot [a]$

THAT IS,

$$b_1 = s_{11} a_1 + s_{12} a_2 + \ldots + s_{1n} a_n$$

$$b_n = s_{n1} a_1 + s_{n2} a_2 + \ldots + s_{nn} a_n$$

WHERE PROPORTIONALITY CONSTANTS ARE DEFINED (NORMALIZATION) SUCH THAT,

$\tfrac{1}{2} a_p a_p^* = \tfrac{1}{2} |a_p|^2$ = POWER INCIDENT AT PORT p

$\tfrac{1}{2} b_p b_p^* = \tfrac{1}{2} |b_p|^2$ = POWER EXITING AT PORT p

THUS,

$|a_p|$ = INCIDENT WAVE PEAK AMPLITUDE AT PORT p

$|b_p|$ = EXITING (REFLECTED) WAVE PEAK AMPLITUDE AT PORT p

AND

$v_p = a_p + b_p$ = NORMALIZED VOLTAGE AT PORT p

$i_p = a_p - b_p$ = NORMALIZED CURRENT AT PORT p

HENCE

$$a_p = \tfrac{1}{2}(v_p + i_p)$$

$$b_p = \tfrac{1}{2}(v_p - i_p)$$

Figure VI-14 **The Scattering Matrix**

and 2. If, instead, total reflection occurs with reflection coefficient, Γ, at ports 2 and 4, the direct and coupled waves thereby are re-directed into the coupler. Because of the 90° phase quadrature introduced between direct and coupled waves each time a signal travels through the coupler, the resultant waves exiting the coupler add in phase at the normally decoupled port 3 and 180° out of phase at the input port 1. The net result is that power which originally enters the coupler at port 1 exits totally at port 3, having been reflected from the symmetric terminations at ports 2 and 4. Thus, the hybrid's properties permit conversion of a symmetric pair of one-port reflective networks into a perfectly matched two-port transmission network. Actually, this function is achieved only when the direct and coupled waves have the same magnitude. But since this magnitude varies slowly as a function of frequency, fairly good transmission can be obtained over an octave bandwidth, as is seen from the *s* matrix demonstration to follow.

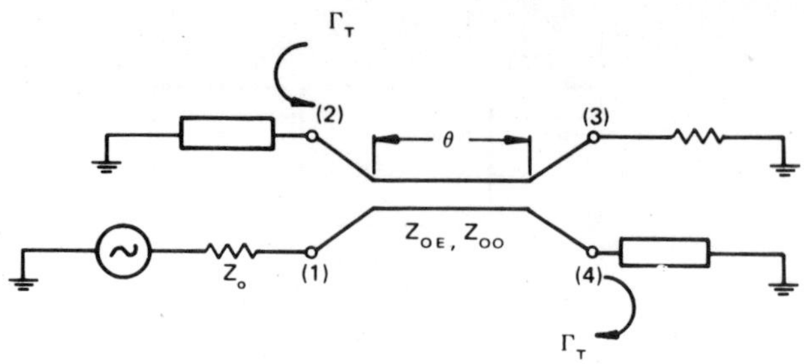

$$Z_{OE}\, Z_{OO} = Z_o^{\,2}$$

$$[b] = [s] \cdot [a]$$

$$\begin{bmatrix} b_1 \\ b_2 \\ b_3 \\ b_4 \end{bmatrix} = \begin{bmatrix} 0 & f_1 & 0 & f_2 \\ f_1 & 0 & f_2 & 0 \\ 0 & f_2 & 0 & f_1 \\ f_2 & 0 & f_1 & 0 \end{bmatrix} \cdot \begin{bmatrix} a_1 \\ \Gamma_T b_2 \\ 0 \\ \Gamma_T b_4 \end{bmatrix}$$

where $f_1 = \dfrac{j\,k \sin\theta}{\sqrt{1-k^2}\cos\theta + j\sin\theta}$ $\qquad f_2 = \dfrac{\sqrt{1-k^2}}{\sqrt{1-k^2}\cos\theta + j\sin\theta}$

Figure VI-15 *s* **Matrix Representation for the Reflectively Terminated Backward Wave Hybrid Coupler**

s Matrix Analysis of Reflectively Terminated Hybrid

The analysis is begun by first evaluating the terms of the *s* matrix shown in Figure VI-15. Basically, each s_{ij} parameter is equal to one (complex) ratio of the wave amplitude, b_i which emerges from the i port due to an excitation, a_j, at the j port with all other ports match terminated to that excitation. For example, the first row of *s* parameters for the hybrid is evaluated by explicitly writing the equation this row represents. From Figure VI-14

$$b_1 = s_{11}a_1 + s_{12}a_2 + s_{13}a_3 + s_{14}a_4 \qquad \text{(VI-23)}$$

With all ports match terminated except port 1, and with an input a_1, $b_1 = 0$ since the input is matched. Hence, $s_{11} = 0$. An input a_2 produces a coupled wave at port 1; the value of the resulting wave, given by Equation (VI-20), is abbreviated f_1a_2. Thus $s_{12} = f_1$. An input a_3 with the other ports matched produces a zero output at port 1 because of the perfect directivity of the coupler. Finally, an input a_4 produces a direct wave given by Equation (VI-21) and abbreviated f_2a_4 at port 1; hence, $s_{14} = f_2$.

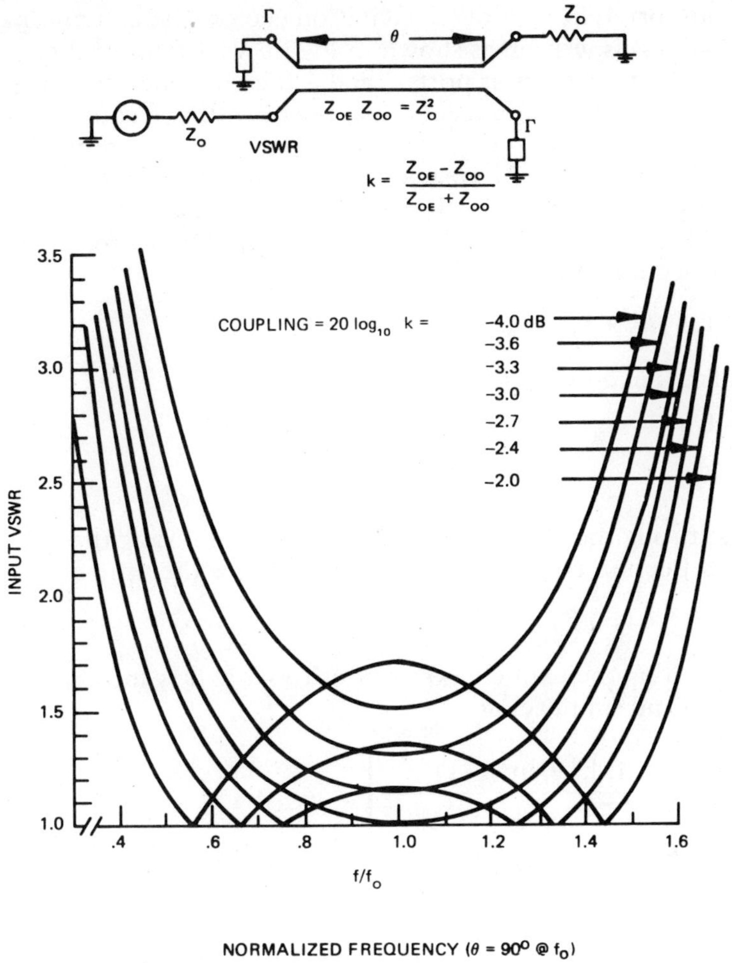

Figure VI-16 **Theoretical VSWR Performance vs. Frequency for Reflectively Terminated Hybrids of Various Couplings**

The remaining *s* matrix terms are evaluated in similar fashion. Because an ideal hybrid coupler for which $Z_{OE} Z_{OO} = Z_0^2$ has perfect input match, infinite directivity, and four-port symmetry, its complete *s* matrix requires only the f_1 and f_2 functions for definition.

The operation with reflective terminations as defined in Figure VI-16 corresponds to an input wave a_1 at port 1. The direct result is two output waves, b_2 and b_4. These re-enter the coupler as $a_2 = \Gamma_T b_2$ and $a_4 = \Gamma_T b_4$, respectively. Since there is a matched load at port 4, $a_4 = 0$; the a excitation vector (the column matrix of a values) is written as shown in Figure VI-15. Since there are reflecting terminations at ports 2 and 4 there is generally not a perfect match at port 1, except when the operating frequency and coupling length are such as to give exactly equal outputs at ports 2 and 4. The degree of mismatch, as a function of coupling and frequency, can now be estimated using the s matrix. Performing the indicated matrix multiplication in Figure VI-15 for b_1, the net reflection at the input gives

$$b_1 = f_1 \Gamma_T b_2 + f_2 \Gamma_T b_4 \tag{VI-24}$$

Similarly, expressions for b_2 and b_4 are

$$b_2 = f_1 a_1 \tag{VI-25}$$

$$b_4 = f_2 a_2 \tag{VI-26}$$

Substituting these into Equation (VI-24) and noting that the complex reflection coefficient, Γ_1, at the input is the ratio b_1/a_1 gives

$$\Gamma_1 = (f_1^2 + f_2^2)\,\Gamma_T \tag{VI-27}$$

Substituting the explicit expressions for f_1 and f_2 into Equation (VI-27) and simplifying

$$\Gamma_1 = \left[\frac{1-k^2\ (\sin^2\theta+1)}{(\sqrt{1-k^2}\cos\theta + j\sin\theta)^2}\right]\cdot\Gamma_T \tag{VI-28}$$

From Equation (VI-28) it can be seen that the matched impedance, backward wave, hybrid shown in Figure VI-16 with symmetric reflections at its direct and coupled ports has a perfect input match ($\Gamma_1 = 0$) when the numerator expression in Equation (VI-28) goes to zero (since the denominator is always finite). For finite Γ_T, the numerator is zero when

$$k^2\ (\sin^2\theta+1) = 1 \tag{VI-29}$$

We note that with a 3 dB coupler (to be precise, 3.0103 dB for exactly equal outputs) the direct and coupled waves have equal amplitude when the coupling length θ is 90° at the coupler's center frequency, f_0. For this condition, cancellation at port 1 of the wave reentering the coupler from the Γ_T terminations should be total, and the match at port 1 should be perfect. Equation (VI-29) shows this assumption to be true since $k = 1/\sqrt{2}$. However, tighter coupling ($k > 0.707$) produces a perfect match at pairs of frequencies symmetrically disposed about f_0. This situation can be understood physically by noting that, if the coupled wave is larger than the direct wave at f_0, it decreases in amplitude for frequencies removed symmetrically from f_0. The coupler's perfect 90° phase difference between direct and coupled waves *at all frequencies* then insures the perfect cancellation at those two frequencies (spaced symmetrically about f_0) at which the amplitude balance occurs.

Equation (VI-28) is shown graphically in Figure VI-16 for various coupling values, as a function of normalized frequency, f/f_0, assuming that the terminations at ports 2 and 4 are lossless (i.e., totally reflecting, $\Gamma_T = 1$). The voltage standing wave ratio

$$\text{VSWR} = \frac{1+|\Gamma_1|}{1-|\Gamma_1|} \qquad \text{(VI-30)}$$

has been used in place of Γ_1 in Figure VI-16 for ease of comparison with typical component specifications. Notice that for a 2.7 dB coupler ($k = 0.732$) the input VSWR is below 1.2 for $0.65 \leqslant f/f_0 \leqslant 1.3$, an octave frequency bandwidth.

This reflectively terminated coupler evaluation was chosen, not only for its *s* matrix demonstrative value, but also because of its importance in control circuitry, as is seen in the following chapters. This use of *s* parameters demonstrates the computational power of the proper choice of matrix definition for a given network analysis. If there is any doubt, an attempt to derive these results for the hybrid by alternative methods may heighten one's appreciation of the *s* matrix*.

In the preceding example, it was possible to find a general closed form solution to a problem of considerable intricacy. The *s* matrix is equally useful when it is evaluated numerically from laboratory measurements for multi-ports, in which case the *s* parameters must be specified for each analysis frequency. The use of a computer for data reduction of a multi-port device is then particularly helpful.

C. FORTRAN Computer Programming

1. *Scope*

It is not the intent of this section to serve as a complete reference to FORTRAN IV programming, the computer language used for

*Other analyses are possible for the backward wave hybrid, however. See, for example, Chen Y. Ho; "Analysis of dc Blocks Using Coupled Lines," *IEEE Transactions on Microwave Theory and Technique, Vol. MTT-23, No. 9,* September 1975, pp. 773-774.

the examples which follow. On the other hand, a circuit designer not already using FORTRAN may find texts on programming intensely detailed and therefore may postpone (indefinitely) the beginning of the acquisition of this important language. Developing a fluency in a language proceeds at a rate proportional to the motivation for learning the language. This section is intended to introduce or review FORTRAN IV and to show how it is effective in the evaluation of microwave control circuitry.

The treatment is short enough that with a brief reading, the useful elements and technical details of writing a FORTRAN IV program to solve a microwave circuit problem can be seen in perspective. The program examples to follow can be read without any previous programming experience; having seen them, the reader should be more than prepared to use a standard text [1] for any further detailed information that may be needed for his own programming.

2. *FORTRAN IV Summary*

Variable Types and Accuracy

One of the most useful, yet initially confusing, aspects of FORTRAN IV programming is the variety of variable types available. The principle computational variable types are *integer, real,* and *complex.* It is further possible to define *logical* variables, which take on "true" or "false" values, and *string* variables, which have alphanumeric codes or names as their values. We discuss string variables later in relation to variable topology network analysis programs, but for now the integer, real, and complex variables suffice. All FORTRAN variables are defined with a name having one to six alphanumeric characters, the first of which must be a letter. Sample variable names are X, Y, VOLTS, A25, and so forth.

Integer variables are those which have no fractional part. If the variable name begins with any of the letters I, J, K, L, M, or N, the FORTRAN compiler (which generates from the *source* program you write a *compiled* program in suitable notation to be run on the particular computer equipment being used) automatically interprets the variable as integer, unless a specific statement at the beginning of the program indicates otherwise. Thereafter, the computer truncates any decimal part which the variable might have or could obtain from calculations performed within the program. Integer variables usually are used for indexing repetitive program sequences (DO loops), for subscripts, and for certain logical statements.

Real variables are those which can have a decimal part; they can be expressed in decimal notation, such as with 1.2965, 3.2, 0.519, or floating point (scientific notation in which 129.1 = 1.291E2, -111. = -1.11E2, 0.015 = 1.5E-2, and so forth). Variables defined within a FORTRAN program whose names do not begin with I, J, K, L, M, or N are automatically treated as real variables. It is customary to define real variables with an A letter prefix if their names might otherwise begin with an integer letter. For example, a real variable "LOSS" defined as "ALOSS" eliminates the need for specification as a real variable and the chance of misinterpretation as an integer.

Complex variables must be so defined at the beginning of the program; thus, there is no possible confusion with integer and real variables. Complex variables are those which have real and imaginary parts. Both parts can be expressed in terms of real quantities in that they have whole and fractional parts. The availability of complex notation is a major advantage of the FORTRAN IV language for electrical engineering.

Accuracy obtained with most computers using FORTRAN IV is commensurate with arithmetic performed using exponential notation to eight significant figures, with number sizes between 10^{-40} and 10^{+40}. If more significant figures are required a statement calling for *Double Precision* for designated variables allots double the usual storage space for variable quantities giving, typically, 16 significant figure accuracy. The author has tested a sample calculation with 99 cascaded matrix multiplications and results were accurate to better than 1% using standard single precision variables. Thus, in most cases, single precision can be expected to suffice.

Undefined and improperly defined variables represent one of the most vexing FORTRAN programming bugs to find. *An undefined variable is automatically assigned a value of zero and most compilers give no fault indication in this instance. A complex variable not specifically designated as such (at the beginning of the program) has its imaginary part set equal to zero,* and is treated throughout the program as a real variable.

Always run sample calculations with known solutions through a new program. Keep in mind that, while in practice, computers are nearly fault free, programmers are not. *No program is ever debugged with one hundred percent certainty,* and any "strange" results should be subjected to questioning and independent verification.

Any of the three variable types described – integer, real, and complex – can be defined as up to a three dimensional *array* permitting the handling of matrix operations in very convenient form. This definition is accomplished through the use of *Subscripted Variables* which are defined at the beginning of the *Main* program (before the first executable statement) by a DIMENSION statement of the form

DIMENSION NAME (n_1, n_2, n_3)

where the appropriate variable name is followed by a parenthesis with one number listing the maximum extent of the array for each dimension used. Examples of subscripted variable DIMENSION statement are

DIMENSION X (10)
DIMENSION Y (10, 20)
DIMENSION Z (100, 2, 15)

Once defined, a call of the variable name above (without parentheses) is sufficient to move the entire array between the main program and subprograms, defined later. Reference to an individual *element* of an array is made simply by calling out its coordinates. For example, with the arrays defined above, we might wish to define an addition

X (9) = Y (9, 1) + Z (9, 1, 5)

The enormous value of the FORTRAN subscripted variable notation becomes clear with the matrix algebra examples which are presented later.

If a real or complex specification is required of a subscripted variable, then the word "DIMENSION" is omitted. Thus, for example

COMPLEX Z (2, 2)

is sufficient to define a set of four values located in a 2 x 2 array, each of which is itself a complex number.

Mathematical Operations

With FORTRAN there are only five basic arithmetic operations, as shown in Table VI-1.

Symbol	Operation	Fortran Statement Example
+	*Addition*	A = B + C
–	*Subtraction*	A = B – C
*	*Multiplication*	A = B * C
/	*Division*	A = B/C
**	*Exponentiation*	A = B ** C, (A = B^C)

Table VI-1 **Basic FORTRAN Mathematical Operations**

Parentheses must be used to avoid having two operation symbols together. Thus A = B*–C is incorrect while A = B*(-C) is correct. Integer and real variables must not be used in the same expression; the one exception is that a real expression can be raised to an integer power.

The operation carried out within a statement follows the hierarchy:

1) Operations within parentheses.
2) All exponentiations.
3) All multiplications and divisions.
4) All additions and subtractions.

In general, it is good practice to use parentheses whenever there is a question about the sequence of operations. For example, A = B+C**2 produces A = B + C^2, while A = (B+C)**2 produces A = $(B+C)^2$.

Sequence of Operations and Statement Numbers

The FORTRAN compiler generates a program sequence in the same order as the *Statements* are entered regardless of the *Statement Number* assignments. The first five spaces of the statement are reserved for a statement number, but the use of the number is optional. Generally it is used only if the statement must be referenced somewhere else in the program. No sequence need be kept in the assignment of statement numbers, but each numbered statement must have a unique number, as we demonstrate later. *FUNCTION* and *SUBROUTINE* subprograms are compiled separately from the main program; it is permissible to assign the same statement numbers within these subprograms as are assigned within the main program.

The sixth space is used if the statement cannot be fitted in the 72 remaining spaces. Any character (except 0) in this space causes the compiler to continue the previous statement beginning with the seventh space. In FORTRAN IV, a single statement can be extended over 19 additional lines, although particular compilers may have some different limit to the numbers of lines useable for a statement. Of course, only the first line of a continued statement may be assigned a statement number.

A "C" in the first space (of the five spaces reserved for the statement number) is used to tell the compiler that the statement is a *Comment* – used for note keeping within the source program and ignored by the compiler in the generation of a machine language program.

The 72 spaces (or columns) available for statements are labelled on IBM cards (and source program writing pads) to facilitate keeping track of the column allocations for statement numbers or comment designation, continuation, and statement context. When generating a source FORTRAN program using an on-line, time-sharing terminal, special "editor" programs have been designed to keep track of the spacing automatically. Typical use of comments, program statements, and statement numbers are shown in the sample program to calculate N factorial in Figure VI-17. The columns 73-80 of the IBM card format are not read by the standard FORTRAN IV compiler; therefore, a program assembled on cards can use these columns for number identification of the cards to facilitate reorganizing the deck should it become shuffled inadvertently.

DO Loops

The utility of a computer language comes from the facility to write brief statements, the execution of which effects the completion of what would otherwise be complicated and/or laboriously tedious programming tasks. The DO statement, when skillfully employed, can make short work for the programmer of repetitous calculations. The following sequence of program steps in Table VI-2 calculates the value of N factorial for a variable named "N" and stores the result as a variable "NFACT."

The DO statement example given stated simply means "perform the indicated operations to statement 100, starting with I = 1 and stopping at I = N, increasing N in steps of 1." The last 1 could have been omitted since a unity step is assumed if no step size is specified.

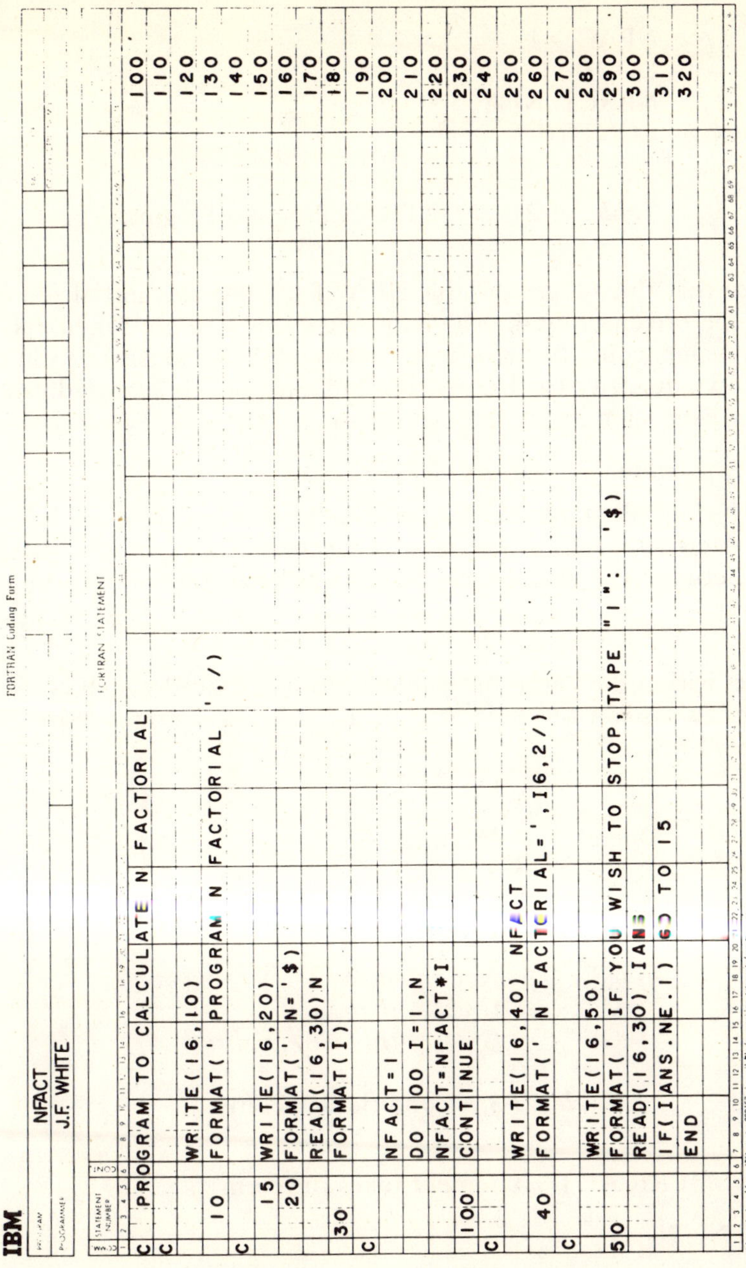

```
C     PROGRAM TO CALCULATE N FACTORIAL                                  100
C                                                                       110
      WRITE(16,10)                                                      120
10    FORMAT(' PROGRAM N FACTORIAL ',/)                                 130
C                                                                       140
15    WRITE(16,20)                                                      150
20    FORMAT(' N= '$)                                                   160
      READ(16,30)N                                                      170
30    FORMAT(I)                                                         180
C                                                                       190
      NFACT=1                                                           200
      DO 100 I=1,N                                                      210
      NFACT=NFACT*I                                                     220
100   CONTINUE                                                          230
C                                                                       240
      WRITE(16,40) NFACT                                                250
40    FORMAT(' N FACTORIAL= ',I6,2/)                                    260
C                                                                       270
      WRITE(16,50)                                                      280
50    FORMAT(' IF YOU WISH TO STOP, TYPE "1": '$)                       290
      READ(16,30) IANS                                                  300
      IF(IANS.NE.1) GO TO 15                                            310
      END                                                               320
```

Figure VI-17 Sample Program on FORTRAN Coding Form to Calculate N Factorial

```
      NFACT=1
      DO 100 I=1, N, 1
      NFACT=NFACT*I
100   CONTINUE
```

Table VI-2 **FORTRAN Program Fragment**

Notice that the integer variable NFACT has its value initialized to unity before entering the DO loop. Were this step not taken, its initial value would be zero (if not referenced previously) or some unknown quantity (if this variable had a previously assigned value during execution in the same program).

IF Statements – Control Transfer

The IF (logical) statement allows branching of the program following a decision, based on the relative algebraic or magnitude difference between two quantities. The IF statement takes the form

IF (   ) statement

where, within the parentheses, a condition is placed which consists of two quantities separated by a condition operator from Table VI-3.

Operator	Meaning
.LT.	*Less than*
.LE.	*Less than or equal to*
.EQ.	*Equal to*
.NE.	*Not equal to*
.GT.	*Greater than*
.GE.	*Greater than or equal to*

Table VI-3 **Condition Operators**

The NUMERICAL IF statement takes the form

IF (n) N_1, N_2, N_3

which causes the program sequence to go to

statement N_1 if $n < 0$ $(n = -)$

N_2 if $n = 0$

N_3 if $n > 0$ $(n = +)$

The quantity n can be a single valued variable such as X, Y, or DELTA, or it can be an expression such as (A–B), (A+C/B), and so forth. The double periods separate the operator from the characters which might be used to define the variables being compared. Sample IF statements and related transfer of control are shown in Table VI-4.

1) IF (X.GT.Y) X = Y + 2
2) IF (A.LT.B) GO TO 100
3) IF (A.LT.B. AND .X.LT.Y) (statement)
4) GO TO n (n = statement number)
5) GO TO (n_1, n_2, n_3. . .n_m) i

Table VI-4 **Transfer of Control Examples**

The related GO TO statement can be used to change the execution of program steps from the usual sequential order. Any numbered program step can serve as a point to which control is transferred. As noted earlier, numbering steps is optional except when the step is to be referenced in a *Transfer of Control* statement. Transfer of control can be unconditional, as in example 4) above, or conditional, as in examples 2) and 5). Conditional transfer is called *branching* and represents the "decision making aspect" of computer programming. Example 5) is a computed branch step. The program execution is transferred to step n_1 or n_2, or . . . n_m according-as the integer variable i is equal to 1, 2. . . m.

FORMAT Statements – Writing a Simple Program

In the execution of a computer program, it is necessary for the computer to communicate with the operator in order to request data, to indicate that a certain point has been reached in the program, to output data, and so forth. Such "remarks" by the computer are of course built into the program and stored in FORMAT statements, *which must be assigned a number* so they can be called up by the program when they are needed. These statements may contain English phrases, with or without numbers. Such *alphanumeric* expressions are called *Hollerith fields* and are separated by single quotation or apostrophe (') marks. They may be interspersed with number *Field specifications*, through which data can be read or written by the computer. These expressions are demonstrated by the complete program to determine N factorial shown in Figure VI-17. The means for intercommunications between the computer and the programmer can be readily understood by an examination of this program. A summary of common FORMAT text commands is given in Table VI-5.

	Description	Example	
Fm.n	*Fixed point variable field of m characters (including decimal point and sign) with n characters following the decimal point*	F5.2	(1.45)
Em.n	*Floating point variable field of m characters (including the sign, a leading 0, decimal point, and 4 positions for the exponent) with n significant figures*	E9.2	(0.12E+05)
Im	*Integer variable field of m characters (including sign)*	I5	1246
X	*Blank Space*		
/	*Start a New Line*		
mH	*Hollerith Field*	5HHELLO	
m ( )	*Repeat m Times*	2F3.0	
,	*Separator*	F3.0, F10.2	
$	*Stop (Without Advancing)*	X = '$	

Table VI-5 **Some Format Text Commands**

To the extent possible the sample programs given are in standard FORTRAN IV; however, minor variations can be expected among the "FORTRAN IV" compilers available from various time sharing services, particularly with respect to input and output commands. Nevertheless, the reader can expect that the programs given here should be, with some assistance from the service representatives, easily made to be compatible with any FORTRAN IV system.

A complete program NFACT to calculate N factorial is shown in Figure VI-17. Main and subprograms must be assigned a name which begins with a letter and has no more than six alphanumeric characters. The program is shown as it would be written initially by the programmer on a FORTRAN coding form. Use of the form makes it easy to keep the statement numbers and statement text in the assigned columns.

Although each FORTRAN statement does not require a *statement number,* it is useful to assign *line numbers* in order either that the *deck* of cards can be put in sequence or, if the program is to be read in on a time-shared terminal, that corrections can be made by identifying the *line* which is to be changed. For this reason we have assigned an identification sequence beginning with 100 and increasing 10 for each step, "leaving space" for the insertion of additional statements if needed later on. These identification step numbers, of course, play no part in the compilation of the FORTRAN program and are merely for the programmer's convenience in generating his source program. We now describe the significance of the steps in the program shown in Figure VI-17 to illustrate the use of comments and READ, WRITE, and FORMAT statements.

Lines 100-110 – The first two statements are comments and are so interpreted by the compiler because a "C" appears in the first column. The compiler ignores the rest of the line so designated. The insertion of comments to introduce either blank lines or descriptive remarks permits the programmer to separate operations and to provide reminders of his initial intentions when he wrote the program. In particularly large programs the use of comments greatly aids in both program debugging and in making program modifications.

Step 120 – The WRITE statement is the first executable statement of this program. It requires no statement number since no other statements in the program refer to it. The parentheses following "WRITE" contain two number assignments; the first defines the computer peripheral instrument which is used for the writing (in this case, 16 defines a Teletype terminal), the second defines the numbered statement in which the FORMAT for the writing is defined (statement 10).

Line 130 – This statement is numbered since it defines a FORMAT and must therefore be referred to elsewhere in the program by either a READ or WRITE statement. This particular FORMAT statement contains only a printed message which is defined between the single quotation marks. When the material to be typed begins a line, as in this case, it is necessary to leave a blank space. Most FORTRAN compilers do not print the first character in a line definition; rather, that character is used as a code to determine line spacing in the event that the WRITE FORMAT specification calls for multiple lines. The code for line spacing is given in Table VI-6. In the present case, a blank space suffices.

Blank	*Single line spacing*
0	*Double line spacing*
1	*Advance to top of next page*
+	*Suppress spacing (overprint)*

Table VI-6 **Line Spacing for WRITE FORMATS**

The slash or virgule (/) indicates that a line will be skipped by the program after the message is printed. The various elements within the FORMAT statement are separated from one another by commas; the entire contents of the FORMAT statement are contained within a set of parentheses.

Line 160 – This FORMAT prints a message, then waits for the value of "N" to be typed in. The dollar sign ($) causes the teletype terminal to wait for a data entry by the operator following the printed statement.

Line 170 – The READ statement is similar in format to the WRITE statement. The first number in parentheses describes the computer peripheral – in this case it is again the Teletype terminal, 16 – and the second identifies the FORMAT statement in which the data is to be read. Following the READ statement is the definition of the variables which are to be provided to the computer *in the order that they will be presented.* In this case only one variable is to be read; since it is an integer, it will be read using the *Integer Format* described by statement 30.

Lines 200-230 – This sequence is the actual computational order of the program described earlier as a DO loop example. A DO loop must terminate on a numbered statement. It is good practice to make this statement a simple CONTINUE statement to avoid possible misrepresentation by the compiler. Notice that line 220 contains what is mathematically a false statement; in computer language, however, it is a very useful operation. The interpretation of the statement at line 220 is "replace the value for NFACT with the quantity NFACT multiplied by the current value of the index 'I'." It should be kept in mind that the last computed value for NFACT remains in the computer until a new value is assigned for the variable named. For this reason, NFACT was initialized with the value 1 in line 200 prior to entry into the DO loop.

Lines 290-310 – After line 260, the desired value for N factorial has been computed and printed out; the program could be terminated. However, the use of a simple IF statement permits us to recycle through the program for a new value of N, if so desired. There are many ways in which this branch point can be handled. The method demonstrated in this program favors the assumption that in typical usage the program would be executed for a number of different N values. After the previous value for N factorial is written by the WRITE statement in line 250, the WRITE statement in line 290 is printed on the Teletype terminal. An answer in the form of a 1 is required if the operator wishes to terminate the execution of the program. Any other typed character (or no typed character) before the carriage return is activated causes a return to statement 15, requesting a new value for N.

Line 320 – All programs must have an END statement. Subprograms may follow END, but the collection of statements of the main program is terminated at this point.

Figure VI-18 shows this program as it appeared after being typed into the computer via a time-share terminal using the FORTRAN editor mode; below it is a sample execution of the program. Actually, when this program was first entered, mistakes had been made in programming. These errors were corrected by typing appropriate comments to the editor; a final version of the corrected program was typed out using a LIST statement. We do not discuss the editor function in detail since it is specific to each particular computer system; however, many satisfactory editors are available. Most would require only a few hours of practice to develop sufficient programming familiarity.

FORTRAN IV Supplied Functions

In addition to the basic five mathematical operations listed in Table VI-1, most FORTRAN IV compilers can be expected to supply the commonly used mathematical functions for integer, real, and complex described respectively in Tables VI-7, VI-8, and VI-9.

```
C PRØGRAM TØ CALCULATE N FACTØRIAL
C
        WRITE(16,10)
 10     FØRMAT(' PRØGRAM N FACTØRIAL ',/)
C
 15     WRITE(16,20)
 20     FØRMAT(' N='$)
        READ(16,30) N
 30     FØRMAT(I)
C
        NFACT=1
        DØ 100 I=1,N
        NFACT=NFACT*I
 100    CØNTINUE
C
        WRITE(16,40) NFACT
 40     FØRMAT(' N FACTØRIAL=',I6//)
C
        WRITE(16,50)
 50     FØRMAT(' IF YØU WISH TØ STØP,TYPE "1":   '$)
        READ(16,30) IANS
        IF(IANS.NE.1) GØ TØ 15
        END
```

```
RUN NFACT

PRØGRAM N FACTØRIAL

N=3

N FACTØRIAL=      6


IF YØU WISH TØ STØP,TYPE "1":

N=6

N FACTØRIAL=    720


IF YØU WISH TØ STØP,TYPE "1":  1

CP UNITS   0

EXIT


SYSTEM?..
.
```

Figure VI-18 Time-Share Terminal Printout of Program NFACT With Sample Execution

Function	Definition	Symbol	Function Type (Resulting Variable)
Absolute Value	$\lvert m \rvert$	IABS (m)	*integer*
Choose largest value	Max $(m_1, m_2 \ldots)$	MAX0 $(m_1, m_2 \ldots)$	*integer*
		AMAX0 $(m_1, m_2 \ldots)$	*real*
Choose smallest value	Min $(m_1, m_2 \ldots)$	MIN0 $(m_1, m_2 \ldots)$	*integer*
		AMIN0 $(m_1, m_2 \ldots)$	*real*
Float	Integer to real	FLOAT (m)	*real*
Sign Transfer	Sign of m_2 times $\lvert m_1 \rvert$	ISIGN (m_1, m_2)	*integer*
Positive Difference	m_1 − Min (m_1, m_2)	IDIM (m_1, m_2)	*integer*
Remaindering	$m_1 - \left(\frac{m_1}{m_2}\right) m_2$	MOD (m_1, m_2)	*integer*

Table VI-7 **FORTRAN IV Function with Integer Arguments [1] (All of These Functions are Instrinsic)**

The *External* functions can be listed directly as an argument in a subroutine call statement, while *Intrinsic* functions cannot. However, this distinction does not affect the programming methods to be described.

3. Dedicated Programs

When To Write Them

The need to analyze certain circuits may recur frequently for a particular field. For example, a diode engineering department might regularly wish to assess the loss and isolation bandwidth over which the various PIN diodes can be used in a SPDT switch and to evaluate how various diode packages affect this performance. The evaluation could be addressed with a general analysis program, but a dedicated program written explicitly to handle this situation can be much quicker to use and, as we see below, not at all difficult to write.

In general, there is no simple rule for when it is more practical to write a dedicated program. A judgment is required based on how much time and expense is saved with the specialized program compared with the investment of personal and machine time required to write and debug it. This evaluation, in turn, depends upon both the skill and the value placed on the programmer's time. In the author's view, writing dedicated programs is preferable when a

COMPUSERV FREE MODE
must start program with → 1 FREEIN ← 1st line no. (1, 5, 10 etc.)
→ 10C ← must have comment line, even if blanks
Similar to card mode, except 6 — actual program starts here → 20 WRITE(...
30 20 FORMAT...
≥1 blank space, min.

Function	Definition	Symbol	Function Type (Resulting Variable)
***External* Functions**			
Exponential	e^x	EXP (x) DEXP (x)	*real* *double*
Natural Logarithm	ln (x)	ALOG (x) DLOG (x)	*real* *double*
Common Logarithm	$\log_{10}$ (x)	ALOG10 (x) DLOG10 (x)	*real* *double*
Trigonometric Sine	sin (x)	SIN (x) DSIN (x)	*real* *double*
Trigonometric Cosine	cos (x)	COS (x) DCOS (x)	*real* *double*
Hyperbolic Tangent	tanh (x)	TANH (x)	*real*
Square Root	$x^{1/2}$	SQRT (x) DSQRT (x)	*real* *double*
Arctangent	arctan (x)	ATAN (x) DATAN (x)	*real* *double*
Arctangent (with quadrant specification)	arctan (y/x)	ATAN2(y, x) DATAN2 (y, x)	*real* *double*
***Intrinsic* Functions**			
Absolute Value	$\|x\|$	ABS (x) DABS (x)	*real* *double*
Truncate fractional part	Sign of x times largest integer $\leq \|x\|$	AINT (x) INT (x) IDINT (x)	*real* *integer* *double*
Choose largest value	Max(x_1, x_2. . .)	AMAX1(x_1, x_2. . .) MAX1(x_1, x_2. . .) DMAX1 (x_1, x_2. . .)	*real* *integer* *double*
Choose smallest value	Min(x_1, x_2. . .)	AMIN1(x_1, x_2. . .) MIN1(x_1, x_2. . .) DMIN1(x_1, x_2. . .)	*real* *integer* *double*
Fix	Real to integer	IFIX (x)	*integer*
Sign Transfer	Sign of x_2 times $\|x_1\|$	SIGN(x_1, x_2) DSIGN(x_1, x_2)	*real* *double*
Positive Difference	x_1-Min(x_1, x_2)	DIM(x_1, x_2)	*real*
Take most significant part of Double Precision *argument*		SNGL (x)	*real*
Express Single Precision *in* Double Precision *form*		DBLE (x)	*double*
Create complex number from two real arguments	z = x + jy	CMPLX (x, y)	*complex*

Table VI-8 **FORTRAN IV Functions With Real (or *Double Precision* Real) Arguments [1]**

Function	Definition	Symbol	Function Type (Resulting Variable)
***External* Functions**			
Exponential	e^z	CEXP (z)	*complex*
Natural Logarithm	ln (z)	CLOG (z)	*complex*
Complex Sine	sin (z)	CSIN (z)	*complex*
Complex Cosine	cos (z)	CCOS (z)	*complex*
Square Root	$z^{1/2}$	CSQRT (z)	*complex*
Complex Magnitude	$\lvert z \rvert$	CABS (z)	*real*
***Intrinsic* Functions**			
Complex Real Part	Real (z)	REAL (z)	*real*
Complex Imaginary Part	Imag (z)	AIMAG (z)	*real*
Form Complex Conjugate	z^* (if $z = x+jy$, $z^* = x-jy$)	CONJG (z)	*complex*

Table VI-9 **FORTRAN IV Functions with Complex Arguments [1]**

continuing use for them can be anticipated. They should be viewed as an investment in the engineer's analytic library. Not only are they valuable in their direct use, but their generation improves a programmer's skill, ensuring the ability to develop programs efficiently for the future needs which cannot practically be addressed with general purpose, network analysis programs.

SPDT Switch Analysis Example

A *dedicated* or *fixed topology* circuit analysis program is one for which the layout and type (inductors, capacitors, resistors, etc.) of the circuit elements are already known. Thus, only the specific values and frequencies of interest need be provided to execute the program. Consider how this arrangement would be handled for the SPDT switching circuit shown in Figure VI-2. The basic equations relating the source-to-load voltages at the two output ports are given in Equations (VI-3), (VI-4), (VI-5), and (VI-6). From these equations, the loss, VSWR, and isolation are readily determinable. The complete FORTRAN IV program is shown in Figure VI-19.

The brevity of the listing demonstrates the advantage of a dedicated program to handle simple problems. The program is described as follows:

```
C     SPDT  A PROGRAM TO CALCULATE SPDT TEE SWITCH PERFORMANCE
C
```

(IJKLMN Real) & Complex

```
      REAL LINT,LEXT,LOSS,ISOL
      COMPLEX ZDIODE,YDIODE,ZF,ZR,ZIN,VTEE,VLOADF,VLOADR
```

Input Data

```
C
      WRITE(16,10)
 10   FORMAT(' F1,F2,STEP,Z0='$)
      READ(16,50) F1,F2,STEP,Z0
C
      WRITE(16,20)
 20   FORMAT(' CJ,CP,LINT,LEXT,RF,RR='$)
      READ(16,50) CJ,CP,LINT,LEXT,RF,RR
C
 50   FORMAT(6G)
```

Starting Values

```
C
      TWOPI=6.2831853
C
      F=F1
C
```

Forward Bias

```
100   C=999999
      R=RF
      GO TO 150
```

Z_F, Z_R

```
C
120   C=CJ
      R=RR
C
150   ZDIODE=CMPLX(R,(TWOPI*LINT*F-1000/(TWOPI*F*C)))
      YDIODE=(1/ZDIODE)+CMPLX(0,(TWOPI*F*CP/1000))
      ZDIODE=(1/YDIODE)+CMPLX(0,TWOPI*LEXT*F)
C
      IF(C.EQ.999999) ZF=ZDIODE
      IF(C.EQ.999999) GO TO 120
      IF(C.EQ.CJ) ZR=ZDIODE
```

Load Voltages

```
C
      ZIN=(ZF+Z0)*(ZR+Z0)/(ZF+ZR+2*Z0)
      VTEE=2*ZIN/(Z0+ZIN)
      VLOADF=VTEE*(Z0/(ZF+Z0))
      VLOADR=VTEE*(Z0/(ZR+Z0))
C
```

Loss Isolation VSWR

```
      LOSS=-10*ALOG10(VLOADF*CONJG(VLOADF))
      ISOL=-10*ALOG10(VLOADR*CONJG(VLOADR))
      RHO=CABS((ZIN-Z0)/(ZIN+Z0))
      VSWR=(1+RHO)/(1-RHO)
C
      IF(F.EQ.F1) WRITE (16,200)
```

Freq. Cycle

```
200   FORMAT(' FREQ (GHZ)',6X,'VSWR',5X,'LOSS(DB)',4X,
     +  'ISOLATION (DB)',/,/)
      WRITE (16,210)F,VSWR,LOSS,ISOL
210   FORMAT(2X,F6.2,8X,F5.2,5X,F5.2,7X,F5.2)
C
```

Stop

```
      IF(F.EQ.F2) STOP
      F=F+STEP
      GO TO 100
      END
```

Figure VI-19 **A Sample FORTRAN IV Program to Evaluate the SPDT Switch Shown in Figure VI-2 (See Sample Program Execution in Figure VIII-43)**

Line 3 – A specific statement REAL defines real variables which would otherwise (since their names begin with IJKLMN) be treated by the compiler as integer variables.

Line 4 – Complex variables must be so specified before the first executable step of the program.

Lines 5-15 – Input data is read in this sequence. Note that a general FORMAT (statement 50) can be used. Fixed or floating point numbers of any length are recognized using this FORMAT (with the specific time-shared system being used).

Line 16 – This statement is the first executable one. The value for 2π is used in reactive calculations. Therefore, to save machine computations and to simplify programming, this quantity is calculated initially and identified as a program "variable."

Lines 20-43 – This sequence calculates the diode impedance in both forward (ZF) and reverse (ZR) bias states, for a given frequency (F) in gigahertz. Statement 100 "shorts out" the junction capacitance, CJ, by setting it equal to 99999 pF. Notice the form of the complex definition for DIODE in statement 150 and the simplicity of the subsequent complex arithmetic in the following two statements, with lines 35-38 reading exactly as Equations (VI-3), (VI-4), (VI-5), and (VI-6), respectively.

Lines 40-43 – These statements are directly recognizable for the loss, isolation, reflection, coefficient magnitude, and VSWR of the device. The FORTRAN IV-supplied complex notations used in these statements are familiar mathematical expressions to electrical engineers and are listed in Table VI-9.

4. Subroutines and Functions

Defining and Using Them

One of the most attractive features of the FORTRAN IV language is the ease with which recursive procedures can be handled. There are three methods for defining such programmer generated operations (separate from system supplied functions such as sin, log, etc.):

1) Statement Function
2) FUNCTION Subprogram.
3) SUBROUTINE Subprogram.

A *Statement Function is a function which can be defined using a single FORTRAN statement and which computes only one value.* Statements can be continued over several lines but there can be only one equal sign; thus, this operation is limited to what can be described by a single equation. *A statement function must preceed the first executable program statement.* For example

```
ALOSS (P2, P1) = 10*ALOG10(P2/P1)
```

is a statement function which might be useful in electrical engineering programs. Whenever, within a program, there are two defined power values, their ratio in decibels could be determined quickly by using this function. The variables P2 and P1 are dummy variables; any two variables or values could be used in the execution of the statement function. Thus the sequence shown below is possible

```
POWER = 1
REF   = 0.001
PDBM  = ALOSS (POWER, REF)
```

which, the reader recognizes, converts the assigned value of POWER in watts to dBm.

The single statement definition limitation of the statement function is overcome by using the FUNCTION subprogram. *FUNCTION subprograms are listed at the end of the main program and require an identification statement, a defined operation (which may occupy many statements), a RETURN (to the main program), and an END statement.* As such they require more programming labor; however, considerably increased generality is acquired. FUNCTION subprograms can have their own DIMENSION statements (should subscripted variables be used), complex variable definitions, and so forth; *all of the variables' names used within the subprogram are treated as separately defined from those used in the main program* (unless a COMMON statement is used in the main program to specify otherwise). FUNCTION subprograms can have more than one independent variable in their specifications; for example, we could define

```
FUNCTION SUM (A, B, C)
SUM = A + B + C
```

but *only one calculated value is returned to the main program by a FUNCTION subprogram.*

Figure VI-20 shows sample FUNCTION subprograms used to calculate the complex hyperbolic cosine and sine; these are useful in performing the lossy line Smith Chart Transformation to be de-

OPERATIONS PERFORMED:

$$\text{CCOSH (Z)} = \frac{e^{+Z} + e^{-Z}}{2}$$

$$\text{CSINH (Z)} = \frac{e^{+Z} - e^{-Z}}{2}$$

SUBROUTINE

```
C
C
      CØMPLEX FUNCTIØN CCØSH(Z)
      CØMPLEX Z
      CCØSH=(CEXP(Z)+CEXP(-Z))/2.0
      RETURN
      END
C
C
      CØMPLEX FUNCTIØN CSINH(Z)
      CØMPLEX Z
      CSINH=(CEXP(Z)-CEXP(-Z))/2.0
      RETURN
      END
C
C
C
```

Figure VI-20 **Sample Function Subprograms Used to Calculate Complex Hyperbolic Cosine and Sine for Real Function Subprograms ("Complex" is Omitted in the Title)**

scribed next. The five digit statement identification number seen in the left column is, of course, not part of the FORTRAN IV program. We use it later to connect the subprograms together to form a general network analysis program.

The FUNCTION subprograms are listed at the end of the main program which uses them. Thereafter, specification of the FUNCTION by name is sufficient to signal the computer to use it. A sample use of the FUNCTIONS in Figure VI-20 is seen in the Smith Chart Transformation subroutine to be described. The FUNCTION subprogram requires a *single executable statement* for its definition and *returns only a single value to the main program.* This limitation is eliminated by using the SUBROUTINE *subprogram,* for which *the number of executable steps and functional relationships is essentially unlimited.*

Smith Chart Subroutine

A subroutine to perform the Lossy Smith Chart Transformation given by Equation (VI-2) is shown in Figure VI-21. The complex TANH function has been broken up into SINH and COSH for which the functions shown in Figure VI-20 are used. Notice that subroutines *can call other* subroutines and functions. The TANH was reevaluated using SINH and COSH to avoid division by zero (when the line loss is zero and $\gamma \ell = 90°$).

Like the function subprogram, subroutines are listed at the end of the main program which uses them. Because they have an essentially unlimited number of input and/or output variables, a special

OPERATION PERFORMED

$$Y_{IN} = \frac{Y_L \text{ COSH } (\gamma \ell) + (1/Z_C) \text{ SINH } (\gamma \ell)}{\text{COSH } (\gamma \ell) + Y_L Z_C \text{ SINH } (\gamma \ell)}$$

SUBROUTINE:

```
C
C   SMITH CHART TRANSFØRMATIØN ALØNG A LØSSY LINE
          SUBRØUTINE SMITH(YL,GAMMAL,ZC,YIN)
          CØMPLEX YIN,GAMMAL,CCØSH,CSINH,YL,A,B
          IF(ZC.LT.1.0E-12) ZC=1.0E-12
          A=YL*CCØSH(GAMMAL)+(1/ZC)*CSINH(GAMMAL)
          B=CCØSH(GAMMAL)+YL*ZC*CSINH(GAMMAL)
          IF(CABS(B).LT.1.0E-12) B=CMPLX(1.0E-12,0.0)
          YIN=A/B
          RETURN
          END
C
```

Figure VI-21 **Subroutine to Perform the Lossy Line Smith Chart Transformation (Equation (VI-2))**

call is used to direct the main program to use them. The subroutine name followed by a parentheses listing the input and output variables *in the same order* as they are listed in the subroutine subprogram name is used in a call within the main program. For example, the Smith Chart Transformation can be executed within a main program by the statement

CALL SMITH (A, B, C, D)

where the dummy variables A, B, C, D (suitably defined as real, integer, complex, subscripted, etc. variables) have defined values in the main program and are to be interpreted as complex load admittance, complex propagation constant, characteristic impedance, and the calculated complex input admittance, respectively. The use of subroutine calls and their variable specification are demonstrated in the general network analysis program for which we are developing these particular subprograms.

Matrix Multiplication Subroutine

Analyses of networks using the ABCD matrix require the repetitive multiplication of 2 x 2 complex matrices. Rather than program this operation each time it is needed, we store it as the subroutine shown in Figure VI-22. This subroutine accepts two complex matrices, A and B; multiplies them together according to the rules for complex matrix multiplication; and returns to the main program the complex matrix product, C. As can be seen from the program, the use of subscripted variables permits a terse subroutine call, involving only the subscripted variable names assigned to the computer matrices. The subscript numbers are not listed explicitly. The variable identification adequately describes where the subscripted array begins in memory.

In examining this subroutine one may wonder why the operation first sets a matrix D equal to the product A x B and then equates this value to the desired matrix solution C for return to the main program. This precaution is a subtle but important one. Frequently, we want to multiply one matrix, B, by another matrix, A, *and then return the resultant matrix as B.* This overwriting of values is a common programming technique, minimizing the amount of computer storage and variable definition required. Thus, we want subroutine MAT to be able to handle a call of the form

CALL MAT (A, B, B)

OPERATION PERFORMED:

$$[C] = [A] \cdot [B]$$

$$\begin{bmatrix} C_{11} & C_{12} \\ C_{21} & C_{22} \end{bmatrix} = \begin{bmatrix} A_{11} & A_{12} \\ A_{21} & A_{22} \end{bmatrix} \cdot \begin{bmatrix} B_{11} & B_{12} \\ B_{21} & B_{22} \end{bmatrix}$$

SUBROUTINE:

```
01880          SUBRØUTINE MAT(A,B,C)
01890   C
01900   C    CØMPLEX ELEMENT, 2X2 MATRIX MULTIPLICATIØN
01910   C
01920          CØMPLEX A(2,2),B(2,2),C(2,2),D(2,2)
01930          D(1,1)=A(1,1)*B(1,1)+A(1,2)*B(2,1)
01940          D(1,2)=A(1,1)*B(1,2)+A(1,2)*B(2,2)
01950          D(2,1)=A(2,1)*B(1,1)+A(2,2)*B(2,1)
01960          D(2,2)=A(2,1)*B(1,2)+A(2,2)*B(2,2)
01970          DØ 10 I=1,2
01980          DØ 10 J=1,2
01990          C(I,J)=D(I,J)
02000   10     CØNTINUE
02010          RETURN
02020          END
02030   C
```

Figure VI-22 **Subroutine for Complex 2 x 2 Matrix Multiplication**

To perform this multiplication four steps (lines 1930-1960 in Figure VI-22) are required, for each of which the *old value of the B matrix must be preserved* if the correct answer is to be obtained for the *new* B matrix. Hence the need for an intermediate *scratch pad* matrix D.

PERF Subroutine to Evaluate Loss, VSWR and Phase From the ABCD Matrix

The Smith Chart subroutine example could have been written as a FUNCTION subprogram since only one value, YIN, is returned to the main program. Likewise, most compilers would probably permit the matrix array C to be returned from a FUNCTION subprogram written for the 2 x 2 matrix multiplication. Frequently, however, we wish to have a number of separate results calculated by the subprogram; a subroutine must then be used.

For example, any two-port network for which the overall ABCD matrix has been evaluated can have its input impedance, Z_{IN}; input VSWR; its transmission LOSS; and transmission PHASE de-

NETWORK:

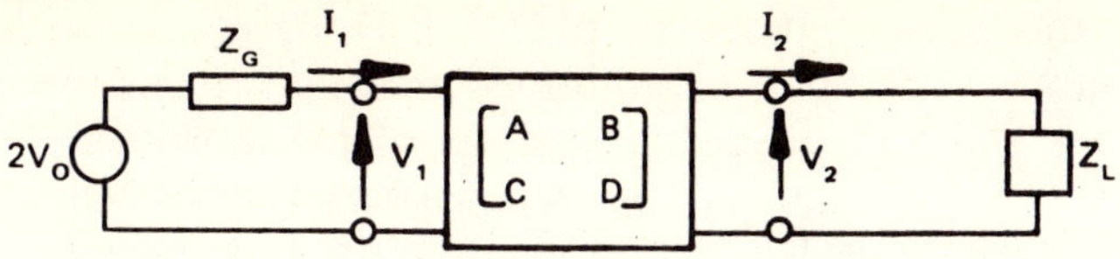

CALCULATIONS:

$$Z_{IN} = \frac{AZ_L + B}{CZ_L + D} \qquad \text{LOSS} = \left|\frac{V_2}{V_0}\right|^2 \cdot \frac{\text{REAL}(Z_G)}{\text{REAL}(Z_L)}$$

$$\text{REFL.} = \frac{Z_{IN} - Z_G}{Z_{IN} + Z_G} \qquad \text{PHASE} = \text{ARG}(V_2/V_0)$$

$$\text{VSWR} = \frac{1 + |\text{REFL}|}{1 - |\text{REFL}|} \qquad \frac{V_0}{V_2} = \frac{Z_G + Z_{IN}}{2Z_{IN}} \cdot \left(A + \frac{B}{Z_L}\right)$$

SUBROUTINE:

```
      C
            SUBRØUTINE PERF(ABCD,ZG,ZL,VSWR,ALØS,PHASE,ZIN)
            CØMPLEX ABCD(2,2),ZG,ZL,ZIN,REFL,RATIØ,A,B,C,D
            A=ABCD(1,1)
            B=ABCD(1,2)
            C=ABCD(2,1)
            D=ABCD(2,2)
      C   INPUT IMPEDANCE
            ZIN=(A*ZL+B)/(C*ZL+D)
      C   REFLECTIØN CØEF.
            REFL=(ZIN-ZG)/(ZIN+ZG)
      C   CALCULTIØN ØF INPUT VSWR
            VSWR=(1.0+CABS(REFL))/(1.0-CABS(REFL))
      C   RATIØ ØF MAXIMUM AVAILABLE TØ ACTUAL LØAD VØLTAGE (VO/V2)
            RATIØ=0.5*((ZIN+ZG)/ZIN)*(A+(B/ZL))
      C   INSERTIØN LØSS IN DB
            ALØS=10*ALØG10(CABS(RATIØ*RATIØ)*REAL(ZL)/REAL(ZG))
      C   TRANSMISSIØN PHASE (-ARG(VO/V2))
            X=REAL(RATIØ)
      C   PRECAUTIØN TØ AVØID DIVISIØN BY ZERØ
            IF(ABS(X).LT.(1.0E-8)) X=1.0E-8
            Y=AIMAG(RATIØ)
      C   CALCULATIØN ØF TRANSMISSIØN PHASE (RADIANS)
            PHASE=-ATAN(Y/X)
      C   CØNVERSIØN ØF PHASE FRØM RADIANS TØ DEGREES
            PHASE=PHASE*57.2957804
            RETURN
            END
      C
```

Figure VI-23 **Subroutine to Calculate Performance of a Two-Port from its ABCD Matrix**

termined using subroutine PERF, shown in Figure VI-23. This subroutine directly accepts a complex ABCD matrix array along with complex generator and load impedances; it returns the calculated input and transmission parameters. This subroutine, along with the other subprograms presented, forms a part of our network analysis program. It can, however, be used directly in a variety of special circuit evaluation programs.

5. *General Network Analysis Program*

General Network Analysis Plan

Computer programming represents an investment of effort – a kind of capitalization in paperwork – commonly termed *software.* As an engineer develops programming skill he generates increasingly complex programs the generation and debugging of which can take considerable amounts of his own, as well as the computer's, time. It is desirable, therefore, that these programs have maximum generality so that they can be used long after the particular initial jobs for which they were originated have been completed. Building-in generality requires creativity on the programmer's part, but a general purpose program need not be formidably complicated, particularly if *skillful use* of subprograms is made. We can demonstrate this fact with a general topology network analysis program, both because this large program illustrates construction using subprogram building blocks, and because the general network analysis is itself a good addition to the microwave circuit designer's program library.

Consider the general network shown in Figure VI-24. It consists of a number, N, of two-port circuit elements connected in cascade between a generator and load. The desired performance parameters, we have already seen, can be evaluated using subroutine PERF, provided that the overall ABCD matrix for the cascade of N elements is known. This matrix can be obtained by the cascade multiplication of the individual ABCD matrices for each of the N elements.

The general network programming problem is one of bookeeping. How is the cascade multiplication programmed when the kinds of N circuit elements which make up the cascade are only specified at the time of the execution of the program? To meet this requirement we first need a subroutine which can accept coded names for circuit elements; we then assign the proper ABCD matrix to each of them.

SCHEMATIC OF GENERAL CIRCUIT:

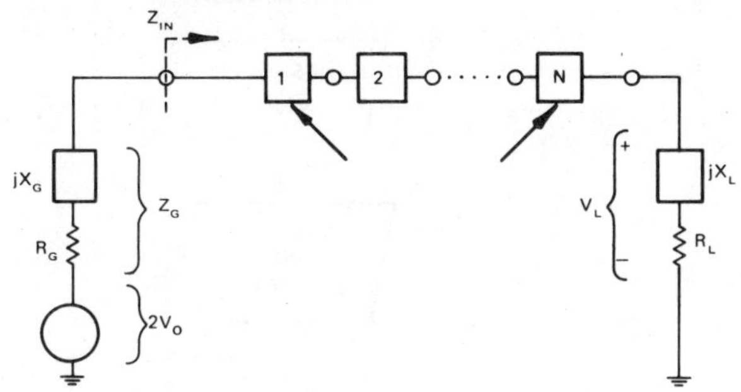

CALCULATED PERFORMANCE PARAMETERS:

$$\text{LOSS} = 10 \log_{10} \left[\left| \frac{V_O}{V_L} \right|^2 \bullet \frac{\text{Re}(Z_L)}{\text{Re}(Z_G)} \right]$$

$$\text{PHASE} = \text{ARGUMENT}\,(V_O/V_L)$$

$$\rho = \left| \frac{Z_{IN} - \text{Re}(Z_G)}{Z_{IN} + \text{Re}(Z_G)} \right|$$

$$\text{VSWR} = \frac{(1 + \rho)}{(1 - \rho)}$$

Figure VI-24 **A General Two-Port Circuit Cascade and its Performance Parameters**

The basic *flow chart* of the network analysis program is shown in Figure VI-25; each of the major steps can be isolated as a subroutine. Not only does this approach permit the complete program to be written in easier-to-visualize steps, it also creates subroutines which are likely to be useable separately for future programming needs.

READ (English Reading Subroutine)

The first subroutine required is one which translates the element list into an array of data intelligible to the computer. We use a name code *mnemonic* which is easy to remember for element types. For this demonstration the subprogram can recognize four different element types; however, from the principles presented, the element library can be made as extensive as desired.

1) A length of transmission line.
2) A series LCR circuit.

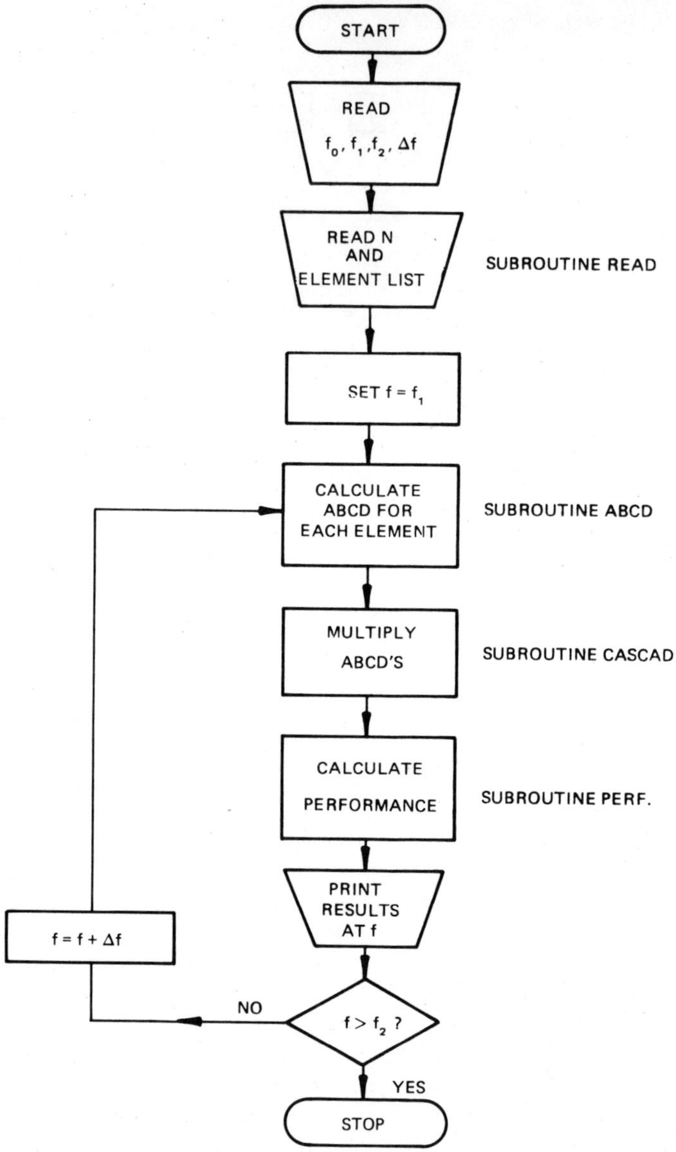

Figure VI-25 Flow Chart for General Network Analysis Program

3) A parallel LCR circuit.
4) A shunt stub terminated in a capacitance, C.

These types are called LIN, SLC, PLC, and STB, respectively. Figure VI-26 shows the schematic circuits and parameter specifications for these two-port elements which are to be recognized by the network analysis program. Although limited in extent, this brief element library suffices for many practical analyses problems, since judicious choice of the parameter value assignments permits removal of unwanted L, C, and R parts of the elements. For each element it is necessary to input a string of values to the program which serves to identify the element kind and assign values to its descriptive parameters. For example

LIN 50 90 0.1

ELEMENT SCHEMATIC	CODE NAME	PARAMETERS (1)	(2)	(3)	(4)
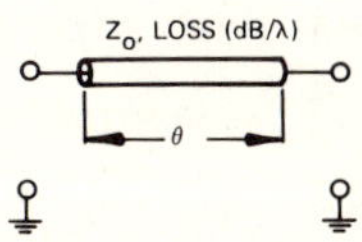	LIN	Z_o (Ω)	θ (° @ f_o)	LOSS (dB/λ @ f_o)	–
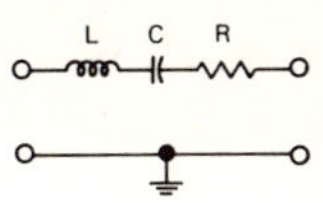	SLC	L (nH)	C (pF)	R (Ω)	–
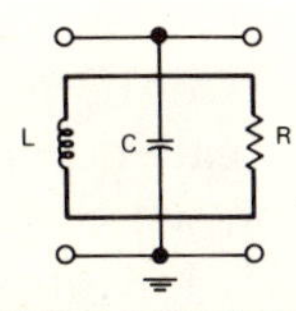	PLC	L (nH)	C (pF)	R (Ω)	–
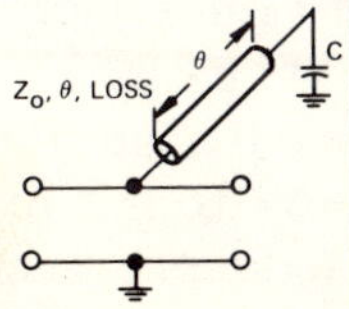	STB	Z_o (Ω)	θ (° @ f_o)	LOSS (dB/λ @ f_o)	C (pF)

Figure VI-26 **Element Kinds and Descriptive Codes for General Network Analysis Program**

```
C
C
          SUBRØUTINE READ(N,IST,ST)
          DIMENSIØN IST(20),ST(20,4),NEM(4)
          DATA NEM/'LIN','SLC','PLC','STB'/
100       FØRMAT(' HØW MANY ELEMENTS? '$)
110       FØRMAT(1G)
120       FØRMAT(' ELEM ',I3,' TYPE & PARAMETERS = '$)
130       FØRMAT(A3,5G)
          WRITE(16,100)
          READ (16,110)N
          DØ 200 I=1,N
140       WRITE(16,120)I
          READ (16,130)IST(I),(ST(I,J),J=1,4)
          DØ 150 K=1,4
150       IF(NEM(K).EQ.IST(I)) GØ TØ 200
          WRITE(16,160)
160       FØRMAT(' NØ SUCH ELEMENT, TRY AGAIN.')
          GØ TØ 140
200       CØNTINUE
          RETURN
          END
C
C
```

Figure VI-27 **Subroutine READ to Input a List of Two-Port Element Types (Described with String Variables) and Their Parameter Values**

is translated as a transmission line with 50 Ω characteristic impedance, 90° electrical length, and 0.1 dB loss per wavelength at the center frequency, f_0, for which the analysis is to be made. Electrical length and loss are recalculated for each analysis frequency at a later time. Figure VI-27 shows the subroutine READ which requests N, the number of elements, and their parameter descriptions.

This subroutine contains a *string variable* DATA list containing the mnemonic names of the element types. After being read, this list is stored in a string variable integer array, called IST, which lists up to 20 element types in the sequence in which they appear in the array. A second real variable array ST(I, J) stores the element number as I and the four parameter values which apply to it as J1, J2, J3, and J4. Because an error could easily be made in the spelling of the name (even hitting the space bar on the teletype prior to the name would cause the computer to misinterpret the name), the READ subroutine contains a sequence to compare the name entered with the DATA list, print the error message in FORMAT 160, and redemand the element description if an unrecognized name is entered.

IST(I) = name of element no. I
ST(I,J) = parameter values of element no I, for J=1 to 4.
NEM(K) = (Element type)$_k$; k=1 to 4

The string variables used for names may consist of arbitrary alphanumeric combinations (including spaces) of up to six characters. The input/output is performed using the A FORMAT shown in the 130 FORMAT statement. Data is entered using parentheses, commas, and "single quotations," as demonstrated by the DATA Statement. Notice that the call statement for this subprogram

SUBROUTINE READ (N, IST, ST)

contains only variable information for *return* to the main program, since this subroutine generates its inputs directly from its DATA and READ statements.

ABCD (Subroutine Library of Element ABCDs)

Having converted the engineering descriptions of the elements into subscripted array data interpretable by the computer program, the next step is to assign proper ABCD matrix descriptions to the elements. This assigment is performed using the subroutine ABCD shown in Figure VI-28. The call for ABCD specifies N, IST, and ST as inputs. It also requests FO, the frequency (in gigahertz) at which the element parameters were specified, and F, the actual frequency for which the transmission matrix, ABCDS, for each element of the string is to be calculated.

This subroutine identifies the type of element using the DO loop in lines 730-760 and assigns the proper sequence, beginning with respective statement number 100, 200, 300, and 400 for the LIN, SLC, PLC, and STB elements. An overall DO loop beginning at line 710 cycles through the identification and matrix specification sequence for each of the N elements. The resulting three dimensional array, ABCDS, contains as its first subscript the number of the position of the element in the cascade, reading from I=1 at the generator to I=N at the load end of the cascade. Since all of the elements listed are symmetric with respect to the input/output ports it follows that A=D in their transmission matrices; this equality is performed for all elements at statement 950. The specific calculations of ABCD parameters for each element type follow the requirements in the ABCD library given in Figure VI-4.

The matrix evaluation steps for SLC and PLC elements contain special provisions for those cases when the input of a zero value (or no value) for a parameter would render a trivial result for the analysis. With the SLC element this situation occurs when C=0, totally disconnecting the input and output ports of the cascade.

```
C
        SUBRØUTINE ABCD(N,IST,ST,F,FO,ABCDS)
        CØMPLEX ABCDS(20,2,2),Y,Z,GAMAL,CCØSH,CSINH
        DIMENSIØN IST(20),ST(20,4),KIND(20),NEM(4)
        DATA NEM/'LIN','SLC','PLC','STB'/
        TPI=6.2831853
        DØ 960 I=1,N
C  DETERMINE ELEMENT KIND
        DØ 50 K=1,4
        IF(IST(I).EQ.NEM(K)) KIND(I)=K
50      CØNTINUE
        GØ TØ (100,200,300,400) KIND(I)
C "LIN" ELEMENT(K=1), (ZC-ØHMS,DEG-AT FO,LØSS-DB/WL,--)
100     BETAL=(F/FO)*ST(I,2)/57.29570
        ALPHAL=(ST(I,3)/8.6858896)*(BETAL/TPI)*SQRT(FO/F)
        GAMAL=CMPLX(ALPHAL,BETAL)
        ABCDS(I,1,1)=CCØSH(GAMAL)
        ABCDS(I,1,2)=ST(I,1)*CSINH(GAMAL)
        ABCDS(I,2,1)=(1.0/ST(I,1))*CSINH(GAMAL)
        GØ TØ 950
C  "SLC" SERIES LCR ELEMENT (K=2), (L-NH,C-PF,R-ØHMS,--)
C  SHØRT ØUT A ZERØ CAPACITY SERIES C
200     IF(ST(I,2).LT.(1.0E-12)) ST(I,2)=1.0E+12
        Z=CMPLX(ST(I,3),(TPI*F*ST(I,1)-159.15494/(F*ST(I,2))))
        ABCDS(I,1,1)=CMPLX(1.0,0.0)
        ABCDS(I,1,2)=Z
        ABCDS(I,2,1)=CMPLX(0.0,0.0)
        GØ TØ 950
C  "PLC"  PARALLEL LCR ELEMENT (K=3), (L-NH,C-PF,R-ØHMS,--)
C  ØPEN CIRCUIT ZERØ VALUE SHUNT L ØR R ELEMENTS
300     IF(ST(I,1).LT.(1.0E-12)) ST(I,1)=1.0E+12
        IF(ST(I,3).LT.(1.0E-12)) ST(I,3)=1.0E+12
        Y=CMPLX((1.0/ST(I,3)),((-1.0/(TPI*F*ST(I,1)))+
     +  (F*ST(I,2)/159.15494)))
        ABCDS(I,1,1)=CMPLX(1.0,0.0)
        ABCDS(I,1,2)=CMPLX(0.0,0.0)
        ABCDS(I,2,1)=Y
        GØ TØ 950
C "STB" CAP. TERM. SHUNT STUB (I=4),(ZC,DEG,LØSS,C)
C  MAKE FINITE A ZERØ C
400     IF(ST(I,4).LT.(1.0E-12)) ST(I,4)=1.0E-12
C  STUB SUSCEPTANCE
        Y=CMPLX(0.0,TPI*F*ST(I,4))
C  STUB PRØPAGATIØN CØNST.
        BETAL=(F/FO)*ST(I,2)/57.295780
        ALPHAL=(ST(I,3)/8.6858896)*(BETAL/TPI)*SQRT(FO/F)
        GAMAL=CMPLX(ALPHAL.BETAL)
C  TRANSFØRM C ALØNG STUB
        CALL SMITH(Y,GAMAL,ST(I,1),Y)
        ABCDS(I,1,1)=CMPLX(1.0,0.0)
        ABCDS(I,1,2)=CMPLX(0.0,0.0)
        ABCDS(I,2,1)=Y
950     ABCDS(I,2,2)=ABCDS(I,1,1)
960     CØNTINUE
        RETURN
        END
```

Figure VI-28 **Subroutine ABCD to Evaluate the ABCD Matrices for a Cascade of N Elements Described According to Figure VI-26**

Statement 200 replaces a C value of less than 10^{-12} pF with a 10^{+12} pF capacitance, interpreting the lack of a specified C value as indicative that the series L-C-R circuit consists of, at most, only an L-R series combination. Since no difficulty arises with zero values for L or R, this provision is not required for them. Similarly, with PLC elements, zero valued parallel L and R are converted to large magnitudes, effectively open circuiting and, hence, removing them from the cascade.

Not only can some zero valued circuit parameters produce trivial results; in some cases they cause *computer overflows* (i.e., division by zero) when used as divisors in arithmetic operations. The programmer must provide for circumventing such occurrences; an example of such a precaution is contained in statement 400. Here, although a zero C at the end of the stub is physically meaningful, the eventual division by zero which could occur in the SMITH subroutine to translate the admittance of C along the stub's length would give a division *overflow*. Hence, the C value is assigned a minimum magnitude of 10^{-12} (we could use even smaller magnitudes, but this value is both sufficiently small for our purposes and large enough that any FORTRAN IV compiler will have no difficulty dividing with it).

```
C
C
C
              SUBRØUTINE CASCAD(N,ABCDS,ABCDT)
              CØMPLEX ABCDS(20,2,2),ABCDT(2,2),NEXT(2,2)
              ABCDT(1,1)=ABCDS(N,1,1)
              ABCDT(1,2)=ABCDS(N,1,2)
              ABCDT(2,1)=ABCDS(N,2,1)
              ABCDT(2,2)=ABCDS(N,2,2)
              IF(N.EQ.1) GØ TØ 100
              DØ 50 I=1,(N-1)
              NEXT(1,1)=ABCDS((N-I),1,1)
              NEXT(1,2)=ABCDS((N-I),1,2)
              NEXT(2,1)=ABCDS((N-I),2,1)
              NEXT(2,2)=ABCDS((N-I),2,2)
              CALL MAT(NEXT,ABCDT,ABCDT)
 50           CØNTINUE
 100          RETURN
              END
C
C
```

Figure VI-29 **Subroutine CASCAD to Form the Resultant ABCD Matrix From a Cascade of N Matrices**

AS( ) for ABCDS( )
AT( ) " ABCDT( )

After the determination of the string of cascaded element matrices, ABCDS, they must be multiplied to form a resultant matrix, ABCDT, representing the terminal characteristics of the complete N element cascade. This procedure can be done using subroutine CASCAD shown in Figure VI-29. The procedure consists of equating ABCDT to the last (Nth) element of the cascade. If N=1, the case when the circuit under analysis has but one element, no multiplication is necessary and line 1300 ensures the immediate return of ABCDT to the main program. For N greater than one, the next to last ((N−1)th) elements matrix is set equal to a scratch pad matrix, NEXT, which in turn is used to multiply ABCDT. The result of the product NEXT·ABCDT is returned by the previously developed subroutine MAT (Figure VI-22) and is used to overwrite ABCDT. This procedure is repeated through N−1 multiplications by the DO loop beginning on line 1310, after which the proper value for ABCDT is achieved; this value is returned to the program calling CASCAD.

The Complete Program

At this point we have prepared subprograms for all of the major steps of the general two-port network analysis program described by the flow chart in Figure VI-25. A main program to link them together is simple and short, as can be seen by the sample program, WHITE, in Figure VI-30. Besides linking the subprograms together, the main program inputs frequency data and the generator and load impedances, then outputs the calculated circuit performance.

To attach the subprograms, it is necessary merely to list them at the end of the main program. The sequence in which they are appended need not correspond to that in which they are called. The complete listing of program WHITE with subprograms comprises line numbers 100-2150, as summarized below in Table VI-10.

Program/Subprogram	Line Numbers	Figure Reference
WHITE	100-400	VI-30
READ	410-640	VI-27
ABCD	650-1200	VI-28
CASCAD	1210-1410	VI-29
PERF	1420-1700	VI-23
CCOSH	1710-1780	VI-20
CSINH	1790-1870	VI-20
MAT	1880-2030	VI-22
SMITH	2040-2150	VI-21

Table VI-10 **Program WHITE Summary**

```
10     FØRMAT(' PRØGRAM WHITE - CIRCUIT ANALYSIS (REV.A)',/)
20     FØRMAT(' ZL (RL,XL--ØHMS) = '$)
30     FØRMAT(' ZG (RG,XG--ØHMS) = '$)
40     FØRMAT(' F0,F1,F2,DELTA (IN GHZ.) = '$)
50     FØRMAT(100G)
       CØMPLEX ZL,ZG,ABCDS(20,2,2),ABCDT(2,2),ZIN
       DIMENSIØN IST(20),ST(20,4)
       WRITE(16,10)
       WRITE(16,20)
       READ (16,50)ZL
       WRITE(16,30)
       READ (16,50)ZG
       WRITE(16,40)
       READ (16,50)F0,F1,F2,DELTA
       CALL READ(N,IST,ST)
       WRITE(16,200)
       F=F1
100    CALL ABCD(N,IST,ST,F,F0,ABCDS)
70     CØNTINUE
       CALL CASCAD(N,ABCDS,ABCDT)
       CALL PERF(ABCDT,ZG,ZL,VSWR,ALØS,PHASE,ZIN)
       WRITE(16,210)F,VSWR,ALØS,PHASE,ZIN
200    FØRMAT(' FREQ-GHZ      VSWR     LØSS-DB',3X,
     + 'PHASE-DEG     ZIN-(R),    (JX)',/)
210    FØRMAT(F7.3,F8.2,F9.1,F10.1,F11.1,F10.1)
       F=F+DELTA
       IF(F.GT.(F2+DELTA/2)) GØ TØ 300
       GØ TØ 100
300    END
C
C
```

Figure VI-30 **WHITE, a General Two-Port Cascade Network Analysis Program**

As a demonstration of the execution and utility of such a general program, we use it to verify the performance similarity between a quarter wavelength line section and the lumped π (CLC) equivalent which we derived earlier in Figure VI-5.

The first program execution shown in Figure VI-31 specifies a load of 12.5 Ω and a generator of 50 Ω. Without a suitable matching circuit this combination would produce a VSWR of 4, reflecting 36% of the available generator power for an insertion loss of 1.4 dB. By using a "quarter wave transformer" of 25 Ω characteristic impedance, a perfect match (VSWR = 1) is obtained at the center frequency (1 GHz), and the VSWR remains below 2 over the 0.7-1.3 GHz bandwidth.

The second program execution shown in Figure VI-31 shows the same mismatched generator-load combination with a shunt C-series L-shunt C, π equivalent of the quarter wave transformer. Each element value has 25 Ω reactance (C = 6.4 pF and L = 4 nH),

RUN WHITE

PRØGRAM WHITE - CIRCUIT ANALYSIS (REV.A)

ZL (RL,XL--ØHMS) = 12.5

ZG (RG,XG--ØHMS) = 50

FO,F1,F2,DELTA (IN GHZ.) = 1 .5 1.5 .1

HØW MANY ELEMENTS? 1

ELEM 1 TYPE & PARAMETERS = LIN 25 90

FREQ-GHZ	VSWR	LØSS-DB	PHASE-DEG	ZIN-(R),	(JX)
0.500	2.76	1.1	-38.7	20.0	15.0
0.600	2.35	0.8	-47.8	24.6	17.5
0.700	1.95	0.5	-57.5	30.9	18.7
0.800	1.58	0.2	-67.9	38.9	17.1
0.900	1.26	0.1	-78.8	46.6	10.8
1.000	1.00	0.0	90.0	50.0	-0.0
1.100	1.26	0.1	78.8	46.6	-10.8
1.200	1.58	0.2	67.9	38.9	-17.1
1.300	1.95	0.5	57.5	30.9	-18.7
1.400	2.35	0.8	47.8	24.6	-17.5
1.500	2.76	1.1	38.7	20.0	-15.0

RUN WHITE

PRØGRAM WHITE - CIRCUIT ANALYSIS (REV.A)

ZL (RL,XL--ØHMS) = 12.5

ZG (RG,XG--ØHMS) = 50

FO,F1,F2,DELTA (IN GHZ.) = 1 .5 1.5 .1

HØW MANY ELEMENTS? 3

ELEM 1 TYPE & PARAMETERS = PLC 0 6.3661976

ELEM 2 TYPE & PARAMETERS = SLC 3.9788737

ELEM 3 TYPE & PARAMETERS = PLC 0 6.3661976

FREQ-GHZ	VSWR	LØSS-DB	PHASE-DEG	ZIN-(R),	(JX)
0.500	3.08	1.3	-36.3	16.6	7.0
0.600	2.70	1.0	-44.7	19.2	8.7
0.700	2.27	0.7	-54.0	23.2	10.3
0.800	1.83	0.4	-64.5	29.4	11.2
0.900	1.39	0.1	-76.5	38.7	9.3
1.000	1.00	0.0	90.0	50.0	-0.0
1.100	1.47	0.2	75.1	53.7	-19.9
1.200	2.27	0.7	59.6	40.8	-37.0
1.300	3.60	1.7	44.6	24.2	-40.0
1.400	5.70	2.9	31.2	13.6	-36.1
1.500	8.92	4.4	19.8	7.8	-31.2

Figure VI-31 Sample Executions of WHITE to Show Equivalence of Quarter Wave Transformer and its π Equivalent with Frequency (See Figure VI-5 for Derivation of Equivalence)

as specified by the equivalence derivation in Figure VI-5. The program execution verifies that at 1.0 GHz the lumped element π circuit gives exactly the same performance as the distributed quarter wave transformer. It can also be seen that the lumped circuit matching deteriorates somewhat more rapidly with frequency excursions from f_0.

The two executions took only about a minute to print out yet the same data calculated by hand to this accuracy could take hours of laborious calculation. This program shows only a sample of what can be accomplished with general analysis programs. More comprehensive treatments can be found in the literature [9].

References

[1] McCracken, Daniel D.: *A Guide to FORTRAN IV Programming,* John Wiley and Sons, New York, 1965 (August 1968 printing).

[2] Altman, Jerome L.: *Microwave Circuits,* D. Van Nostrand Co. Inc., New York, 1964.

[3] Collin, Robert E.: *Foundations for Microwave Engineering,* McGraw-Hill, Inc., New York, 1966.

[4] Rubin, Stanley: "Analyzing Four-Terminal Networks Using Matrix Algebra – Part I," *The Electronic Engineer's Design Magazine,* December 1966.

[5] Beatty, R.W.; and Kerns, D.M.: "Relationships between Different Kinds of Network Parameters, Not Assuming Reciprocity or Equality of the Waveguide or Transmission Line Characteristic Impedances," *Proceedings of the IEEE, Vol. 52,* p. 84, January 1964.

[6] Matthei, Young, and Jones: *Microwave Filters, Impedance Matching Networks, and Coupling Structures* (Chapter 13 – "Hybrid Couplers"), McGraw-Hill, Inc., New York, 1964.

[7] Montgomery, C.G.; Dicke, R.H.; and Purcell, E.M.: *Principles of Microwave Circuits* (Vol. 8, MIT Radiation Laboratory Series), McGraw-Hill, Inc., New York, pp. 146-156, 1948.

[8] White, J.F.: "Semiconductor Microwave Phase Control," *NEREM Record,* pp. 106-107, 1963.

[9] Green, Peter E.: "General Purpose Programs for the Frequency Domain Analysis of Microwave Circuits," *IEEE Tranactions on Microwave Theory and Techniques, Vol. MTT-17, No. 8,* pp. 506-526, August 1969.

The author especially thanks John Kostriza for the matrix algebra, Seymour Cohn for the backward wave coupler analysis, Nancy Gillis and Frank Arrington for the FORTRAN IV programming techniques, and Harold Stinehelfer for reviewing the chapter.

Questions

1. If a perfect short circuit ($Z_L = 0$) is placed at the end of a quarter wavelength of Z_C characteristic impedance transmission line, show that at the other end of the line the input impedance, Z_{IN}, is purely resistive and has a value given approximately by

   $$Z_{IN} \approx \frac{34.7\ Z_C}{\alpha(\mathrm{dB}/\lambda)}$$

   where $\alpha(\mathrm{dB}/\lambda)$ is the attenuation of the line in decibels per wavelength.

   Hint: Use Equation (VI-2), expand tanh ($\gamma\ell$) in exponential form, and approximate $e^x \approx 1 + x$ for $x \ll 1$.

2. Using the lossless equation for the transformation of impedances along a transmission line (Equation (VI-1)), show that a quarter wave transformer of characteristic impedance Z_T matches a resistive load of impedance RZ_0 to a generator of impedance Z_0 when

   $$Z_T = \sqrt{R}\, Z_0$$

3. Derive an approximate expression to calculate how much insertion loss is introduced in a matched Z_0 system if the shorted quarter wavelength stub in Question 1 is connected across the main line?

   Use this expression to calculate the loss when the stub has Z_0 impedance and 1 dB loss per wavelength.

4. Derive the ABCD matrix given in Figure VI-4 for a length of lossless transmission line using the evaluative conditions given in Figure VI-3. Show that this matrix gives the same input impedance expression for the line when it is terminated in a load Z_L as does the Smith Chart Transformation given in Equation (VI-1).

5. Make a flow chart (example in Figure VI-25) and write a FORTRAN IV program to evaluate the input impedance of a lossy line terminated in a load impedance Z_L. The subroutines in Figures VI-20 and VI-21 may be used.
6. Load the network analysis program in Figure VI-30 into a computer. Debug it by observing its "signature" as it reproduces the sample evaluations shown in Figure VI-31. Does this action verify that the program you've loaded is identical to that given in the text?
7. Use the program in Question 6 to design by trial and error a bandpass filter consisting of two 90° shorted stubs separated by 90° on a transmission line between matched generator and load. Use as initial specifications 1 dB maximum loss over a ±5% bandwidth about $f_0 = 1$ GHz and 10 dB minimum isolation for frequencies separated more than ±20% from f_0. Can "better" performance be achieved? What response is obtained at $2f_0$? At $3f_0$?
8. Modify the network analysis program in Figure VI-30 so that a PIN diode equivalent circuit of the type shown for a diode in package style 30 (Table III-2) can be added to the element analysis list in Figure VI-26. Add both series and shunt diode options to the element list. What changes must be made to the program and subroutines in order for the larger number of input parameters needed to define the diode to be accommodated?

Chapter VII

Limiters and Duplexers

A. Introduction to Practical Circuit Designs

This chapter and the two which follow discuss the practical design of microwave control circuits using the PIN diode. From an organizational point of view, though, it would seem proper to have a chapter preceding these three device chapters in which the general methods of circuit construction would be addressed. In converting an equivalent circuit with computer generated theoretical performance into a practical device, a host of questions must be answered. The kind of circuit medium – waveguide, coax, stripline, microstrip, and so forth – must be determined. The method of introducing bias to the diodes, while separating it from the RF circuitry, may be with lumped conductors and capacitors or with distributed transmission line chokes. The diodes themselves may be used in chip form and bonded directly to the circuit ground plane, or packaged diodes may be employed. In short, there are a great many practical decisions to be made once the basic device equivalent circuit has been selected.

The question is – how can these constructional questions be handled in a logical systematic fashion? From the author's point of view, they can't. The weighting given by various practical constructional alternatives is difficult to define *a priori* in textbook fashion. For example, it is all very well to decide that waveguide circuitry offers the lowest practical insertion loss medium, but how is this benefit to be weighted against the fact that the designer may have readily accessible to him facilities for making stripline circuits. In this case, the use of waveguide hardware, even if, in principle, it is better suited to the task at hand, would require the acquisition of waveguide component manufacturing equipment and techniques. In practice, designers use circuit media which are familiar and readily accessible.

Therefore rather than attempt in this book to summarize the characteristics and tradeoffs of various circuit media and circuit realization tricks, we try to demonstrate a wide variety of device design approaches, leaving the reader to adapt them to the particular needs as is seen fit. For example, post mounting of diodes is treated in conjunction with limiter circuits but the same mounting configuration can, of course, be applied to switches and phase shifters. A variety of bias circuits, including lumped capacitors, conductors, and coaxial and printed circuit chokes, is presented in the course of describing limiters, switches, and phase shifters throughout the next three chapters. Thus, even if the reader is interested only in limiters, examination of the circuit realization techniques used for limiters, switches, and phase shifters is recommended.

B. How Limiters Function

We saw in Chapter II that PIN diodes used at microwave frequencies cannot be expected to rectify in the low frequency sense. However, a whole class of semiconductor control components — self-actuated limiters — do perform an important radar function because their impedance changes under the influence of a high level RF signal. *A limiter is a two-port network which passes low power but attenuates high power signals.* There are three phenomena through which this non-linear *limiting* performance can be achieved. In an actual device, more than one principle may be applicable:

1) *Some diodes are capable of rectification* at microwave frequencies, such as point contact and Schottky Barrier diodes (which, of course, cannot be classed as PIN diodes).
2) *Varactor diodes have a capacitance which is voltage variable* and which responds rapidly enough to change characteristics at microwave frequencies. It is this ability that permits varactor diodes to multiply microwave frequencies. Varactors are chemically identical to PIN diodes, as was seen in Chapters I and II, the PIN representing a special doping profile extension of the graded PN junction, or varactor, diode.
3) *PIN diodes undergo RF conductivity modulation* within the I region when subjected to a large magnitude microwave current.

We examine briefly the first two methods of microwave semiconductor limiting so as to put the conductivity modulation limiting obtainable with the PIN diode into perspective. Consider first the I-V characteristic and rectification performance of a diode capable

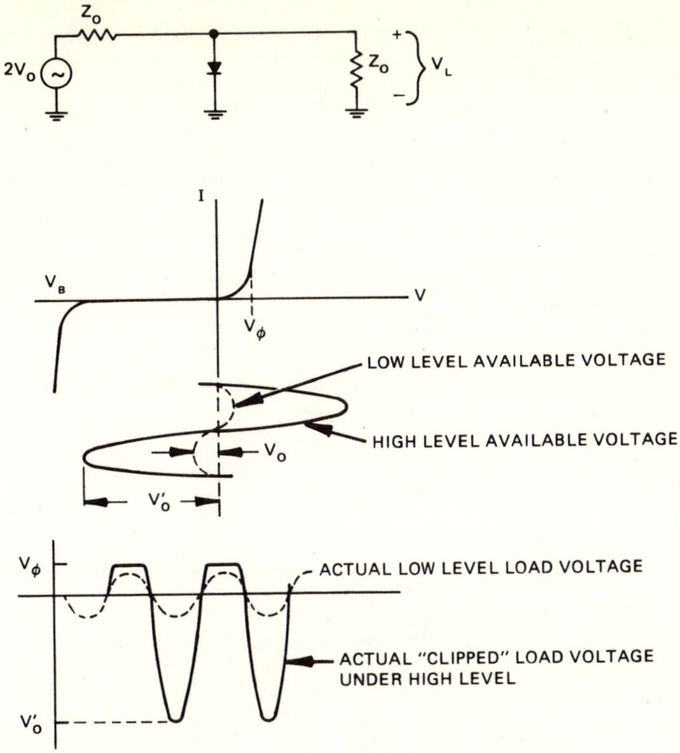

Figure VII-1 **Ideal Half Wave Clipper**

of responding to an RF waveform at microwave frequencies, shown in Figure VII-1. If the RF voltage applied does not exceed the junction potential of the diode mounted in shunt with the transmission line, practically no diode conduction occurs and half the generator voltage – the *available voltage* – reaches the matched load. As the available voltage is increased, partial conduction occurs during the forward-going portion of the RF sine wave, resulting in a clipped voltage waveform appearing at the load. The device described represents a kind of power limiter in that it passes with little attenuation RF signals with voltages which do not exceed the junction potential of the diode, but it clips forward-going voltages which are greater than that value. Such a device would offer to the receiver in a radar system some protection from high power signals.

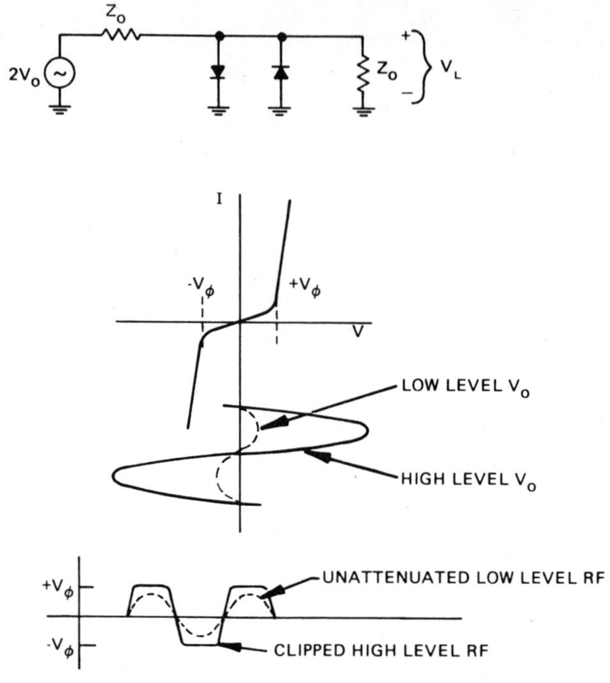

Figure VII-2 **Back-To-Back Ideal Limiter**

At best, however, this device limits only under the forward-going portion of the RF waveform, and so would produce somewhat less than 3 dB of isolation. This problem could be overcome by mounting an additional diode of the opposite polarity in parallel with the first diode, producing a symmetrically clipped waveform, as shown in Figure VII-2. With the resulting *back-to-back* limiter, the potential isolation theoretically is unlimited. The limiter clips the incident voltage, limiting to a peak RF voltage approximately equal to the junction potential of the diodes. If the clipping action is quite sharp, then the RF signal which passes to the receiver contains harmonics of the fundamental RF frequencies; these, however, can be eliminated with a low pass filter. The practical problem is that those diodes which have a sufficiently rapid turn-on time at microwave frequencies must, of necessity, have a very thin depletion layer. Furthermore, to keep their microwave capacitance small, their junction area must likewise be made small, and the resulting small volume diode cannot protect against very high power microwave signals without burning out.

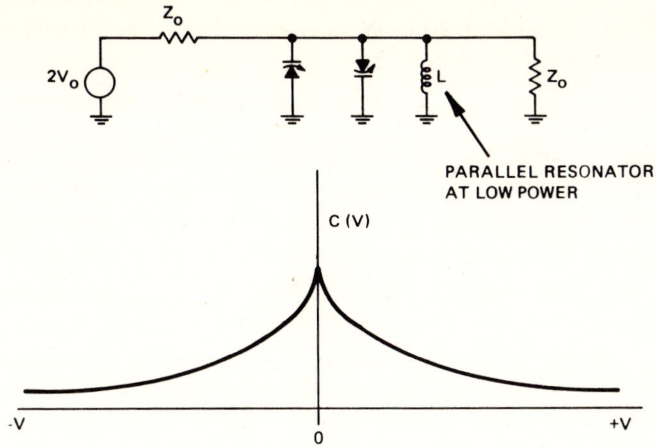

Figure VII-3 **A Varactor Limiter Which Produces C-Change to Reflect High Level Signals**

The second phenomenon, capacitance change with microwave voltage, also can be used for microwave power limiting. For example, reexamining the C(V) law characteristic shown for a tuning diode in Figure I-12, a capacitance change of about 10 to 1 with RF voltage could be used to provide high reflection to high power signals when the diode is used in a parallel resonant circuit, as shown schematically in Figure VII-3. This limiter produces a variable reflection as a function of incident power. A pair of back-to-back mounted varactors on the transmission line have a relatively large total capacitance for low power signals. This capacitance can be parallel resonated by a shunt coil. At high power levels the average capacitance decreases, detuning the parallel resonant circuit and reflecting more of the power. Additionally, since varactor diodes have a very fast response time, some rectification also occurs at high power, providing additional limiting via the clipping mode described in Figure II-2. However, the varactor diode suffers from the same limitation as the point contact and Schottky Barrier diodes in that, to achieve reasonable operating bandwidth, small capacitance diodes must be used and power handling is consequently very limited.

The third mechanism for semiconductor limiting is the RF actuation of a PIN diode. Recall that in Chapter II, we demonstrated with the charge control model that practical PIN diodes *do not rectify* in the microwave spectrum. While this statement is true, the PIN operated at zero bias with a dc short circuit across its terminals nevertheless *does provide limiting action.* How does this limiting occur and how can it be reconciled with the previous charge control model description? The answer to both these questions was presented in the classic paper by Leenov [1]. We now reexamine the charge control model under the condition that a PIN diode is zero biased but has a high level RF current signal incident upon it.

Recall that in the Chapter II derivation for I region resistance, Equation (II-34), it was assumed that holes and electrons are uniformly distributed throughout the I region. While this assumption is a good one when the diode is operated under heavy forward bias, it must be reexamined when the diode is zero biased but has a high level microwave signal applied. Before doing so, let us consider the assumption of uniform electron and hole density in the I region under forward bias. The sources for electrons and holes are the N+ and P+ regions, respectively. Under forward bias, each carrier type is injected from the appropriate boundary into the I region where its presence as a charged mobile carrier provides microwave conductivity and results in low resistance to a microwave signal. However, because of recombination, its *lifetime* as a current carrier is limited. Since there are no sources for electrons and

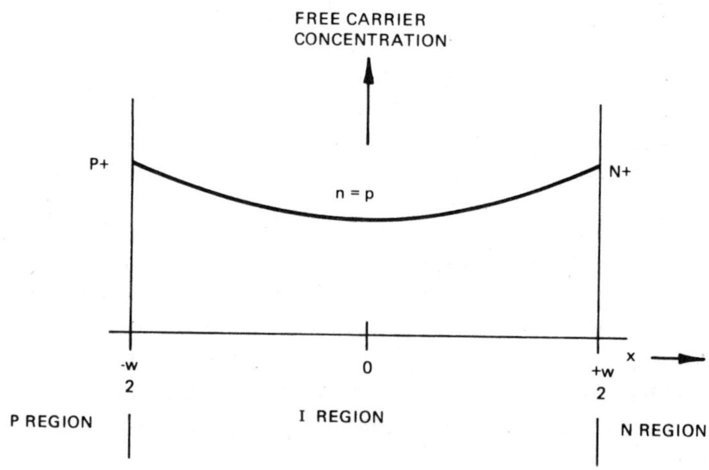

Figure VII-4 **I Region Carrier Concentration in the Forward Biased PIN Diode**

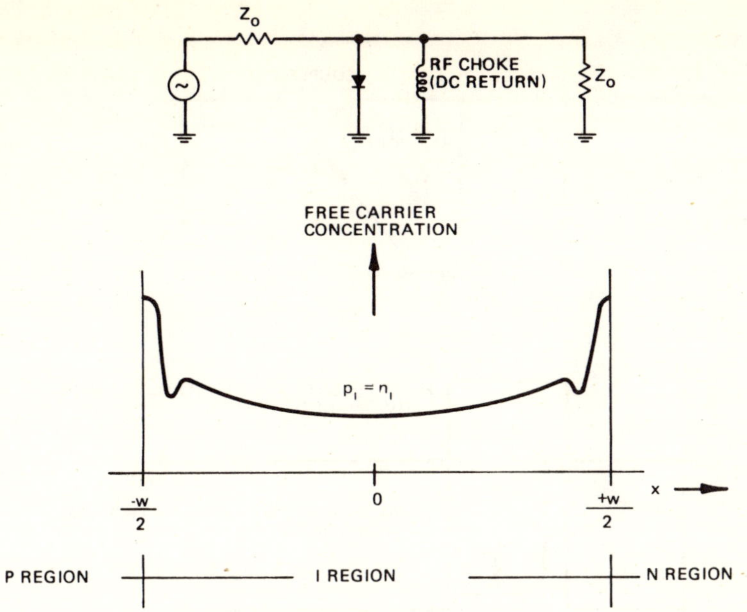

Figure VII-5 **I Region Carrier Concentration in a PIN Diode Under High Level RF Excitation but with dc Path Short Circuited**

holes within the I region (neglecting intrinsic thermal generation), it follows that the concentration of them is somewhat less in the center than it is at the boundaries of the I region. This differentiation is shown schematically by the carrier profile distribution in Figure VII-4. Of course, the distribution of carriers with distance in the I region is well enough behaved that the use of an average carrier concentration is adequate for the resistance calculated using Equation (II-34). But we note the actual profile to emphasize that the carriers move toward the center by diffusion from the I region boundaries where the concentrations are greater.

Suppose now that the dc bias is removed and replaced by a short circuit path, the need for which is described later. The diode is installed in a transmission line carrying a high level microwave signal, as shown in Figure VII-5. We already have seen in Chapter II that there is not sufficient time in the duration of the RF forward-going half cycle for holes injected from the P+ side or electrons injected from the N+ side to completely traverse the I region

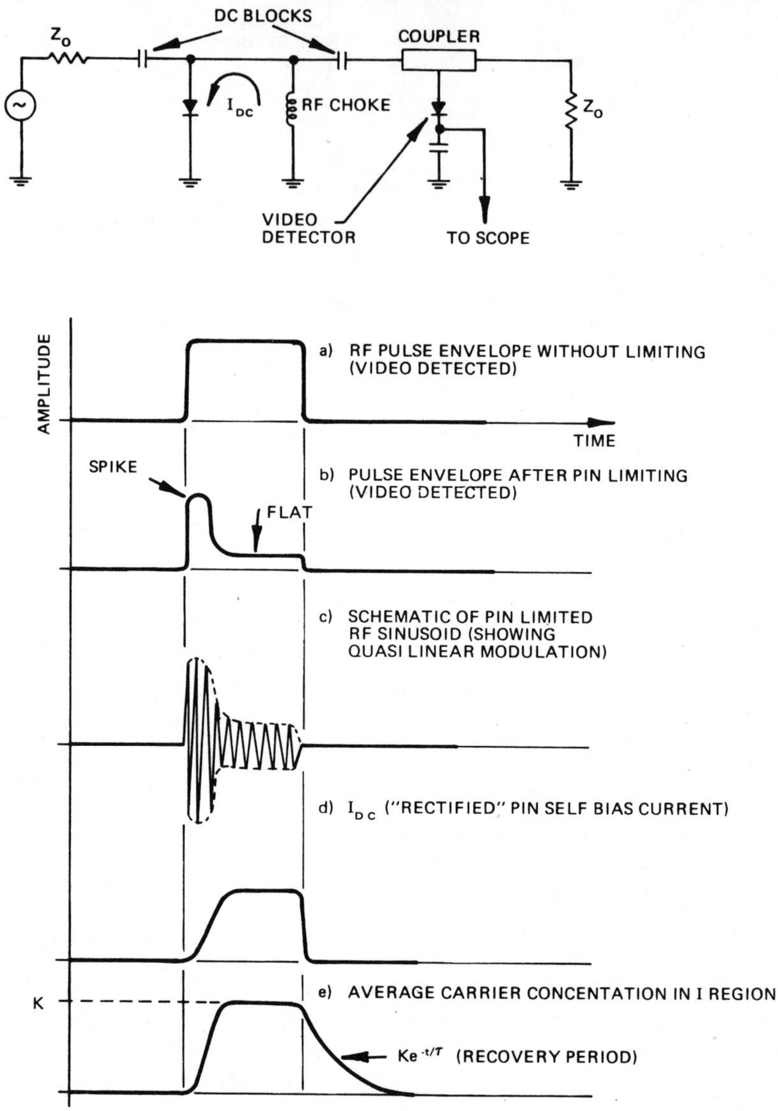

Figure VII-6 Microwave Limiting Obtained with RF Induced Conductivity Modulation of PIN Diode.

width. However, a large microwave current, once established in the diode (initially through the mechanism of displacement current across the I region) causes some partial carrier injection near the P+ and N+ boundaries during forward-going half cycles of the RF signal. Most of this charge is withdrawn when the RF voltage reverses, *but not all of it.* Some diffuses a little too far into the center of the I region to be recovered totally by the negative-going half of the RF cycle. The result is a trickling of electrons and holes into the I region. These distribute themselves, according to Leenov's analysis, as shown in the carrier distribution profile in Figure VII-5.

After a few RF cycles a steady state distribution is attained and remains static during the remainder of the RF pulse, serving to conductivity modulate the I region. During the transient period in which the charge is built up in the I region, the PIN limiter provides relatively little limiting to the high power RF signal. The RF signal passes through the PIN limiter with little attenuation during this *spike leakage* period, after which a relatively high attenuation is achieved and the power passed is called *flat leakage,* as diagrammed in Figure VII-6.

Before proceeding further, it is important to distinguish carefully this *RF conductivity modulation* limiting from the *RF rectification* limiting described for a single diode in Figure VII-1. As one observes the detected high power pulse envelope after it has passed through a single diode PIN limiter, it is tempting to say that the PIN, being slower, passes a leakage spike initially before it begins to rectify. This statement is wrong! If it were true, the most limiting that could be obtained would be less than 3 dB, since only half the RF wave is affected by rectification, as was shown in Figure VII-1. Rather, with conductivity modulation limiting, a transient spike leakage period occurs during which the average charge within the I region is built-up. Throughout this period and during the steady state, *flat leakage* period which follows it, the diode's impedance to the microwave signal is *quasi-linear.* That is to say, the diode presents substantially the *same conductivity to both forward and reverse-going halves of the RF cycle.* The limiting achieved can be, and experimentally is observed to be, much greater than 3 dB.

A dc current path must be provided in the PIN limiter, as shown in Figure VII-6. The charge, consisting of holes and electrons in equal quantities (to preserve charge neutrality), injected into the I region is continually undergoing recombination. This recombination produces a dc current in the external circuit which has the same polarity as would a forward bias applied current. In order to achieve microwave conductivity modulation, it is necessary to provide an external path for dc current continuity. The equivalent circuit of this limiter is shown with a single diode and dc return to emphasize its quasi-linear nature. That is, the same limiting is provided to both halves of the RF cycle, in distinction from the half wave rectified performance obtained with a fast carrier diode type limiter described by Figure VII-1. However, in principle, the dc continuity could be realized with an additional PIN diode of the opposite polarity replacing the RF choke and mounted in the back-to-back configuration shown in Figure VII-2. In such installations the diodes must be well matched in order to have nearly identical recombination current waveforms.

Leenov has shown that the resistance of the I region of a PIN diode activated through a microwave current is given by

$$R_I = \frac{W}{\sqrt{D_{AP}/2\pi f}} \cdot \frac{1}{(e/kT)} \cdot \frac{1}{I_{RF}} \qquad \text{(VII-1)}$$

when

$W \ll L = \sqrt{D_{AP}}$ (i.e., I region width narrow compared to carrier diffusion length)

where (in silicon)

$L = 1.7\sqrt{\tau}$ (mils) $= 0.004\sqrt{\tau}$ (centimeters)

τ = average carrier lifetime (microseconds)

W = I region width (centimeters)

D_{AP} = ambipolar diffusion constant (15.6 centimeters^2/second)

f = RF frequency (hertz)

(e/kT) = 40/volt (for T = 273 Kelvin)

I_{RF} = RF current (amperes rms) through the PIN

For silicon diodes used at room temperature Equation (VII-1) can be simplified to

$$R_I = \frac{w\sqrt{F}}{20\, I_{RF}} \qquad \text{(VII-2)}$$

where R_I is in ohms, w is the I region width (in microns), F is the operating frequency (in gigahertz), and I_{RF} is the microwave current (in amperes rms).

The action of a high level microwave signal is much less effective in producing conductivity modulation of the I layer of the diode than is a dc current. For example, at a microwave frequency of 1 GHz, with a diode having 50μ (2 mil) I region width, the RF current required to reduce R_I to a value of 1 Ω is calculated to be

$$I_{RF} = \frac{50\sqrt{1}}{(20)(1)} = 2.5 \text{ amperes (rms)} \qquad \text{(VII-3)}$$

However, we can see from Table III-1 that a typical 50 μ I region width PIN diode requires only 50 mA of dc bias to conductivity modulate its I region resistance to 1 Ω or less. Thus, although a PIN can be conductivity modulated with an RF excitation, this mechanism is orders of magnitude less efficient as a means for reducing the resistance of the diode. It does, nevertheless, have the advantage that a receiver protector using self actualization is less likely to fail in a system, because it requires no driver. Furthermore, a duplexer design for self-biased operation protects the radar from adjacent radars, for which no foreknowledge of the time of arrival of high power RF pulses is available as a control for an external driver circuit.

After the high power microwave signal, the concentration of holes and electrons in the I region does not disappear immediately, but decays exponentially with a time constant equal to the average carrier lifetime — the definition for lifetime. During this *recovery period* the limiter has a relatively high insertion loss, the radar receiver it protects is therefore less sensitive to signals. By convention the recovery time of a limiter usually is defined as the time required for the limiter's insertion loss to return to within 3 dB of its low level insertion loss following the cessation of the high power pulse. The radar is generally useable even within this recovery period because early radar echoes correspond to nearby targets for which the received signal level is relatively high.

The length of the recovery time is related to the diode's (minority carrier) lifetime and to the extent of conductivity modulation realized. Measurements made by Brown [2], shown in Figure VII-7, indicate that, for a given diode lifetime, recovery time is linearly proportional to the peak RF power applied up to some

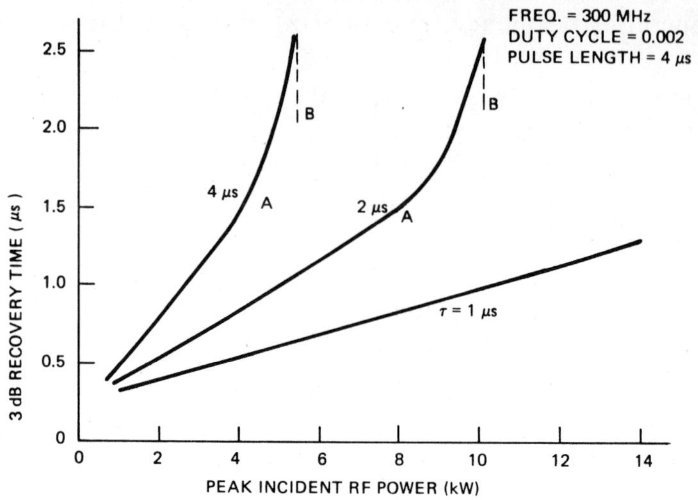

Figure VII-7 **Recovery Time for PIN Diodes with Various Lifetimes as a Function of Incident RF Peak Power (After Brown [2])**

peak power level (point A on the graph) at which a change in slope of the recovery time versus incident peak power is experienced. This change is due to thermal heating of the diode.* For further RF power increases, the recovery time increases more rapidly, with burnout of the diode being encountered when the graph approaches a vertical slope (point B in Figure VII-7).

Thus, *measurement of recovery time can be used for nondestructive monitoring of the power handling capacity of a self-actuated PIN limiter.* The recovery time measurement is performed by injecting a constant level, low power, microwave signal into the high power test set, as shown in the block diagram in Figure VII-8.

Leenov's work shows that any PIN diode can be conductivity modulated by a strong enough microwave current, but this finding does not mean that all PIN diodes make satisfactory microwave limiters if the incident RF power level is high enough. Unless the PIN does reach a sufficiently high conductivity condition at relatively low RF power, it will burn out as the incident RF power is

*Heating generates electron-hole pairs intrinsically, as described in Chapter I. Thus, recovery time proceeds, not at a rate proportional to carrier lifetime, but as the slower thermal cooling rate of the junction.

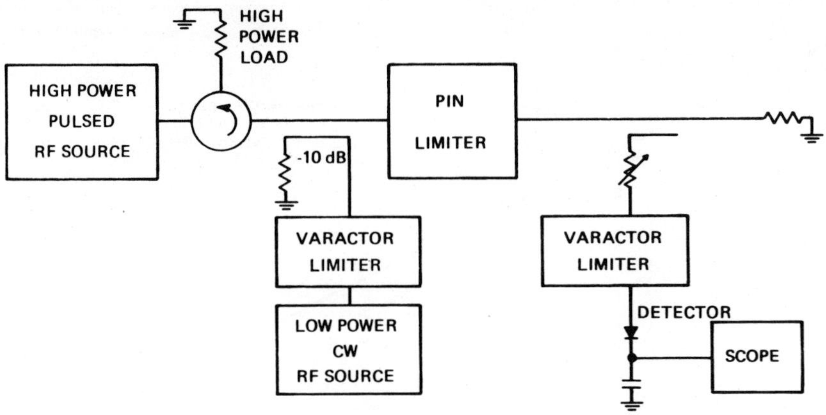

Figure VII-8 **Measurement of PIN Limiter Recovery Time**

increased. Practically speaking, there is a maximum I region width useable at each operating frequency. Equivalently, given a PIN diode, its I region width determines the maximum frequency at which it can be used for practical limiting. Because failure due to excessive absorbed power cannot be precisely reproduced, Brown [2] chose an experimental method for determining the maximum frequency of use for a selection of PIN diodes having four representative junction thicknesses 2.5, 12.7, 25, and 36 μ (0.1, 0.5, 1.0, and 1.4 mils). He mounted these diodes in shunt with a 50 Ω coaxial line, and subjected them to high power microwave pulses of one μs length with rise time of about 50 ns.

The criterion for "successful operation" was the selection of the maximum frequency at which a barely observable spike was obtained on an oscilloscope used to monitor the pulse after it passed through the diode. During the onset of this leakage spike, the RF power absorbed within the diode was high and power handling was uncertain. Figure VII-9 shows Brown's experimental data. He concluded that diodes with a junction width of approximately 1 mil (25 μ) could be used up to about 150 MHz; 0.5 mil (13 μ), up to 1 GHz; and 0.1 mil (2.5 μ), up to 10 GHz. To date, there has been no published significant improvement in the understanding* of the selection of PIN diodes for microwave limiting beyond Brown's experimental data, which describes the diode choice made for most practical limiter designs very well.

*However, Garver [3] observed that Brown's data fits a straight line (as shown in Figure VII-9) when plotted on log-log axes.

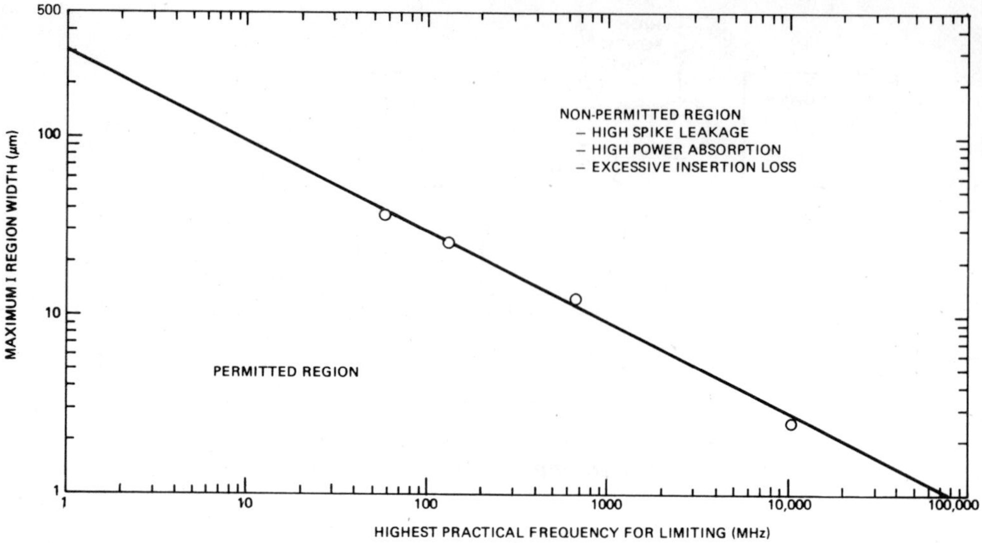

Figure VII-9 **Brown's [2] PIN Limiter Data Showing I Region Width vs. Highest Practical Frequency of Use**

C. Coaxial Duplexers

At microwave frequencies below one GHz, fairly wide I region diodes with large junction areas provide practical limiter operation. Consequently it is practical to design totally passive duplexers. *A radar duplexer is a three-port which permits use of a single antenna for transmission and reception.* It isolates the receiver from the transmitter during high power transmission by virtue of the high conductivity in PIN diodes used to shunt the transmission line, as shown in Figure VII-10. The duplexer can be made self-actuated using PIN diode limiters, or it can be switched using injected dc bias currents. In turn, the dc bias may be rectified by other diodes exposed to a relatively small portion of the high power microwave signal, or the bias may be separately applied prior to each RF high power pulse.

To gain an appreciation for the design approach followed in the development of a high power PIN diode duplexer, we review the performance requirements and design considerations for a practical high peak power, long pulse length, UHF duplexer. From the radar system requirements, the general performance specifications for this duplexer were established, as shown in Table VII-1.

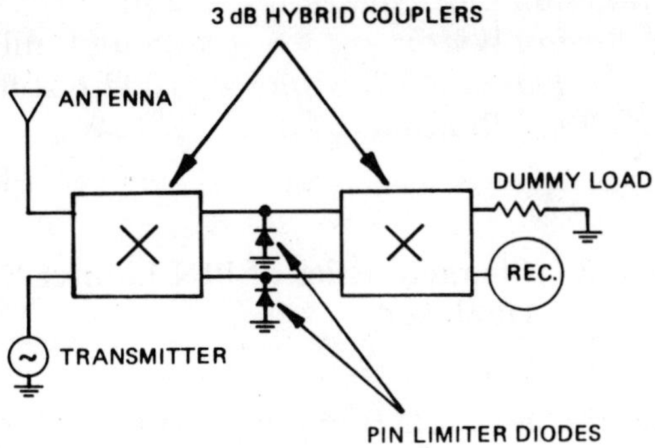

Figure VII-10 **Balanced Duplexer Circuit Using Self-Actuated PIN Limiter Diodes**

Frequency Bandwidth	200-225 MHz
Peak Power (max)	200 kW
Average Power (max)	12 kW
Pulse Length (max)	200 μs
Receiver Protection (min)	60 dB
Recovery Time (max)	10 μs
Transmit Loss (max)	0.5 dB
Receiver Loss (max)	1.0 dB

Table VII-1 **UHF Duplexer Requirements**

At the time this duplexer was designed, the most probable candidate for the diode choice was a 2 pF diode having the characteristics listed in Table VII-2.

In view of the high peak power and long pulse length requirements of the duplexer, the first question is how many diodes are required. At the low RF frequency, self-actuated limiter performance is practical. In order to determine how much power can be controlled by a single PIN diode with the characteristics listed in Table VII-2, it is necessary to specify a maximum allowable temperature rise for the diode junction. The maximum power dissipation within the diode can be estimated from this figure and related to the maximum allowable RF current, as follows.

Junction Capacitance	2 pF
I Region Width	25 μ (1 mil)
Zero Biased RF Resistance	10 Ω (estimated)
Thermal Resistance (max)	8°C/W
Package	Axial Prong Ceramic

Table VII-2 **Characteristics of PIN Limiter Diode for UHF Duplexer**

Suppose that the peak junction temperature rise resulting from the incident 200 μs RF pulse is to be held to a maximum of 60°C. The diode thermal resistance is specified in Table VII-2 to have a maximum value of 8°C/W. Since the junction capacitance is 2 pF, the heat sinking capacitance of the silicon chip can be estimated from Table III-1. Although this diode type is not specifically listed in the table, a 25 μ, 0.2 pF diode has heat sinking capacity (HC) of 2 μJ/°C; thus, a 2 pF diode, having the same I region width but ten times the junction area, has 20 μJ/°C of heat sinking capacity. The thermal time constant of the diode is then HC x θ = 160 μs. Equation (III-12) can now be applied directly to solve for the maximum power dissipation allowed within a 200 μs pulse, as shown below.

$$60^\circ\text{Celsius} = P_D\ (8^\circ\text{Celsius/watt})\ (1 - e^{-200/160}) \qquad \text{(VII-4)}$$

$$P_D = 10 \text{ watts}$$

With a self-actuated limiter the allowable power dissipation is not directly related to the forward resistance of the diode, R_F, since this resistance depends upon the RF current through the diode which in turn is related to the allowable dissipation. The interrelation of these factors can be taken into account by combining the equation for the high power dissipation in a SPST switch, Equation (V-20), with Leenov's equation for the I region resistance, R_I, of a self-actuated PIN diode as a function of microwave current, I_{RF}. These expressions, written in common notation are, respectively

$$P_D = (I_{RF})^2 \cdot R_F \qquad \text{(VII-5)}$$

$$R_F = R_C + \frac{w\sqrt{F}}{20\ I_{RF}} \tag{VII-6}$$

Equation (VII-6) is Leenov's equation with an extra term, R_C, added to account for the contact and package resistance which must be expected in a practical diode. Combining these expressions gives

$$I_{RF} = \frac{1}{2R_C}\sqrt{\frac{w^2F}{400} + 4R_CP_D} - \frac{w\sqrt{F}}{40R_C} \tag{VII-7}$$

Allowing 0.2 Ω for contact resistance and applying the values w = 25 μ and P_D = 10 W to this expression gives

$$I_{RF} = \frac{1}{2(0.2)}\sqrt{\frac{(25)^2(0.2)}{400} + 4\ (0.2)\ (10)} - \frac{25\sqrt{0.2}}{40\ (0.2)} \tag{VII-8}$$

I_{RF} = 5.8 amperes (rms)

As previously noted, the diode's forward resistance is a function of the RF current, which must not exceed 5.8 A if dissipation is not to exceed 10 W peak. Under this RF current excitation, the forward resistance of the diode (from Equation (VII-6)) is 0.3 Ω. There is now sufficient information to apply the fundamental power limit equation for duplexer operation to determine how much power can be switched per diode and, accordingly, the minimum number of diodes required to build the duplexer. Applying this expression (Equation (V-39))

$$P_{MAX}/\text{diode} = \frac{(0.2)\ (10\ \text{watts peak})}{4\ (0.3\ \text{ohm})\ (10\ \text{ohms})\ (2\pi \times 200 \times 10^6\ \text{hertz} \times 2\ \text{picofarads})^2}$$

$$P_M = 26\ \text{kilowatts peak/diode} \tag{VII-9}$$

From this result it follows that the complete duplexer to handle 200 kW of peak power theoretically could be built using a total of 8 diodes, or 4 diodes in each of two SPST shunt type limiters mounted between hybrid couplers, as shown in the balanced duplexer configuration in Figure VII-10. However, before judging the advisability of this design, let us determine at what impedance level, Z_0, the SPST switches must operate in order for the RF current not to exceed 5.8 A rms per diode during the 200 μs pulse. With the shunt configuration shown in Figure VII-10, the 4 diodes in each SPST switch carry twice the line current commensurate with a power flow of 100 kW in each switch. That is

$$I_L{}^2 Z_0 = P$$

$$\left(\frac{4 \times 5.8}{2}\right)^2 Z_0 = 100{,}000 \qquad \text{(VII-10)}$$

$$Z_0 = 743 \text{ ohms}$$

This calculation indicates that, while it is possible to control the 200 kW with only 8 diodes, the required single pole single throw limiters must be designed about a characteristic impedance of 743 Ω. This impedance level is impractical to obtain and so it is necessary to consider a circuit which, although requiring a larger number of diodes, is realizable. In this case a lower characteristic impedance level, and hence a higher value of RF line current, must be switched. For this duplexer, any characteristic impedance for the SPST switches other than 50 Ω is inconvenient to realize, since it would require transforming sections* consisting of quarter wavelengths of transmission line; at 200 MHz, a quarter wavelength is 150 cm (59 in). If, instead, of a 743 Ω characteristic impedance, 50 Ω is used, the total current carried by the parallel diodes in each SPST limiter increases by the square root of the impedance ratio – by approximately a factor of 4. Accordingly, to design the limiters with a 50 Ω characteristic impedance, 16 rather than 4 diodes are needed in each switch.

To accommodate 16 diodes at a plane in the transmission line, a large transmission line cross section must be used. Figure VII-11 shows a 3-1/8 in (8 cm) coaxial line cross section with a typical diode installation. The axial prong ceramic packaged PIN is mounted in shunt with the line using beryllium-copper spring fingers to assure good thermal and electrical continuity. In series with the diode is a disk bypass capacitor which serves to series resonate the inductance of the diode, insuring a low impedance across the line at high RF power. In addition, the bypass capacitor permits dc checking of the diode without removing it from the

*Alternatively, lumped circuit transformers could be built as demonstrated in Figures VI-5 and VI-31; however, for such high power operation these circuits are equally inconvenient to realize.

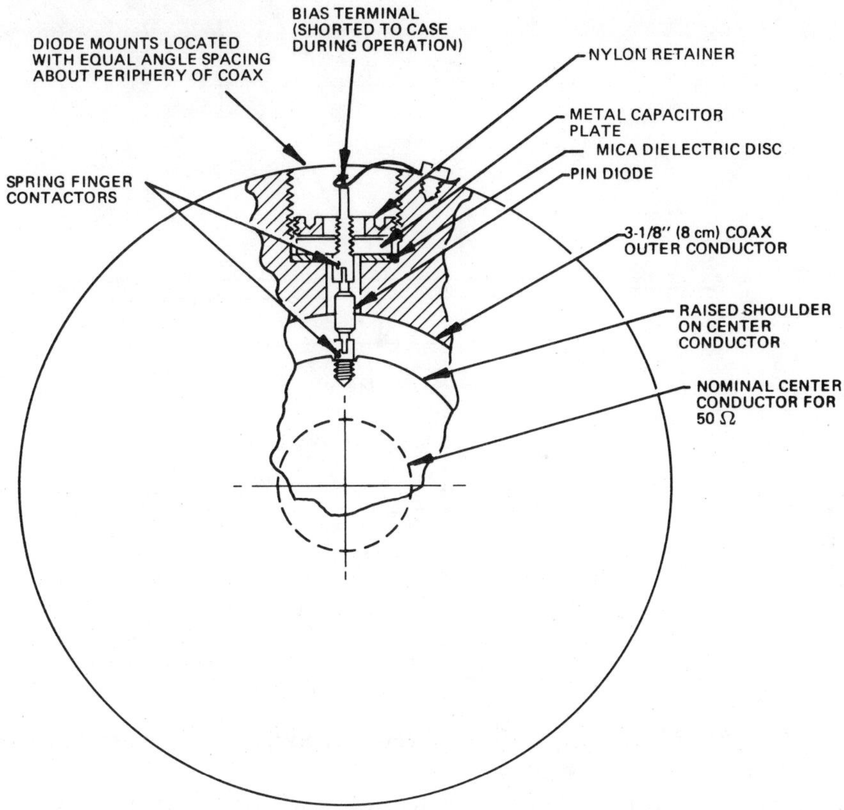

Figure VII-11 **Typical Shunt Diode Mount for High Power Coaxial Duplexer**

switch. In operation, the bias terminal is shorted to the transmission line case to provide a dc return path for the "rectified" current. The completion of this dc path is established between the center conductor and outer conductor by using a shorted transmission line stub. Not only is the dc return provided, but RF parallel resonant tuning of the total capacitance of the 16 diodes in the switch in the low loss, receive condition state is supplied as well.

The equivalent circuit of each SPST limiter for the high power coaxial duplexer is shown in Figure VII-12; a photograph of the completed unit is in Figure VII-13. The large value of series tuning

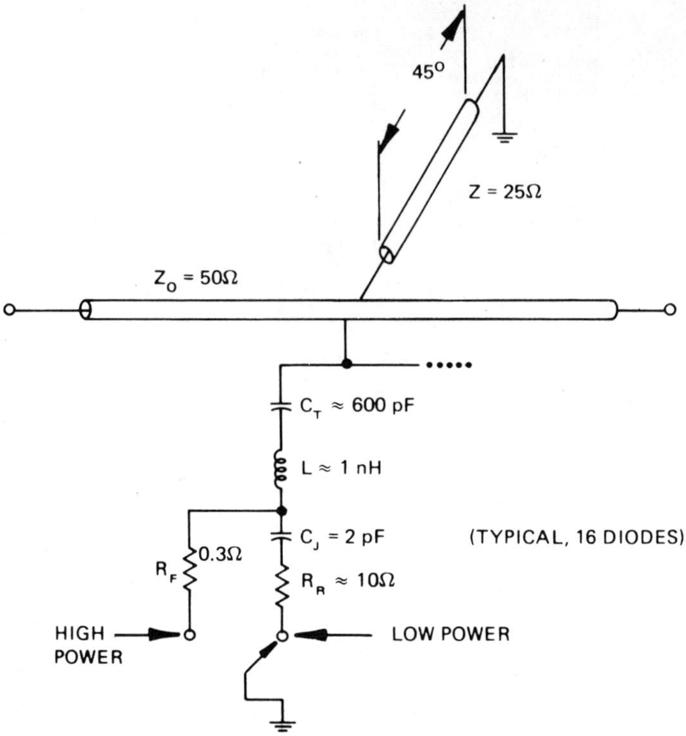

Figure VII-12 **Equivalent Circuit of SPST Limiter used in UHF Duplexer**

capacitance, 600 pF, required the use of a multiple disk capacitor. The effect of package capacitance, on the order of 0.2 pF, has been neglected in the equivalent circuit since it is less than one tenth of the 2 pF junction capacitance. The parallel combinations of the sixteen diodes result in a total line shunting capacitance of 32 pF, producing a capacitive reactance of -j25 Ω at 0.2 GHz. This reactance can be parallel resonated using a short circuited 25 Ω transmission line stub having a length of 45°. In practice, a pair of 50 Ω stubs, each 45° long, are used; they can be seen in Figure VII-13 following the high power limiter diode installation. To effect the parallel tuning without introducing a real impedance transformation the tuning stubs must be electrically close to the plane of the diodes; however, placing them slightly behind the plane of the diode limiters shields them from the high incident power and permits their realization in small diameter coax. The use of two 50 Ω stubs in each SPST switch also permits maintenance of electrical symmetry in the large diameter coaxial line cross section;

Figure VII-13 **UHF Duplexer Using Thin Base PIN Diodes for RF Activated Switching**

the stubs are mounted on opposite sides of the line diameter, and thus introduce little asymmetric loading. Accordingly, less higher order mode excitation is caused than would be by a single stub.

The duplexer shown in Figure VII-13 was evaluated in a functioning radar having 150 kW peak power, 10 kW average power, and pulse width of 200 μs. The loss on transmit through the duplexer was found to be only 0.15 dB, but, with zero bias, the receive loss was 2 dB. It was found that this receive loss could be reduced to 1 dB by the application of 1 V of reverse bias to the diode bias terminals. The high power duplexer gives about 40 to 50 dB of protection (depending upon the antenna mismatch); a low power varactor limiter provides the necessary 20 dB of additional receiver protection.

Considering the equivalent circuit in Figure VII-12 in light of these measured insertion loss values, the following conclusions can be made about the estimates of forward and reverse bias diode resistance:

1) The low insertion loss on transmit of only 0.15 dB is mainly attributable to ohmic losses in the input hybrid coupler and the SPST limiter housings; the figure is consistent with theoretical expectations. The calculated diode loss under high power (*arc loss*) is calculated from Table V-1 as

$$\text{Arc Loss} = \frac{4(R_F)/(N/2)}{Z_0}$$

$$= \frac{4\ (0.3\ \text{ohm})/(32/2)}{50\ \text{ohms}} \qquad \text{(VII-11)}$$

$$= 0.0015 = 0.15\% = 0.006\ \text{decibel}$$

which is negligible.

2) The measured low power insertion loss of 2 dB is about twice that which would have been expected had the eight-diode power-optimized circuit been used. However, with 16 diodes and the equivalent circuit shown in Figure VII-12, even less receive loss would be anticipated. From this equivalent circuit, converting the series RC model for the diode junction consisting of a 2 pF capacitor and a 10 Ω series resistor into the RC parallel equivalent, gives a line loading resistance of about 16 kΩ at 0.2 GHz, or 1 kΩ total for the parallel combination of 16 diodes. The insertion loss for 1 kΩ across a 50 Ω line is only 0.2 dB. Thus, it can be seen that the measured loss is an order of magnitude larger at zero bias than what would be expected from the estimated equivalent circuit in Figure VII-12. This result occurred because the diodes used were not completely punched-through at zero bias.* Thus we can conclude that the series resistance at reverse bias is about 100 Ω, dropping to about 50 Ω with -1 V of applied bias. Despite this fact, the duplexer functioned better in the actual radar application than the gas tube duplexer which it replaced; both the long and short range performances were consistently better with the semiconductor device.

Note that, in the assignment of the effective switching impedance, Z_0, for the SPDT switches, it is better to use R_F (i.e., from the arc loss formula in Table V-1) rather than R_R (in the receive loss formula) because reverse resistance, particularly when operated near zero bias, is generally less precisely known than is forward resistance.

*See Chapter II for a discussion of diode punchthrough phenomena.

D. High Frequency, Waveguide Limiters

1. Stub Mounted Diodes

At microwave frequencies above 3 GHz, using a series capacitor to resonantly tune the diode inductance and a shunt inductor to parallel tune the total capacitance of the diode and its package becomes impractical. At these high frequencies the actual reactance values are more difficult to estimate and, furthermore, the addition of the tuning elements becomes increasingly troublesome to accomplish with circuit components the dimensions of which are small compared to the operating wavelength. Instead, the diode may be embedded within a coaxial stub which, in turn, shunts the transmission line in which switching is required.

To appreciate how switching can be effected in this manner, consider the equivalent circuit shown in Figure VII-14. For simplicity, assume that the diode is represented as a simple switched capacitance and that the shunt connected transmission line has a total electrical length of 180°. Let us examine the impedance in shunt with the main line which results for the two switch positions of the diode. If the diode is short circuited to represent the forward bias condition, the net electrical length of the stub is the sum of θ_1 and θ_2, or 180°; a short circuit shunts the main transmission line, providing the isolation condition. In the second case, let the switch shown in parallel with the capacitor C be open, representing the reverse bias condition of the diode. Further, suppose that

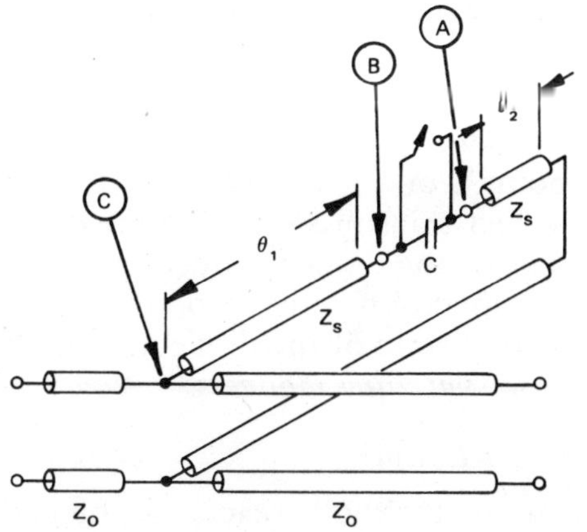

Figure VII-14 **Diode Embedded in a Shunt Stub**

θ_1 is equal to 135° and θ_2 is equal to 45° and that the characteristic impedance of the stub, Z_S, is selected to have a magnitude equal to half of the capacitive reactance of the diode capacitance, C. Consider the impedances which then result at points A and B. At point A the impedance looking toward the short circuit end of the stub is equal to $+jZ_S$, representing an inductive reactance having a magnitude equal to the characteristic impedance of the stub. At point B, the net impedance has this same magnitude but with the opposite sign, since the diode's reactance is capacitive and, by selection of the stub impedance, has been made equal to twice the value of Z_S. The reactance of $-jZ_S$ at point B is transformed to an open circuit at point C on the main transmission line by the line length, $\theta_1 = 135°$. Thus, perfectly matched transmission occurs on the line in the reverse bias state of the diode.

It can be seen that by proper choice of the characteristic impedance of the stub and the electrical lengths θ_1 and θ_2 the capacitive reactance of the diode can be tuned out at any desired center frequency. To the extent that the diode has series inductance, this reactance likewise can be tuned out since it forms a part of the distributed impedance of the stub itself. The conditions described in this example are not unique and there are other combinations of stub impedance and electrical lengths θ_1 and θ_2 which give perfect isolation and perfect match (neglecting dissipative losses in the stub and diode) for the diode's two bias states.

The stub circuit offers another advantage in that the switching sense can be changed merely by adding or subtracting 90° to θ_1. For example, if all of the parameters described are maintained except that θ_1 is changed from 135° to 45°, then under the forward bias condition, with the switch closed, the net electrical length of the stub is 90° and an infinite impedance or open circuit is presented at point C on the main transmission line. This situation results in perfect transmission when the diode is forward biased; reverse biasing the diode produces a short circuit on the transmission line. Operation which produces an isolation condition under the reverse bias state of the diode in a shunt mounted stub is called *the reverse switching mode.*

2. *Waveguide Post Coupling*

The method of tuning the diode reactances by embedding the diode in a post is particularly well suited for switches and limiters which must be built in waveguide. Figure VII-15 shows a mechanical schematic diagram of the method for coupling a coaxial line

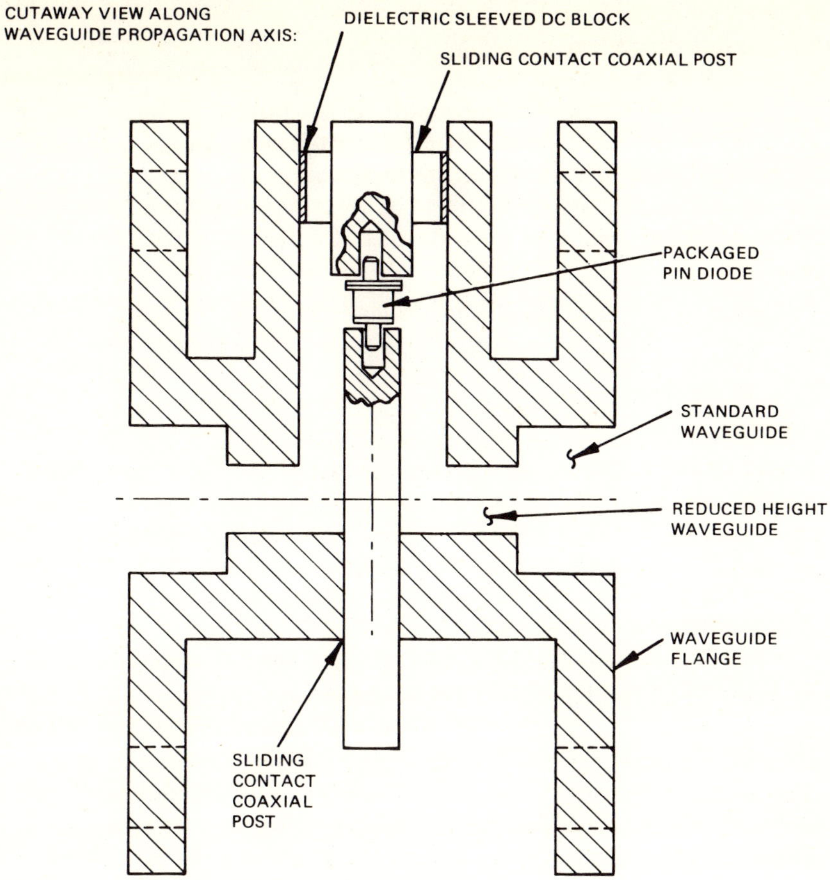

Figure VII-15 **Mechanical Schematic of Post Coupled PIN Limiter in Waveguide**

containing a packaged PIN diode to a rectangular waveguide transmission system. Typically for limiters and switches, the post is centered in the broad wall of the waveguide; for phase shifters, it may be close to a narrow wall for light coupling. The post is oriented parallel to the electric field of the dominant TE_{10} mode. The waveguide may be stepped in height to enhance the broadband tuning of the low impedance coaxial structure to the relatively high impedance associated with the propagating TE_{10} mode. The electrical lengths, θ_1 and θ_2, can be separately adjusted by positioning of the sliding dc block and the coaxial posts connected to the two prongs of the diode. Different impedances can be used for θ_1 and θ_2.

Generally the circuit is tuned empirically, but an analysis can be made after the fact by measuring the lengths θ_1 and θ_2. This information is useful in estimating the peak power capability, in performing tolerance studies, and in evaluating possible improvements which might be realized were diodes of different characteristics to be used.

The post coupling of diodes in waveguide is a common technique not only for limiters, switches, and phase shifters, but also for active sources, such as Gunn and avalanche oscillators. One of the first questions which arises in the analysis of the post coupling is – what value of characteristic impedance should be assigned in the waveguide? [4-6] Because of the importance of this basic post coupling structure in waveguide semiconductor control and source circuits, let us digress briefly to examine the rationale behind this important *waveguide impedance concept.*

3. Waveguide Impedance and Green's Function

Because of the distributed nature of the fields in waveguide, no unique values can be assigned to voltage and current associated with the power flow. Consequently, the characteristic impedance of rectangular waveguide has no unique definition until the use of that definition is specified. For example, we could define the impedance of the waveguide as the ratio of the vertical electric field to the horizontal magnetic field in the center of the waveguide cross section. This definition would be particularly appropriate if we were to desire to match the waveguide to free space, but it is conceptually invalid (and experimentally less accurate [4]) for calculating the current induced on a post when it shunts the broadwalls of the waveguide. The calculation of this current is, in fact, related to the present problem of the post coupling of diodes in the guide. *We use as a defining criterion that our definition for waveguide impedance gives the correct value for induced current on a thin post when it is connected directly across the broadwalls of the waveguide.*

The situation for the impedance definition is shown schematically in Figure VII-16. A conductor having small cross sectional dimensions compared with the a dimension of the waveguide is connected between the broadwalls of the guide at a distance, d, from one

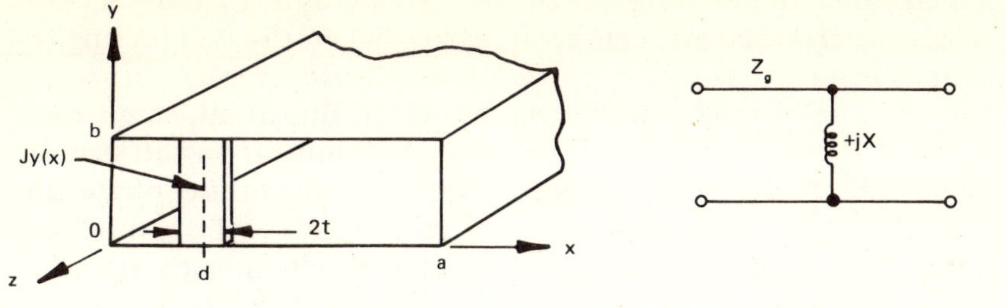

(a) THIN (IN Z DIRECTION) INDUCTIVE STRIP

(b) EQUIVALENT CIRCUIT

Figure VII-16 **The Inductive Strip Model Used to Define the Waveguide Characteristic Impedance for Post Coupled Devices**

of the narrow walls. To simplify the mathematical evaluation of the current, I, induced on this conductor, we assume that the conductor is realized as a thin conducting strip rather than a round post. This assumption, however, results in no loss of generality since, even with the post, we assume that the cross sectional dimensions are small compared to the waveguide width in order for the resulting current to be confined to the immediate region in the x-z plane a distance, *d*, from the narrow wall of the waveguide. For most waveguide obstacles the solution for distributed current is very complex and often impractical to perform; but for a thin post or strip, our assumptions permit a simple solution.

Physically, the current on the strip model is induced when a TE_{10} *(dominant)* mode wave traveling in the -z direction is incident at the reference plane z = 0, where the strip is located. Except for the implicit sinusoidal variation with time, the expression for the TE_{10} mode is

$$E_1 = E_0 \sin(\pi x/a) \cdot e^{-\Gamma_1 z} \qquad \text{(VII-12)}$$

where the propagation constant Γ_n for the infinite set of TE_{n0} modes is given by

$$\Gamma_{n0} = \sqrt{(n\pi/a)^2 - (2\pi/\lambda_0)^2} \qquad \text{(VII-13)}$$

Equation (VII-12) describes the electric field in the cross section of an *unperturbed* waveguide; however, when the conducting foil strip is installed, *the electric field must vanish over its entire surface,* S. The physical mechanism by which this disappearance occurs is the establishment of a current distribution on the strip by the incident TE_{10} wave. This current has a magnitude, phase, and spatial distribution such that it excites a reflected TE_{10} mode and other TE_{n0} modes, the sum of the fields of which just serve to cancel the incident TE_{10} mode over the surface of the conducting strip. Elsewhere in the cross section of the waveguide the electric field is not cancelled, however, and the energy which is not reflected continues to propagate beyond the obstacle.

Impinging on the metal strip, the incident TE_{10} wave induces a current distribution, $J_y(x)$, the magnitude and spatial variation in x of which is such as to excite higher order mode fields

$$E_{n0} = E_{n0}\,(z = 0)\,\sin\,(n\pi x/a) \cdot e^{-\Gamma_n z} \qquad \text{(VII-14)}$$

which cancel the incident electric field at the surface of the strip, satisfying the boundary condition that the tangential E field at the surface of a perfect conductor is zero. This mode set can be thought of as a Fourier series; that is, an infinite number of sinusoidal terms the magnitudes of which add up to the incident TE_{10} over the x range $(d-t) \leqslant x \leqslant (d+t)$. An infinite number of such TE_{n0} waves is excited, but for waveguide used in its dominant mode, all of them except the TE_{10} are evanescent (i.e., *cut off*), and only the TE_{10} component of the induced mode set is able to propagate *(be scattered)* from the conducting strip obstacle. The remaining mode amplitudes fall off rapidly with distance (on the z axis) from the obstacle; their presence accounts for energy storage and imparts the reactive character to the obstacle, in this case that of a shunt inductor.

The general expression which relates the scattered TE_{n0} mode amplitudes, $E_S(x, z)$ to a filament of induced current $J_y(x')$, at x equal to x′ and z equal to 0, is called a *Green's function.* For a uniform vertical (y direction) current filament, the Green's function for rectangular waveguide is [7 (Section 8.5)]

$$G(x, y, x') = \frac{-j\omega\mu_0}{a} \sum_{n=1}^{\infty} \frac{1}{\Gamma_n} \sin\left(\frac{n\pi x}{a}\right) \sin\left(\frac{n\pi x'}{a}\right) e^{-\Gamma_n |z|} \qquad \text{(VII-15)}$$

and the scattered field, $E_S(x, z)$, anywhere in the waveguide can then be evaluated from

$$E_S(x,z) = \int_S G(x, z|x')\, J_Y(x')\, dx' \qquad \text{(VII-16)}$$

where S is the surface of the obstacle.

From the Green's function the normalized reactance of the inductive strip can be determined; happily, though, we need not carry the analysis this far. For our purposes we assume that the normalized inductive reactance of the thin post can be "looked-up" [8 (pp. 258-263)] or measured in the laboratory. From the equivalent circuit normalized reactance, $\bar{x}$, we can determine the reflection coefficient of the post obstacle to the incident wave. But the Green's function formula *also* gives a value for the amplitude of the scattered TE_{10} mode in terms of the current on the obstacle, $J_Y\,(x')$. Combining these expressions, we solve for the total current on the post. Knowing the simultaneous values for incident power, P; post current, I; and normalized reactance, $\bar{x}$, permits us to say *how the effective characteristic impedance, Z_G, of the waveguide must be defined* to have produced the current I on the post. This definition is that of characteristic impedance which we seek to use in the analysis of circuits which couple diodes to rectangular waveguide by means of posts. This derivation of Z_G is carried out as follows.

The dominant mode wave reflected from the obstacle, E_R, is identical to the incident wave given by Equation (VII-12) except for a complex amplitude factor, R, (the reflection coefficient) and the fact that it propagates toward the generator, (in the +z direction) rather than toward the load (in the -z direction). Thus, by definition

$$E_R = RE_0 \sin(\pi x/a) \cdot e^{\Gamma_1 z} \qquad \text{(VII-17)}$$

where $R = E_R/E_0$ is the complex reflection coefficient.* The complex amplitude, E_R, is found from the Green's function in Equation (VII-16) for the case where n is equal to 1; that is

$$E_R = \frac{-j\omega\mu_0}{a\Gamma_1} \sin(\pi x/a) e^{\Gamma_1 z} \int_{d-t}^{d+t} J(x') \cdot \sin(\pi x'/a) \cdot dx' \qquad \text{(VII-18)}$$

*We use R here for reflection coefficient rather than Γ, as elsewhere in this book, to distinguish from the mode propagation constants, Γ_N.

For our analysis the post is considered thin compared to the width of the waveguide. Applied to the thin strip representation of the post, this assumption means that $2t \ll a$. Since the strip is narrow, the current density is virtually constant with x. Then $J(x) \approx J_0$, and $J_0 \cdot 2t \approx I$, the *peak RF current induced on the strip.* The integration in Equation (VII-18) then produces

$$E_R = \frac{-j\omega\mu_0}{a\Gamma_1} \cdot \sin(\pi d/a) \cdot I \cdot \sin(\pi x/a)\, e^{\Gamma_1 z} \qquad \text{(VII-19)}$$

Now the *average* incident propagating power, P, related to the incident TE_{10} wave the *peak* amplitude of which is E_0 is given by

$$E_0 = 2\sqrt{\frac{P}{ab} \cdot \frac{\lambda_G}{\lambda_0} \cdot \sqrt{\frac{\mu_0}{\epsilon_0}}} \qquad \text{(VII-20)}$$

Equations (VII-17), (VII-19), and (VII-20) together with $\Gamma_1 = 2\pi/\lambda g$ and $\omega = 2\pi/\lambda_0\sqrt{\mu_0\epsilon_0}$ can be combined to give

$$4R^2P = \sqrt{\frac{\mu_0}{\epsilon_0}} \cdot \frac{\lambda_G}{\lambda_0} \cdot \frac{b}{a} \cdot \sin^2(\pi d/a) \cdot I^2 \qquad \text{(VII-21)}$$

However, from transmission line theory, when a reactance which produces a reflection coefficient R is connected in shunt across a transmission system of characteristic impedance Z_G as shown in Figure VII-17, a peak current, I, is induced which is related to the incident average power P according to

$$8R^2P = Z_G \cdot I^2 \qquad \text{(VII-22)}$$

Examining Equations (VII-21) and (VII-22) indicates that for these equations to be consistent with each other, the waveguide impedance which we seek must be defined as

$$Z_G = 2 \cdot \sqrt{\frac{\mu_0}{\epsilon_0}} \cdot \frac{\lambda_G}{\lambda_0} \cdot \frac{b}{a} \cdot \sin^2(\pi d/a) \qquad \text{(VII-23)}$$

Readers familiar with waveguides should recognize that Equation (VII-23) is just the "voltage-power" definition for waveguide impedance; that is, it is the definition of waveguide impedance which results from relating the waveguide voltage between the waveguide broadwalls at x equal to d (were there no post present) to the

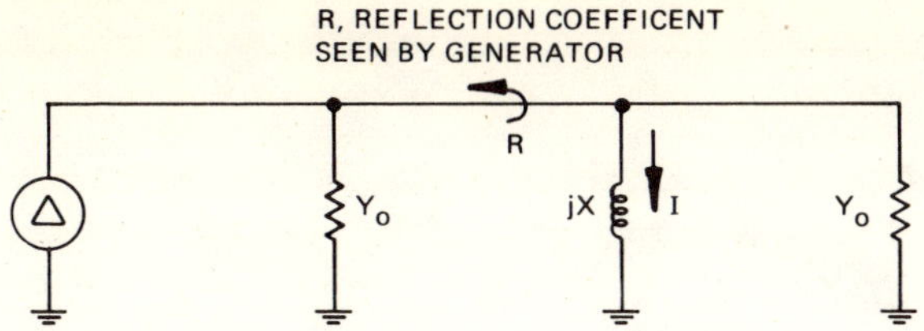

Figure VII-17 **Lumped Circuit Used to Derive Absolute Impedance Equivalent for the Post Obstacle in Waveguide**

propagating power. This fact provides an easy way to remember the Z_G definition appropriate for post coupling, *but it was not used a priori as a justification for this definition because, with the post present, the waveguide voltage at d is zero.*

Implicit in this derivation for Z_G is the assumption that the current on the post is constant (i.e., has no y variation). For some coupling circuits this assumption may not be valid; however, to the extent that the post current couples to the TE_{10} propagating mode in the waveguide, only the constant component of I is significant; the TE_{10} mode itself has no y variation and therefore can couple only to a uniform y directed current. For a more general solution, see Reference 6.

The elaborate steps taken here are to provide a basis for the important concept of waveguide impedance as it applies to post coupling. Such coupling is one of the most, if not *the* most, practical way to couple the two terminal PIN diode to the distributed wave propagation in rectangular waveguide. The use of this definition for Z_G has been used [5,9] with effective results; we shortly demonstrate its usefulness for the waveguide limiter. The derivation for Z_G from first principles is presented both to prevent a feeling of analytical uncertainty in its use by the practicing circuit designer and to highlight the assumptions which govern its use, namely

1) The coupling post is thin compared to the waveguide width.
2) The current on the post is essentially constant along the y direction.*

*See Reference 6 for a more complete analysis by Eisenhart when the post current is not assumed constant along its length.

Figure VII-18 **Photograph of Post Coupled Waveguide Limiter. The Protruding Wire Lead Couples RF Field to a Detector Diode Whose Rectified Current Speeds the PIN Turn-On**

4. Practical Application of Post Coupling

A Ku-band PIN limiter is shown in Figure VII-18. A detector diode in an axial lead glass package detects a portion of the RF energy. The rectified current from the detector is applied to the bias terminal of the post coupled PIN diode, increasing the speed with which the PIN reaches the isolation condition.

The PIN limiter stage was empirically tuned and the resulting circuit analyzed using a computer program; Figure VII-19 shows the equivalent circuit appropriate to its analysis.* With a practical limiter, the post is connected in shunt with the main transmission line through the reactance, jX_A, of the portion of the post within

*This analysis, including the programming, was performed by Roland Ekinge at Microwave Associates, Inc., Burlington, Massachusetts during his sabbatical leave from Chalmers University, Gothenburg, Sweden.

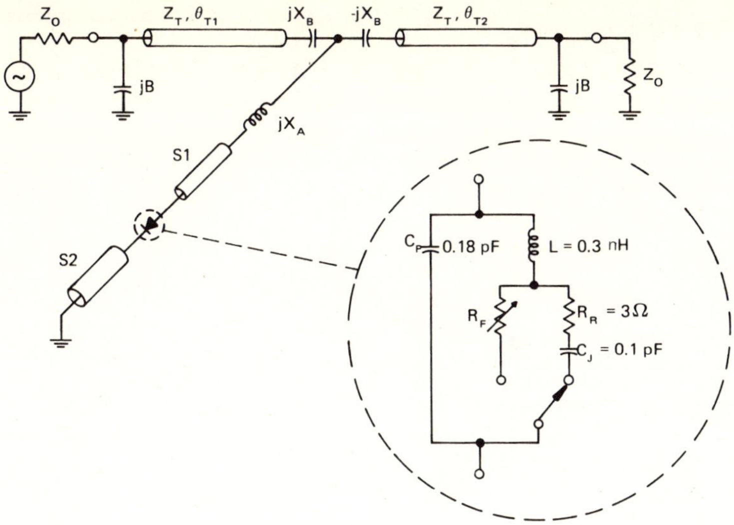

Figure VII-19 **Schematic Diagram for a Practical Ku-Band, Post Coupled, Waveguide PIN Limiter**

the waveguide. Application of the operating principles described for the ideal circuit shown in Figure VII-14 is different in this case only to the extent that the reactance presented by the coaxial stub at the entrance to the waveguide broadwall must be the negative of $+jX_A$ (the post reactance in the waveguide), rather than zero, in order for the isolation state to be realized.

The circuit in Figure VII-19 contains discontinuity reactance elements represented by jB, for the shunt capacitance encountered at a change in waveguide height, and $-jX_A$, for the series capacitance energy storage introduced near a shunt post. The values of these elements depend upon the waveguide and post structure dimensions. Their values for the specific limiter shown were evaluated within the computer program (because of the complexity of the analytic expressions) using the formulation by Marcuvitz for the post [8 (pp. 258-263)] and discontinuity susceptance due to a change of waveguide height [8 (pp. 307-308)].

Waveguide impedance Z_0 and Z_T were evaluated from Equation (VII-23) at each analysis frequency based upon the actual circuit dimensions. Standard WR-62 waveguide for which a = 1.57 cm (0.620 in) and b = 0.79 cm (0.310 in) was used. Z_T was a sym-

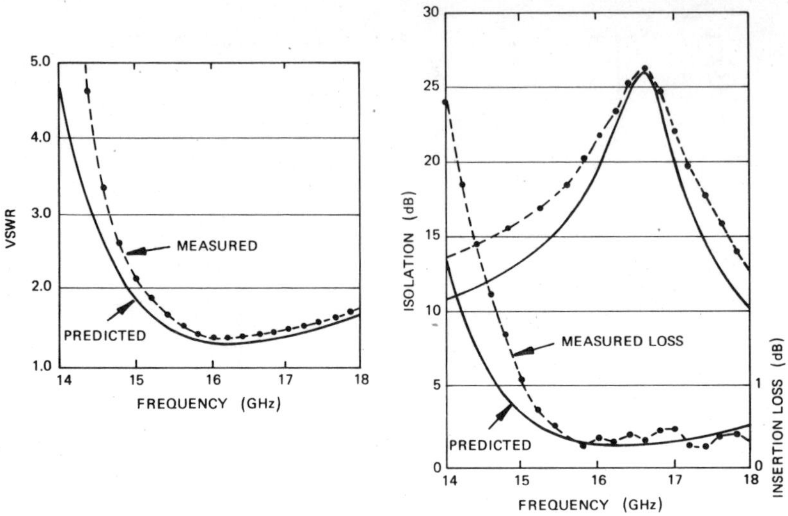

Figure VII-20 **Computer Calculated Specifications Compared with Measured Properties of a Ku-Band Limiter.**

metrically reduced height section in which b = 0.4 cm (0.160 in). Other circuit parameters were:

	S1	*S2*
Characteristic Impedance (ohms)	50	19
Electrical Length at 16.5 GHz	84°	49°
Loss (decibels/wavelength)	0.1	0.1
Post Diameter	0.28 centimeter	(0.111 inch)
θ_{T1} *at 16.5 GHz*	105°	
θ_{T2} *at 16.5 GHz*	85°	

Table VII-3 **Ku-Band Limiter Circuit Parameters**

The calculated and measured values of isolation, insertion loss and VSWR for the limiter are shown in Figure VII-20. Since measured values of the diode resistances R_F and R_R at Ku band were not available, the values used, $R_F = 2.3\ \Omega$ and $R_R = 2\ \Omega$, were selected to force the calculated values of isolation and insertion loss respectively to agree with those measured at the limiter's center frequency of 16.5 GHz. The effectiveness of the equivalent circuit can be assessed from the similarity of the calculated and measured

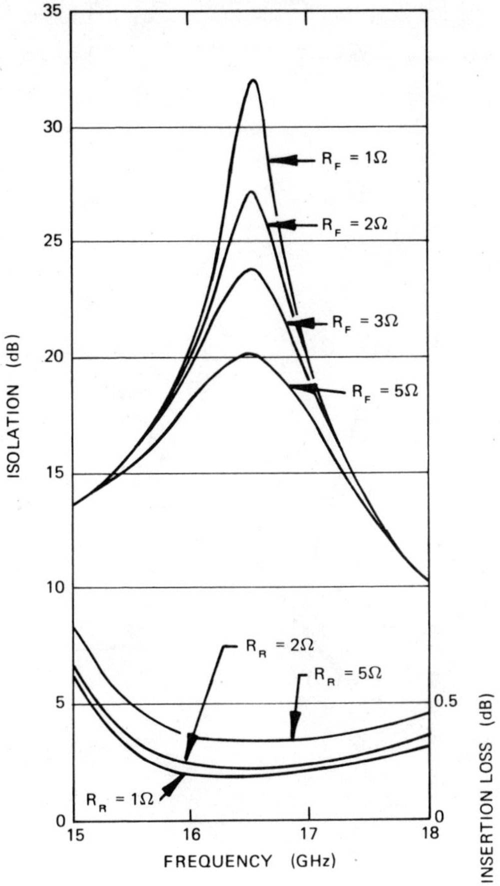

Figure VII-21 **Computer Estimated Changes in Limiter Performance Produced with Varying R_F and R_R**

isolation, loss, and VSWR performance "signatures" over the 14-18 GHz bandwidth.

The utility of the analytical approach performed after a model has already been designed by experimental methods can be seen in Figure VII-21, showing computer calculated isolation and insertion loss performance which can be expected for various values of diode resistances R_F and R_R. Controlled experiments and sufficiently accurate measurements to determine this data empirically would be extremely tedious to perform in this high operating frequency band.

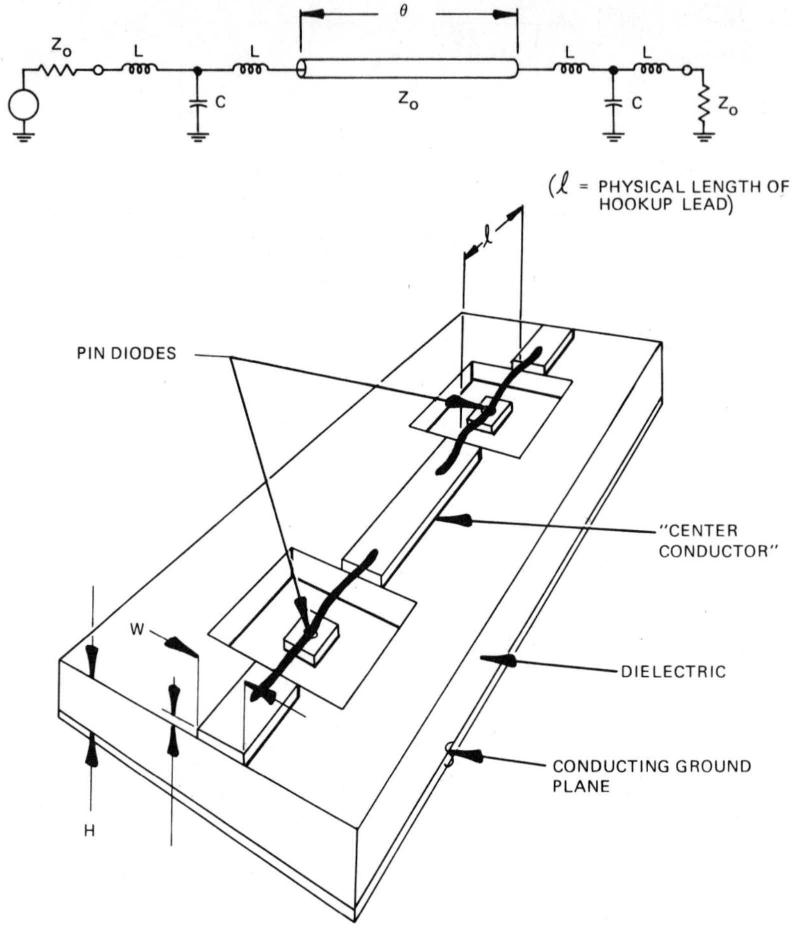

Figure VII-22 **Microstrip Limiter and Equivalent Circuit Using Packageless "Chip" Diodes**

E. Integrated Circuit Limiters

1. Mechanical Layout

One of the simplest and broadest bandwidth approaches to the design of limiters and switches dispenses with the diode package as such – the *hybrid integrated circuit.* The term "hybrid" is used because the diode and circuit are not both made of the same material. Hybrid integrated circuits are usually made with *microstrip circuitry,* TEM transmission line consisting of a conductor pattern above a single ground plane. Printing the circuit pattern on a dielectric such as teflon fiberglass (TFG) or alumina ceramic provides a reproducible means for realizing the miniature circuit geometry. Figure VII-22 shows a typical, two stage limiter having a pair of diodes mounted on the ground plane and spaced along the transmission line. Flexible wire leads provide mechanical stress relief in the connection of the center conductor to the diode. In addition, by proper selection of the lead lengths and sizes, series inductances are realized which can be used to form a matched tee filter and thereby tune out the reflection that would otherwise be introduced by the diode's capacitance.

2. Matched Tee Diode

If the total inductance, 2L, of both hookup straps satisfies

$$Z_0 = \frac{\sqrt{2L}}{C} = \frac{1}{Y_0} \tag{VII-24}$$

then the diode installation will resemble a length of transmission line and its transmission match will be frequency independent at frequencies for which both the reactance, $j\omega L$, has a magnitude which is small compared to Z_0 and, correspondingly, the susceptance, $j\omega C$, is small compared with Y_0.

The performance of this circuit can be expressed in a general way through normalizing. If the input VSWR of the network is graphed as a function of ωCZ_0, where $\omega = 2\pi f$, the resulting curve can be used to estimate the VSWR for any capacitor C and load Z_0 which is tuned according to the criterion given by Equation (VII-24). This method of tuning produces a *maximally flat* low pass response; that is, the VSWR (or reflection insertion loss) is minimum at zero frequency and has a maximally flat rate of change with frequency for the number of elements used. The ratio of the total inductive reactance to Z_0 is equal to the ratio of the capacitive susceptance to Y_0, specifically $2\omega L/Z_0$ is equal to $\omega C/Y_0$, at all frequencies. At the frequency for which ωL equals Z_0, the total inductive and capacitive reactances are equal to each other and to Z_0. This point is unity on the ωCZ_0 axis in the normalized performance graph shown in Figure VII-23. As an illustrative example, the graph indicates that a 0.32 pF diode (the reactance of which is 50 Ω at 10 GHz) using matched tee tuning has a VSWR equal to 1.28 at 10 GHz when installed in a 50 Ω system and it

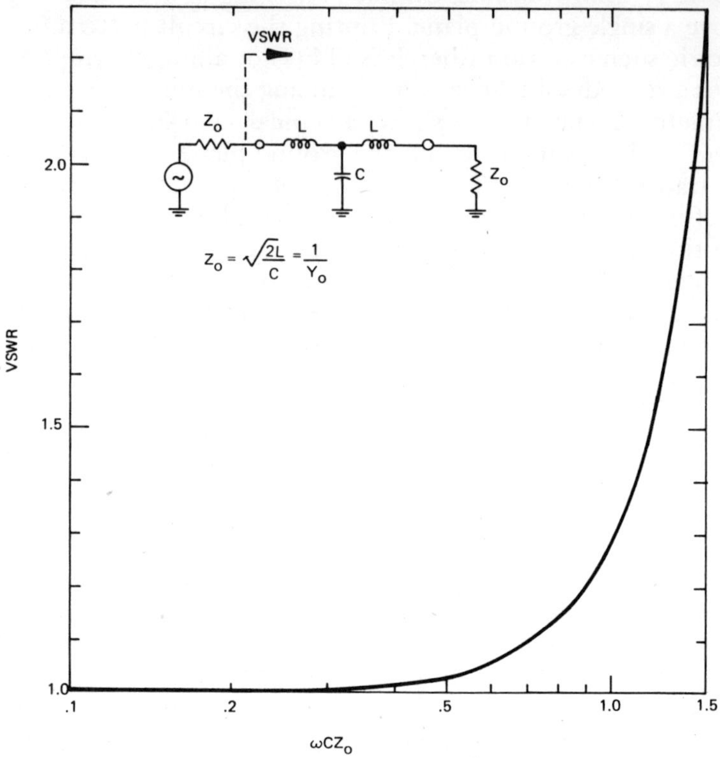

Figure VII-23 **VSWR vs. Normalized Frequency for a Maximally Flat, Low Pass Tee Diode Installation**

has lower VSWR for all frequencies below 10 GHz, specific values being estimable directly from the graph.

In conventional filter theory the low pass tee circuit is usually described in terms of its cutoff frequency, at which half the incident power is reflected. This frequency corresponds to $\omega CZ_0 = 2$ and would occur at 20 GHz for the C = 0.32 pF example. Clearly, this definition for cutoff frequency is not very useful for switch designs, however, since the input VSWR for 3 dB of reflection loss is 5.8, well off the scale in Figure VII-23 and much too large a mismatch for practical control device design.

It might appear that selecting the value for L to satisfy the maximally flat condition in Equation (VII-24) is advantageous only if operation beginning at very low and extending continuously to

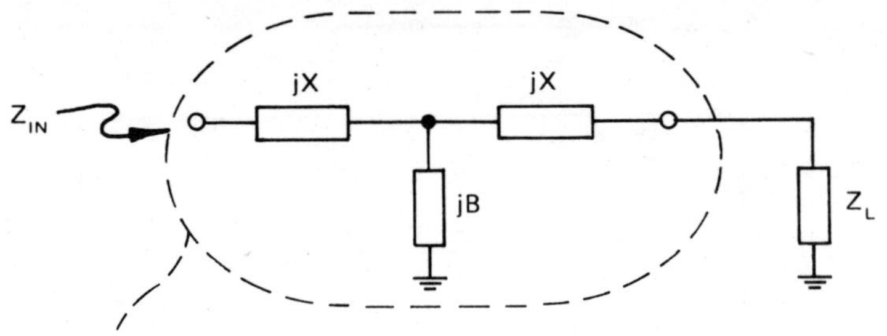

$$\begin{bmatrix} A & B \\ C & D \end{bmatrix} = \begin{bmatrix} (1 - BX) & jX(2 - BX) \\ jB & (1 - BX) \end{bmatrix}$$

$$Z_{IN} = \frac{AZ_L + B}{CZ_L + D}$$

IF WE REQUIRE THAT $Z_{IN} = Z_O$ WHEN $Z_L = Z_O$
(i.e. MATCHED INPUT FOR MATCHED OUTPUT)

THEN,

$$\frac{X}{Z_O} = \frac{1 \pm \sqrt{1 - (BZ_O)^2}}{BZ_O} \qquad \text{(VII-25)}$$

Figure VII-24 **ABCD Matrix Derivation of the Condition for a Perfectly Matched Tee Diode Installation**

high frequencies is desired. In particular, if L is chosen to give a perfect match at some higher frequency, then a broader overall bandwidth of operation might be expected. We now explore this tuning criterion.

To determine the choices of L under which a perfect match can be obtained, we write the overall ABCD matrix for the tee network following the procedure used for the π network in Figure VI-5. Imposition of the matching condition (Z_{IN} equals Z_0 when Z_L equals Z_0) gives Equation (VII-25) in Figure VII-24, from which the proper value for $X(\omega_0)$, the reactance of one of the hookup straps, can be determined from $B(\omega_0)$, the susceptance of C at the desired matching frequency, $f_0 = \omega_0/2\pi$.

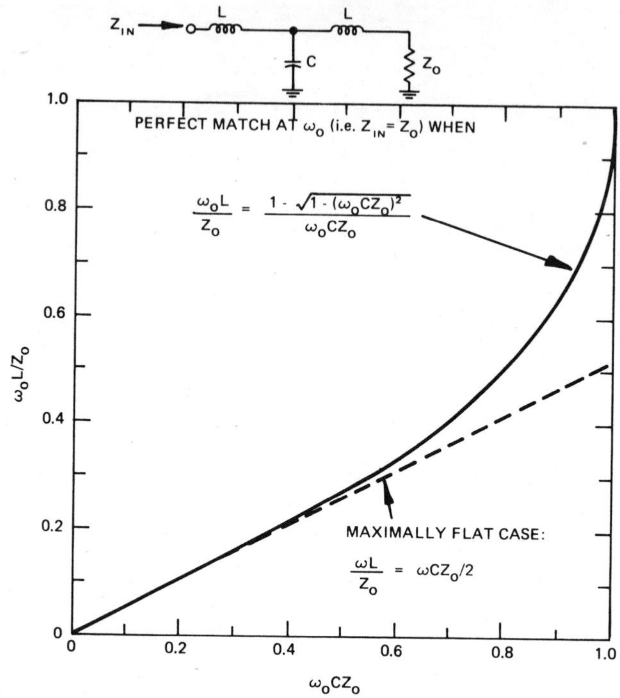

Figure VII-25 **Condition for a Perfectly Matched Tee Diode Installation**

Equation (VII-25) has two solutions and both give a perfect match at the same frequency, f_0; however, the smaller solution, obtained with the (−) sign for X/Z_0, is preferable, since it gives the lowest VSWR at frequencies above and below Z_0. The resulting relationship is shown graphically in Figure VII-25. It can be seen that:

1) If the circuit is tuned at a frequency for which the capacitive reactance magnitude is less than half of Z_0, (i.e., $\omega_0 C Z_0 < 0.5$) the required $\omega_0 L/Z_0$ is approximately equal to $\omega C Z_0/2$; clearly, this condition is identical to the maximally flat case (for which $\omega_0 = 0$).

2) As the matching frequency is increased, $\omega_0 L/Z_0$ approaches $\omega_0 C Z_0$.

3) It is not possible to achieve a perfect match with the tee circuit shown in Figure VII-25 above the frequency at which the capacitive reactance equals Z_0; that is, above the frequency for which $\omega C Z_0 = 1$.

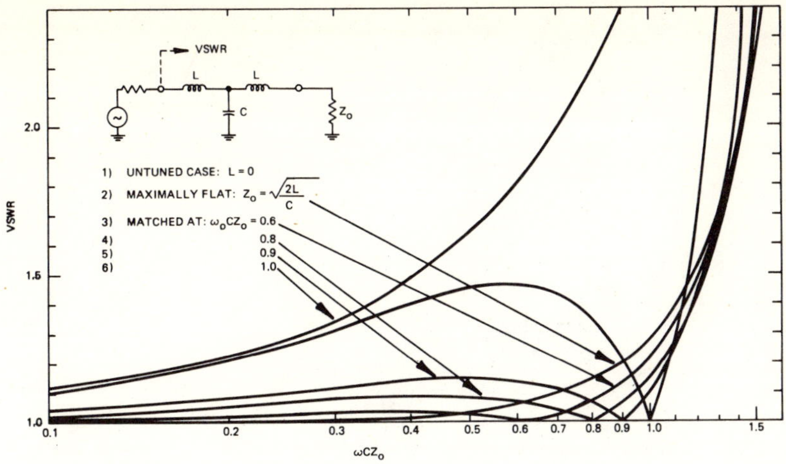

Figure VII-26 **Comparison of Tee Circuit Matching Methods**

Figure VII-26 shows a comparison of the various tee circuit matching methods. Curve 1 shows the VSWR obtained when L = 0, and no matching is performed. Curve 2 is the maximally flat performance, clearly providing a great improvement over the unmatched case. However, curves 3 through 6, respectively representing the performance when a perfect match is obtained at $\omega_0 CZ_0$ equal to 0.6, 0.8, 0.9, and 1.0, show that the off resonance VSWR performance is progressively deteriorated, compared with the maximally flat case, as the circuit match is made to approach the frequency at which $\omega_0 CZ_0$ is equal to 1. For most practical applications, therefore, it is likely that the maximally flat response is the most desirable.

3. Microstrip Impedance and Inductance Estimates

A convenient means of estimating the inductance of the short hookup leads is by relating their geometry to that of a microstrip transmission line. Not only is this representation a computational convenience, it is a more realistic approximation at the higher microwave frequencies, where lead lengths become an appreciable part of a wavelength (as is seen in the example to follow).

Microstrip consists of a "center" conductor above a single ground plane, as shown in Figure VII-27. If the center conductor can be supported without dielectric, the transmission line is homogeneous, having the same dielectric constant above the center conduc-

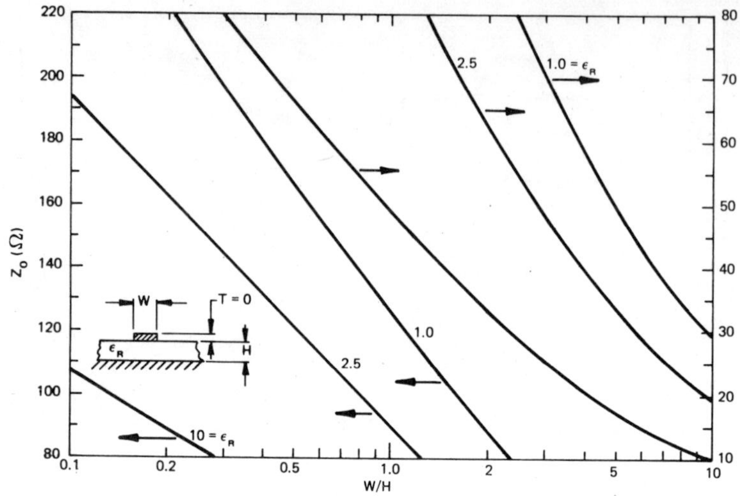

Figure VII-27 **Wheeler's [10] Microstrip Characteristic Impedance for Zero Thickness Center Conductors**

tor as below. Usually, though, the center conductor strip is printed and etched on a dielectric substrate, producing an *inhomogeneous* transmission line in which part of the propagating electric fields is within a dielectric, ϵ_R, and part is in air. The resulting solution for Z_0 in the inhomogeneous case requires some approximation, the most widely used reference being the work of Wheeler [10] from which the curves in Figure VII-27 were generated.

For inhomogeneous microstrip the calculation of the propagating wavelength is complicated by the fact that the RF energy travels in the compound dielectric medium which consists of air and the substrate dielectric; the resulting *effective* relative dielectric constant $\epsilon_{EFFECTIVE}$ is somewhere between 1.0 and ϵ_R. Wheeler's work anticipated this computational difficulty as well, and the ratio $\epsilon_{EFFECTIVE}/\epsilon_R$ is shown graphically in Figure VII-28.

Recently, the equations of Wheeler were published in a convenient FORTRAN IV computer program by Kwon [11] which is contained in Appendix E. With this program the characteristic impedance and effective dielectric constant can be accurately calculated for any value of substrate dielectric constant, ϵ_R, and strip dimensions, W and T.

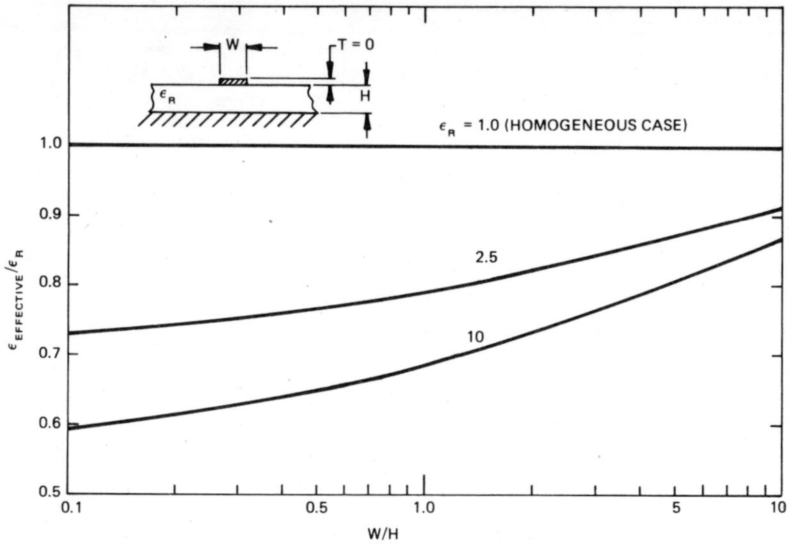

Figure VII-28 **Effective Dielectric Constant for (Inhomogeneous) Microstrip**

To estimate hookup lead inductance based on microstrip impedance, the inductive reactance magnitude is equated to the product of the microstrip line impedance and its electrical length in radians

$$Z \cdot \theta \approx 2\pi f L \qquad \text{(VII-26)}$$

The following example demonstrates the design procedure.

4. Design Example

Suppose it is desired to design a 50 Ω two stage limiter, of the form shown in Figure VII-22, which is useable over the 1-10 GHz bandwidth. From Figure VII-23 it is seen that if the VSWR of a single diode is to be kept below 1.5, then ωCZ_0 must not exceed approximately 1.15 at the highest frequency of operation, corresponding to a maximum value for C of about 0.2 pF. From Equation (VII-24), the proper value for 2L is 0.5 nH, or 0.25 nH per hookup lead. Furthermore, let us assume that a teflon fiberglass substrate is used with T = 1 mil (25 μm) copper center conductor pattern.

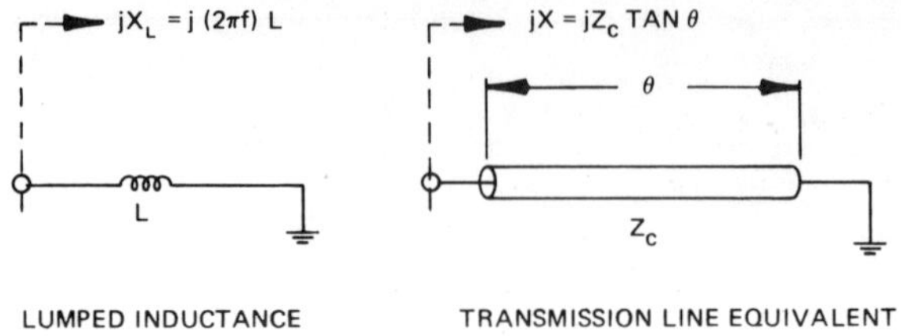

$$(2\pi f)\, L = Z_C \tan\theta$$

FOR SMALL θ,

$$Z_C\theta \approx 2\pi fL$$

Figure VII-29 **Approximating a Short Line Length by a Lumped Inductor, and Vice Versa**

Since the diode chip dimensions are small, it is necessary to use a small ground plane spacing, H, to avoid a large discontinuity at the diode installation. Furthermore, the use of small circuit dimensions inhibits the propagation of higher order modes within the microstrip circuit housing. Accordingly, we choose a substrate thickness, H, equal to 10 mils (0.25 mm). From Figure VII-27, the proper center conductor width, W, for Z_0 equal to 50 Ω, is 28 mils (7.1 mm). Assume that a 2 mil wide (5 mm) strap is used as a hookup lead from the diode to the 50 Ω line center conductor, and that its average height above the ground plane is 5 mils (0.12 mm). Since the lead is essentially surrounded by air, $\epsilon_R = 1$ and W/H = 0.4; from Figure VII-27, the lead's characteristic impedance as a microstrip line is about 180 Ω. The equivalence given in Figure VII-29 between a lumped inductor and a short length of high impedance transmission line can be estimated at any convenient frequency. Actually, the frequency term can be eliminated from Equation (VII-25), but greater physical insight is obtained by comparing the inductive reactance to an electrical length of line at a specified frequency. Thus, at 10 GHz, a 0.25 nH inductance has a reactance magnitude of 15.7 Ω; from Figure VII-29, this value is realized by a section of 180 Ω line the electrical length of which at 10 GHz is 0.087 rad (5°).

Figure VII-30 shows the VSWR performance of the 0.2 pF matched tee diode with frequency using both the lumped and distributed models. This data was calculated using program WHITE, developed

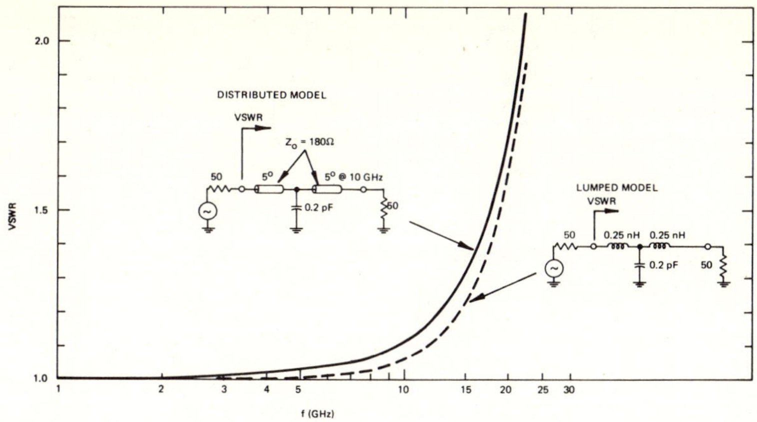

Figure VII-30 **VSWR of Single 0.2 pF Diode in Maximally Flat, Low Pass Tee Installation, Calculated Using Program WHITE (Chapter VI)**

in Chapter VI. Figure VII-31 shows the calculated performance for both diode tees when separated by a 50 Ω line 90° long at 10 GHz. Notice that although the VSWR for the single diode distributed model was higher than for the lumped model, the reverse is true for the two diode circuit performance shown in Figure VII-31, which has a match resonance near 7 GHz contributed by the distributed nature of the circuit.

Depending upon the application, dc returns and blocks [12] may be required in the circuit shown in Figure VII-22; they, too, can be modeled using program WHITE. Lumped and distributed bias elements are discussed in Appendix I, pp. 534-536.

5. Practical IC Limiter Data

The two stage limiter shown in Figure VII-22 can be conveniently packaged in a modular form which provides mechanical protection, hermetic sealing, and compatibility with stripline circuitry, as shown in Figure VII-32. A glass seal feedthrough designed as a 50 Ω section, replaces the circuit connector. Figure VII-33 shows the test fixture used for performing RF measurements. The measured low power VSWR and insertion loss performance over X-band for a practical two stage limiter using the design principles described is shown in Figure VII-34. Reflective contributions

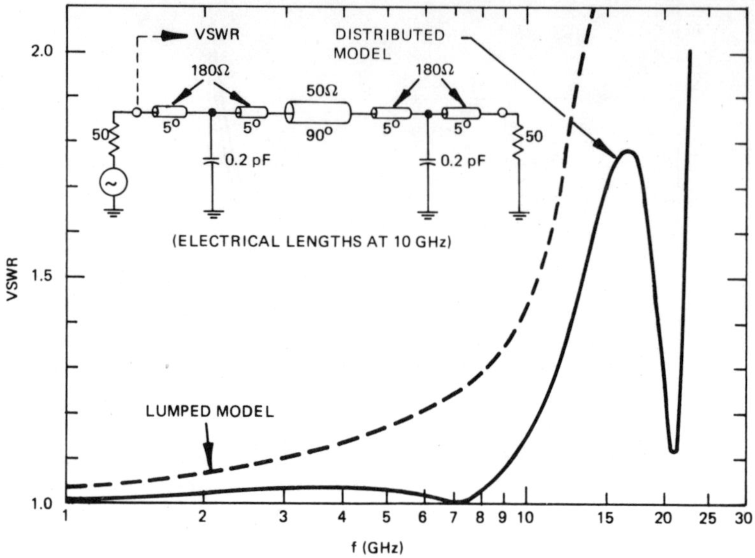

Figure VII-31 **VSWR of Two Stage Limiter with Maximally Flat, Low Pass Tee Matching, Calculated Using Program WHITE (Chapter VI)**

from the SMA connectors, glass feedthroughs, and bias elements contribute to the VSWR performance anticipated for the basic circuit evaluated in Figure VII-31. The limiter provides protection from 1 kW peak RF pulses with 1 μs pulse length and 0.001 duty cycle, limiting this incident level to a maximum output of 0.25 W peak, which corresponds to a self-actuated, high power isolation of about 36 dB.

F. Bulk Limiters

1. Basis of Operation

Gas tubes have been used for decades in radar duplexers, operating on the principle of ionization to separate high and low power signals. Their main disadvantage comes from their deterioration with time as the electrodes, bombarded by high energy ions and electrons, are eroded; furthermore, the glass seals eventually develop leaks or are broken. These problems arise because the ionized medium is gaseous. If, however, the ionized medium were a solid, the need for hermetic enclosure could be eliminated and the device made more controllable and reproducible.

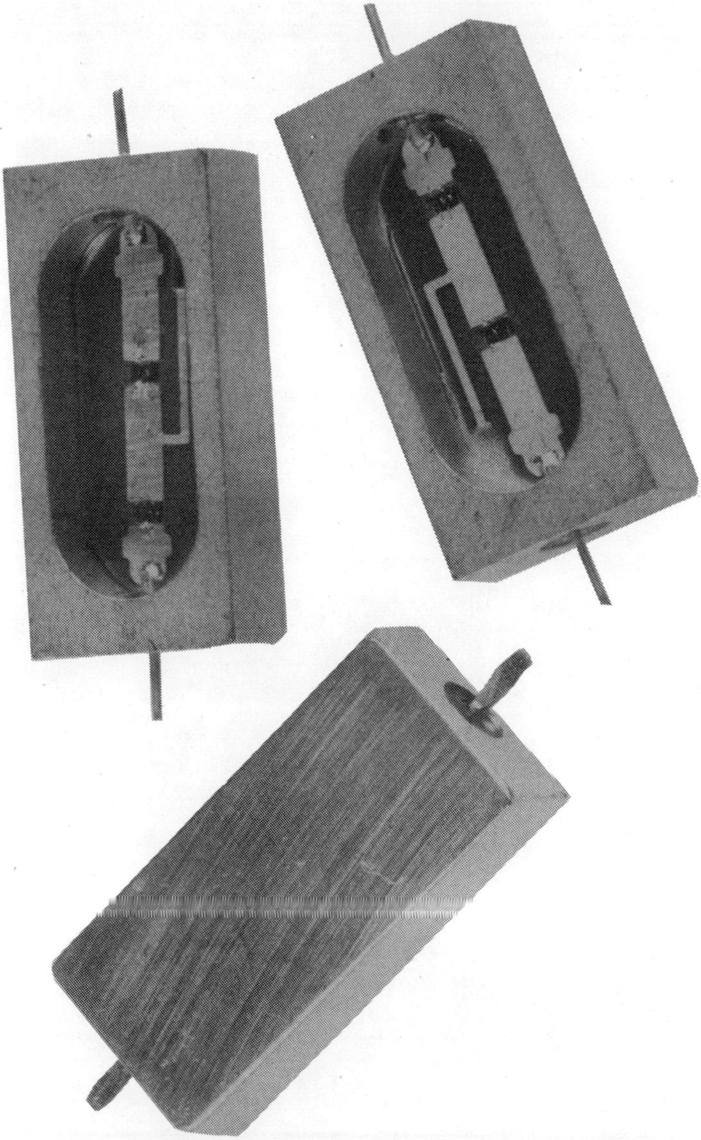

Figure VII-32 Photograph of a Two Stage Integrated Circuit Limiter in a Stripline Modular Package, Measuring 0.53 in (1.35 cm) Long, 0.25 in (0.64 cm) Wide, and 0.125 in (0.32 cm) High. The Glass RF Feedthroughs and Soldered Top of the Package Provide an Hermetic Enclosure.

Figure VII-33 Test Fixture Showing Installation of Integrated Circuit Module in Stripline

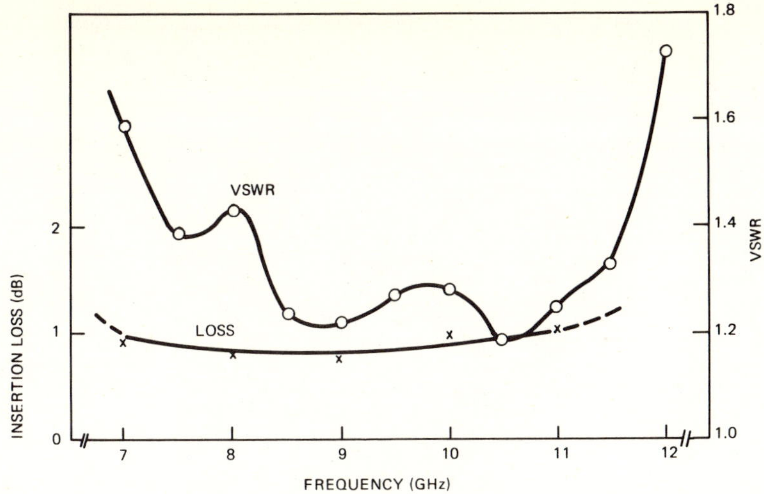

Figure VII-34 **Measured VSWR and Insertion Loss Performance at Low Power for the Two Stage X-Band Limiter Shown in Figure VII-29**

An ionizable dielectric medium is available with silicon and other semiconductor materials. In Figure I-13, the breakdown field in silicon is seen to be around 300 kV/cm for low impurity material. Thus, by installing a piece of silicon in a high RF field circuit location, a form of *avalanche limiter* can be obtained. In principle the thickness of the silicon is limited only by the magnitude of the RF field at which limiting is to be performed – an advantage over diode limiting, as in Figure VII-9, with which the useable diode thickness diminishes with frequency. Such limiting utilizing a material like silicon has been shown practical [13, 14, 15] and is called *bulk limiting*, not because the devices are bulky compared with with diodes, but *because the operation essentially is dependent only on the bulk properties of the semiconductor.* In fact, the bulk limiters are actually more compact than corresponding diode waveguide limiters. As with gas tube duplexers, the semiconductor bulk limiters are placed in the high electric field region of a resonant structure in order for them to limit at suitably low incident RF power levels. Figure VII-35 shows how this arrangement is made by mounting a silicon chip at the center of a waveguide resonant iris. For narrow height waveguide irises which are thin in the direction of propagation (z direction), the resonance occurs at

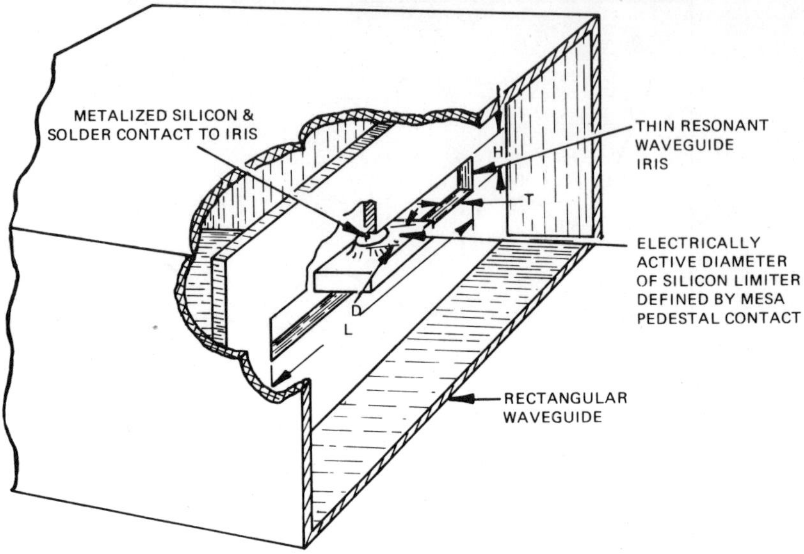

Figure VII-35 **Sketch of Bulk Limiter Arranged by Placing Silicon at High Electric Field Region of Resonant Waveguide Iris**

the frequency for which the iris horizontal slot length is about half the freespace wavelength [16]. With the semiconductor installed in the slot, however, the actual resonant frequency is lower, due to the capacitive loading introduced by the silicon.

Experimentally, it is found that a silicon sample about 0.25 mm (0.010 in) thick produces satisfactory limiting performance when mounted in a resonant iris for which L ≈ 1 cm (0.4 in), H = 0.64 mm (0.025 in), and T = 0.81 mm (0.032 in). A typical silicon chip size is about 1.0 mm (0.040 in) square but the electrically active region is confined to a circular area, typically about 0.76 mm (0.030 in) in diameter. The corresponding silicon thickness is about 0.51 mm (0.020 in). The capacitance of the active region for this example is only about 0.07 pF, lower than that of most of the smallest PIN diodes; yet, because of its thickness — 51 μ compared with 2-20 μ for PIN diodes used as limiters at X-band — its active volume is much greater. Thus, the bulk limiter permits both higher RF power sustaining capacity and broader device bandwidth.

2. Low Power Operation and Equivalent Circuit Modelling

In Chapter V we derived that a SPST switch can be represented as a three-port network, with the switching element terminating one of the ports and generator and load attached to the other two. This representation permitted us to develop the fundamental performance limitations for diodes used as switches. The equivalent circuit may have appeared unduly cumbersome at that time, since practical diode switches are almost always realized simply by mounting the diode directly across the transmission line and tuning out any parasitic reactances. The utility of the general three-port network derivation is more apparent, however, in the analysis of the bulk limiter using the waveguide iris, wherein it is not intuitively obvious just how the equivalent circuit must be handled in order to determine the RF voltage across or RF current through the silicon element.

As derived in Chapter V, the equivalent circuit for any lossless three-port network is as shown in Figure VII-36, under the condition that the reference planes for the three-port have been chosen such that a short circuit at any one decouples the other two. This equivalent circuit is used to represent the metallic portion of the waveguide iris. The assumption of losslessness implicit in the use of this equivalent circuit is reasonable since such iris structures pass energy at resonance with only 0.2 dB or less of insertion loss. The semiconductor element is represented as a lossy capacitive load at port 3, in the same manner as used to analyze SPST diode switches.

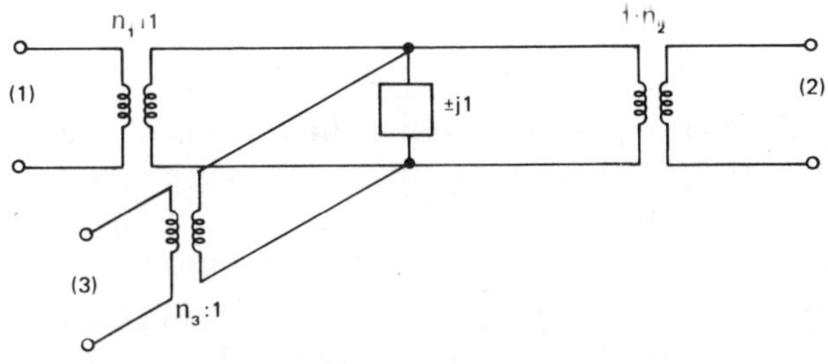

Figure VII-36 **The General Equivalent Circuit for a Lossless Three-Port Under the Reference Plane Selection Which Causes a Short at any Port to Decouple the Other Two**

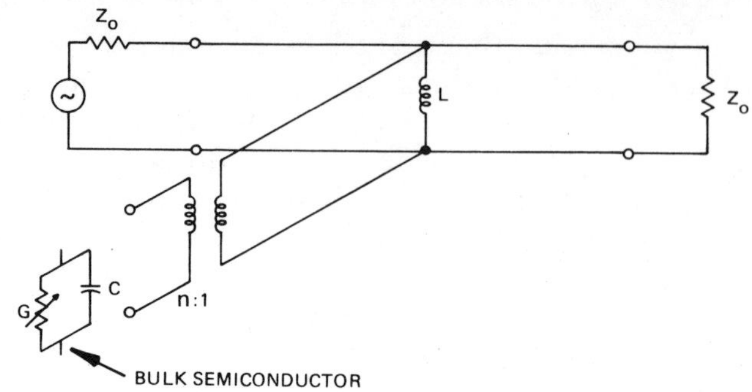

Figure VII-37 **Equivalent Circuit of the Resonant Iris Bulk Limiter, with Matched Generator and Load**

Physical arguments can be used to deduce the simplest form for the equivalent circuit. For a thin iris, placing a short circuit at its center or within the waveguide immediately adjacent to either side of the slot collapses the RF electric fields at all "three ports." Thus, by its very nature, the waveguide iris satisfies the reference plane selection (at and around the resonant frequency) needed to apply the equivalent circuit representation in Figure VII-36. Furthermore, if the same size waveguide is used at both sides of the iris (ports 1 and 2), the resulting symmetry dictates that n_1 equals n_2 and the effects of these transformation ratios can be absorbed within a new value for n_3, which we designate simply as n. Finally, since the reactance of the silicon element at X-band is predominantly capacitive, it follows that, at the resonant frequency, f_0, of the silicon loaded iris, the metallic portion of the iris must be inductive in nature for its parallel combination with the capacitive silicon chip to have zero net susceptance at resonance. The equivalent circuit near resonance then reduces to that shown in Figure VII-37. Of course, this equivalent circuit has so far only been demonstrated to apply at *a single frequency.* Since the decoupling condition for the iris clearly applies over a modest bandwidth about the resonant frequency, the circuit configuration consisting of a single transformer is likewise applicable. However, there is no assurance *a priori* that the turns' ratio, n, can be treated as a frequency invariant constant. We observe shortly that for this simple iris, the value for n does indeed remain essentially constant over the 3 dB transmission bandwidth about resonance, permitting the use of this model. Thus, some very useful deductions can be made

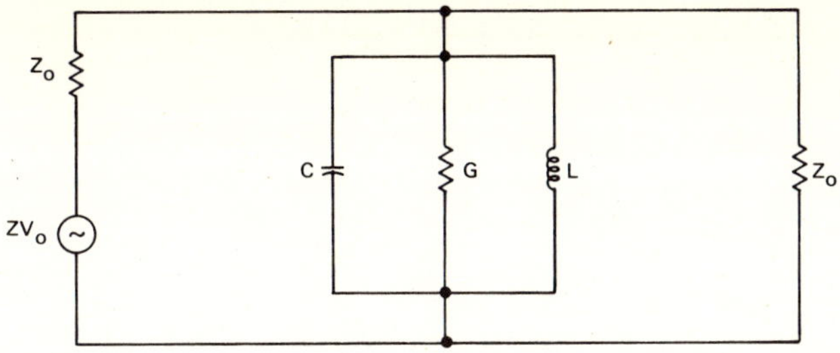

Figure VII-38 **The Simplified Equivalent Circuit for the Bulk Effect Iris Limiter Near Resonance**

about both the RF electric field magnitudes and resulting silicon conductivity values which occur as the incident microwave power is varied.

Since we base the evaluation of the turns' ratio, n, on the measured 3 dB bandwidth of the resonant iris, it is essential that we verify that all of the elements of the equivalent circuit essentially are frequency invariant over the 3 dB bandwidth. Absorbing the turns' ratio, n, into the definition for the effective characteristic impedance, Z_0, of the transmission system, the equivalent circuit reduces to a simple parallel LC resonator, as shown in Figure VII-38. It can be noted that the characteristic impedance of the waveguide itself, its appropriate definition, and its variability with frequency, as well as the turns' ratio and its frequency variation, have all been lumped into the system impedance, Z_0, of this simplified resonance model.

Figure VII-39 shows the measured low level insertion loss of a typical limiter compared with the theoretical universal resonance curve of the simplified LCG parallel equivalent circuit shown in Figure VII-38. The theoretical curve is fitted to the measured data by first adjusting G to give the same loss at resonance as actually measured, then adjusting the L and C values to have the same resonant frequency and 3 dB bandwidth as measured for the actual iris. As can be seen, the measured and calculated loss-versus-frequency contours match closely enough to suggest that the equivalent circuit parameters, Z_0, G, L, and C, indeed can be treated as constants – at least within the 3 dB bandwidth about resonance. Thus, taken as a whole, the fixed turns' ratio transformer model and the parallel resonant circuit model are seen to represent a valid approximation over the 3 dB bandwidth.

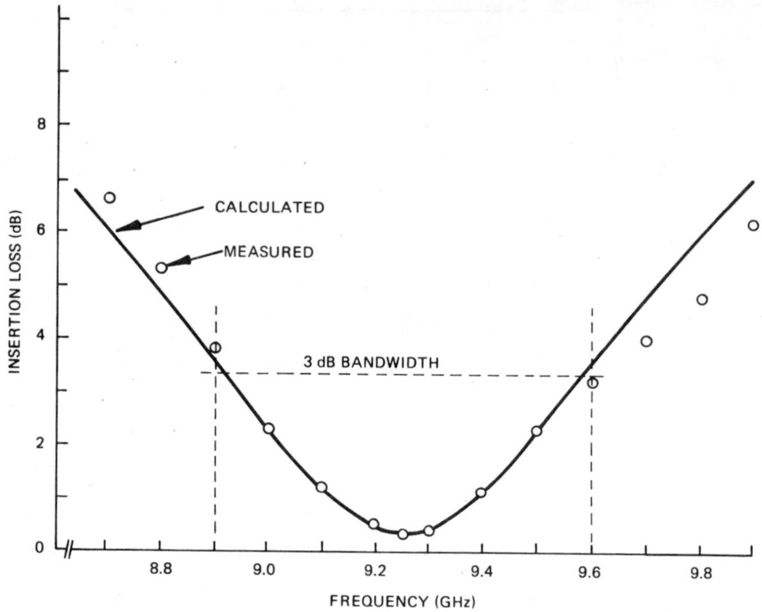

Figure VII-39 **Measured Low Level Insertion Loss of the Bulk Effect Limiter Compared to a Universal Resonance Curve (Sample π-Si-22)**

Our purpose in verifying this assumption for the equivalent circuit arises because of the need to estimate Z_0 from the 3 dB bandwidth and then calculate the RF voltage across the silicon (represented by the capacitor C) as a function for the incident RF power. From this calculation, the limiter circuit can be assessed with respect to its capacity to induce avalanche breakdown within the silicon. In a coaxial system, where an absolute specification of Z_0 is straightforward, this verification would not be necessary. However, with the resonant waveguide iris, the determination of the RF fields at the center of the silicon loaded iris from first principles using a Green's function analysis, to determine Z_0 by the method used for the post coupled diode limiter, would be impractically difficult.

The assignment of specific values to Z_0, G, L, and C in the circuit of Figure VII-38 are determined from measured data as follows. From Equation (II-40) the insertion loss produced by an admittance shunting the transmission line is

$$\text{Insertion Loss} = \left| 1 + \frac{GZ_0}{2} + \frac{jBZ_0}{2} \right|^2 \quad \text{(VII-27)}$$

For the iris limiter the data of which is shown in Figure VII-39, resonance occurs at 9.35 GHz where the insertion loss is 0.35 dB. The net susceptance, B, at resonance is zero and, hence

$$\text{Loss}\,(f_0) = 1.08 = \left(1 + \frac{GZ_0}{2}\right)^2 \quad \text{(VII-28)}$$

At 3 dB band edges, $f_1 = 9.0$ GHz and $f_2 = 9.7$ GHz, the loss increases by 3 dB to 3.35 dB. Then, at these band edges

$$\text{Loss} \approx 2.16 = \left(1 + \frac{GZ_0}{2}\right)^2 + \left(\frac{BZ_0}{2}\right)^2 \quad \text{(VII-29)}$$

from which we can conclude that the contribution to the right side of the equation from the term $(BZ_0/2)^2$ is approximately equal to one and therefore

$$BZ_0 \approx 2 \quad \text{(VII-30)}$$

at the band edges.

It is useful at this point to develop an approximate expression for the net susceptance of a parallel LC circuit at a frequency, f, which is equal to the sum of f_0 and Δf, where f_0 is the resonance frequency and Δf is a small departure in frequency from resonance. The exact expression for B is

$$B = \omega C - \frac{1}{\omega L} \quad \text{(VII-31)}$$

which can be written in terms of the resonance radian frequency, ω_0, and the departure from resonance, $\Delta\omega$, as

$$B = (\omega_0 + \Delta\omega)C - \frac{1}{(\omega_0 + \Delta\omega)\,L} \quad \text{(VII-32)}$$

Defining B_0 as equal to $\omega_0 C$, which in turn is equal to $2\pi f_0 C$, and noting that the capacitive and inductive susceptances are equal at resonance

$$B = B_0 \left(\frac{\omega_0 + \Delta\omega}{\omega_0} - \frac{\omega_0}{\omega_0 + \Delta\omega} \right) \tag{VII-33}$$

$$B = B_0 \left(\frac{\omega_0^2 + 2\Delta\omega\omega_0 + \Delta\omega_0^2 - \omega_0^2}{\omega_0 (\omega_0 + \Delta\omega)} \right)$$

Simplifying and dropping second order terms (on the assumption that $\Delta\omega \ll \omega_0$) gives

$$B \approx B_0 \left(\frac{2\Delta\omega}{\omega_0} \right) = 2 \frac{\Delta f}{f_0} B_0 \tag{VII-34}$$

Equation (VII-34) is of general utility and can also, by duality, be used to determine the net reactance, X, near resonance of a series LC circuit by replacing B and B_0 with X with X_0, respectively.

Returning to the evaluation of the iris limiter, if the approximation for B derived in Equation (VII-34) is substituted into Equation (VII-30),

$$Z_0 = \frac{f_0}{\Delta f_{3\,dB} B_0}$$

$$Z_0 = \frac{1}{2\pi C (f_0 - f_1)} \tag{VII-35}$$

where capacitance is measured in farads and frequency is in hertz. This result is very useful. Equation (VII-35) indicates that the effective switching, or "socket" impedance can be calculated from the 3 dB bandwidth and capacitance. The 3 dB bandwidth is measured easily and the device capacitance can be estimated using Equation (II-38).

For the bulk limiter sample (π-Si-22) the low level loss of which is plotted in Figure VII-38, the electrically active cylindrical volume has a calculated parallel plate capacitance, neglecting fringing, of 0.07 pF. From the loss versus frequency graph, the 3 dB

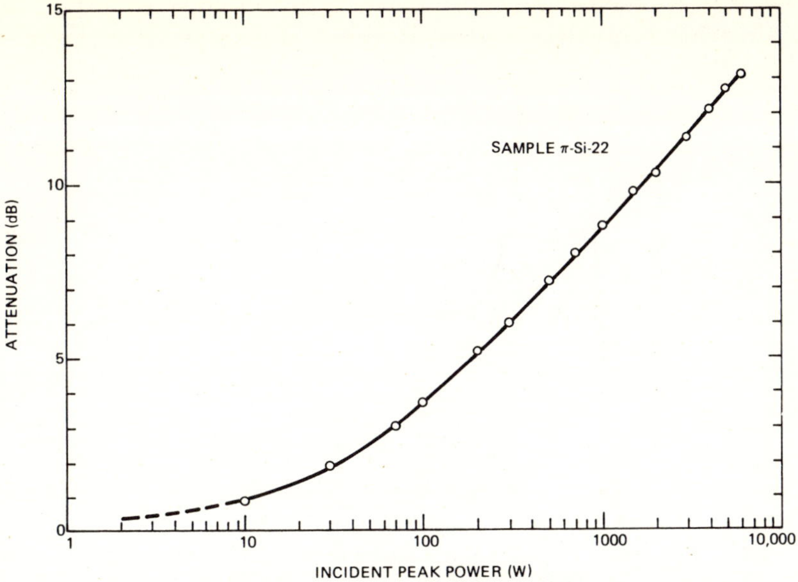

Figure VII-40 **Attenuation vs. Peak Incident Power Measured at 9.35 GHz Using 0.001 Duty Cycle and 1 μs Pulse Length**

bandwidth is 0.7 GHz and thus f_0-f_1 = 0.35 GHz. The effective characteristic impedance in which the limiting is performed is then, from Equation (VII-35), equal to 6500 Ω. This value is a remarkably high impedance level which would be very difficult to realize with such low insertion loss by other circuit impedance transformation methods. The metallic portions of the iris introduce very little loss, since with the silicon chip installed the center frequency loss is only 0.35 dB.

3. High Power Operation

The very high effective impedance results in a relatively large RF voltage across the silicon for a given incident RF power on the iris; accordingly, a high electric field within the iris precipitates the desired avalanche condition for limiting. Figure VII-40 shows the measured isolation versus incident peak power for this iris sample.

An RF pulse length of 1 μs, duty cycle of 0.001, and test frequency of 9.35 GHz were used for these measurements. The limiter in-

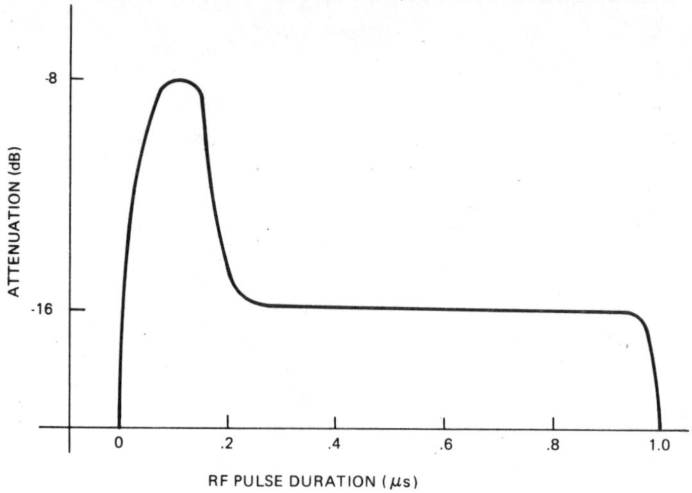

Figure VII-41 **Typical Waveform Observed as Leakage Through Bulk Limiter**

curred an irreversible increase in low power insertion loss after exposure to the 8 kW peak power, and therefore is judged to have failed at this level. The isolation described by Figure VII-39 is the average obtained over the 1 μs pulse; however, the limiting varies within the pulse due to the spike leakage period of approximately 0.2 μs, as shown in Figure VII-41. In a typical application the bulk limiter would be followed by a lower power diode limiter to attenuate the spike and flat leakage to sufficiently low levels to prevent receiver damage.

4. Conductivity versus E Field Estimate

We have, with the equivalent circuit model developed so far, enough information to calculate the absolute value of conductivity, σ, of the silicon chip as a function of the applied RF electric field. This calculation is based upon the measured isolation with applied peak power shown in Figure VII-39, and is performed by estimating both the conductivity which corresponds to each isolation value and the electric field which corresponds to each value of applied power.

Consider first the conductivity calculation. For a cylindrically shaped conductor, the capacitance, C, and conductance, G, measured between its top and bottom surfaces are related to the cross-sectional area, A, and thickness, t, by

$$C = \epsilon_0 \epsilon_R A/t \tag{VII-36}$$

$$G = \sigma A/t \tag{VII-37}$$

Combining, an expression for σ in terms of G and C is

$$\sigma = \epsilon_0 \epsilon_R G/C \tag{VII-38}$$

But the normalized conductance, GZ_0, is related to the insertion loss at resonance (by Equation (VII-28) or the graph in Figure II-17). If Equation (VII-28) is solved for the normalized conductance value, then

$$GZ_0 = 2(\sqrt{\text{Isolation}} - 1) \tag{VII-39}$$

In associating the total conductance GZ_0 with the RF dissipation in the silicon sample, we implicitly ignore dissipation in the metallic portion of the iris; in view of the low insertion loss of the iris, this assumption is reasonable. For example, from Figure VII-39, the isolation reaches 6 dB at an incident RF pulse power of 300 W. From Equation (VII-39), the corresponding value for GZ_0 is 1.99. Since $Z_0 = 6500\ \Omega$, G = 0.0003 mhos, and the corresponding level of conductivity reached within the silicon is, from Equation (VII-38)

$$\sigma = 0.0885 \frac{\text{picofarad}}{\text{centimeter}} \times 11.8 \times \frac{0.0003 \quad \text{mho}}{.07 \quad \text{picofarad}}$$

$$= 0.005\ (\text{ohm centimeter})^{-1}$$

$$= 5\ (\text{kilohm-centimeters})^{-1} \tag{VII-40}$$

Next, let us evaluate the RF electric field within the silicon which produces this level of conductivity. At resonance the net susceptance of the iris is zero. The iris, the generator, and the load form a simple voltage divider network as shown in the simplified equivalent circuit in Figure VII-42. The RF voltage across the silicon chip, V_S, is readily calculated from the available source voltage, V_L, using Equation (VII-41) in Figure VII-42, since the value of GZ_0 is known. From the incident power, P, the source voltage is

$$V_L = \sqrt{PZ_0} \tag{VII-42}$$

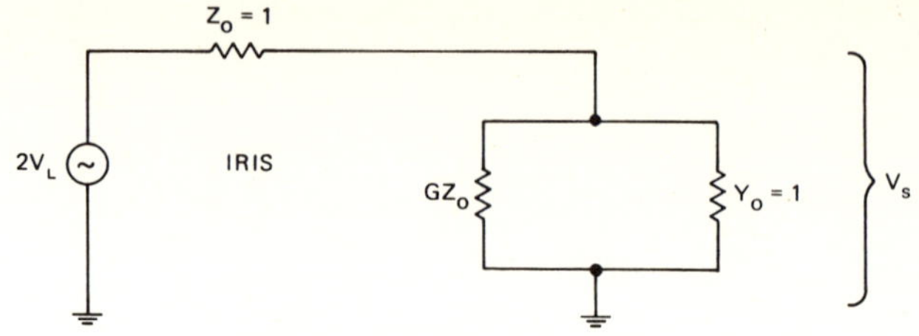

$$V_s = 2V_L \frac{\left(\frac{1}{1 + GZ_0}\right)}{\left(1 + \frac{1}{1 + GZ_0}\right)} = \frac{2V_L}{2 + GZ_0} \qquad \text{(VII-41)}$$

Figure VII-42 **Simplified Normalized Equivalent Circuit for the Bulk Limiter Iris at Resonance used to Estimate Electric Field in the Silicon**

and hence the field in the sample

$$E = V_S/t \qquad \text{(VII-43)}$$

where t is the cylindrical height of the electrically active silicon. Combining Equations (VII-41) and (VII-42) gives

$$E = \frac{2\sqrt{PZ_0}}{t\,(2 + GZ_0)} \qquad \text{(VII-44)}$$

For the present example, for which the isolation is 10 dB at P equal to 1700 W

$$E = \frac{2\sqrt{1700 \text{ watts x } 6500 \text{ ohms}}}{(.05 \text{ centimeter})\ (2 + 4.32)} = 20 \text{ kilovolts (rms)/centimeter} \qquad \text{(VII-45)}$$

In this manner, the conductivity versus applied RF electric field characteristic can be determined from the isolation versus incident RF peak power measurements. For the iris sample (π-Si-22) presently discussed, the resulting characteristic is shown in Figure VII-43. This information is remarkable for the fact that the RF limiting which we presume to be avalanche induced occurs at electric field strengths which are about an order of magnitude lower

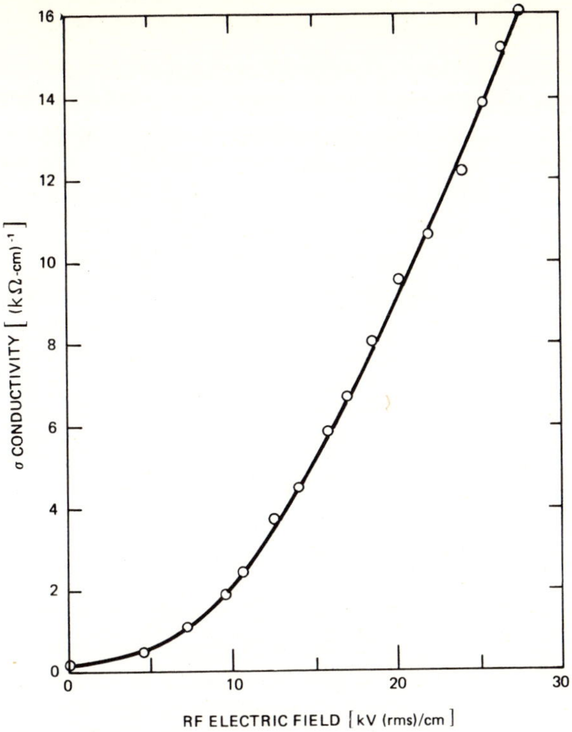

Figure VII-43 **Conductivity versus RF Electric Field for Bulk Iris Sample π-Si-22 Calculated from Isolation vs. Incident Power (Figure VII-39)**

than those observed for the low frequency breakdown in silicon shown in Figure I-13. The reason for this reduction in the field required for avalanche when the excitation is at RF frequencies is not apparent.

5. Blocking Contacts

Until now, silicon has been presented merely as a high resistivity medium with ohmic connections. As noted in Chapter II, however, even the highest resistivity silicon is either weakly P-type (π-material) or N-type (ν-material). Moreover, the sintering of a metal such as nickel (which is used as a base for metallic contacts because of its good mechanical adherence properties) into the silicon creates an N + contact. Thus the ohmic behavior is obtained only with

ν-type silicon; with π-type, a blocking (PN junction) contact results.

The blocking and ohmic contacts at high microwave fields would be expected to have nearly the same behavior since RF conduction is induced by avalanche multiplication and is not principally dependent upon the contacts. But at low RF fields, the predicted insertion loss would be less with a blocking contact resulting from the use of π-material, since the high resistivity silicon is swept free of any residual free charge by the built in junction potential. Experimentally this conclusion is borne out; lower insertion loss is found with the π-type silicon samples. Consequently, experiments have been conducted with π-type material.

6. Return Loss, and Isolation

If the bulk limiter is used only to *protect* a radar receiver from high power, then the specific value of isolation obtained need only be sufficiently large to permit a following lower power varactor or thin base PIN diode to sustain the RF leakage through the bulk device. If, however, the bulk limiter is to serve as the switching element in a radar *duplexer,* such as the example shown in Figure VII-10, it must reach a high enough conducting state to reflect the power incident upon it with little absorption (with high *return loss*), since *power absorbed,* P_{AB}, in the limiter and in the load behind it appears as insertion loss in the radar's transmission mode. Figure VII-44 shows this situation schematically.

The fraction of the incident power dissipated in the limiter is called *arc loss.* The remaining portion of the absorptive loss results from leakage of power to the load. *The reciprocal of the ratio of the leakage power dissipated in the load to the incident power is the isolation.* As with insertion loss, the reciprocal is used so that isolation in decibels is a positive quantity.

We have already evaluated isolation with Equation (VII-28) for a single element, purely conductive limiter. Arc loss is expressed using the approximate relationship in Equation (II-40), which is valid when the limiter's normalized conductance is large compared to unity. However, with the bulk limiter (and with some self-actuated diode limiters), the conductance is not always large enough to justify this approximation. Let us therefore distinguish among, and evaluate, *absorptive loss, return loss,* and *arc loss* (ratios) using exact relationships.

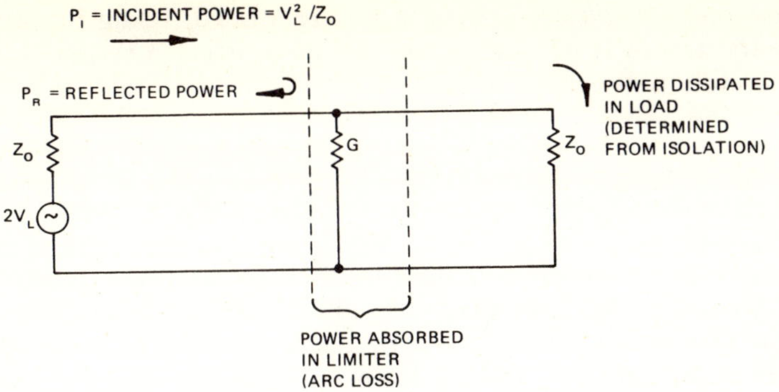

Figure VII-44 **Representation of Limiter Circuit Showing Return Loss, Arc Loss, and Isolation**

The return loss ratio is best calculated from the reflection coefficient evaluated at the generator side of the limiter.

$$\Gamma = \frac{1 - y}{1 + y} \tag{VII-46}$$

where

$$y = 1 + GZ_0 \tag{VII-47}$$

The *return loss is the fraction of incident power which is reflected back toward the generator,* and is given by

$$\text{Return Loss Ratio} = \frac{P_R}{P_I} = |\Gamma|^2$$

$$= \left(\frac{GZ_0}{2 + GZ_0}\right)^2 \tag{VII-48}$$

The *power absorbed ratio, P_{AB}/P_I which represents the fraction of the incident power absorbed by G and the matched load, Z_0,* is then

$$\text{Power Absorbed Ratio} = \frac{P_{AB}}{P_I} = 1 - |\Gamma|^2 \tag{VII-49}$$

The *power absorbed ratio represents insertion loss on transmit for the balanced duplexer* shown in Figure VII-10. Using Equations (VII-46) and (VII-47), this ratio can be written in terms of normalized conductance directly as

$$\text{Absorptive Loss (transmit)} = \frac{P_{AB}}{P_I} = \frac{4\,(1 + GZ_0)}{(2 + GZ_0)^2} \qquad \text{(VII-50)}$$

The arc loss and the power dissipated in the load (as fractions of the incident power) are easily calculated as fractions of the absorptive loss (which is just their sum). These separate fractions of the power absorbed by G and Z_0 are proportional to their respective conductances divided by the total conductance. Thus

$$\text{Arc Loss Ratio} = \text{Absorptive Loss Ratio} \times \left(\frac{GZ_0}{1 + GZ_0}\right) \qquad \text{(VII-51)}$$

$$= \frac{4\,GZ_0}{(2 + GZ_0)^2}$$

and the *leakage* which is the fraction of the incident power that is dissipated in the load is

$$\text{Leakage} = \text{Absorptive Loss Ratio} \times \frac{1}{1 + GZ_0}$$

$$= \frac{4}{(2 + GZ_0)^2} \qquad \text{(VII-52)}$$

It can be seen that, as expressed by Equation (VII-52), *leakage is equal to the reciprocal of the isolation* given by Equation (VII-28).

The return loss, absorptive loss, and arc loss ratios (in percent) are shown graphically as functions of the normalized conductance in Figure VII-45. The variations in these functions can best be understood by considering the limiter's operation as its conductance increases from zero to its maximum value.

When GZ_0 is small compared to unity, practically all of the incident power travels by it and is absorbed in the load. For the duplexer circuit shown in Figure VII-10, this situation is desirable in the *receive* condition, but would be disastrous on *transmit* because the high power from the transmitter would burn

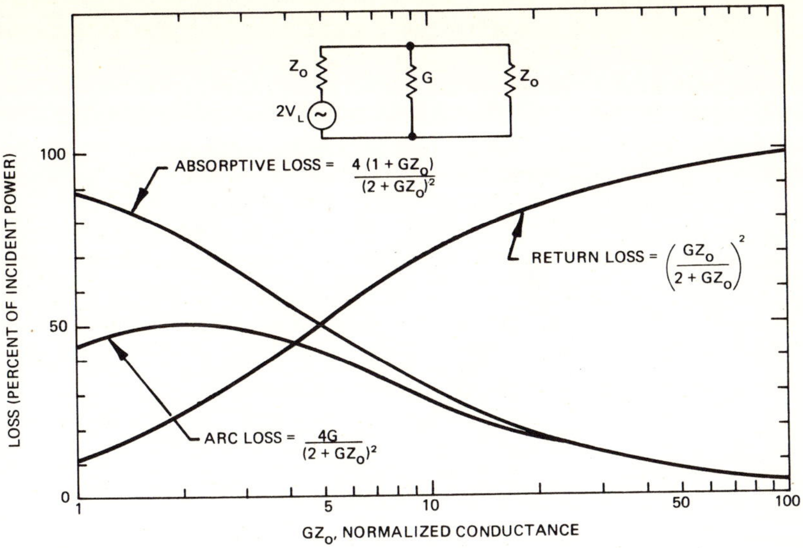

Figure VII-45 **Return Loss and Arc Loss in Percent of Incident Power vs. Normalized Conductance Values**

out the receiver. Even if the receiver could sustain the power, the transmit loss would be infinite since transmitter power would not be directed to the antenna. The return loss with $GZ_0 \approx 0$ is nearly 100%. In the receive condition, however, this figure is what we want, since the low level received signal at the antenna (Figure VII-10) must not be reflected from the limiters and so be directed to the transmitter rather than the receiver.

As the limiter conductance increases (with high incident power), corresponding to the transmit condition, it absorbs some of the incident power, producing arc loss. *A maximum arc loss occurs when GZ_0 equals 2; at that point, 50% of the power incident is absorbed in the limiter, 25% is absorbed in the load, and 25% is returned.* Expressed in decibels, the corresponding values are

Return Loss = 6 dB
Arc Loss = 3 dB
Leakage (Isolation) = 6 dB

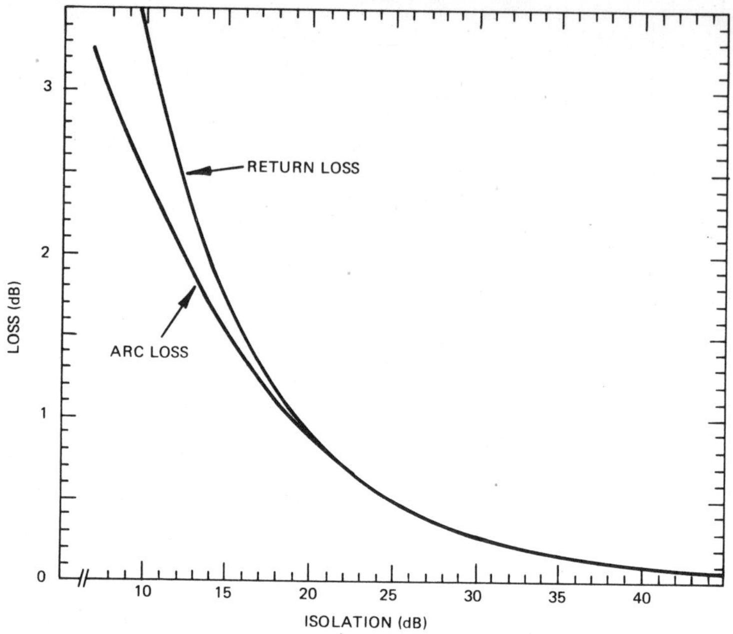

Figure VII-46 **Return Loss and Arc Loss as Functions of the Isolation Produced by a Single Line Shunting Conductance**

As the normalized conductance increases further the arc loss accounts for an increasingly greater fraction of the absorptive loss, *but the total power absorbed is less,* since more of the incident power is reflected. Figure VII-46 shows absorptive and arc losses versus related isolation, in decibels as they would be measured practically. Notice, for example, that for the duplexer to have no more than 0.5 dB of insertion loss on transmit, the absorptive loss must not exceed 0.5 dB and, from Figure VII-45, the isolation must be at least 25 dB. Using Equation (VII-39), the minimum value for GZ_0 corresponds, in this example, to

$$GZ_0 = 2(\sqrt{316} - 1) \qquad \text{(VII-53)}$$

$$= 34$$

Thus, it can be seen that for duplexer operation the achievable conductance is more constrained than for limiter use alone, and must satisfy both the requirements for receive protection and for low transmission loss under high power.

I wish to thank H. Griffin, N. Brown, and C. Genzabella for the design assistance and the Navy Bureau of Ships for sponsoring much of the development of the coaxial duplexer. R. Ekinge and F. Jellison provided the waveguide limiter analysis and experimental results. D. Higgs, M. Litchfield, and J. Kim contributed the integrated circuit limiter. K. Mortenson, A. Armstrong, P. Bakeman, and J. Park performed most of the development of the bulk limiter.

References

[1] Leenov, D.: "The Silicon PIN Diode as a Microwave Radar Protector at Megawatt Levels," *IEEE Transactions on Electron Devices, Vol. ED-11, No. 2,* pp. 53-61, February, 1964.

This paper is a classic on the subject of PIN diode conductivity modulation with dc and/or RF excitations.

[2] Brown, Norman J.: "Design Concepts for High-Power PIN Diode Limiting," *IEEE Transactions on Microwave Theory and Techniques, Vol. MTT-15, No. 12,* pp. 2-742, December 1967.

[3] Garver, Robert V.: *Microwave Diode Control Devices,* Artech House, Inc., Dedham, Massachusetts, 1976.

[4] Walker, R.M.: "Waveguide Impedance — Too Many Definitions," *Electronic Communicator, Vol. 1,* p. 13, May-June 1966.

[5] White, J.F.: "Simplified Theory for Post Coupling Gunn Diodes to Waveguide," *IEEE Transactions on Microwave Theory and Techniques, Vol. MTT-20, No. 6,* pp. 372-378, June 1972.

[6] Eisenhart, R.L.; and Khan, P.J.: "Theoretical and Experimental Analysis of a Waveguide Mounting Structure," *Transactions on Microwave Theory and Techniques, Vol. MTT-19, No. 8,* pp. 706-719, August 1971.

[7] Collin, R.E.: *Field Theory of Guided Waves,* McGraw-Hill, Inc., New York, 1960.

[8] Marcuvitz, N.: *The Waveguide Handbook* (Vol. 10, Radiation Laboratory Series), McGraw-Hill, Inc., New York, 1951, pp. 257-258.

[9] Deloach, B.C.: "A New Microwave Measurement Technique to Characterize Diodes and an 800 GHz Cutoff Frequency Varactor at Zero Bias," *1963 IEEE MTT Symposium Digest,* pp. 58-69.

[10] Wheeler, Harold A.: "Transmission Line Properties of Parallel Strips Separated by a Dielectric Sheet," *IEEE Transactions on Microwave Theory and Techniques. Vol. MTT-13, No. 3,* pp. 172-185, March 1965.

[11] Kwon, Andrew H.: "Microcomputer Program," *1976 Microwave Journal Handbook and Buyers Guide.* A FORTRAN IV computer program to calculate microstrip line characteristic impedance based on the work of Wheeler, Reference 10 (reprinted in Appendix E).

[12] Howe, Harlan Jr.: *Stripline Circuit Design,* Artech House, Inc., Dedham, Massachusetts, 1974.

[13] Mortenson, K.E.; and White, J.F.: "Nonrefrigerated Bulk Semiconductor Microwave Limiters," *IEEE Journal of Solid State Circuits, Vol. SC-3,* pp. 5-11, March 1968.

[14] Mortenson, K.E.; et al.: "Bulk Semiconductor Microwave Limiters," *TR-ECOM-0186-F* (Parts I and II, Report on Study Funded by US Army Electronic Command, Fort Monmouth, New Jersey, at Rensselaer Research Corp., Troy, New York, August 1970; 215 pages total).

[15] Armstrong, A.L.: "Bulk Semiconductor Limiters," *ECOM-0292-F* (Final Report for US Army Electronics Command, Fort Monmouth, New Jersey under Contract 72-C-0292 for Limiter Study at Rensselaer Research Corp., Troy, New York, March 1974; 54 pages total).

[16] Moreno, Theodore: *Microwave Transmission Design Data,* Dover Publications, Inc., New York, 1948.

Chapter VIII

Switches and Attenuators

A. Broadband Coaxial Switches

1. Introductory Remarks

In the design of microwave switches — where the diode control element is expected to switch between nearly open and nearly short circuit impedances — it would seem that only the use of packageless diodes is practical since the parasitic inductance and capacitance associated with diode packages cannot help but have an adverse effect on switching performance. Though this observation is correct, packaged diodes offer many advantages. Switches can be built with replaceable diode sockets, permitting simple maintenance; packaged diodes can be used, eliminating the need for the skilled labor required for chip assembly. The interchangeability of diodes in packaged form affords the possibility of special fine tuning which is impractical with diode chips that are soldered or bonded into the switch device.

In this section, the techniques for broadbanding both the transmission and the isolation states of a SPDT switch are examined. Much of the theory presented also applies to chip diode circuit designs, since it involves handling of the spacing requirements for the diodes in the isolate state and tuning of the diode capacitance in the transmission state.

The stimulus for the particular approach to be described came from a correspondence by Fisher [1] in which he suggested that the capacitance of a diode could be tuned out with a short-circuited length of transmission line less than a quarter wavelength long, and that the resulting parallel resonant circuit could be made to have properties similar to those of a simple quarter wave resonant stub. These results are achieved when the tuning is performed with a stub of the proper characteristic impedance and length. This method of simulating a quarter wave stub using a diode and its parallel tuner has proved to be practical over an octave bandwidth [2].

However, in order for the switch to have a practical octave bandwidth, it must have good performance in the isolation state, as well as in the transmission state. Packaged diodes, particularly, have series inductances which limit the amount of isolation that can be obtained over a given bandwidth even if the diode's resistance could be reduced to zero. Package inductance can be series resonated using a series capacitor; the question becomes at what frequency (or frequencies, if two different diodes are used to obtain the necessary isolation) should the resonance be set to achieve the highest minimum value of isolation throughout the bandwidth? In other words, how should the diodes be series tuned for the minimum value of isolation obtained anywhere in the desired passband to have the largest value?

In answering these questions, the elements of design theory that are applicable to many switching applications are explored. These topics include the application of bandpass filter theory to the transmission state of the diode switch, the isolation to be obtained from "quarter wave spaced" diodes (i.e., what constitutes a good approximation to quarter wave spacing), the effect of series resonant inductive tuning on the transmission state capacitance of a diode, and the effect of stagger series tuning on maximum isolation bandwidth.

2. The Transmission State

Fisher's Equivalent

Figure VIII-1 shows the equivalent circuit of an idealized diode represented as a simple capacitance tuned with an inductive stub to parallel resonance at some desired frequency. The resultant capacitor-stub combination, of course, has zero net susceptance at the resonant frequency and a susceptance-versus-frequency characteristic which is dependent upon C, Y_T, and θ_T. Near resonance, this characteristic can be approximated graphically as a straight line. The susceptance-versus-frequency characteristic of a quarter wave stub about its resonant frequency is a tangent function which likewise graphs as a relatively straight line. Fisher's equivalence involves choosing the diode tuning stub admittance length so that the slopes of these two characteristics in the vicinity of resonance are matched. The equivalence, then, represents an

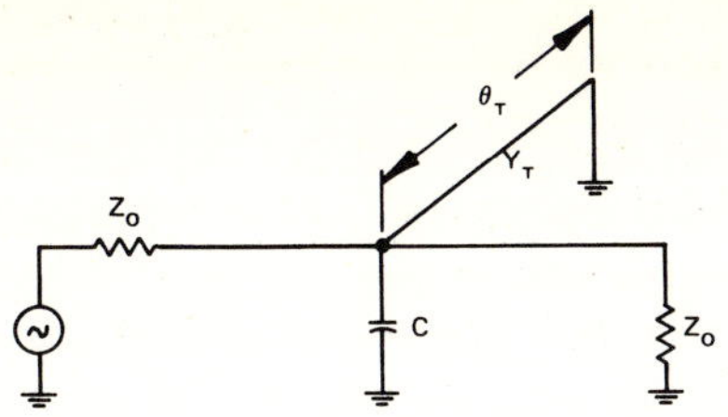

a) TUNED DIODE IN SHUNT WITH THE LINE

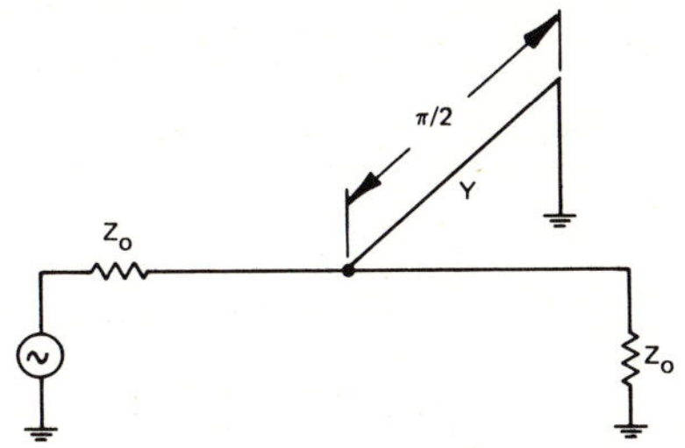

b) EQUIVALENT 90° STUB IN SHUNT WITH THE LINE

Figure VIII-1 **The Parallel Resonated Diode and its Quarter Wavelength Stub Equivalent Circuit**

effective Taylor simulation of a quarter wave stub by the capacitor and shunt tuner; that is, at the resonant frequency, the susceptance and its derivative with frequency are equal to that of the quarter wave stub. The advantage gained is that there is a large body of information available about the characteristics of bandpass filters which use quarter wave stubs as their elements. By finding the equivalence of a tuned diode and quarter wave stub, a composite bandpass filter design which uses quarter wave stubs can be formulated, increasing the achievable bandwidth and making the particular performance to be obtained directly calculable from filter theory.

The conditions to be met are found by requiring that the diode and its parallel tuner be resonant at some specified frequency, ω_0, and that its susceptance-versus-frequency derivative matches that of the quarter wave stub (also resonant at ω_0) which it replaces. The resulting expressions are obtained as follows. Figure VIII-2 shows the equivalent circuit models of the diode capacitance with parallel tuning stub, Y_T, and the equivalent quarter wave stub, Y_E. The admittances looking into these two networks are totally

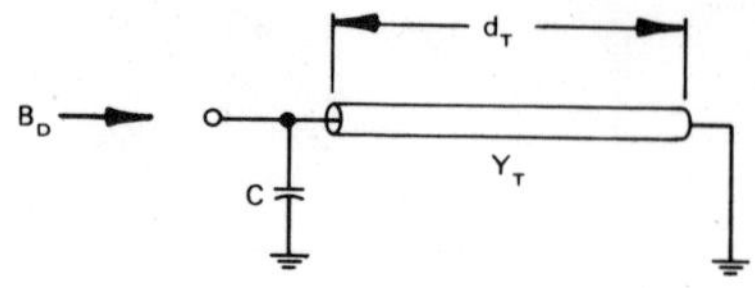

a) DIODE CAPACITANCE PARALLEL RESONATED USING SHORTED INDUCTIVE STUB

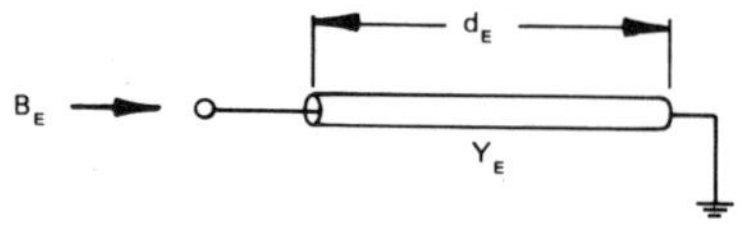

b) EQUIVALENT QUARTER WAVELENGTH STUB

Figure VIII-2 **Circuit Models used for Fisher Equivalent Circuit**

susceptive since the effects of dissipative insertion loss are neglected for this analysis. The susceptance values are represented as B_D and B_E, respectively. The susceptance of the parallel tuned diode is given by

$$B_D = j\omega C - jY_T \cot \frac{\omega d_T}{V}$$

where V*, the velocity of propagation on the tuning stub, is

$$V = \frac{3 \times 10^8}{\sqrt{\epsilon_R}} \quad \text{meters/second} \qquad \text{(VIII-1)}$$

The susceptance for the equivalent quarter wave stub similarly is

$$B_E = -jY_E \cot \frac{\omega d_E}{V} \qquad \text{(VIII-2)}$$

The corresponding radian frequency derivatives are then

$$\frac{dB_D}{d\omega} = j\left(C + Y_T \frac{d_T}{V} \csc^2 \frac{\omega d_T}{V}\right) \qquad \text{(VIII-3)}$$

*See Appendix A, p. 501, for a more precise value of the velocity of light.

$$\frac{dB_E}{d\omega} = jY_E \frac{d_E}{V} \csc^2 \frac{\omega d_E}{V} \tag{VIII-4}$$

Equating the derivatives at the center frequency, ω_0, and noting that at ω_0

$$d_E = \frac{\lambda_0}{4} \tag{VIII-5}$$

$$\text{hence } \csc^2 \frac{\omega d_E}{V} = 1$$

gives

$$Y_E \frac{\lambda_0}{4V} = C + Y_T \frac{d_T}{V} \csc^2 \frac{\omega_0 d_T}{V} \tag{VIII-6}$$

noting that

$$V = f\lambda$$
$$\omega = 2\pi f$$

hence

$$\frac{\lambda_0}{4V} = \frac{\pi}{2\omega_0}$$

Furthermore at ω

$$\omega_0 C = Y_T \cot \frac{\omega_0 d_T}{V}$$

$$\text{hence } \frac{\omega_0 d_T}{V} = \text{arc cot } \frac{\omega_0 C}{Y_T}$$

But if $\cot x = B/A$; $\csc^2 x = 1 + (B/A)^2$

$$\text{therefore,} \qquad \csc^2 \left(\frac{\omega_0 d_T}{V}\right) = 1 + (\omega_0 C/Y_T)^2 \tag{VIII-7}$$

With these expressions it is possible to rewrite Equation (VIII-6) as

$$Y_E = \frac{2}{\pi} \left\{ \omega_0 C + Y_T \left[\text{arc cot} \left(\frac{\omega_0 C}{Y_T} \right) \right] \cdot \left[1 + \left(\frac{\omega_0 C}{Y_T} \right)^2 \right] \right\} \tag{VIII-8}$$

This result is the original by Fisher; it relates the equivalent stub's admittance to the capacity and admittance of the tuning stub used to parallel resonant the capacitor. Since the result is transcendental, it is desirable to graph it for convenient application. In order to obtain a single graph in two dimensions it is necessary to reduce the three variables — Y_E, Y_T, and C — to two. This result can be achieved by normalizing any two of the variables to the third. It is most convenient to normalize the capacitance susceptance, $\omega_0 C$, and the admittance, Y_T, to the desired equivalent stub admittance, Y_E. Then Equation (VIII-8) can be rewritten as

$$\frac{\pi}{2} = \frac{\omega_0 C}{Y_E} + \frac{Y_T}{Y_E} \left[\text{arc cot} \left(\frac{\omega_0 C}{Y_E} \cdot \frac{Y_E}{Y_T} \right) \right] \left[1 + \left(\frac{\omega_0 C}{Y_E} \cdot \frac{Y_E}{Y_T} \right)^2 \right] \tag{VIII-9}$$

This equation is plotted, along with the condition for resonance given in Equation (VIII-7), in Figure VIII-3. For more accurate estimation, Table VIII-1 gives the corresponding variable values for common steps in the evaluation of $\omega_0 C/Y_E$.

Examination of the two curves in Figure VIII-3 shows that for small normalized susceptance, $\omega_0 C/Y_E$, the required tuning stub is nearly a quarter wavelength long ($\theta \approx 90°$) and its normalized admittance, Y_T/Y_E, is nearly unity. These values are as expected

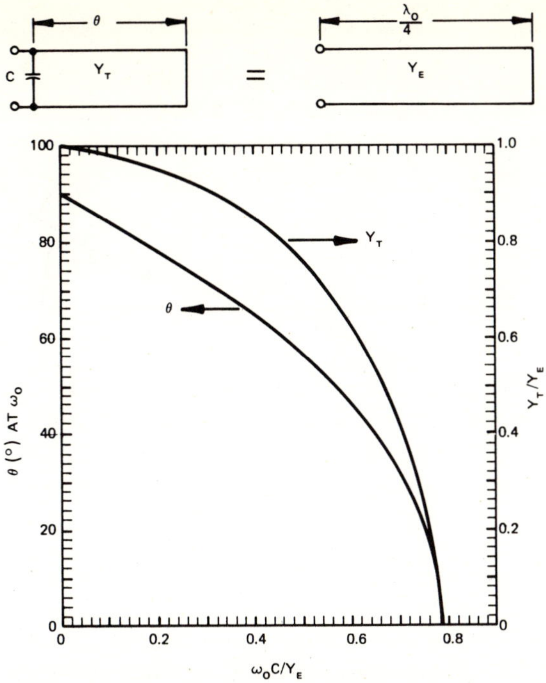

Figure VIII-3 Conditions for Simulating a Quarter Wave Stub with a Tuned Diode

$\omega_0 C/Y_E$	$Y_T Y_E$	θ (° at f_0)
0.	1.000	90.0
0.1	.985	84.2
0.2	.956	78.4
0.3	.915	71.8
0.4	.850	64.8
0.5	.760	56.7
0.6	.627	46.2
0.7	.436	31.9
0.750	.300	21.8
0.780	.120	8.75
0.7854 (π/4)	.000	.00

Table VIII-1 Tuned Capacitor Values for Simulating a Quarter Wave Stub

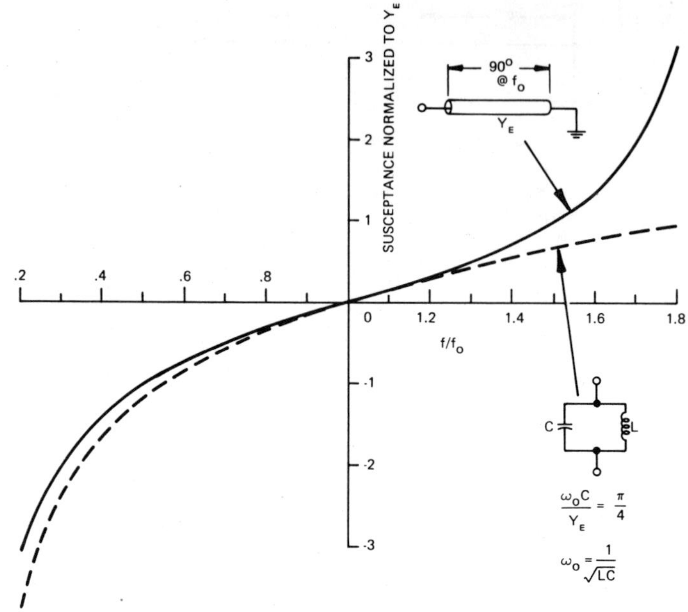

Figure VIII-4 **Comparison of Susceptance Behavior of Quarter Wave Stub with Tuned Capacitance Simulated Stub**

since a small capacitive susceptance requires that almost the entire job of simulation is borne by the tuner, Y_T. Since the tuner and simulated device are identical under zero susceptance, the approximation can be expected to be best for small $\omega_0 C$ values. But small C implies low power handling capability; thus, it is often desirable to use the largest possible C.

The largest allowable value for $\omega_0 C/Y_E$ is $\pi/4$. As can be seen from the curves, θ and Y_T/Y_E are both equal to zero at this value. This situation can be interpreted as the limiting case in which the tuning is performed by a *lumped inductor* rather than a distributed line length. Although the equivalence between the quarter wave stub and the tuned diode is assured by design only in the vicinity of f_0, it turns out that, even for the limiting case when $\omega_0 C/Y_E$ has its maximum value, the susceptance-versus-frequency contours of the two circuits agree within 20% over the range $0.65 < f/f_0 < 1.3$, *a bandwidth ratio of 2 to 1.* This fact is shown graphically in Figure VIII-4 along with the defining equations for C and L in this limiting case.

Building the Diode into a Filter

Having found a means for simulating a quarter wave stub with a tuned diode, the next consideration is how this component can be built into a bandpass filter, thereby achieving a better transmission bandwidth than could be obtained with an ordinary tuned diode. The simplest bandpass filter to employ is a maximally flat one, which has perfect match at a center frequency and a gradual monotonic degradation with frequency deviation from the band center.

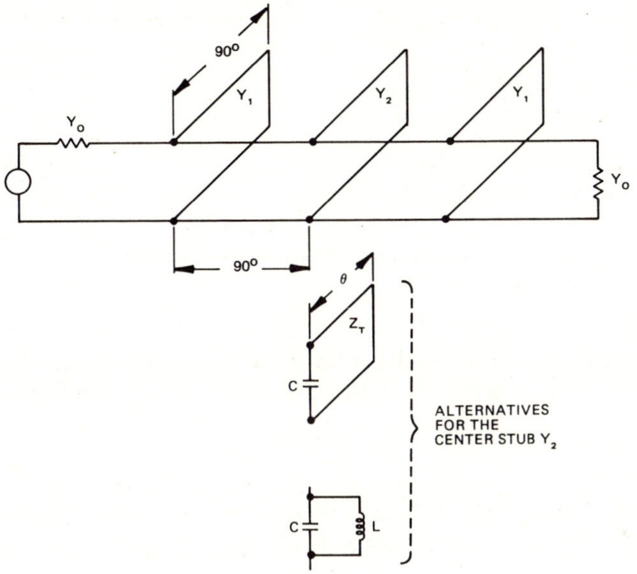

Figure VIII-5 **Three Element Bandpass Filter Showing Replacement of Center Stub with Tuned Diode**

A three element bandpass filter is shown in Figure VIII-5. All stub lengths and spacings are 90° at the center frequency, f_0. The choice of $Y_1 = Y_0$ and $Y_2 = 2Y_0$ yields a maximally flat transmission response about f_0. The performance with frequency, calculated using program WHITE (derived in Chapter VI), is shown in Figure VIII-6, together with the response obtained when the center stub is replaced with a tuned capacitor (diode) for two conditions, $\omega_0 C/Y_2$ equal to 0.4 or 0.7854 ($\pi/4$). In practical terms, if the filter is designed for a 50 Ω system ($Z_0 = 50\ \Omega$, $Y_0 = 0.02$ mhos), the Y_1 stub has a characteristic impedance of 50 Ω and the Y_2 stub has one of 25 Ω.

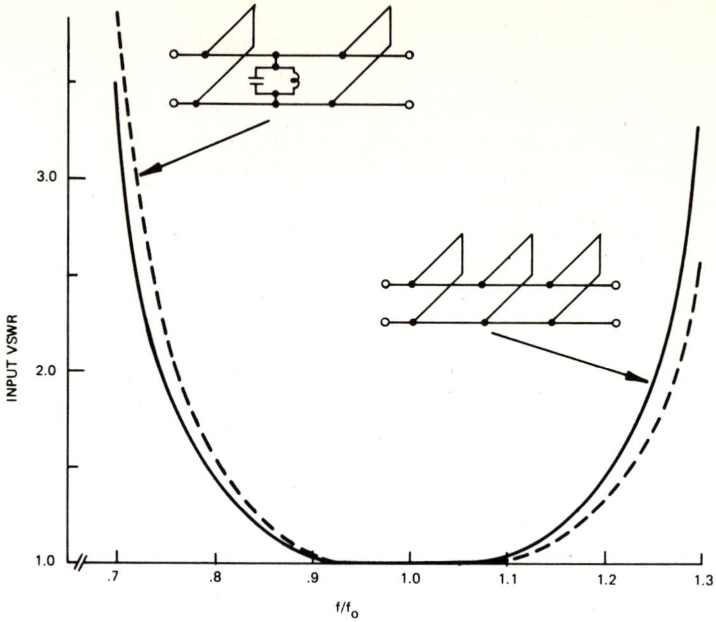

Figure VIII-6 **Maximally Flat Stub Filter Response Compared with the Same Filter using a Tuned Diode Center Stub**

For $\omega_0 C/Y_2 = 0.4$, the reactance of C must be 62.5 Ω. If, for example, the center frequency is chosen to be 1 GHz, the value for C equals 2.55 pF, a fairly large diode capacitance. The corresponding tuning stub (from Table VIII-1), has a normalized admittance, Y_T/Y_2, equal to 0.85 and length, θ, equal to 64.8°. The characteristic impedance, Z_T, of the tuning stub is then 25 Ω/0.85 = 29.4 Ω.

For $\omega_0 C/Y_2 = \pi/4$, the reactance of C is 31.8 Ω, corresponding at 1 GHz to a 5.0 pF capacitance. The shunt inductance, L, which resonates at 1 GHz is 5.07 nH.

The passband VSWR response of the three stub filter is shown in Figure VIII-6 with the response obtained when the center stub is replaced with the inductively tuned diode. As can be seen, the simulated stub performance closely approximates that of the stub filter; of course, even closer approximation can be expected when a somewhat lower diode capacitance is tuned with the appropriate inductive stub.

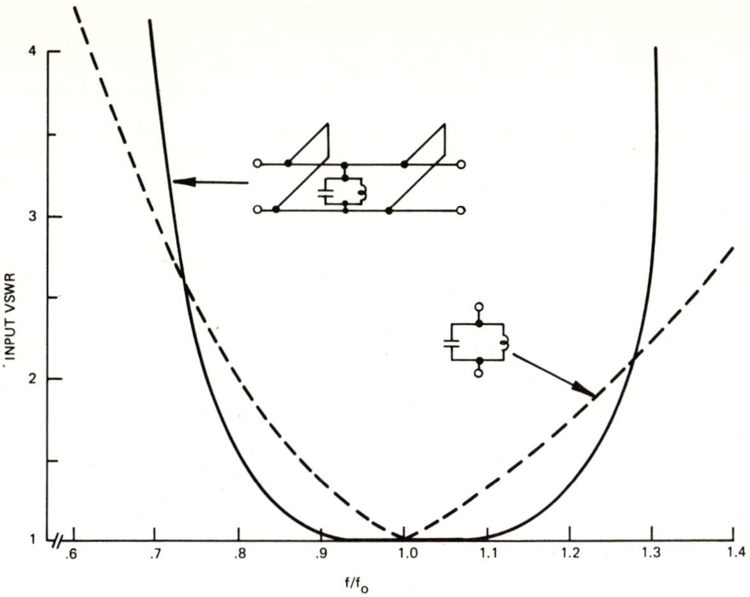

Figure VIII-7 **Response of Filter Tuned Diode Compared with that for Induction Tuning**

The next potential question is how much better is the transmission match obtained when the tuned diode is incorporated within a filter than that which would be achieved using the tuned diode alone? Figure VIII-7 shows the VSWR performance with frequency for the filter tuned diode compared with the response obtained when tuning is realized with only the inductor. It can be seen that, rather than increase the transmission VSWR bandwidth, the use of an added filter *redistributes the response.* That is to say, VSWR is improved, over the mid-section of the band while it is degraded beyond some frequencies. In the example shown in Figure VIII-7, VSWR is reduced (compared to simple inductive tuning) by the added filter stubs within $0.75 < f/f_0 < 1.25$ but increased outside of this bandwidth. A filter design using a greater number of elements would permit some further reduction of VSWR within the passband at the expense of steeper *skirts* at the band edges and, of course, at the expense of greater structural complexity, as well.

Overall reduction in VSWR for wider bandwidth is accomplished only by reducing the magnitude of the normalized capacitive sus-

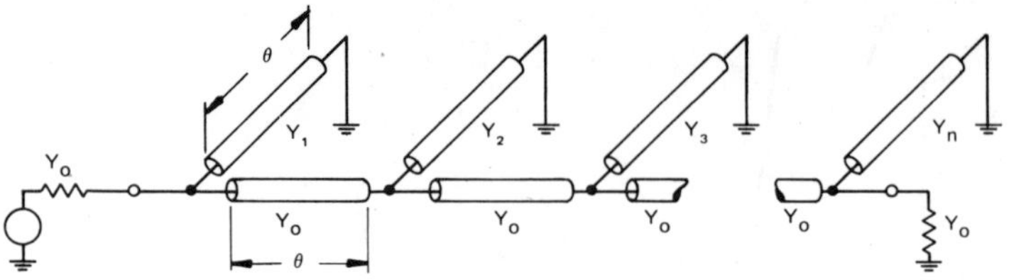

Figure VIII-8 **Equivalent Circuit for Mumford's Maximally Flat Filter Design Tables**

ceptance*. For this example the largest capacitance associated with stub simulation, 5 pF with $f_0 = 1$ GHz, is used to demonstrate the relative tuning effectiveness. In practice, a lower capacitance normally would be used and a broader bandwidth would be obtained for a given allowed maximum VSWR and/or reflective insertion loss.

Mumford's Bandpass Filter Tables

Virtually any filter design criterion which utilizes either stub or capacitor elements can be used to optimize the transmission characteristics of a diode switch. For a treatment of filter theory and design data, Matthei, Young, and Jones' work is very comprehensive [3]. The maximally flat filter consisting of shorted quarter wavelength stubs with quarter wavelength spacing has been summarized in tabular form by Mumford [4] and is particularly easy to apply. The form of the filter is shown in Figure VIII-8.

The insertion loss of the filter (measured when it is placed between a matched resistive generator and a load having the same characteristic admittance, Y_0, as that of the main line of the filter) can be found at any frequency from

$$\frac{P_0}{P_L} = 1 + K_N \frac{\cos^{2N}\theta}{\sin^2\theta} \qquad \text{(VIII-10)}$$

where N is the number of shorted stubs of length l and

$$\theta = 2\pi l/\lambda \qquad \text{(VIII-11)}$$

*Normalized susceptance must be used to define bandwidth limits since use of a lower main line impedance, of course, always reduces the reflection caused by a given capacitive obstacle.

$$K_N = \left[\frac{k_1 (k_2 + 2) \ldots\ldots (k_N + 2)}{2} \right]^2 \qquad \text{(VIII-12)}$$

and

$$k_R = Y_R / Y_0 \qquad \text{(VIII-13)}$$

is the admittance of the Rth stub normalized to the main line.

Mumford's values for filters with 3, 4, and 5 elements are shown in Table VIII-2; the original reference [4] can be consulted for designs using 6 to 10 stubs. The filters are symmetrical; therefore, only the normalized admittances for the first stub to the center are listed. As can be seen from the tabulated values, the largest admittance stubs are located at the center of the filter; accordingly, this point is where the diode should be embedded if the largest possible capacitance is to be accommodated. This fact has a special significance with respect to SPDT switches, as is illustrated later.

Table VIII-2 does not contain two-stub filter values since, by symmetry, the characteristic admittances of both stubs must be equal ($k_1 = k_2$) so no tables are necessary. Equation (VIII-10) is applicable to the two-stub filter. In this case, the choice of k_1 (or k_2) is arbitrary, with larger values of k permitting the use of larger capacitance diodes, at the expense of a smaller bandwidth for given VSWR. Thus, for the two-stub filter, the loss is

$$\frac{P_0}{P_L} = 1 + \left[\frac{k_1 (k_1 + 2) \cos^2 \theta}{2} \right]^2 \qquad \text{(VIII-14)}$$

Alternatively, the insertion loss versus frequency can be computed using program WHITE. The stubs are entered as STB elements having sufficiently large terminating capacitance values to simulate short circuits. Tuned diodes can be realized using the PLC and STB elements in parallel to simulate the shunt tuning. Data for Figure VIII-6 were obtained in this manner.

Filters Suitable for SPDT Switches

So far, only the use of filter design as it applies to the two-port SPST switch has been examined. Fisher also suggested that a SPDT shunt diode tee switch can be designed using the stub filter approach; such a method is illustrated in Figure VIII-9(a). Diodes

	10 Log K_N	k_1	k_2	k_3
Three-Stub Filters				
1)	−12.728	0.100	0.200	
2)	− 0.944	0.300	0.600	
3)	+ 5.46	0.500	1.000	
4)	+10.138	0.700	1.400	
5)	+15.56	1.000	2.000	
6)	+21.156	1.400	2.800	
7)	+27.604	2.000	4.000	
8)	+31.904	2.5	5.0	
9)	+35.563	3.0	6.0	
Four-Stub Filters				
10)	− 5.17	0.1	0.292	
11)	+ 3.253	0.2	0.571	
12)	+13.329	0.4	1.109	
13)	+25.668	0.8	2.141	
14)	+35.909	1.3	3.395	
15)	+44.873	1.9	4.877	
16)	+56.734	3.0	7.568	
Five-Stub Filters				
17)	+ 3.452	0.100	0.366	0.532
18)	13.577	0.200	0.694	0.989
19)	20.523	0.300	1.005	1.410
20)	26.002	0.400	1.304	1.808
21)	30.601	0.500	1.596	2.193
22)	38.16	0.700	2.166	2.933
23)	44.324	0.900	2.724	3.648
24)	54.172	1.300	3.819	5.038
25)	66.970	2.000	5.702	7.403
26)	77.874	2.800	7.829	10.058

Table VIII-2 **Mumford's Maximally Flat Stub Filters [4]**

are located a quarter wavelength from the input in each of two output arms. Switching the diodes between forward and reverse bias directs the input power alternately to the two outputs. The *off* arm has a near short circuit produced by the forward biased diode, which results in a quarter wave stub in shunt with the transmission path to the *on* arm. This stub must have the characteristic admittance, Y_0, of the through line since it represents a transmission path when it becomes an *on* arm. The resultant equivalent circuit for the *on* path for a three-stub filter design is shown in Figure VIII-9(b).

A filter using as few as two stubs can be used, in which case both stubs have Y_0 characteristic admittance. Design #5 in Table VIII-2 gives a maximally flat response when the center stub has admittance of 2.0 and the end stubs have the required $Y_1/Y_0 = 1$ relationship. Examination of Table VIII-2 shows only one other filter having a unity normalized admittance – #19, a five-stub design. The VSWR versus frequency response curves for these three-filters are shown in Figure VIII-10. These curves emphasize the characteristic increase in steepness obtained as more elements are used; however, they cannot be interpreted as a direct comparison of the absolute VSWR bandwidth performances obtainable for various

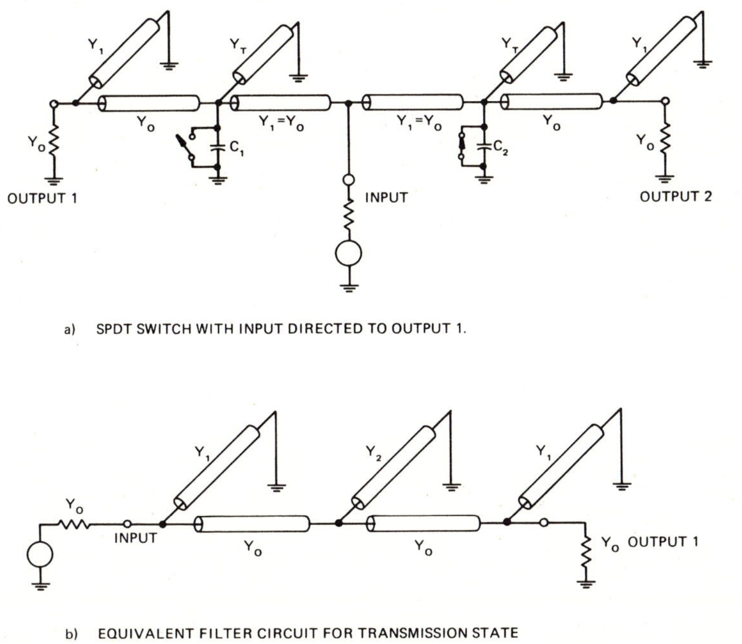

Figure VIII-9 **SPST Switch Designed using Filter Circuit Approach with Equivalent Two-Port Stub Filter Representing Transmission State**

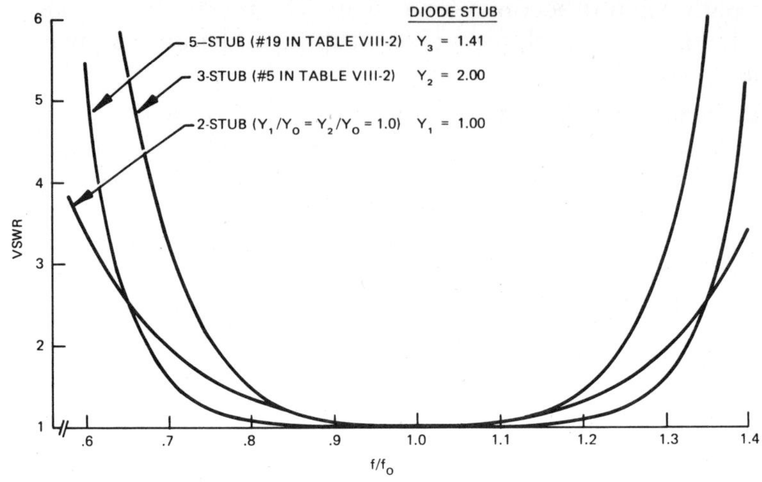

Figure VIII-10 **Frequency Response of Maximally Flat Filters Suitable for SPDT Switches**

numbers of stubs since they do not all incorporate the same admittance magnitude center stub. The two-stub permits a diode to be imbedded in a $Y_1/Y_0 = 1$ stub, the three-stub allows the largest diode capacitance to be simulated within a $Y_2/Y_0 = 2.0$ stub, and the five-stub permits an intermediate capacitance value to be simulated within its center stub, $Y_3/Y_0 = 1.41$.

Actually, it is not necessary to limit the filter designs usable for SPDT switching to those for which the stub adjacent to the tuned diode, simulated stub has a normalized characteristic admittance of unity. Should the design require that the adjacent stub have a normalized admittance, Y_{ADJ}/Y_0, greater than unity, an auxiliary shunt stub can be added at the tee junction, as shown in Figure VIII-11, the admittance, Y_{AUX}, of which is

$$Y_{AUX}/Y_0 = Y_{ADJ}/Y_0 - 1 \qquad \text{(VIII-15)}$$

Similarly, as shown in Figure VIII-11, a pair of series stubs could be added at the tee junction of the SPDT switch to give a smaller net admittance for the stub created by the switch's off arm. The series method is, of course, much less practical to implement. While the designs discussed so far can be realized practically in microstrip, stripline, or coaxial line media, the series stub is difficult to develop in any medium except coax.

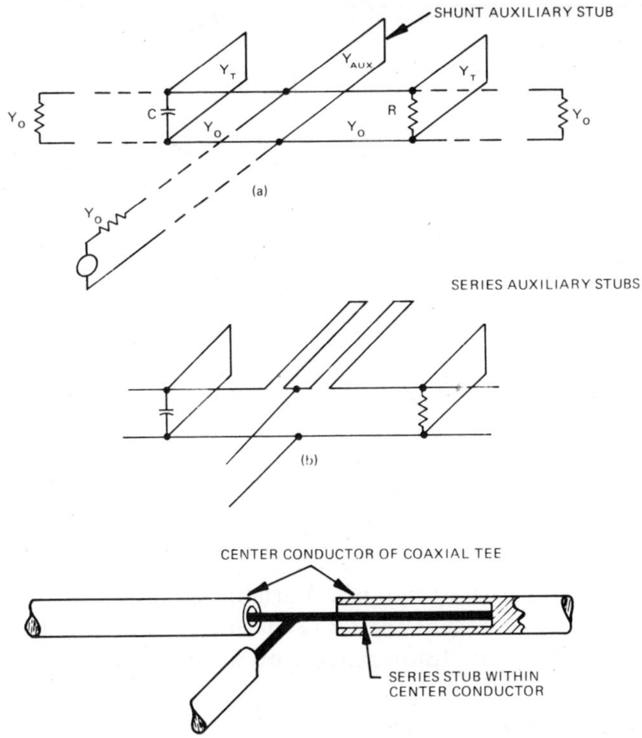

Figure VIII-11 **Methods for Adding Auxiliary Stub(s) at the SPDT Switch Tee Junction to Make the Off Arm Stub Admittance Appear (a) Larger or (b) Smaller**

Effect on Capacitance Introduced by Series (Isolation) Tuning

Until now, the diode in its pass state has been modeled as a simple capacitor, though it actually has series inductance which must be series resonated (usually by a tuning capacitor) to obtain maximum isolation in the forward biased state. This series tuning, of course, neutralizes the series inductive reactance only at the series resonance frequency. Over an octave bandwidth the effect of series reactance on the pass (reverse biased) state is to change the net susceptance, ωC. It is necessary to assure that this change is small enough that the anticipated pass state filter response can be realized with reasonable approximation. The situation can be visualized as one in which the diode capacity, C, is replaced by an equivalent capacity, C_E, which has a frequency dependence to account for the net reactance of the diode series inductance, L_S, and

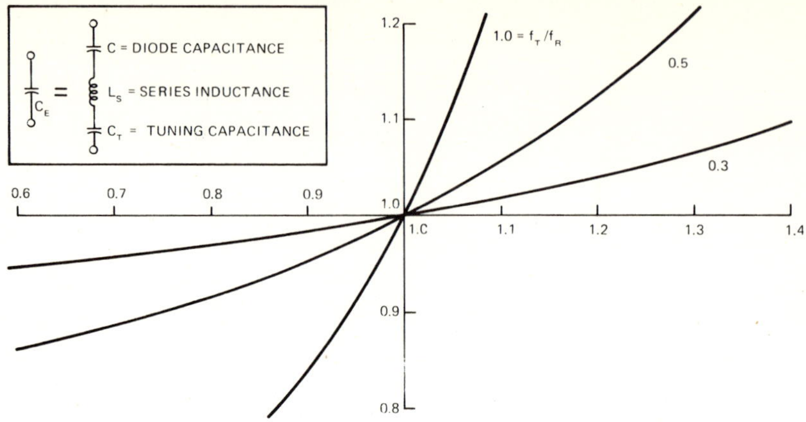

DEFINE: $f_T = \dfrac{1}{2\pi\sqrt{L_S C_T}}$ AND $f_R = \dfrac{1}{2\pi\sqrt{L_S C}}$

THEN: $C_E = \dfrac{C}{1-(f^2-f_T^2)/f_R^2}$

Figure VIII-12 **A Frequency Variable Capacitor Model to Represent the CLC Circuit Obtained with a Diode, its Inductance, and a Tuning Capacitor**

tuning capacitance, C_T. An expression for C_E is given in Figure VIII-12 along with curves showing how C_E/C varies with frequency normalized to the series resonance frequency, f_T.

The performance can be related to the ratio of f_R (the self resonant frequency of the diode capacity, C, with its own series inductance, L_S) and C_T. For octave bandwidth, C_E/C must remain close to unity over the range $0.65 < f/f_T < 1.3$, which gives a 2 to 1 bandwidth. From Figure VIII-12 it can be seen C_E/C remains within about 5% of unity over this bandwidth when $f_T/f_R = 0.3$ or less. This figure corresponds to an inductive reactance at f_0 which is 0.09 or less of the diode's capacitive reactance; equivalently, $C/C_T = 0.09$. Figure VIII-13 shows how the three-stub filter shown in Figure VIII-7 is affected by L_S and C_T.

3. The Isolation State

Single Diode Tuning

It is clear from Figure VIII-10 that a SPDT switch used in a 5-stub filter design could have an input VSWR of no more than 1.2 for $0.75 < f/f_0 < 1.35$ or 2.6 for $0.65 \leqslant f/f_0 \leqslant 1.35$. These values

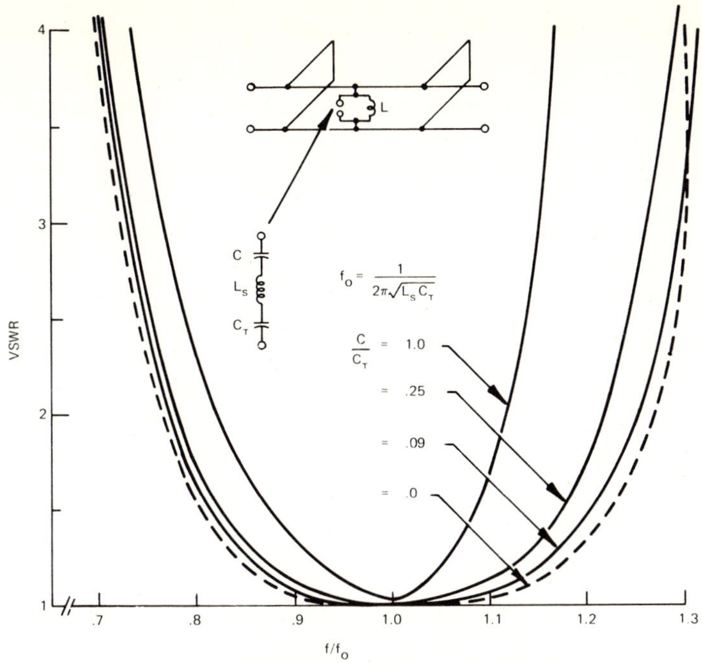

Figure VIII-13 **Effect on Filter Response Caused by Series (Isolation) Diode Tuning**

correspond respectively to reflective insertion loss performance below 0.1 dB over a 1.66 to 1 bandwidth and below 1 dB over a 2 to 1 (octave) bandwidth. Of course, for such switch designs to be practical it is necessary to maintain satisfactory isolation performance over the same bandwidth. With packaged diodes the isolation bandwidth is limited by the amount of series inductance and the manner in which it is tuned out.

Typically, the forward resistance of a PIN diode is less than 1 Ω; when shunt mounted across a 50 Ω line, this diode would provide, using Equation (II-41) and adding 6 dB for SPDT, an isolation of 34 dB as a SPDT switch. However, series inductance greatly reduces the short circuiting quality of the diode; an inductance of about 1 nH is typical. In a switch with center frequency of 1.5 GHz, this fact results in nearly 10 Ω of inductive reactance, 20% of the impedance of a 50 Ω line system. Figure VIII-14 shows the isolation obtainable with various untuned reactance magnitudes. The presence of an untuned reactance which is 20% of Z_0 and which produces less than 10 dB of isolation at midband is unac-

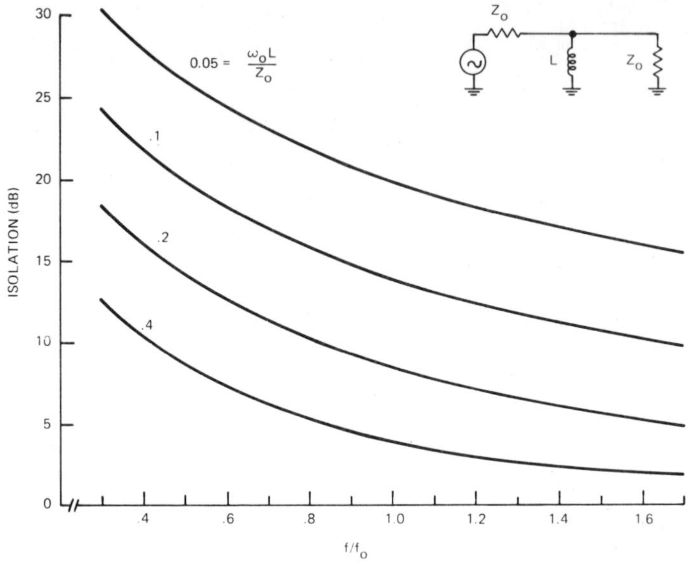

Figure VIII-14 **Isolation Limited by Untuned Inductor**

ceptable. Figure VIII-15 shows the isolation obtained when the same inductances are series resonated at the center frequency; their effect is eliminated at resonance, and the isolation bandwidth determined by the magnitude of inductive reactance which must be neutralized resonantly.

Two Diode Synchronous Tuning

The combination of diode resistance and inductive reactance makes it difficult to obtain more than 20 to 30 dB of isolation from a single diode installation, and a second diode is needed if a greater isolation and/or isolation bandwidth is required. If the second diode is installed directly in parallel with the first, the isolation theoretically increases by about 6 dB since the line shunting admittance is doubled. In practice, somewhat more than a 6 dB improvement is often obtained, since the individual inductances are not independent of each other and are often reduced by the resulting symmetry of the pair (especially in coaxial lines, wherein the diodes are mounted diametrically opposite one another). A further advantage of mounting both diodes at the same plane of propagation is that they share the RF current, decreasing the diode thermal rise. A similar discussion applies for the use of two series diodes, provided that their dimensions are small compared

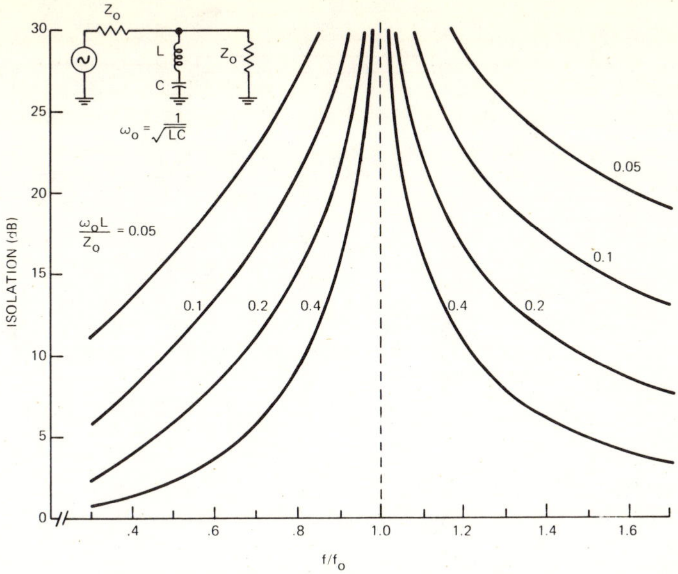

Figure VIII-15 **Isolation Limited by Series Tuned Inductor**

with the operating wavelength. The isolation increases by about 6 dB and the RF voltage applied to each is halved, increasing the voltage limited power capability.

However, a much greater isolation increase is achieved by separating the diode pair by a length of line approximately equal to a quarter wavelength. The resulting isolation is then approximately equal to the sum of the individual isolation values plus 6 dB, provided each isolation value is large enough. How this situation comes about and how close to 90° the spacing must be for the isolation to be estimable in the manner described above are the next topics to be examined.

Figure VIII-16 shows two shunt admittances, Y_1 and Y_2, spaced an electrical distance θ apart between matched generator and load. These admittances represent the two diodes in their isolate states. The overall ABCD matrix between ports (1) and (2) is also evaluated in the figure. The insertion loss, or isolation, of the network when placed between matched resistive generator and load is found from the expression in Figure VI-3. If Z_0 is assigned a value of unity, then Y_1 and Y_2 have the dimensions of normalized admittance, and the resulting expression for isolation is

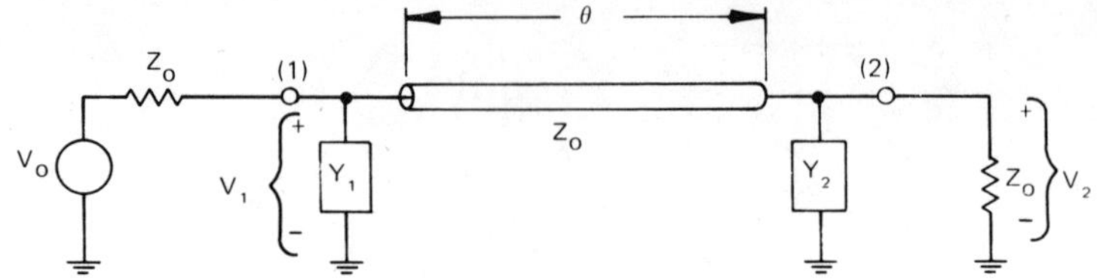

$$\begin{bmatrix} A & B \\ C & D \end{bmatrix}_{12} = \begin{bmatrix} 1 & 0 \\ Y_1 & 1 \end{bmatrix} \cdot \begin{bmatrix} \cos\theta & jZ_0 \sin\theta \\ jY_0 \sin\theta & \cos\theta \end{bmatrix} \cdot \begin{bmatrix} 1 & 0 \\ Y_2 & 1 \end{bmatrix}$$

$$= \begin{bmatrix} \{\cos\theta + jZ_0Y_2 \sin\theta\} & \{jZ_0 \sin\theta\} \\ \left\{\begin{matrix}(Y_1 + Y_2)\cos\theta \\ +j(Y_0 + Z_0 Y_1 Y_2)\sin\theta\end{matrix}\right\} & \left\{\begin{matrix}\cos\theta \\ +jY_1 Z_0 \sin\theta\end{matrix}\right\} \end{bmatrix}$$

Figure VIII-16 **Line Length Loaded by a Pair of Spaced Admittances**

$$\text{Isolation} = 1/4\ |(2+Y_1+Y_2)\cos\theta + j(2+Y_1+Y_2+Y_1Y_2)\sin\theta|^2 \text{ (SPST)} \quad \text{(VIII-16)}$$

For a switch which is to be operated over an octave, if $\theta = 90°$ at midband, $60° \leqslant \theta \leqslant 120°$ throughout the octave bandwidth. The minimum value for $\sin\theta = 0.87$ and the maximum value for $\cos\theta = 0.50$. If it is assumed further that the magnitudes for the normalized Y_1 and Y_2 are 10 or greater (corresponding to isolation values of at least 14 dB for the diodes when they are used individually), Equation (VIII-16) can be approximated by

$$\text{Isolation} \approx 1/4\ |Y_1Y_2|^2 \text{ (SPST)} \quad \text{(VIII-17)}$$

with an error of no more than 30%. The minimum isolation obtained under these conditions is 34 dB and the maximum error of 30% introduced by the approximation in Equation (VIII-17) corresponds to only ±1.5 dB.

The isolation obtained with an individual line shunting normalized admittance is

$$\text{Isolation} = \left|1 + \frac{Y}{2}\right|^2 \approx 1/4\ |Y|^2 \quad \text{(VIII-18)}$$

Thus, expressed in decibels, it can be seen that for the two element SPDT switch

$$\text{Isolation (dB)} \approx 10 \log_{10} \left|\frac{Y_1 Y_2}{2}\right|^2 = 10 \log \left|\frac{Y_1}{2}\right|^2 + 10 \log \left|\frac{Y_2}{2}\right|^2 + 6 \qquad \text{(VIII-19)}$$

Stated simply, *the total isolation in decibels of two diodes spaced a quarter wavelength apart at midband is approximately equal to the sum of their individual isolation values plus 6 dB.* This approximation is valid to within about ±1.5 dB, provided that the spacing is between 60° and 120° and the individual diode isolation values are each at least 14 dB.

When two such SPST switches are used with complementary biases, as in the tee style SPDT switch shown in Figure VIII-9, the isolation is increased by about four times (6 dB) because the current established in the first element of the off biased SPST switch is reduced to about half the value it would experience were it mounted directly between generator and load. Thus the SPDT isolation is given by

$$\text{Isolation} \approx |Y_1 Y_2|^2 \ \text{(SPDT)} \qquad \text{(VIII-20)}$$

The importance of deriving this result is apparent when the stagger tuning of the switching elements is examined, but first consider the isolation to be obtained from a pair of quarter wave spaced diodes over an octave bandwidth when their inductances are series resonated at the same frequency. From Figure VIII-15 it can be seen that the isolation-versus-frequency characteristic drops off more slowly for a given fractional frequency increment above f_0 than below. For example, the isolation obtained at 1.2 f_0 is always greater than that at 0.8 f_0, for a given ratio of $\omega_0 L/f_0$. It is shown later that *to have equal isolation values at the band edges, the series tuning of the inductance values should be performed at the frequency which is the geometric mean of the band edge frequencies.*

Thus, for a switch used over the bandwidth 1.0 to 2.0 GHz, the series tuning should be performed at $f_T = \sqrt{f_1 f_2} = 1.414$ GHz. Figure VIII-17 shows the isolation to be obtained from a pair of diodes tuned for equal isolation at the edges of an octave bandwidth.

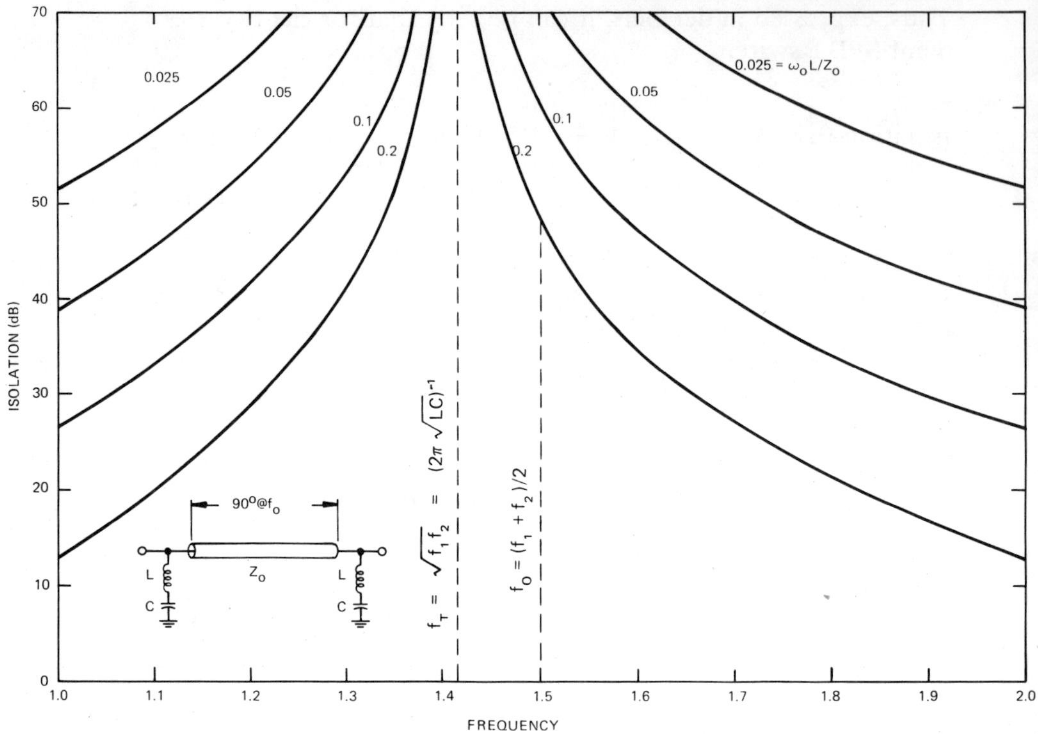

Figure VIII-17 **Isolation for Two Element Octave Band Switch Tuned at $f_T = \sqrt{f_1 f_2}$**

Stagger Tuning

It is clear from an examination of Figure VIII-17 that isolation potential is wasted by tuning both diodes at the same frequency; this procedure produces much greater isolation at the band center than at the band edges. It would be better to tune the diodes separately near the two band edges. In this way the isolation near midband drops to the minimum value obtained at the band edges, but this minimum-isolation level is greater than that achieved with synchronous tuning of both elements. The question is how can the two resonant frequencies, f_A and f_B, be determined such that equal isolation is obtained at the band edges and at a third frequency within the band? To resolve this problem, the approximation for isolation given in Equation (VIII-17) is useful.

Figure VIII-18 shows the switching circuit model to be used and the critical frequency definitions. Although not proved as of yet, it turns out that the frequency within the band at which the minimum isolation is obtained is the geometric mean of the band

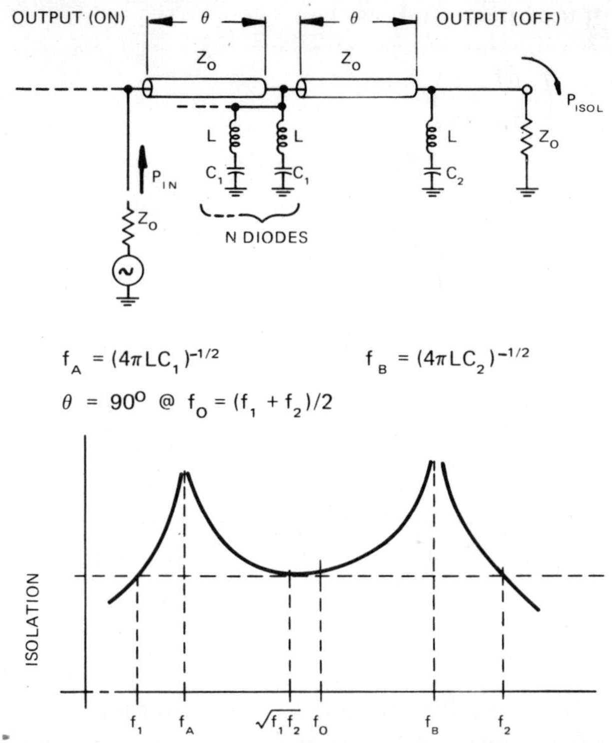

Figure VIII-18 **A Stagger Tuned SPDT Switch**

edge frequencies, f_1 and f_2, as shown in the sketch in Figure VIII-18. For this analysis the switching elements and their tuning capacitors are assumed to be lossless, a condition which causes infinite isolation at the series resonant frequencies, f_A and f_B — clearly an inaccuracy. However, only the *minimum* value of isolation that occurs between f_1 and f_2 is of interest. This value is determined almost completely by the reactance of the elements away from resonance; thus, the lossless assumption is reasonable.

The circuit to be analyzed in Figure VIII-18 contains N parallel diodes in the first position, as might be necessary for high power handling. The isolation as a function of frequency, f, is found from Equation (VIII-20) by expressing the susceptances B_1 and B_2 explicitly. To find this value, it is noted first that the total series reactance of the first set of N diodes is

$$jX_1 = \frac{j2\pi f_A L}{N} \left(\frac{f}{f_A} - \frac{f_A}{f}\right) \qquad \text{(VIII-21)}$$

The susceptance magnitude is then

$$B_1 = \frac{N}{2\pi f_A L}\left(\frac{f f_A}{f^2 - f_A^2}\right) \qquad \text{(VIII-22)}$$

and similarly

$$B_2 = \frac{1}{2\pi f_B L}\left(\frac{f f_B}{f^2 - f_B^2}\right) \qquad \text{(VIII-23)}$$

The isolation is then

$$\text{Isolation} = |B_1 B_2 Z_0^2|^2 = \frac{N^2 Z_0^4}{(2\pi L)^4}\left[\frac{f^2}{(f^2-f_A^2)\,(f^2-f_B^2)}\right]^2 \qquad \text{(VIII-24)}$$

To examine the variation of isolation with frequency, only the factor in brackets need be considered. This factor is defined as

$$F = \frac{f^2}{(f^2-f_A^2)\,(f^2-f_B^2)} \qquad \text{(VIII-25)}$$

Imposition of the condition that the isolation (and hence the value of F) have the same values at f_1 and f_2 gives the result that

$$f_A f_B = f_1 f_2 \qquad \text{(VIII-26)}$$

This equality may be used to calculate f_B once f_A is known.

If it further is required that the same magnitude for F be obtained at some frequency between f_A and f_B, and that this value be the minimum isolation within this frequency range (obtained by setting $dF/df = 0$), the resulting expression for f_A and f_B is

$$f_{A,B} = \left[\left(\frac{f_1+f_2}{2}\right)^2 \pm \sqrt{\left(\frac{f_1+f_2}{2}\right)^4 - (f_1 f_2)^2}\,\right]^{1/2} \qquad \text{(VIII-27)}$$

Equation (VIII-27) has two real roots between f_1 and f_2 for up to octave bandwidths ($f_2 \leqslant 2f_1$), the lower of which may be defined as f_A and the higher as f_B. The set of Equations (VIII-24), (VIII-26), and (VIII-27) can be used for SPST (and SPDT by adding 6 dB to the isolation values calculated) for any bandwidth up to an octave provided the magnitudes of $B_1 Z_0$ and $B_2 Z_0$ are

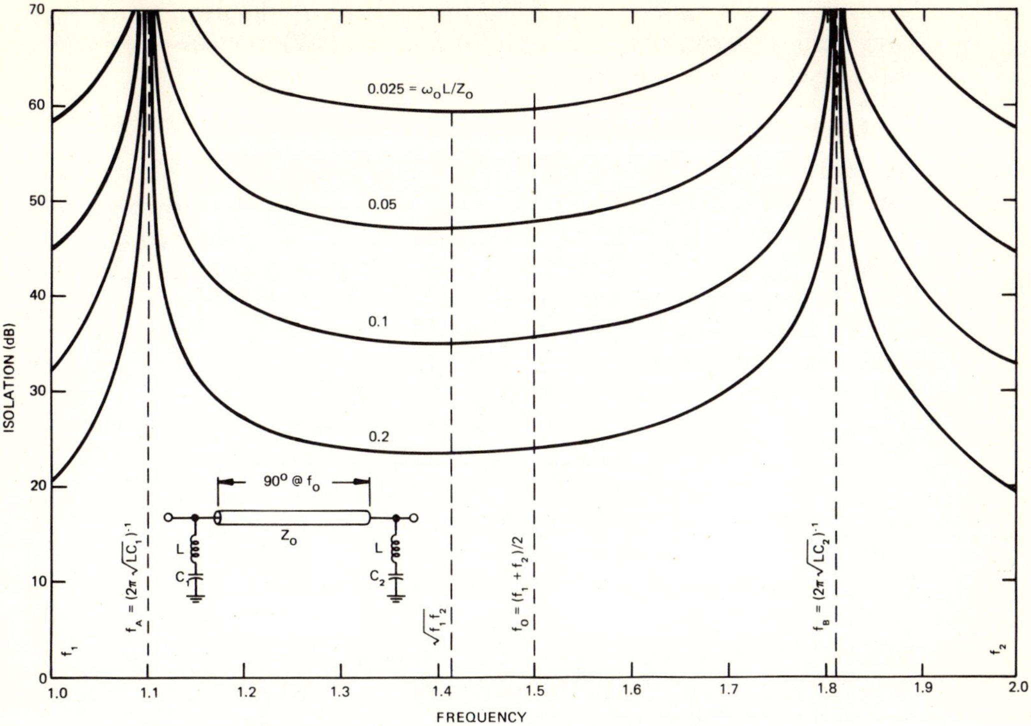

Figure VIII-19 **Isolation of Stagger Tuned SPST Switch**

both at least 10, and the diodes are spaced 90° apart at $f_0 = (f_1 + f_2)/2$. The solution for the octave band stagger tuning gives

$$f_1 = 1.00$$
$$f_A = 1.10$$
$$\sqrt{f_1 f_2} = 1.41$$
$$f_B = 1.81$$
$$f_2 = 2.00$$

For the two element switch under the conditions previously examined in Figure VIII-17 for synchronous tuning, the resulting isolation characteristics are as shown in Figure VIII-19. Note that the improvement in minimum isolation is about 6 dB for all cases of $\omega_0 L/Z_0$ examined. The curves in Figures VIII-17 and VIII-19 were plotted using program WHITE with a subroutine section added to handle the series LC circuit in shunt with the line. As such, the curves do not make the approximation of Equation (VIII-17) to calculate isolation. Nevertheless, nearly equal isolation values at band edges f_1 and f_2, as well as $\sqrt{f_1 f_2}$, were obtained (within a few decibels), indicating that this approximation, which was used to derive the expression in Equation (VIII-27) for f_A and f_B, is reasonably accurate.

4. Diode Dissipative Losses

An exact solution for the diode dissipative losses in the SPDT tee switch shown in Figure VIII-18 would be cumbersome. However, an approximate solution which is based on a small perturbational analysis is usually adequate and can provide insight into the parameters of importance in the selection of diodes and the operating circuit impedance. For this analysis, it is assumed that the voltages and currents in the switch are influenced by the diode resistance values, which are small compared with the other circuit reactances and/or the respective characteristic impedances of the lines in the circuit. In addition to this small perturbation assumption, the circuit is assumed to be perfectly matched. Therefore, diodes in shunt with the line "see" line voltage under reverse bias and line current (in SPDT switches). An analysis of the diode losses in the SPDT switch discussed so far serves to demonstrate the design analysis approach.

Figure VIII-20 shows the simplified equivalent circuit used for loss calculations of the SPDT switch. Multiple (N) diodes may be accommodated at the first position for both high power handling and isolation enhancement. The tuning stubs, diode series inductance, package capacitance, and series tuning capacitors have been omitted since their overall effect on insertion loss is usually small. Finally, in the off arm, only the losses of the switching element in the position nearest the tee junction are included; the RF current through it is approximately the line current, I_L, since it "short circuit" terminates a Z_0 impedance quarter wavelength stub with V_L excitation. In the on arm, the diodes have a reduced equivalent circuit, as shown in Figure VIII-20, composed of C_J and R_R.

If jX is the reactance of a single reverse biased diode with junction capacitance C_J, the dissipative insertion loss of N such diodes is

$$\text{Loss}_R = \frac{P_{DR}}{P_A} = \frac{{V_L}^2}{|R_R + jX|^2} \cdot R_R\,(N+1) \cdot \frac{Z_0}{{V_L}^2} \qquad \text{(VIII-28)}$$

$$\approx R_R Z_0\,(N+1)/X^2 \qquad \text{(VIII-29)}$$

Now, the reverse biased cutoff frequency is

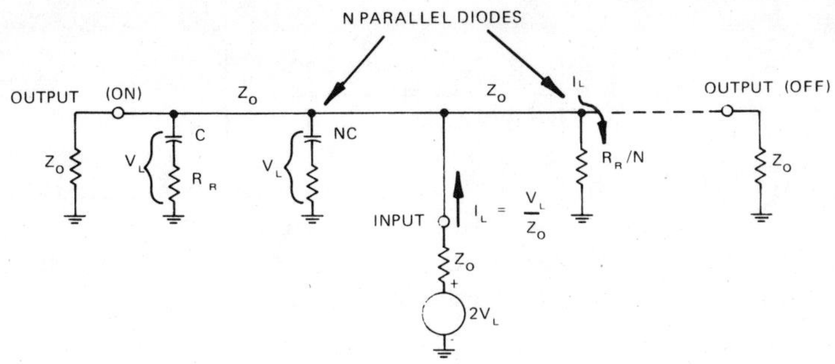

Figure VIII-20 **SPDT Switch Equivalent Circuit used to Estimate Diode Dissipative Losses**

$$f_C = \frac{1}{2\pi R_R C_J} \qquad \text{(VIII-30)}$$

Hence

$$\frac{R_R}{X} = \frac{f}{f_C} \qquad \text{(VIII-31)}$$

Thus, it may be written that

$$Loss_R = \frac{Z_0}{X} \cdot \frac{f}{f_C} \cdot (N+1) \qquad \text{(VIII-32)}$$

The three factors on the right side of Equation (VIII-32) can be identified respectively as the circuit coupling factor, the diode quality at frequency f, and the total number of reverse biased diodes.

Meanwhile, the N forward biased diodes in the leading position of the "off" arm carry line current; their dissipative loss is

$$Loss_F = \frac{P_{DF}}{P_A} = \left| \frac{I_L{}^2 \cdot R_F/N}{I_L{}^2 Z_0} \right| = \frac{R_F}{NZ_0} \qquad \text{(VIII-33)}$$

After algebraic manipulation, the combined losses can be written as

$$\text{Loss}_D = \text{Loss}_F + \text{Loss}_R$$

$$= \frac{f}{f_C}\left[\frac{R_F X}{R_R N Z_0} + \frac{(N+1)\,Z_0}{X}\right] \qquad \text{(VIII-34)}$$

Equation (VIII-34) shows that, for an increase in Z_0, the loss resulting from forward biased diodes (the first term) decreases, while the loss for reverse biased diodes (the second term) increases. A minimum total loss is obtained when

$$Z_0 = X\sqrt{\frac{R_F}{R_R N(N+1)}} \qquad \text{(VIII-35)}$$

The minimum loss value, with this choice of Z_0, is

$$\text{Loss}_{\text{MINIMUM}} = \frac{2f}{f_C}\sqrt{\frac{R_F(N+1)}{R_R N}} \qquad \text{(VIII-36)}$$

The reverse biased cut-off frequency, f_C, is used in order to separate the losses for forward and reverse biased diodes. Note, by comparing Equation (VIII-30) for f_C and Equation (V-23) for f_{CS}, that the switching cut-off frequency defined by Hines is

$$f_C = f_{CS}\sqrt{\frac{R_R}{R_F}} \qquad \text{(VIII-37)}$$

Thus the minimum loss for the SPDT tee switch can be written

$$\text{Loss}_{\text{MINIMUM}} = \frac{2f}{f_{CS}}\sqrt{\frac{N+1}{N}} \qquad \text{(VIII-38)}$$

which is identical to the minimum loss predicted for the general switch (Equation (V-25)) except for the factor $\sqrt{(N+1)/N}$ the presence of which is due to the second diode element added to each arm of the switch.

5. *A Practical Octave Band Switch*

Design Calculations

In this section a practical switch designed in 50 Ω coax to cover the 1-2 GHz band and embodying the principles of broadbanding just presented is described.

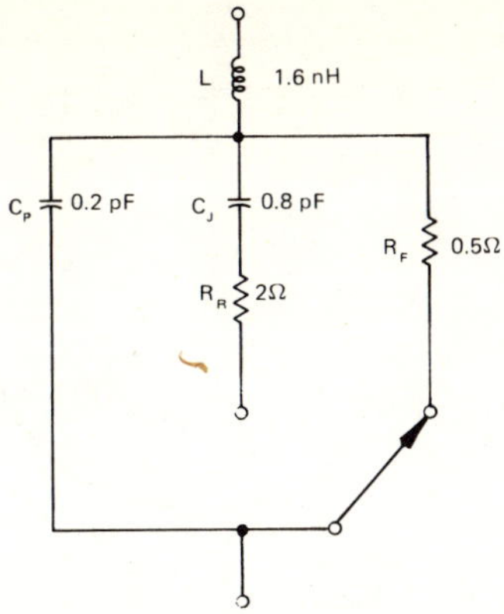

Figure VIII-21 **Equivalent Circuit Model used for Diodes in 1-2 GHz Switch**

From Figure VIII-10, it can be seen that the 5-stub filter (#9 in Table VIII-2) offers the best transmission characteristic of those considered for the SPDT switch. The center element has a moderately high normalized admittance, Y_3, equal to 1.41 (Z_C = 50/1.41 = 35 Ω), permitting adequate allowance for the absorption of diode capacitance. The remaining two stubs have Y_2 = 1.00 (50 Ω), as required for the "off" arm, and Y_3 = 0.3 (166 Ω). The second switching element can be accommodated easily in the Y_2 stub on the output side, providing high isolation. Two diodes, modeled with the equivalent circuit shown in Figure VIII-21 and having a total capacitance of 1 pF, are used in the center element of the filter to simulate a 35 Ω stub. The stubs are arranged so that this pair of diodes forms the first switching element in the off arm. For the simulated center stub, the proper tuning stub is found by evaluating at 1.5 GHz.

$$\frac{\omega_0 C}{Y_E} = 2\pi \frac{\text{radians}}{\text{hertz}} \text{ x } 1.5 \text{ x } 10^9 \text{ hertz x } 2 \text{ x } 1 \text{ picofarad x } 35 \text{ ohms}$$

$$= 0.67 \qquad \text{(VIII-39)}$$

From Figure VIII-3, the corresponding value for Y_T/Y_E is 0.50, indicating that a 70 Ω characteristic impedance 36° length stub should be used to parallel tune the two diodes used in the first element. Similarly, the second element, consisting of one diode,

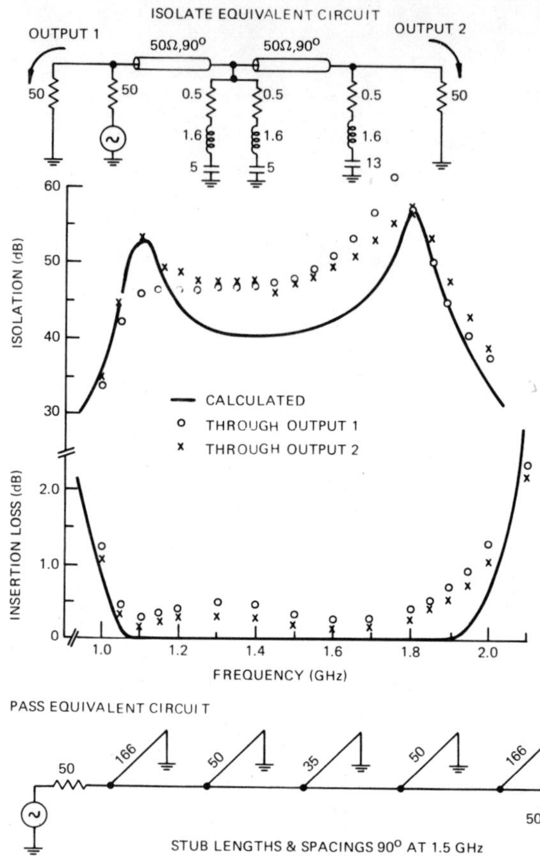

Figure VIII-22 **Measured and Calculated Performance of Stagger Tuned Octave Bandwidth Switch**

simulates a 50 Ω stub. Thus, since $\omega_{0C}/Y_E = 0.47$, a 62 Ω tuning stub 59° long at 1.5 GHz should be used for tuning.

Measured Results

By making preliminary measurements of the isolation characteristics of separate diode installations, the single diode inductance when mounted across the 50 Ω coaxial line was estimated to be 1.6 nH. For a 1-2 GHz octave bandwidth, stagger tuning requires series resonating of the inductance at f_A = 1.10 GHz and f_B = 1.81 GHz. But if a 1.6 nH inductance, which has a reactance of +j 11 Ω at 1.1 GHz, is series resonated at this frequency, the total

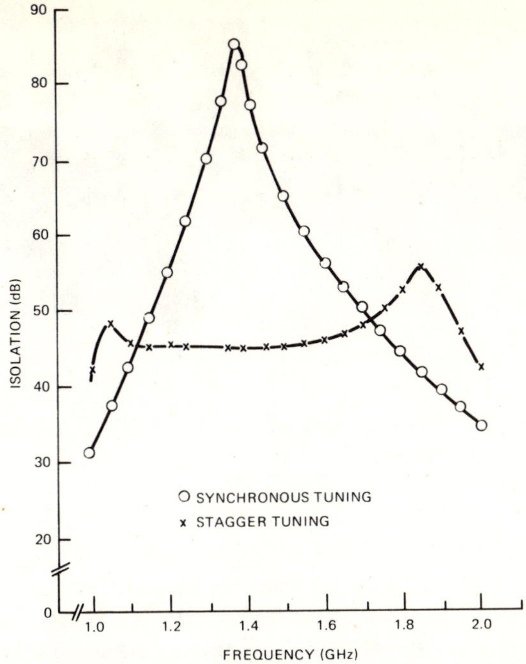

Figure VIII-23 **Measured Isolation with Stagger and Synchronous Tuning**

reactance of the LC combination at the 2 GHz upper band edge of the switch is +j 14 Ω. This evaluation gives a normalized reactance magnitude of only 3 for a 50 Ω system, and violates the assumption of a minimum reactance magnitude of *at least* 10 if the stagger tuning criterion derived is to be accurate. Nevertheless, the switch when so built gave results fairly close to predictions, as can be seen from the measured and calculated loss and isolation parameter shown in Figure VIII-22. Although the initial design criteria were based on the approximations discussed, the calculated performance is exact, having been performed using program WHITE and the equivalent circuits shown in Figure VIII-22 for the isolate and pass states.

Although the stagger tuning does provide about 6 dB greater isolation at the edges of an octave bandwidth switch, this improvement is obtained at the expense of a great reduction in the midband isolation. Figure VIII-23 shows the comparative isolation measured with stagger and synchronous tuning. The synchronous tuning is

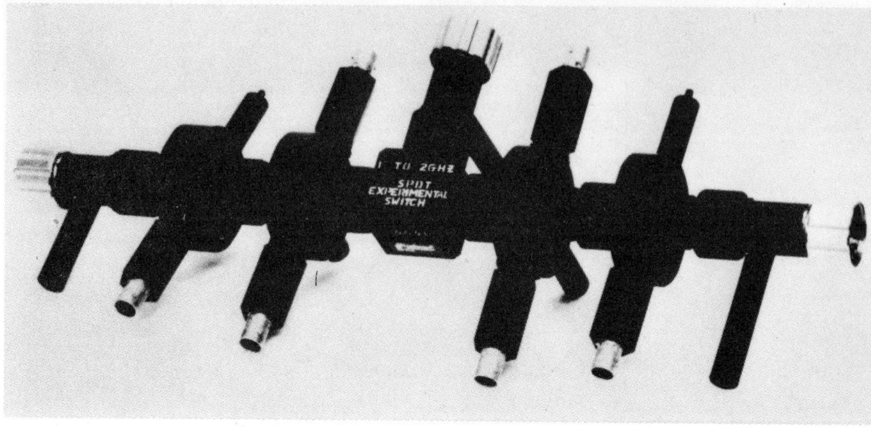

Figure VIII-24 **Photograph of the 1-2 GHz Experimental Switch Model**

performed by series resonating both diode sites at $f = \sqrt{f_1 f_2}$ (1.41 GHz for this 1-2 GHz switch).

A photograph of the coaxial experimental switch model is shown in Figure VIII-24. The four diode sites with tuning stubs and the 166 Ω end stubs were fitted with 7 mm connectors so that they could be tuned separately. The data shown in Figures VIII-22 and VIII-23 were obtained by attaching all of the separate parts together without retuning.

The optimization for minimum insertion loss generally is not practical since the choices of diode capacitance and line impedance are usually geared to achieving the desired bandwidth and power handling capability. However, in this particular design, near minimum diode dissipative loss is achieved coincidentally with the $C_J = 0.8$ pF and $Z_0 = 50\ \Omega$ choice. Thus, applying Equation (VIII-35), the characteristic impedance choice for minimum diode loss at 1.5 GHz (at which $X_J = 199\ \Omega$) is

$$Z_0 = 199 \sqrt{\frac{0.5}{(2)(2)(3)}} \qquad \text{(VIII-40)}$$

$$= 41 \text{ ohms}$$

which is close to the $Z_0 = 50\ \Omega$ actually used. The corresponding loss value is, from Equation (VIII-38),

$$\text{Loss}_{\text{MINIMUM}} = \frac{2(1.5)}{200}\sqrt{\frac{0.5(3)}{2(2)}} \tag{VIII-41}$$

$$= 0.0092$$

$$= 0.04 \text{ decibel}$$

Referring to Figure VIII-22, the measured loss at midband has a minimum of about 0.3 dB which, allowing for dissipation in the circuit itself, is in reasonable agreement with the low expectation above.

High Power Tests

High power tests of the switch at 1.2 GHz indicated that levels of up to 6 kW of peak power could be sustained before the diodes failed. In the tests a 0.001 duty cycle and a 1 μs pulse length were used. The failure at this power was manifest as a permanent increase in the conductance of those diodes which had been reverse biased (those in the on state). The reverse bias voltage applied to the diodes was 150 V; a current of 100 mA was applied to each of the forward biased (off state) diodes.

The RF series resistance of the diodes under forward bias was measured to be about 0.5 Ω; under reverse bias, it is estimated to be 2 Ω. Using these values, at a peak power of 6 kW, the dissipation in each leading forward biased diode is about 15 W; any of the reverse biased diodes dissipates about 30 W. These peak absorption power levels would be expected to cause negligible heating in the 1 μs pulse. The diodes have thermal properties similar to the 0.7 pF, 50 μ (2 mil) I region type described in Table III-1 Thus, with a minimum thermal time constant of 210 μs and thermal resistance of 7°C/W, the temperature rise is estimated to be less than a few degrees centigrade.

This switch was intended for high peak power short pulse applications. The reverse breakdown voltages of the diodes were therefore the determining factor in the maximum power switching capability. Figure VIII-25 shows the dc reverse breakdown characteristic for a typical PIN used. For dc, reverse breakdown at -10 μA is 700-800 V. The peak RF voltage in a 50 Ω line carrying 6000 W is 775 V which, when superimposed on a -150 V reverse bias, definitely exceeds the dc reverse breakdown.

As can be seen, the dc reverse breakdown (if the diode has a sharp reverse I-V characteristic) can provide an approximate indication

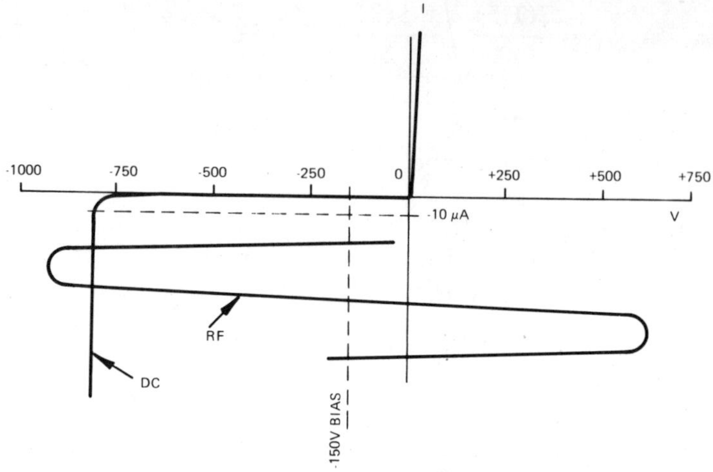

Figure VIII-25 **Comparison of Measured dc Characteristic with RF and Bias Voltages at which RF Burnout is Experienced in Reverse Biased PIN Diode**

of ultimate RF voltage sustaining capability; however, final values must be obtained by measurements made under the particular RF frequency, pulse length, temperature, and control bias which apply in the actual use environment. The safety margin between burn-out and rated power also must be decided. For this particular device, a peak rating somewhat below 1.5 kW would insure safe operation even with a totally reflecting load of any phase.

B. Coaxial High Power

With the octave bandwidth switch just described, it is clear that very high peak power switching requires voltage reduction techniques. There are many ways in which this reduction can be accomplished, one of the most direct being to transform the generator and load impedances to reduced values at the transmission line plane in which the diodes are mounted. This approach is usable for impedance transformations of considerable ratios. A study program to explore the feasibility of diode switching in excess of 100 kW with RF pulses up to 500 μs revealed that low loss transformations to characteristic impedance values of less than 1 Ω are possible. Figure VIII-26 shows a SPDT switch consisting of a tee junction with a large coax diameter SPST switching element. Each SPST effects impedance transformation with

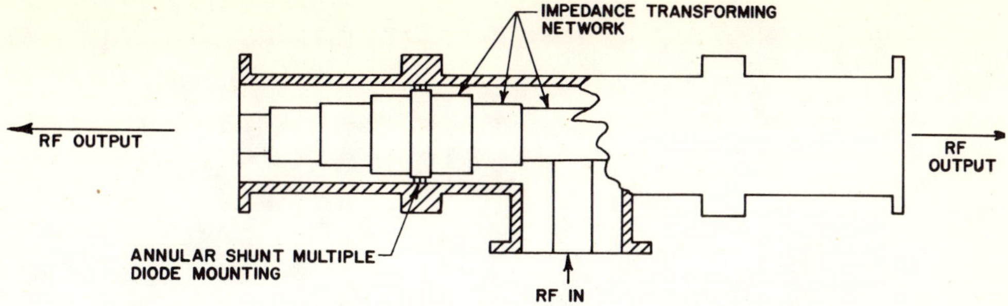

Figure VIII-26 **Plan View of a High Power SPDT Coaxial Switch using Transformers to Achieve a Low Line Impedance at the Plane of the Switching Diodes**

three-section quarter wave transformers. For maximum bandwidth, the SPST switches could be installed between 3 dB hybrid couplers for a balanced switching circuit of the form shown in Figures VII-10 and VII-13.

The diodes used for this evaluation were similar to those described in Figure VIII-21 for the octave switch. Operation over the bandwidth 1200-1400 MHz was desired. Because of the long RF pulse operation, 500 μs, the switch essentially requires as many diodes as would be needed for CW operation at the full peak power. The design of the switch used a 60 to 1 Tchebyschev [5] impedance transformation in 3-1/8 inch (8 cm) coaxial line, transforming the 50 Ω source and load to 0.83 Ω at the diodes. The SPST switch parts can be seen in the photograph in Figure VIII-27.

Identical transformers located at the output sides of the diode annuli transform this low impedance level back to the 50 Ω level of the load. The pair of multi-section transformers mounted back-to-back gave the measured VSWR-versus-frequency characteristic shown in Figure VIII-28. Actually, this graph is quite remarkable considering that a standing wave ratio of about 60:1 is associated with the 0.83 Ω impedance transformation ratio.*

With 48 diodes installed in each switching ring, a minimum of 30 dB of isolation was obtained when each diode was forward biased at 100 mA over the 1.2-1.4 GHz bandwidth for the SPST switching. The insertion loss, including circuit loss, under -100 V

*This transformer is the largest ratio microwave impedance one for which measured performance has been reported.

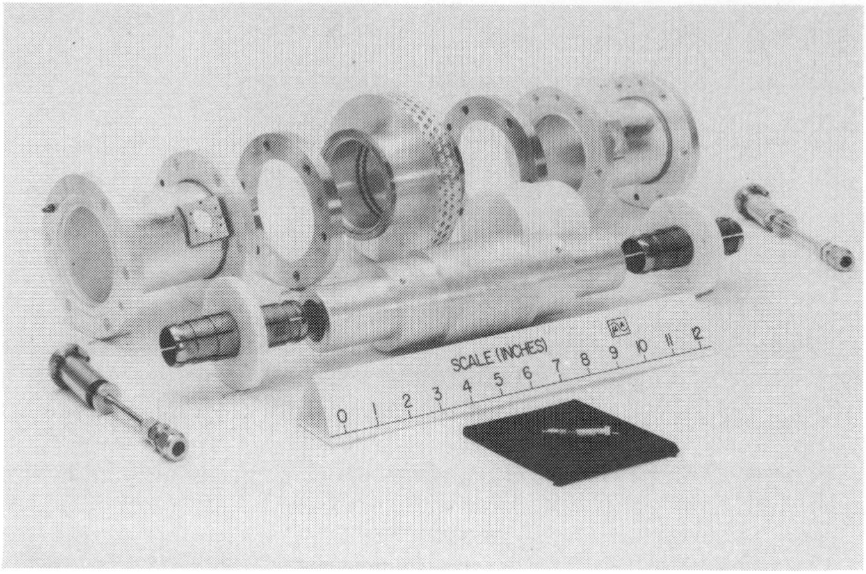

Figure VIII-27 **A 100 kW Switch Built in 3-1/8 in Coaxial Line using Quarter Wave Transformers to f_0 = 0.83 Ω**

reverse bias was below 0.25 dB with one ring filled with diodes; the figure would be below 0.5 dB with both rings filled.

Several factors must be considered in the design of such a high power switch, not the least of which is the fact that the diodes have some effective series inductance when mounted across the transmission line. For this reason, small tuning capacitors, of about 25 pF, were used to series resonate with the inductance in this particular L-band frequency range. Other considerations in such a switch design include requirements that the diodes be separately removable, separately biasable, and spring loaded against the center conductor (so as to assure good contact with slight relative mechanical motion of center and outer conductors), and that there be appropriate filtering in the bias circuit to prevent RF from leaking out the bias terminals. In addition, a diode mount of the smallest possible diameter is required so that as many diodes as possible can be fitted into an annulus.

In this particular design, fabricated in 3-1/8 in coaxial line, diode mounts were made having a maximum diameter of 0.200 in — 48 of them could be fitted on one annular ring. Even so, two switching annuli were required to carry the RF current. A detailed view of a typical diode mount is shown in Figure VIII-29.

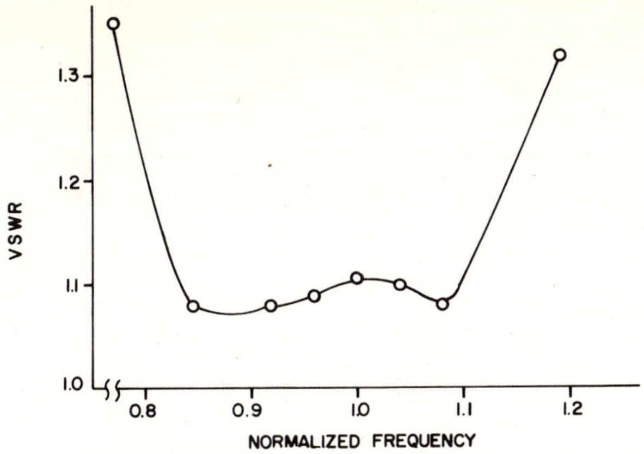

Figure VIII-28 **VSWR Characteristic of Switch with Diodes Removed**

The assembled switch may be seen in Figure VIII-30. Short circuited transmission line stubs mounted in shunt on either side of the diodes provide a dc return path from the center conductor to the outer conductor for biasing. The stubs are hollow and water can be circulated through them to cool the center conductor. Since the 3-1/8 in coaxial line can support high order modes at this frequency, it was found necessary to mount three identical stubs at 90° angular spacings near each primary stub to maintain field symmetry and minimize the excitation of high order modes in the circuit.

By design, it is desired that all diodes carry the same current under forward bias. Since the annular rings in each arm of the switch were separated by about 12 electrical degrees, it was necessary to use compensation to decouple the first ring somewhat from the power so that current sharing between the two rings could be achieved. The method for this compensation used inductive stubs in series with primary ring diodes, as shown in Figure VIII-31. The 1.47 Ω line impedance corresponds to that of the final quarter wave section which transforms 50 Ω to 0.83 Ω; the tuning, to a first order approximation, is frequency independent. There is no need of constructing 0.83 Ω line since this impedance exists only at the end plane of the last transformer section. For experimental purposes, a special biasing circuit was necessary to permit continuous monitoring of the current voltage applied to each diode. The assembled switch, with its bias circuit, is shown in Figure VIII-32,

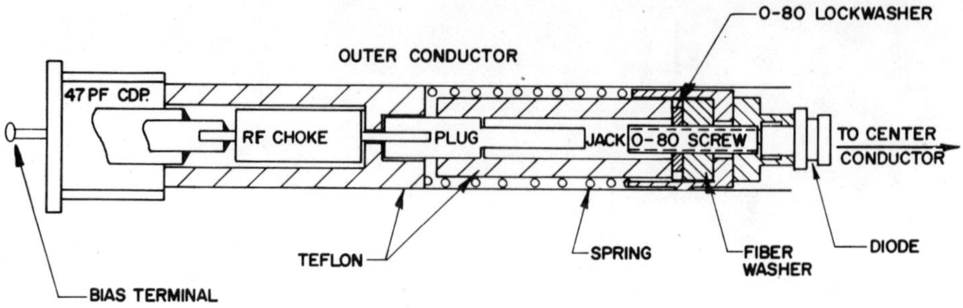

Figure VIII-29 **Typical Diode Assembly**

mounted in a high power circuit used for RF testing above 100 kW.

The test was performed with only one of the annular rings installed. Under -100 V of reverse bias, the power level was increased until the first diode failures were observed. This condition occurred at 120 kW peak power with 500 μs pulse length. Forward bias tests were also made at this level, but no failures were experienced. During the forward bias test, the incident RF power was reduced to one-sixteenth of that applied to the switch when the diodes were reverse biased, since the current strain for a singular annular ring is 16 times that which would occur if the other ring were installed and the other half of the switch were attached in a SPDT configuration. This test demonstrated peak switching at 100 kW and a 500 μs pulse length, corresponding to individual diode stress levels in a SPDT configuration of approximately 300 V rms, with -100 V reverse bias, and 4 A rms with +100 mA forward bias. These results represent the highest RF pulse energy (over 100 kW for 500 μs) ever switched by a diode device designed for full power operation in both forward and reverse biased states.*

There are practical problems which arise due to the use of so many diodes in one device. For this switch, the failure of a single diode excites a *higher order mode* in the large diameter coax, resulting in increased voltage stresses in other diodes. This situation can be handled by suitable derating of the power capability. Figure VIII-33 shows the relative field magnitude at the switching plane with a single shorted diode, A, and shorted diodes diametrically opposite each other, B. Up to 3 dB more stress is experienced with a single

*This distinction is made to exclude switched diode duplexers, which need sustain high power in only one of their bias states.

Figure VIII-30 **Assembled View of the 3-1/8 in Coaxial Switch Containing 98 Diodes**

short circuit diode failure. Since these tests were performed, diodes with higher breakdown voltages have become available, and it is possible that the 100 kW level could be met as a rated level with adequate margin to account for load VSWR and annular higher order mode over voltage stresses. Thus, it is feasible to consider switching the entire output of a high power radar with only a few such high power diode switches.

C. Switched Duplexers

In Chapter VII duplexers which are self actuated were examined. In these circuits the high level microwave signal is sufficient to produce conductivity modulation in the PIN diode and it serves in lieu of a forward bias signal. At that time, however, it was observed that RF is a much less efficient source of carriers in the diode than is dc bias. Without question, a self-actuated duplexer is more desirable because it can protect the receiver, not only from RF high power pulses from the same system, but also from sources of RF power which may enter the antenna due to other radars. On the other hand, for frequencies much above 500 MHz,

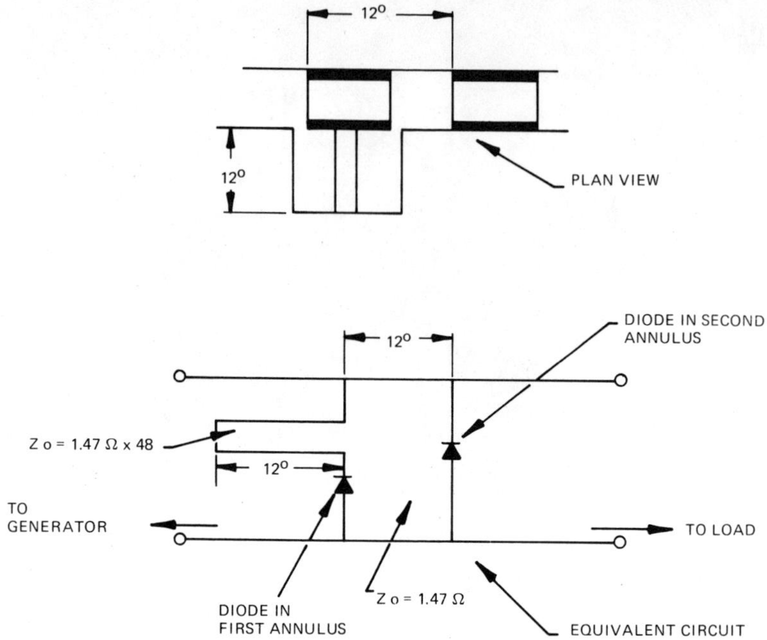

Figure VIII-31 **Current Sharing by the Two Annular Rings of Diodes is Effected by Placing a Short Stub in Series with each Diode in the First Annulus**

self-actuated duplexers capable of sustaining a megawatt of peak power or more become impractical because of the large number of diodes required. If bias can be applied with a driver synchronized to the RF transmitter, much more efficient conductivity modulation of the diode is obtained and very high powered duplexers can be designed using a relatively small number of PIN diodes.

For switching bandwidths of 10-20%, only a single section quarter wave transformer generally is needed to adjust the line impedance to the level in which optimum switching can be realized. Generally, an impedance lower than 50 Ω is desired to permit the use of multiple diodes across the transmission line while maintaining a modest level of insertion loss. A pair of quarter wave transformers form a half-wave line section at the center of which are mounted the shunt switching diodes, as shown in Figure VIII-34. A photograph of a pair of such switching elements designed for a balanced duplexer in the 1250-1350 MHz frequency band [6] is shown in Figure VIII-35.

Figure VIII-32 **The 100 kW Switch with Special Bias Network Installed in the High Power Test Circuit**

Each SPST element contains 8 diodes, each similar to those for which the equivalent circuit is shown in Figure VIII-21. The characteristic impedance of the half wavelength line was 35 Ω, resulting in an effective switching impedance of 25 Ω at the plane of the diodes. The balanced duplexer containing these two switching elements was tested to a peak power of 1.5 MW with 5 μs pulse length and 0.001 duty cycle without failure. The insertion loss was only 0.5 dB with -50 V reverse bias, and the isolation with 100 mA of forward bias per diode was 31 dB for the individual SPST elements. In the balanced duplexer, coupler directivity adds about 20 dB, resulting in a total of about 50 dB of isolation from the main transmitter power, when the radar antenna is well matched to the system.

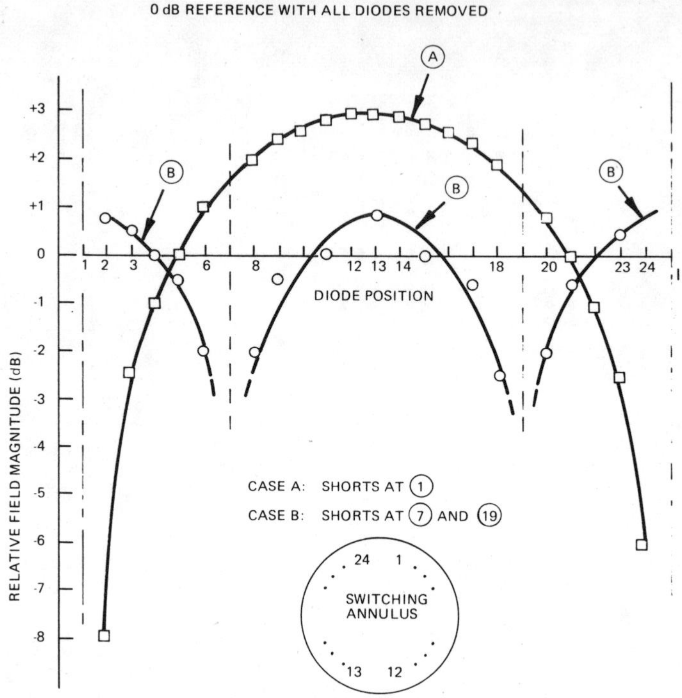

Figure VIII-33 **The Effects on Diode Voltage Stress in a 3-1/8 in Coax Switch Caused by Short Circuits at (A) One Diode Position and (B) Two Diametrically Opposite Diode Positions**

D. Waveguide Switches

Not all switching applications require enormous power handling capability or octave bandwidth. In fact, each application for diode switching generally has a unique set of requirements and its own priority of design features. Through the rest of this chapter, different switching circuit embodiments are examined. One of these is the waveguide switch shown in the photograph in Figure VIII-36. The switch is essentially a dual diode version of the post coupled limiter described in Chapter VII. As was seen with the limiter, the analysis of waveguide control circuits is generally much more difficult than with TEM mode circuits in coax, stripline, or microstrip. In practice, both the design and tuning of the switch usually is done empirically and the lengths of the coaxial sections (S1 and S2 in Figure VII-19) are adjusted with the diode in place to provide minimum loss and maximum isolation. While this approach offers only modest bandwidth, typically about 10%, waveguide circuitry generally contributes the least dissipative insertion loss to the device performance. This situation results because the RF energy is propagated in waveguide, for which the characteristic impedance is very high and losses in the waveguide walls are small. Moreover, for use in microwave systems which already are configured in waveguide, a switch realized directly in waveguide naturally is more compact than one constructed in a TEM medium, requiring transitions to waveguide. The measured performance for the waveguide switch is shown in Figure VIII-37. This switch was designed for use in a pulsed radar system having 1 μs pulse length

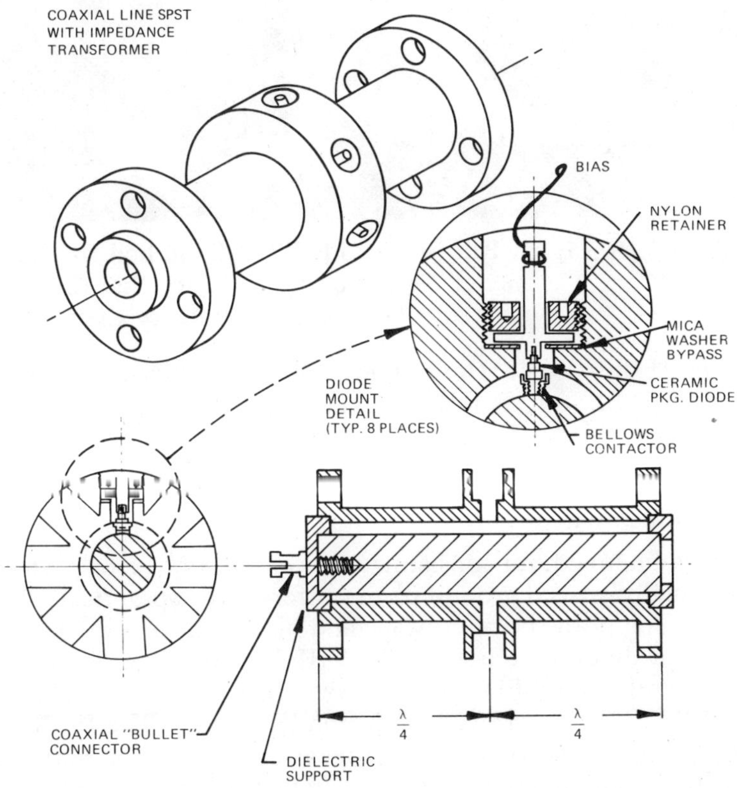

Figure VIII-34 **Plan View of Switched Duplexer SPST Switch with Simple Transformer for 1-5/8 in (4 cm) Diameter Coaxial Line**

Figure VIII-35 **A Pair of 8 Diode SPST Switched Duplexer Elements, Tested under Forward Bias to 1.5 MW with 5 μs Pulse and 0.001 Duty Cycle**

and 0.001 duty cycle. Under these conditions diodes with approximately 0.2 pF junction capacitance and a 3 mil (75 μ) I region survive up to about 5 kW of incident peak RF power when operated with -50 V of reverse bias.

E. Stripline Switches

The stripline medium consisting of a center conductor pattern with equally spaced ground planes offers a very practical means for realizing both the microwave and bias isolation circuitry required for diode switches. Figure VIII-38 shows the assembled view of an L-band SPDT stripline switch, used in the 1,030 to 1,090 MHz, IFF microwave transponder system. The IFF designation is an abbreviation of "Interrogate Friend or Foe;" it was used initially for identifying military aircraft. Such systems now are used commercially to distinguish commercial aircraft in densely travelled traffic control zones. The radar transmits three closely spaced 1 μs pulses; the source alternates from an omnidirectional antenna to a directional one. This stripline switch directs the

Figure VIII-36 Waveguide Switch with Two Post Mounted Diodes; Waveguide Circuitry Generally Provides the Lowest Dissipative Circuit Loss (Switch Photo and Data Courtesy of R. Tennenholtz and D. Conlon, Microwave Associates, Inc.)

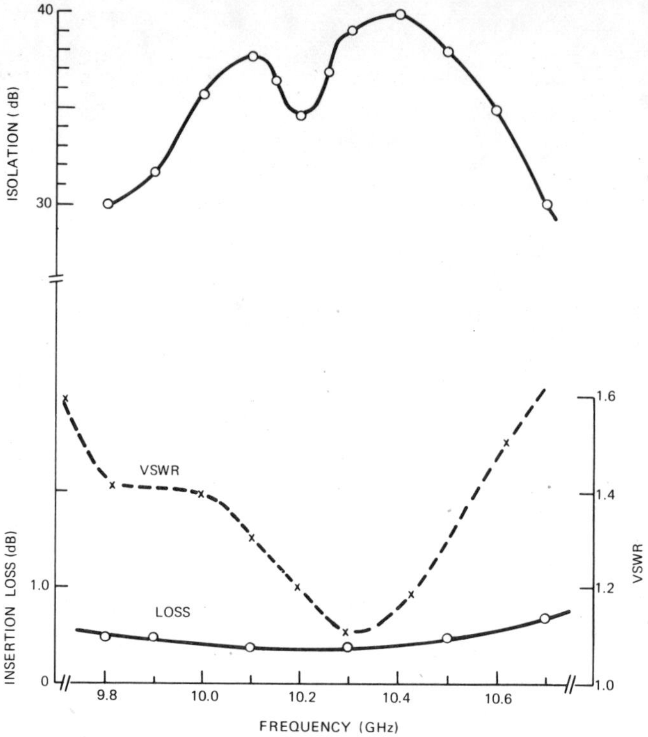

Figure VIII-37 **Measured Performance of the Two Diode, Post Coupled Waveguide Switch**

transmitter power between the antennas; thus, it must have high peak power handling capacity and rapid switching speed. The model shown has a survival level of approximately 3 kW peak when fitted with 4 mil (100 μ) I region diodes and a driver which furnishes -100 V of reverse bias. Occasional use with high incident power combined with high load mismatch can result in diode failure, therefore the circuit has been designed with easily replaceable diodes. Figure VIII-39 shows a closeup view of the removable diode mount consisting of a ceramic package diode soldered into a special O-ring seal screw with a bellows contact for making a reliable, spring loaded ohmic connection with the center conductor of the stripline board.

The schematic diagram of the RF circuit is shown in Figure VIII-40. The two output arms of the switch each have two diodes; these are installed in the "resonant" mode in one arm, and in the

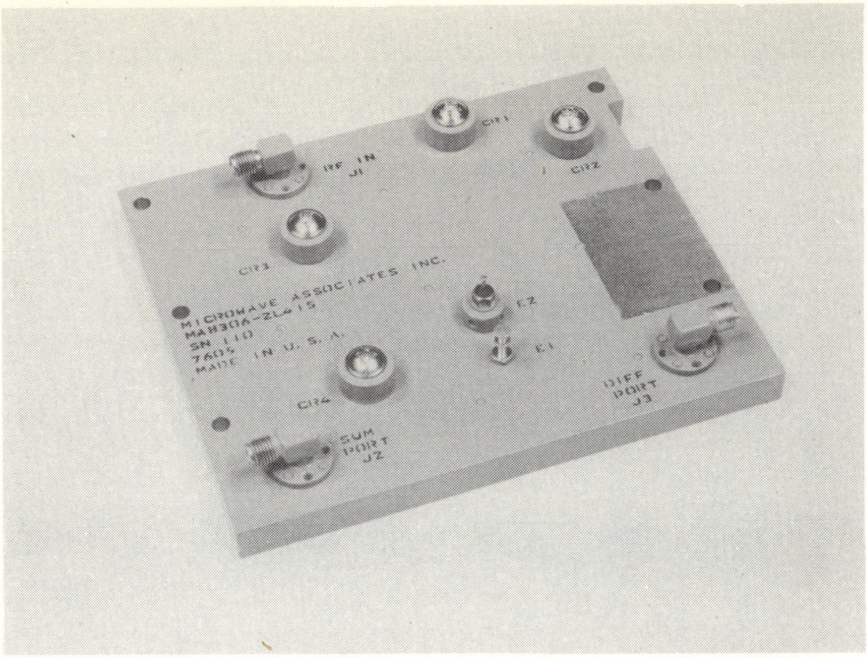

Figure VIII-38 **Stripline L-Band IFF Switch with Two Spaced Diodes per Arm of a SPDT Configuration**

"anti resonant" mode in the other. That is, for Output 1 the diodes are mounted directly across the line – forward bias produces the isolation state; for Output 2 the diodes are mounted at the end of stubs which are about a quarter wave long – forward bias produces the low loss transmission state. The advantage gained is that, should there be a failure in the driver or bias supply, the RF energy is automatically directed to Output 1. Switching to the omnidirectional antenna is mandatory should a failure occur; thus, this "failsafe" configuration is necessary.

Figure VIII-41 shows a photograph of the (center conductor) circuit board. The ground plane spacer has been removed for this view. In this switch the "center conductor" patterns are actually mounted on opposite sides of a thin Teflon fiberglass circuit board. In this way, it is possible to isolate portions of the center conductor from one another with series capacitors. The series capacitance is realized where needed by overlapping the two patterns on opposite sides of the thin center board. Specifically, in this switch, the series capacitance is used for resonantly tuning the inductance of

Figure VIII-39 Removable Diode Detail Showing O-Ring Seal Below Screw Head, Ceramic Package Diode and, Flexible Bellows for Center Conductor Contact

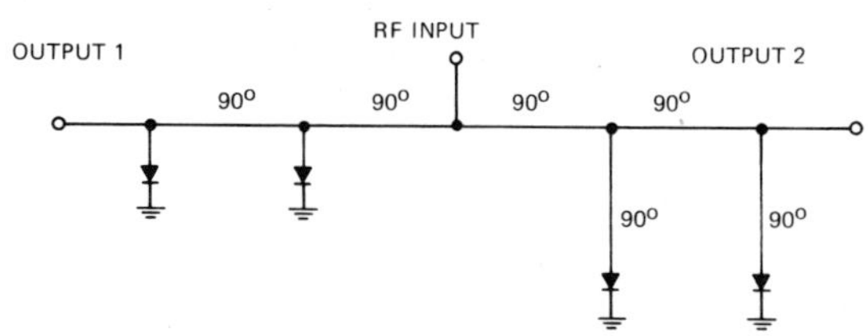

Figure VIII-40 RF Schematic Diagram of the IFF Switch Showing "Failsafe" Configuration; Failure of Bias System Causes Diodes to Assume Capacitive High Impedance State, then RF Power is Directed to Output 1

Figure VIII-41 **View of IFF Switch with Lower Cover and Lower Ground Plane Spacer Removed; Center Board is 0.010 in Thick and has Center Conductor Patterns on Both Sides with Capacitive Coupling Between to Achieve Series Capacitive Elements**

the diodes. It can also be used for achieving bias blocking capacitance elements. The measured performance for the switch is shown in Figure VIII-42.

F. Microstrip Switching

1. "Beam Lead" Diodes

In Figure VI-2, a simple SPDT switch was defined as having two diodes placed at a tee junction to direct generator power alternately between two output loads. This analysis takes into account the effects of package capacitance and inductance as well as the junction capacitance of the diode and its forward and reverse resistance values. A computer program, Figure VI-19, was written to analyze the expected performance of such a switch in terms of diode parameters, operating frequency, and generator characteristic impedance. Clearly, in order to function at high frequency, a

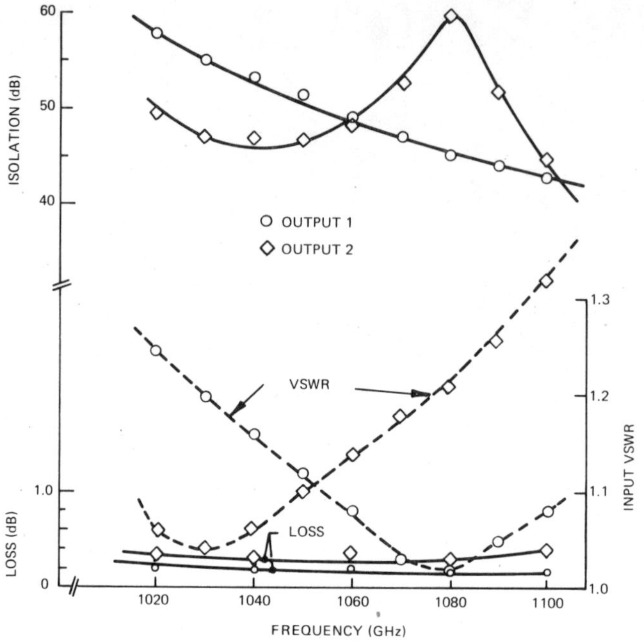

Figure VIII-42 **Measured Performance of the Stripline SPDT Switch**

low capacitance diode must be used to present a high impedance and high isolation in the off arm of the switch. Suppose that a diode having a junction capacitance of .05 pF is selected. Its capacitive reactance at 20 GHz is -j 159 Ω; therefore, in a SPDT configuration, about 10 dB of isolation would be expected for all frequencies below 20 GHz.

Figure VIII-43 shows two executions of the computer program given in Figure VI-19. The first is with the 0.05 pF diode mounted in a typical package having capacitance of .2 pF and internal and external inductances of 0.5 nH. From the figure it can be seen that isolation drops to below 10 dB at only 4 GHz; in fact, the isolation is less than the insertion loss at 8 GHz. The second execution of the program shows the performance obtained if package parasitics are eliminated altogether. The isolation is nearly 30 dB at 2 GHz and drops to the expected 10 dB at 20 GHz. From this figure it can be seen that the package effects far outweigh the reactance of the junction itself for low junction capacitance diodes

```
RUN SPDT

F1,F2,STEP,Z0=2 20 2 50

CJ,CP,LINT,LEXT,RF,RR=.05 .2 .5 .5 5 5

FREQ (GHZ)      VSWR      LOSS(DB)     ISOLATION (DB)


   2.00         1.14        0.57         15.50
   4.00         1.19        1.08          9.09
   6.00         1.20        2.14          5.27
   8.00         1.14        3.93          2.87
  10.00         1.07        6.41          1.51
  12.00         1.12        9.58          0.84
  14.00         1.22       14.31          0.53
  16.00         1.26       20.85          0.49
  18.00         1.16       10.28          1.02
  20.00         1.24        4.65          2.65
CP UNITS    1

EXIT


SYSTEM?..
.RUN SPDT

F1,F2,STEP,Z0=2 20 2 50

CJ,CP,LINT,LEXT,RF,RR=.05 0 0 0 5 5

FREQ (GHZ)      VSWR      LOSS(DB)     ISOLATION (DB)


   2.00         1.10        0.43         29.66
   4.00         1.12        0.45         23.68
   6.00         1.14        0.48         20.21
   8.00         1.16        0.52         17.79
  10.00         1.19        0.57         15.95
  12.00         1.21        0.63         14.43
  14.00         1.24        0.70         13.27
  16.00         1.27        0.77         12.26
  18.00         1.30        0.85         11.40
  20.00         1.32        0.94         10.66
```

Figure VIII-43 **SPDT Switch (Figure VI-2) Performance Calculated using Program SPDT (Figure VI-19) for a Low Junction Capacitance Diode, with and without Package Parasitics**

and that, to achieve broadband and/or very high frequency operation, it is necessary to minimize such package parasitics.

Figure VIII-44 shows the plan view of the *beam lead* diode. The diode is different from the structures described in Chapters I and II in that the geometry is surface oriented. Both the P+ and the N+ zones are diffused into the same surface of the silicon and the resulting diode junction is likewise oriented along the surface. This configuration has the advantage that the capacitance of the junction is very small because the P+ and N+ regions are not arranged parallel to one another as with mesa and planar diodes. A plan view of the PIN beam lead is shown in Figure VIII-44 along with the equivalent circuit. A further advantage of this diode structure is that the metal leads can be plated directly onto the P+ and N+

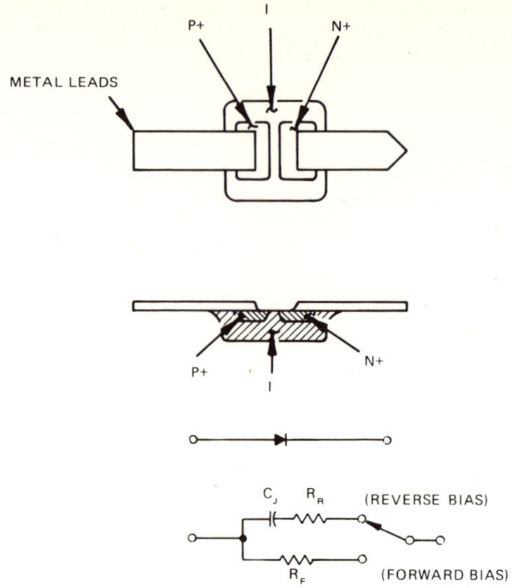

Figure VIII-44 **Plan View and Equivalent Circuit of Beam Lead PIN Diode**

zones while the diode is in batch (wafer) form (that is, with thousands of diodes together on a large silicon wafer). A subsequent etching step separates all of the diodes, leaving the structure shown in Figure VIII-44 in which two leads, or "beams," remain to permit soldering of the tiny structure directly in series with the center conductor of microstrip or stripline transmission systems. It is because of the integral realization of the leads, or beams, which results from the diode processing that the term *beam lead* is derived.

Beam leads have considerably higher R_F when forward biased than do mesa and planar diodes, the P+ and N+ regions of which are cylindrically oriented with respect to one another. The beam lead diode has high resistance because the lifetime associated with carriers is small. Since the junction is formed directly at the surface, recombination of minority carriers injected by forward bias occurs quite rapidly. Lifetime for beam lead diodes is typically 20-100 ns. Forward resistance (R_F) values of 5 to 10 Ω with 100 mA of forward bias are typical. On the other hand, total capacitance of 0.02 to 0.05 pF also is achieved typically, permitting these diodes to serve as extremely broadband switches. In Figure VIII-45 is a photograph of a typical beam lead diode; Figure VIII-

Figure VIII 15 Enlarged Photograph of 0.010 x 0.010 in (250 x 250 μ) Beam Lead Diode Soldered Directly to Center Conductor of 0.007 in, 50 Ω Ground Plane Spaced TFG Microstrip Line (Beam Lead is Model MA-47301, Courtesty of C. McCauley and T. Carr, Microwave Associates, Inc.)

46 shows loss and isolation performance achieved when it is mounted, untuned, in series with a 50 Ω transmission line. The dissipative loss is nearly constant with frequency. Reflective loss arising from reflection interactions between the untuned diode and the test fixture accounts for most of the loss increase observed above 10 GHz; it could probably be eliminated with proper tuning.

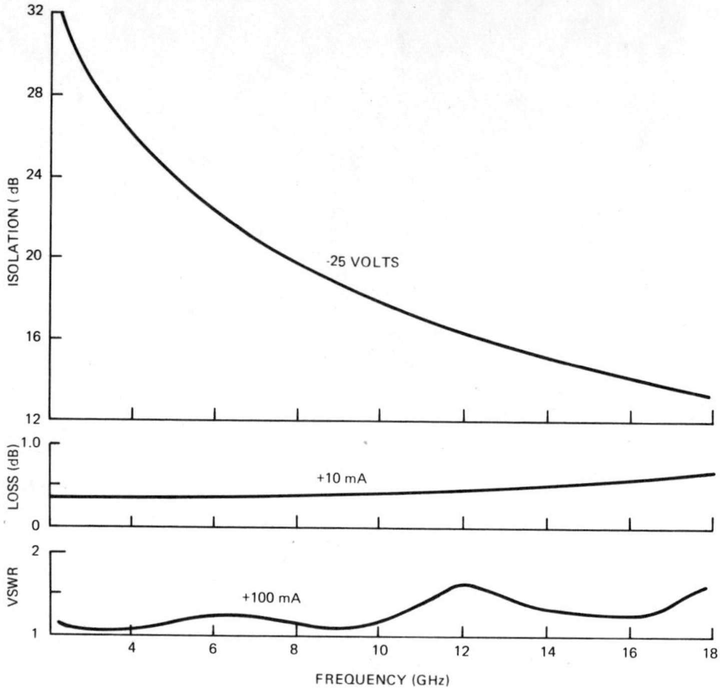

Figure VIII-46 **Typical Measured Performance for 0.02 pF Beam Lead PIN Diode Mounted without Tuning in Series with 50 Ω TFG Microstrip Line (Model MA-47301, Courtesy of C. McCauley and T. Carr, Microwave Associates, Inc.)**

2. *"TR Switch" Subassembly*

At the low frequency boundary of the microwave spectrum, low capacitance, glass passivated PIN diodes can be mounted directly on a ceramic substrate to provide SPDT switching. A photograph of such a diode mounting is shown in Figure VIII-47. This unit is a switch subassembly because it does not include bias components or RF connectors. Measured loss, VSWR, and isolation are shown in Figure VIII-48. Using beryllium oxide as a substrate, thermal resistance values of 15°C/W can be obtained with diodes having about 0.2 pF of capacitance. The dissipative loss of the diode when biased sufficiently to produce less than 1 Ω of forward resistance is only 2% of that incident. Thus, with as much as 100 W of average incident power, only about 2 W would be dissipated in

Figure VIII-47 **A Low Frequency (20 1000 MHz) SPDT Switch Subassembly used for TR (Transmit and Receive) Switching in Communication Radios (Model MA-8334 Series, Courtesy D. Gallagher and V. Philbrook, Microwave Associates, Inc.)**

the forward conducting diode and the associated temperature rise above the heat sink would be only 30°C.

At low frequencies lumped element capacitors and inductors can be utilized to introduce bias to the PIN diodes. Figure VIII-49 shows a typical bias circuit and the values of capacitors and inductors required to provide impedance ratios with the 50 Ω line of 100 to 1 at frequencies up to 1 GHz.

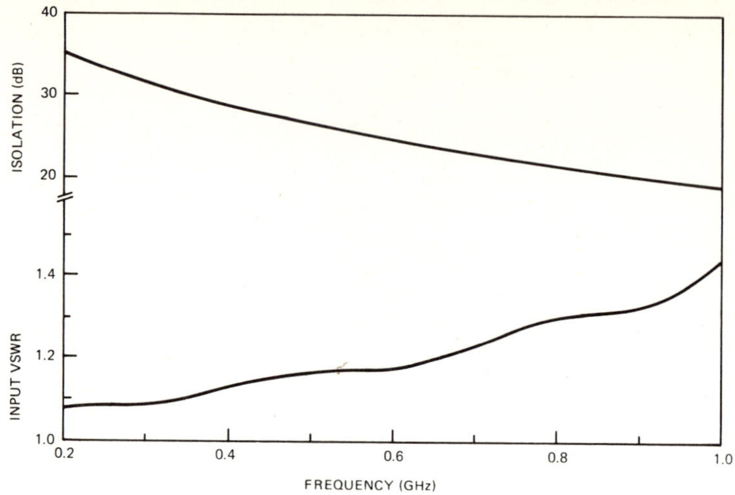

Figure VIII-48 **Typical Isolation and Input VSWR Obtained with TR Switch**

3. *"Fifty Ohm Diode Packages"*

As we have just seen with the beam lead diode and the TR switch subassembly, elimination of the package parasitics permits both broader band operation and greater isolation at high frequencies. In the case of the beam lead, this action necessitates the use of a very low junction capacitance diode in the microwave switching circuit. With the TR switch, although package capacitance is eliminated, series inductance is not; thus, even if low junction capacitance is used, the frequency of use is limited to below 1 GHz. To circumvent the deleterious effects of series inductance, modular microwave packages for PIN diodes are used in which the diode and its hook-up wire are made integral with the microwave circuit, including the series inductance within the center conductor of the transmission line itself, as shown for the microstrip limiter in Figure VII-22. A module of this sort can be considered a special form of diode package. To be broadly useful throughout the microwave spectrum (say, from 1 to 10 GHz), such a package must have a minimum of frequency dependent elements within it. Over such a bandwidth, even the spacing of a pair of diodes produces frequency sensitive behavior.

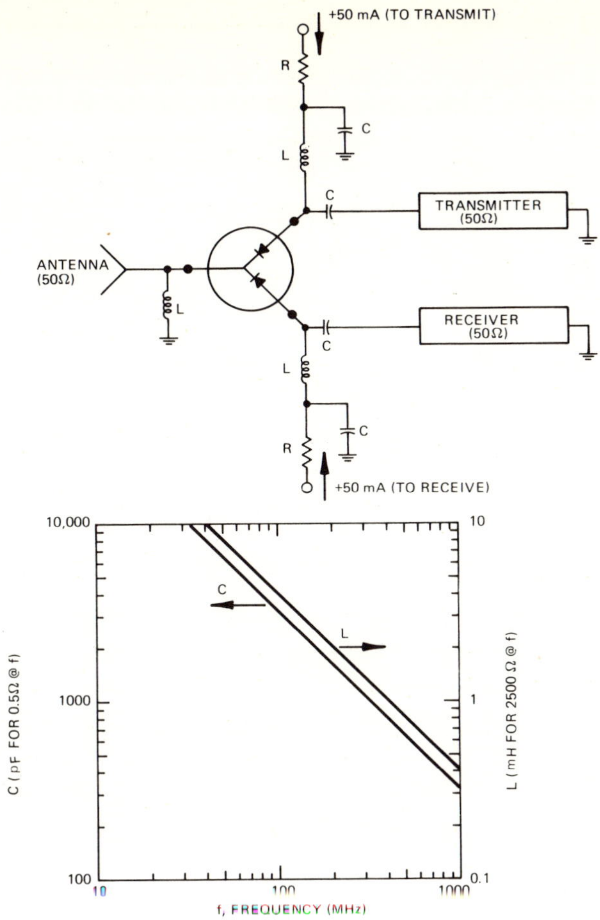

Figure VIII-49 **Typical use of TR Switch Module in Mobile Transmitter-Receiver, Showing use of Lumped Element Bias Circuitry**

The simplest module contains only one diode, with strap lengths and junction capacitance selected to give 50 Ω impedance. This device represents perhaps the ultimate step toward encapsulation of the diode for microwave use. It is for this reason that modules in which the diode strap inductance and junction capacitance have been matched to provide the low pass filter response described in Figure VII-23 are sometimes called the *ideal diode package.* Figure VIII-50 shows practical single diode, 50 Ω packages; Figure VIII-51 shows typical measured performance.

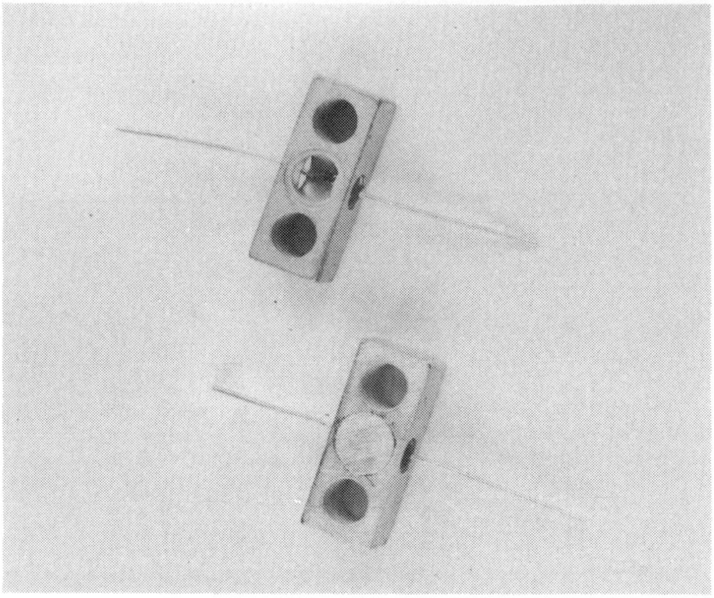

(a) Matched Single Diode Module (Ideal Package)
in *Bolt Channel* Package (MA-47221)

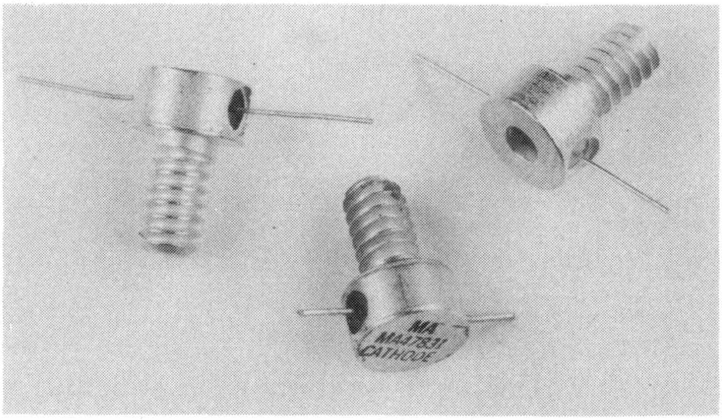

(b) Round Case (Ideal Package) (MA-47202 Series)

Figure VIII-50 The Single Diode Ideal Package Module (Photos and Data Courtesy of D. Gallagher and D. Fryklund, Microwave Associates, Inc.)

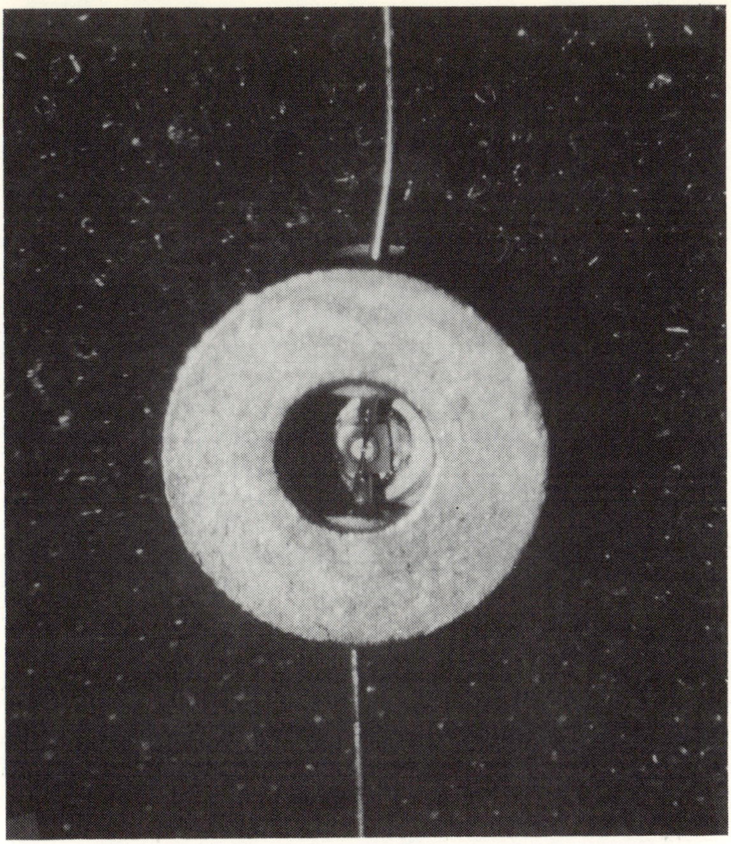

(c) Closeup View Showing Single Shunt Diode Mounted in Round Package

G. Bulk Effect Switching

1. Rectangular Waveguide Switch

The microwave control devices described so far have been "diodes" in the true sense; that is, they have only two terminals. Both the microwave signals applied to them as well as the bias signals used for control must be applied to the *same* two terminals. Such a structure has definite size limitations. For greater power handling capability, large area, wide I region PIN diodes are needed. But a large area results in greater capacitance; for wide bandwidth and high frequency operation, high capacitance at the RF terminals is a disadvantage. Furthermore, with a wide I region, switching speed is reduced; however, even were this reduction not a drawback,

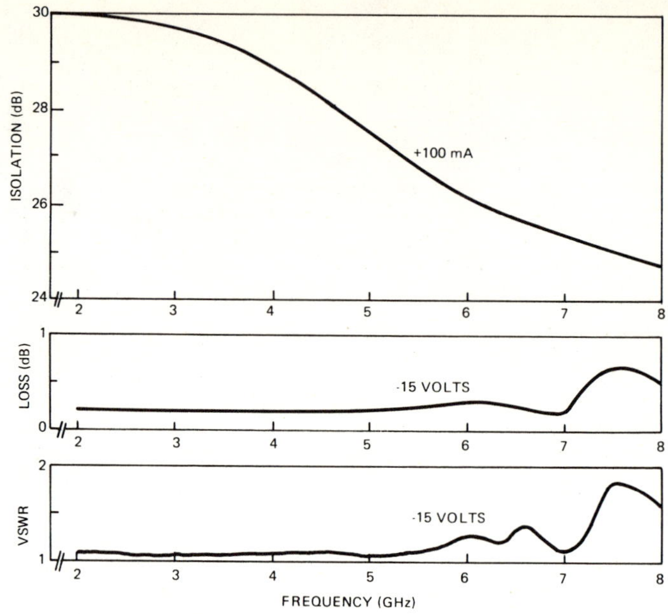

Figure VIII-51 **Typical Performance of Single Diode Module (Ideal Package); Isolation is Resistance Limited at Low Frequency but Inductively Limited at High Frequency (Courtesy of V. Philbrook and D. Fryklund, Microwave Associates, Inc.)**

there is a practical limit to the I region width within which good conductivity modulation is achievable. For this reason, a class of controlled devices having orthogonal bias and RF terminals offers attractive possibilities for high frequency, broad bandwidth, high power handling operation. These devices have been termed *bulk effect* components to emphasize their distributed characteristics.

The concept of a bulk effect microwave *window* is illustrated in Figure VIII-52. A large slab of high resistivity silicon is placed across the waveguide, with P+ diffusions on one side and N+ diffusions on the other side arranged in thin horizontal lines. Metalization patterns over these diffusions allow for biasing the window. Between the two lined structures the device appears as a huge PIN diode which may have, depending on the area of the P+ and N+ diffusions, tens or hundreds of picofarads of capacitance. However, the microwave signal is not applied between the bias terminals, but is oriented at right angles to them. The RF field extends from the

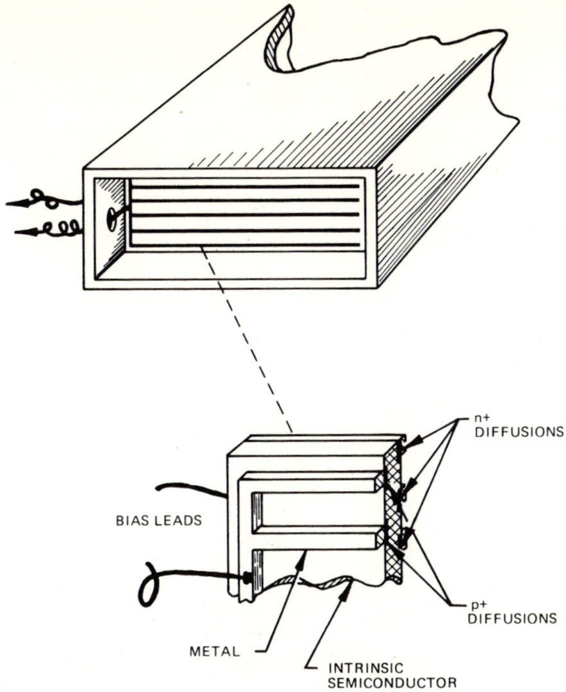

Figure VIII-52 **The Electronic Window Installed in Rectangular Waveguide as a SPST Switch**

top of the waveguide to the bottom, and as such, it "sees" only the dielectric constant contribution of the silicon together with some capacitive reactance contributed by the metallized patterns In practice, the normalized shunt susceptance of such a window mounted in the waveguide is but a small fraction of the characteristic admittance of the guide and hence, even without parallel tuning, the device operates readily over the full waveguide bandwidth with little reflection in the pass state. With this structure the thickness of the window must be kept small so that holes and the electrons injected from the P+ and N+ zones can diffuse throughout the high resistivity portions of the silicon to impart a low resistivity in the forward bias, high isolation state. The diffusion length of these carriers is related to the lifetime, τ, and the diffusion constant, D_{AP}, according to

$$L_D = \sqrt{\tau D_{AP}} \qquad \text{(VIII-42)}$$

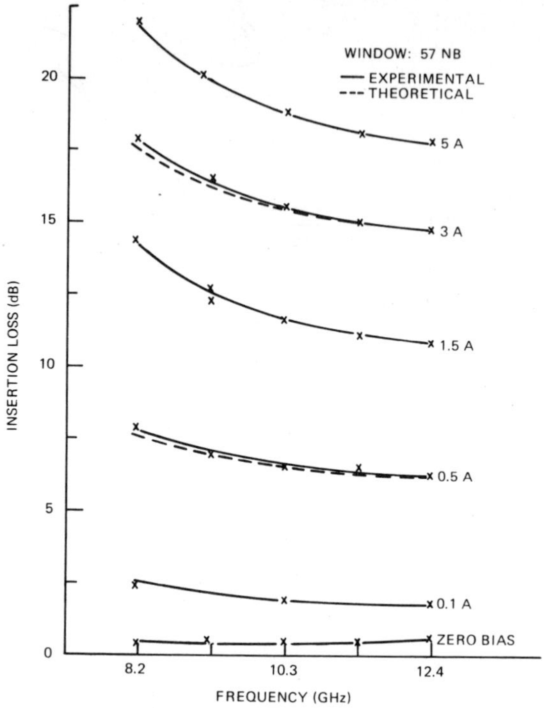

Figure VIII-53 **Measured Performance of Bulk Effect Window in WR-90 Waveguide (Data Courtesy of A. Armstrong, Rensselaer Research Corporation International, Inc., Latham, New York)**

and, since D_{AP} 15.6 cm^2/s

$$L_D \approx 40\sqrt{\tau\,(\text{microseconds})} \quad \text{microns}$$

$$\approx 1.7\sqrt{\tau\,(\text{microseconds})} \quad \text{mils}$$

While a simple effective lifetime definition is useful for estimating the required bias current requirement, it is inadequate to describe the actual distribution of carriers in the element. In fact, the average lifetime of a carrier depends strongly upon whether the carrier's diffusion happens to move it towards the surface and/or contacts, or if it tends to remain within the bulk of the silicon. The recombination of holes and electrons occurs much more rapidly near the diffused contact areas and the surface than it does within the bulk.

In practice, an effective lifetime of about 10 μs can be obtained, thus, the samples must be kept to a thickness of not more than about 100 μ (4 mils) if good conductivity modulation is to be obtained under forward bias.

The measured performance for a window fabricated in WR-90, X-band waveguide is shown in Figure VIII-53. No reverse bias is needed for the transmission state. With increasing forward bias, progressively higher transmission attenuation (isolation) is offered by the window, with isolation of about 20 dB or more obtained at 5 A current. These windows can be operated at essentially the full peak power capacity of the waveguide; typical X-band win-

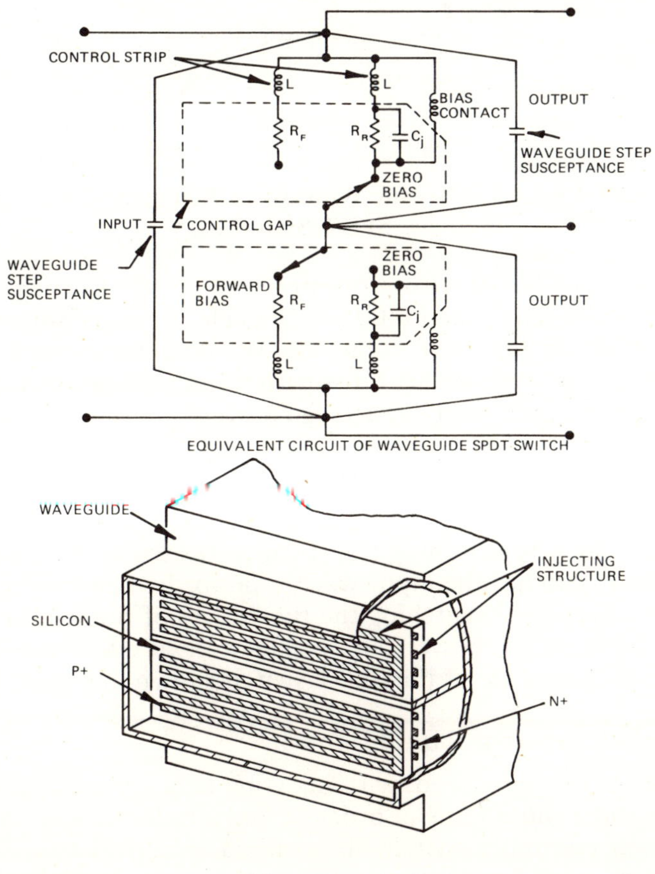

Figure VIII-54 **Waveguide SPDT Switch Structure using Bulk Silicon Window**

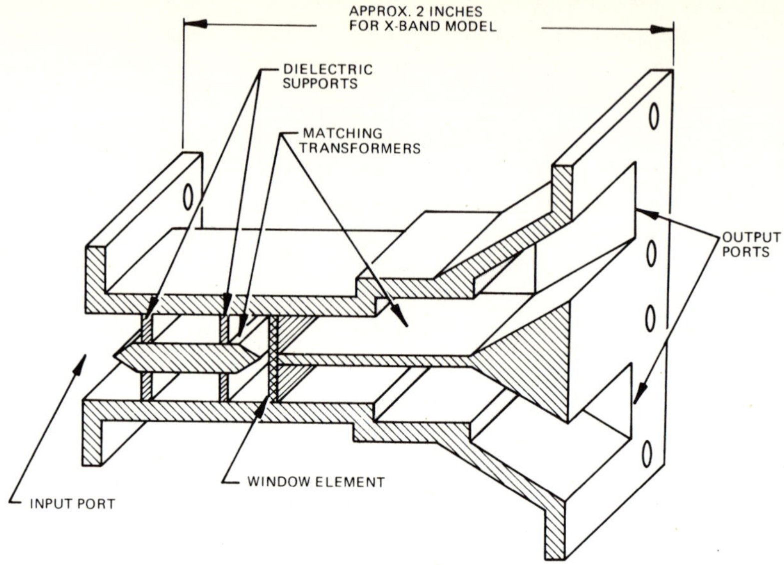

Figure VIII-55 **Electronic Window in Bifurcated Waveguide to Achieve SPDT Switching**

dows have been tested without destruction to 100 kW peak power with 1 μs pulse length and 0.001 duty cycle. Average power handling capability is limited to about 100 W, but ceramic backing of the window can improve heat dissipation, increasing this capability to a few hundred watts of average power.

A configuration to offer SPDT switching is shown in Figures VIII-54 and VIII-55. A section of bifurcated waveguide contains a window element having upper and lower portions which are independently biasable. The matching transformer located at the input of the window matches the input waveguide to the half height guide. Then, by alternately switching the two halves of the window on and off, incident RF power can be made to exit either the upper or lower output ports of the SPDT switch shown in Figure VIII-55.

2. *Surface Oriented Bulk Window*

The window just described, with injecting contacts on opposite faces, provides more than an order of magnitude increase in the peak power capability switchable with practical packaged diode circuits at S- or X-band. As such, it represents a valuable extension in technique for high power switch design. But, with this approach, switching speed is generally in the 1 to 10 μs range, depending up-

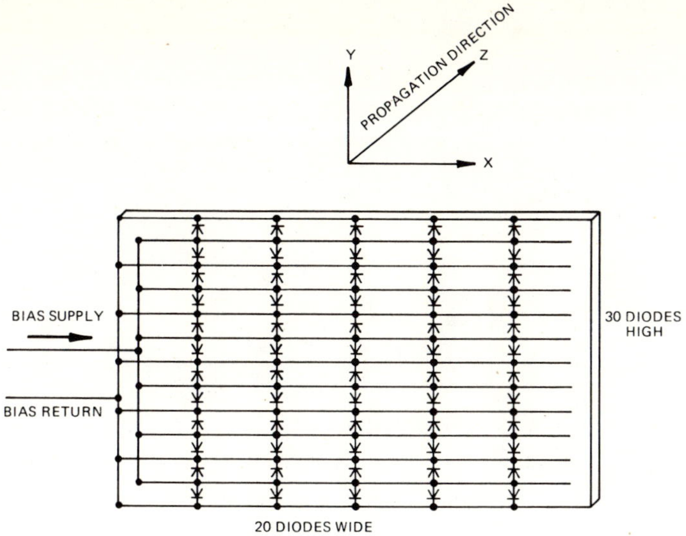

Figure VIII-56 **Schematic of Diode Layout on Silicon Switching Element**

on driver complexity. For some applications, however, it is desirable to combine the high peak power switching capacity with more rapid switching, on the order of 50 ns or less. A means for improving the switching speed of the bulk window is to place both the P+ and N+ diffusions *on the same side of the window,* essentially generating a rectangular array of individual PIN diodes as shown in Figure VIII-56.

The switching speed, T, is related to the current used to produce it, I, and the total charge, Q, stored in the aggregate of the PIN diodes which form the window according to

$$Q < IT \qquad \text{(VIII-43)}$$

The inequality sign is used because some charge is absorbed by recombination during the transition from forward bias to reverse bias. If, for the X-band window, a maximum switching current of 20 A is specified, the total charge must be less than 10^{-6} coulombs to achieve switching in 50 ns. On the other hand, as the silicon switches from a dielectric to a conductive medium, average conductivity, σ, must be greater than about 5 mhos/cm to achieve sufficiently high isolation state conductance. This conductance must be obtained with a total volume of silicon having a low

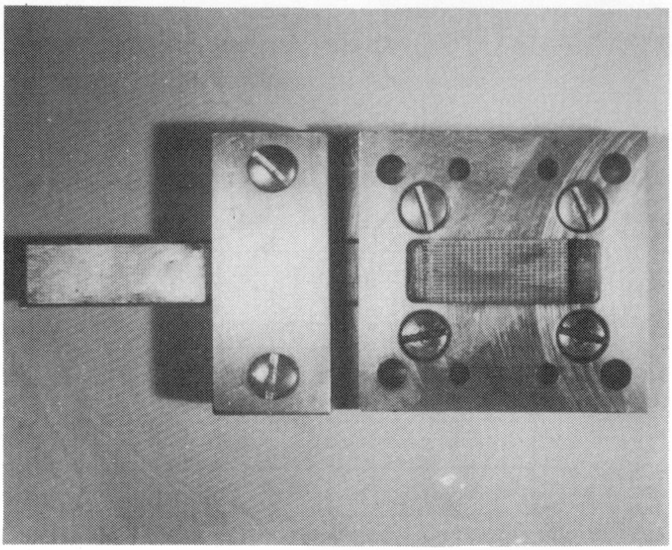

a) Silicon side showing 600 PIN diode matrix array

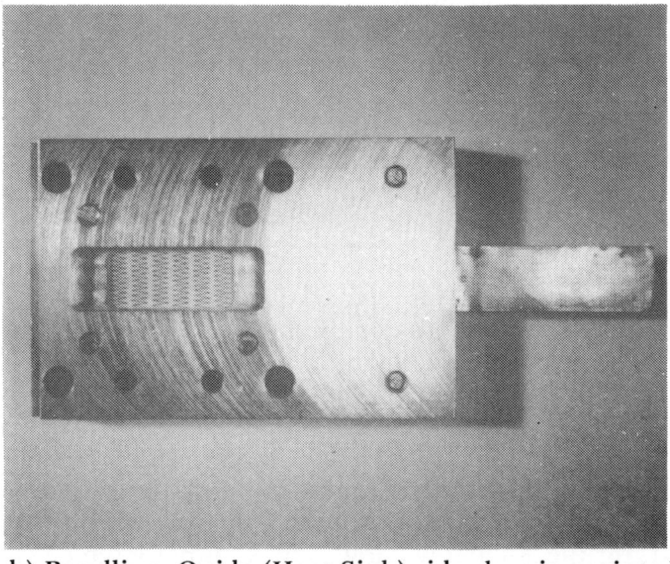

b) Beryllium Oxide (Heat Sink) side showing printed shunt inductive metal pattern to effect parallel tuning

Figure VIII-57 Surface Oriented Bulk Effect Microwave Window (Data and Photographs Courtesy of A. Armstrong and P. Bakeman, Rensselaer Research Corporation International, Inc., Latham, New York)

transmission state susceptance to give low VSWR. From these two conditions, the maximum volume of silicon composing the effective I region of the array of diodes must satisfy

$$V \leqslant \frac{Q\mu_{AP}}{\sigma} = \frac{1 \times 10^{-6} \text{ coulomb} \times 610 \text{ centimeters}^2/\text{volt-second}}{5 \text{ mhos}}$$

$$\leqslant 1.2 \times 10^{-4} \text{ cubic centimeter} \qquad \text{(VIII-44)}$$

The design layout for an X-band window shown in Figure VIII-56 contains 600 individual PIN junctions, each measuring approximately 1 x 1 x 12 mils (25 x 25 x 300 μ) for a total switched volume of 1.1×10^{-4} cm^3.

Photographs of the two sides of an experimental model of the surface oriented bulk effect window are shown in Figure VIII-57. The array of 600 diodes appears on the silicon side. On the back side, a 7 mil (175 μ) beryllium oxide backing was used to improve the heat dissipating characteristics of the window. A metalized inductive pattern on the oxide serves to parallel resonate the shunt capacitance produced in the waveguide by both the silicon and the beryllium oxide.

The device was tested by discharging capacitors into the bias leads to simulate a high speed driver. Pulse bias peaks of approximately 20 A were realized in this manner, and the switching speed of less than 50 ns was obtained for both the forward-to-reverse and the reverse-to-forward bias transitions.

The device was high power evaluated by matching, using an external tuning structure, 50 W of CW microwave power into the window. The resulting flange temperature stabilized at 53°C. This level corresponds to a normal switching capability of over 175 W for the unit tested (which gave only 12 dB of isolation). More than 250 W CW RF power could be switched if sufficient conductivity were achieved to give 19 dB of isolation and a commensurate lower power absorption in the isolate state.

The isolation and insertion loss data obtained for this model are shown in Figure VIII-58. Lower isolation than anticipated (12 dB compared to 20 dB) was obtained because the minority carrier lifetime realized for the PIN diodes was only about 100 ns. It is likely that, with additional fabricational refinements, a switch yielding less than 1 dB of loss and more than 20 dB of isolation could be made, providing an extremely broad bandwidth, high power, and fast switching microwave control component.

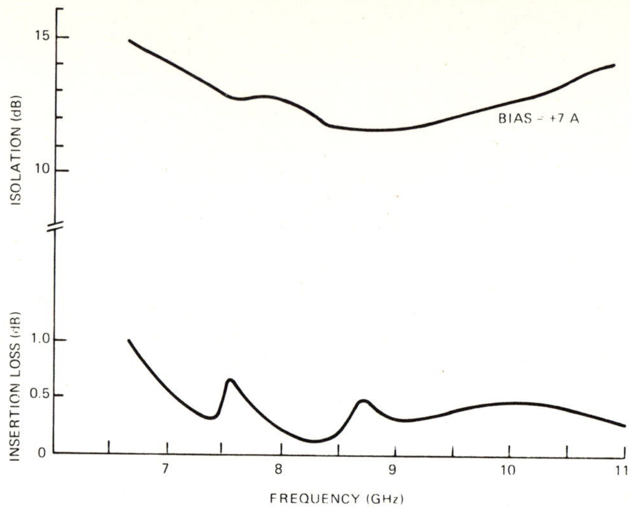

Figure VIII-58 **Isolation and Loss Measured with Surface Oriented Bulk Effect Switch**

3. Inverted Microstrip Bulk Attenuator

By utilizing the orthogonal principles of bulk elements with TEM mode transmission lines, multi-octave bandwidth control devices can be realized. Figure VIII-59 shows the plan view of such a device. This type of transmission line is called inverted microstrip because the substrate which supports the center conductor is above the area in which most of the RF energy propagates. By supporting the line in this manner, a microstrip line with a dielectric that is essentially composed of air is realized, with consequent low dielectric losses. When the dielectric is made of silicon, the prospect for a controllable attenuation function is evident.

By diffusing P+ and N+ channels parallel to the metallized center conductor pattern of the transmission line on the silicon, a variable conductivity path extending from the center conductor to the side walls of the metal channel is created. As forward bias is applied between the P+ and N+ zones, a conducting path is established from the center conductor to ground, creating a transmission line with variable insertion loss along its propagation length.

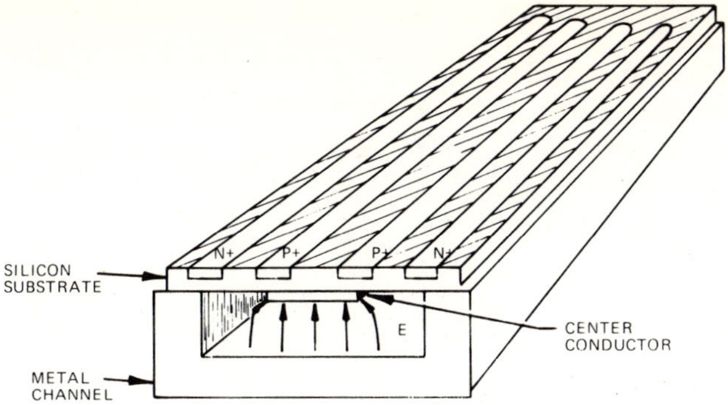

Figure VIII-59 **Inverted Microstrip Bulk Effect Attenuator**

Notice that the control bias can be applied from the top side of the structure, where low microwave field strength exists. Thus, the bias and the microwave fields are decoupled quite naturally. The transmission line itself can propagate signals from dc to arbitrarily high microwave frequencies, limited in practice only by the incidence of higher order modes which can propagate when the width of the metal support channel is equal to a half wavelength.

Figure VIII-60 shows a photograph of an experimental device in which, for fabricational convenience, a microstrip line was supported on an alumina substrate below the silicon. The silicon slice, about 1 cm in length with appropriate diffused and metallized zones parallel to the direction of propagation, is attached ohmically between the center conductor pattern and the metal walls of the transmission line channel.

The measured performance of this attenuator is shown in Figure VIII-61. The performance is remarkably constant considering the broad frequency range over which it was evaluated. The device has largely dissipative properties at frequencies for which its length is long compared to the RF wavelength. The attenuation is somewhat less at high frequencies since there is less reflection, but it is more constant with frequency than it is under low frequency operation. The device shown represents an initial experimental model. As with the other bulk devices, it is likely that further development will produce even more optimum performance over this broad operating frequency range.

Figure VIII-60 Photograph of the Inverted Microstrip Bulk Effect Attenuator (Device Photo and Data Courtesy of A. Armstrong, and P. Bakeman, Rensselaer Research Corporation International, Inc., Latham, New York)

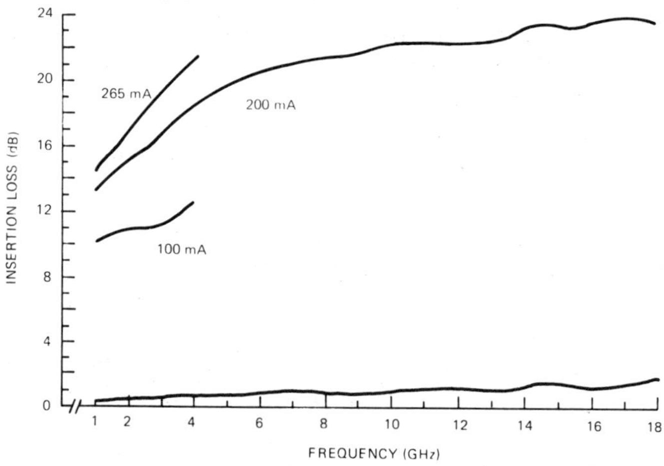

Figure VIII-61 Measured Performance of the Inverted Microstrip Limiter (Courtesy of A. Armstrong and P. Bakeman, Rensselaer Research Corporation International, Inc., Latham, New York)

References

[1] Fisher, R.E.: "Broadbanding Microwave Diode Switches," *IEEE Transactions on Microwave Theory and Techniques, Vol. MTT-13, No. 5,* pp. 706, September 1965.

[2] White, J.F.; and Mortenson, K.E.: "Diode SPDT Switching at High Power with Octave Microwave Bandwidth," *IEEE Transactions on Microwave Theory and Techniques, Vol. MTT-16, No. 1,* pp. 30-36, January 1968.

[3] Matthaei, G.L.; Young, L.; and Jones, E.M.T.: *Microwave Filters, Impedance-Matching Networks, and Coupling Structures,* McGraw-Hill Inc., New York, 1964.

[4] Mumford, W.W.: "Tables of Stub Admittances for Maximally Flat Filters Using Shorted Quarter Wave Stubs," *IEEE Transactions on Microwave Theory and Techniques, Vol. MTT-13 No. 5,* pp. 695-696, September 1965.

[5] Young, L.: "Tables for Cascaded Homogeneous Quarter Wave Transformers," *IEEE Transactions on Microwave Theory and Techniques, Vol. MTT-7, No. 1,* pp. 233-237, January 1959. (*Note* – Corrections to four section designs appear in *IEEE Transactions on Microwave Theory and Techniques, Vol. MTT-8, No. 2,* pp. 243-244, March 1960; also, transformer designs by Young are also described in Reference 3 above.

[6] White, J.F.: "High Power Solid State Switching," *1964 International Solid-State Circuits Conference – Digest of Technical Papers,* pp. 30-31.

[7] Mortenson, K.E.; Armstrong, A.; Borrego, J.; and White, J.: "A Review of Bulk Semiconductor Microwave Control Components," *Proceedings of the IEEE, Vol. 59, No. 8,* pp. 1191-1200, August 1971.

[8] Armstrong, A.L.; and Bakeman, P.E.: "Fast, High Power, Waveguide Bandwidth Microwave Switch," *IEEE MTT Symposium Digest,* 1976.

The author gratefully acknowledges the many who contributed to the realization of the devices described in this chapter. Some who have not been credited explicitly are R. Brunton, D. Conlon, H. Esterly, G. Garas, H.J. Griffin, M.E. Hines, and K.E. Mortenson; the Navy Department, Bureau of Ships; and the Raytheon Company, Wayland, Massachusetts.

Chapter IX

Phase Shifters and Time Delay Networks

A. Introduction

A semiconductor phase shifter is a device the primary function of which is to change, by means of a control bias, the propagation phase of a microwave signal. Though not usually by design, switches, attenuators, limiters, and duplexers also produce phase shift along with their intended functions, but this function is not their intended one.

Any reactance placed in series or shunt with a transmission line introduces phase shift and, therefore, there is a virtually unlimited number of possible phase shifter circuit configurations. However, adding the requirement that the device has minimum insertion and reflective losses, reduces the number of practical circuits to a few. Each offers its own combination of advantages with respect to size, bandwidth, the phase shift obtained per diode, and so forth. One, consisting of alternately switched transmission paths, can provide a phase shift which increases linearly with frequency and thereby represents a special class of phase shifters — *time delay networks.* These circuits have special applicability in microwave phased array antennas which must operate with large array diameter-bandwidth products. Thus, we begin our investigation of phase shifters with these switched line networks. But before doing so, it is necessary to define what is meant by *phase* and *phase shift.*

Figure IX-1 shows the defining convention to be used in this chapter for transmission phase and insertion loss. From Figure IX-1, it can be noted that phase usually is defined using the generator voltage (V_0) phase as a reference, rather than that of the input voltage (V_1) to the network. This preference occurs because, in practice, control networks are not perfectly matched;

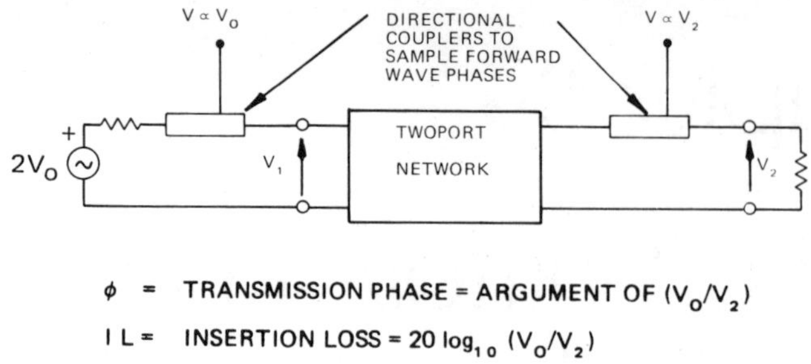

Figure IX-1 **Definition of Transmission Phase and Loss**

consequently, the input voltage phase does not remain constant when the control state of the device is switched. For practical measurements, a phase referenced to the internal voltage of the generator is obtained by using a directional coupler to sample the forward-going wave ahead of the control network, as shown in Figure IX-1. Similarly, an output coupler samples load phase. These couplers give output voltages proportional in amplitude and phase to V_0 and V_2, respectively, even in the presence of mismatches – provided, of course, that the coupler directivity is high enough. *Phase shift* is the *change* in transmission phase of a network. Thus, for example, if a two-port network has two discrete states having respective transmission phases of ϕ_1 and ϕ_2, its *phase shift* is

$$\Delta\phi = \phi_1 - \phi_2 \qquad \text{(IX-1)}$$

The sign to be identified with $\Delta\phi$ depends both upon which state is defined as the reference and how positive phase is defined.

There is a natural tendency to assume that positive values of phase shift should correspond to the "delay" that would be introduced, for example, by inserting extra cable between the generator and the point at which output phase is measured. While the definition of "positive phase" is arbitrary, such a choice would lead to confusion because a steady state sinusoidal voltage waveform usually is defined* in the form $Ve^{j(\omega t+\phi)}$. Since time increases positively, a wave delayed in time should have a *negative phase* as shown in Figure IX-2. In this chapter, we refer only to the magnitude of

*Throughout this book, the ωt portion of complex voltage and current arguments is assumed but not written explicitly.

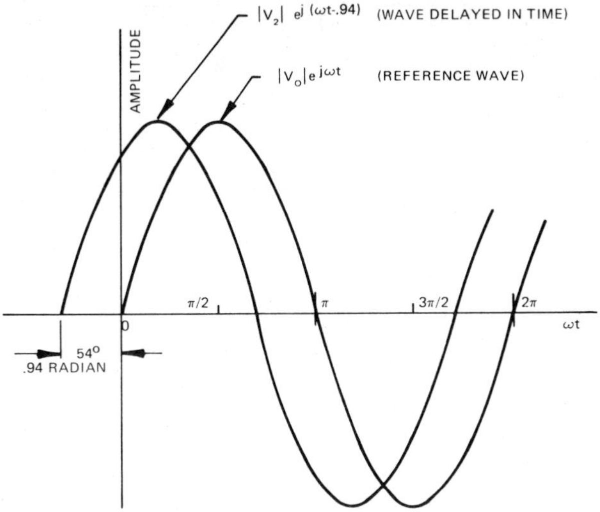

Figure IX-2 **Sample Sinusoid Waveforms Showing that a Time Delayed Signal has Negative Phase**

phase shift and, when necessary, distinguish which network condition(s) corresponds to the shortest electrical length state(s). But, in general, *the phase of a wave delayed with respect to a reference wave* (for example, by a longer electrical length path) *should have a negative phase.*

B. Switched Path Circuits

1. Phase Shift Less than 360°

Conceptually, the most easily visualized means of obtaining insertion phase shift is by providing alternate transmission paths, with their difference in electrical length being the desired phase shift, as shown schematically in Figure IX-3. This approach gives phase shift both proportional to frequency and corresponding to switched time delay. In principle, an array antenna steered with time delay has steering which is frequency independent. However, several wavelengths of delay (approximately equal to the propagation distance across the antenna) are required to effect this frequency independent steering. Providing such controllable delay results in high insertion loss and more costly steering circuitry. Therefore, in practice, time delay of more than one wavelength usually is used ahead of the high power amplifiers driving sectors

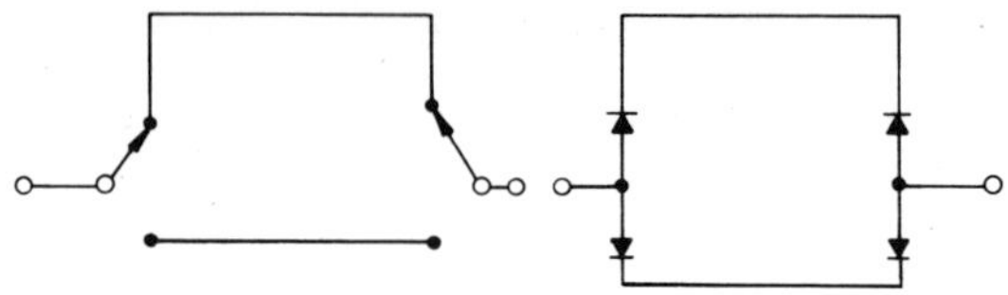

Figure IX-3 **Schematic for Switched Delay Line Phase Shifter**

of large, wide-bandwidth phased arrays; "phase shifters" (providing up to 360° control) are used separately for each radiating element.

In Chapter VI the solution for the input* and output voltages, V_1 and V_2, of the switched path phase shifter described by Figure IX-3 is derived as an example of the even-and-odd mode circuit analysis method. The insertion loss and phase shift for the circuit are determined readily from both these voltages and the defining expressions for loss and phase shift in Figure IX-1. The resulting solution, in terms of diode parameters, is computer programmed using the methods described in Chapter VI and used to calculate results in the following examples.

The practical problem to be avoided in switched path circuits is that of insertion loss resonances. These resonances occur when the electrical spacing between diodes in the *off* path is such as to "tune out" the off diode reactances, causing the off path to be connected directly in parallel with the *on* path. The situation is observed most easily with an example.

Suppose we wish to obtain a 45° phase shift circuit by switching between two 50 Ω line paths with respective electrical lengths of 160° and 205° at 1.5 GHz. Assume that diodes having $C_J = 0.2$ pF and $R_F = R_R = 1\ \Omega$ are used for the switching. Calculated performace in the frequency range 1.2-1.8 GHz, plotted in Figure IX-4, shows that, instead of linearly increasing phase shift with frequency, a phase shift characteristic with abrupt slope changes near 1.25 and 1.6 GHz is obtained. Insertion loss also increases sharply at these *resonant frequencies.* This fact can be understood by noting that at these frequencies the off arm length is such as to provide a "through" path in parallel with the selected path. Thus, with

*Henceforth, phase shift is described as referenced to V_1 for computational simplicity. Any mismatch causing V_1 and V_0 to differ in phase can be treated separately as a phase tolerance perturbation.

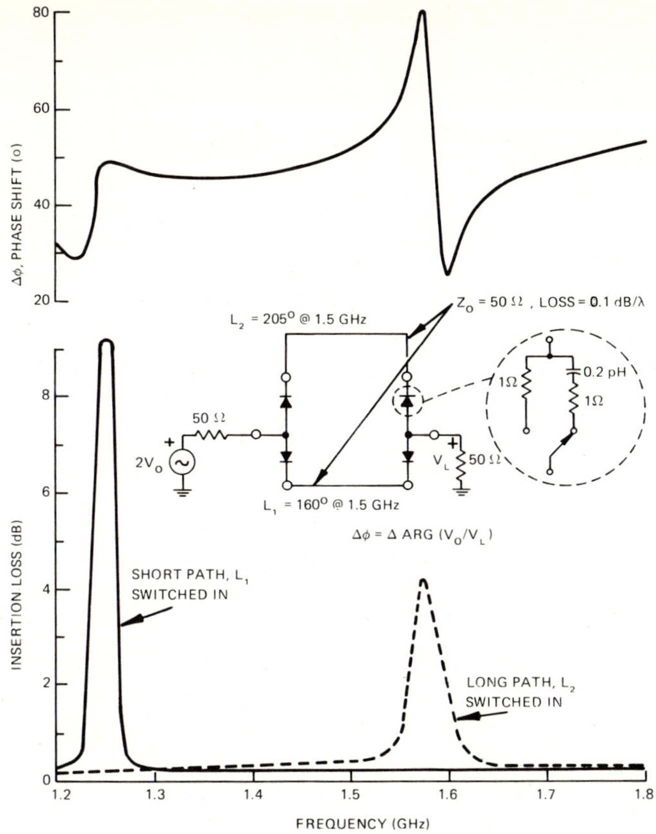

Figure IX-4 **Switched Path Circuit Example with Computed Performance Showing Loss Resonances**

path L_1 switched in, path L_2 should be decoupled, but at 1.25 GHz, path L_2 is 170° long. The two 0.2 pF diodes, which are supposed to isolate L_2 from L_1 each have a normalized reactance of -j13. Reference to the Smith Chart shows that these reactances tune out each other if separated by 172° – approximately the spacing at which the loss resonance is observed in Figure IX-4.

For this example, the means for avoiding the resonance is clear. If the difference $L_2 - L_1$ is maintained at 45° and the absolute magnitudes of L_1 and L_2 are reduced to 45° and 90° (so that neither has an electrical length near 170° within the desired operating bandwidth), the resonance is avoided. Figure IX-5 shows that this approach does serve to eliminate the resonance.

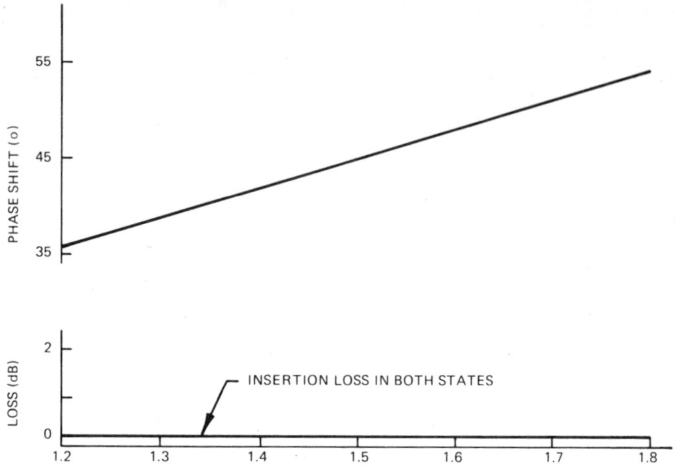

Figure IX-5 **Calculated Performance of Switched Path Circuit in Figure IX-3 with L_1 and L_2 Equal to 45° and 90° Respectively at 1.5 GHz**

To cover a 360° phase shift range with as few two-state phase shifter circuits as is possible, each is made with successive binary steps of the total phase shift range. Thus, a 4-bit phase shifter has four separate two-state circuits; their phase shifts are 180°, 90°, 45°, and 22½° at the design center frequency. With this division, any phase shift value between 0° and 360° can be achieved with an error no larger than one half the smallest bit – in this case, 11 ¼°. The sum of all bits is 348.75°, so 360° cannot be reached; this value, however, is equivalent in the steady state to 0° and a final 22½° phase shift circuit section thus is not needed.

Figure IX-6 shows a practical switched line phase shifter using an alumina (Al_2O_3) substrate microstrip transmission medium. This 4-bit, 16 diode, switched delay line phase shifter, complete with biasing chokes, is printed on a 1 in x 2 in x 0.02 in (2.5 cm x 5.1 cm x 0.5 mm) substrate. Performance is as shown in Figures IX-7 and IX-8. This circuit is used at receiver power levels following a transistor amplifier, and hence the relatively high insertion loss (3 dB) is tolerable. The diodes are similar to the 0.2 pF junction capacitance, 50 μ I region width diodes described in Table III-1. They are biased directly from the TTL logic, operating with zero bias in the isolate state and +5 mA each in the pass state.

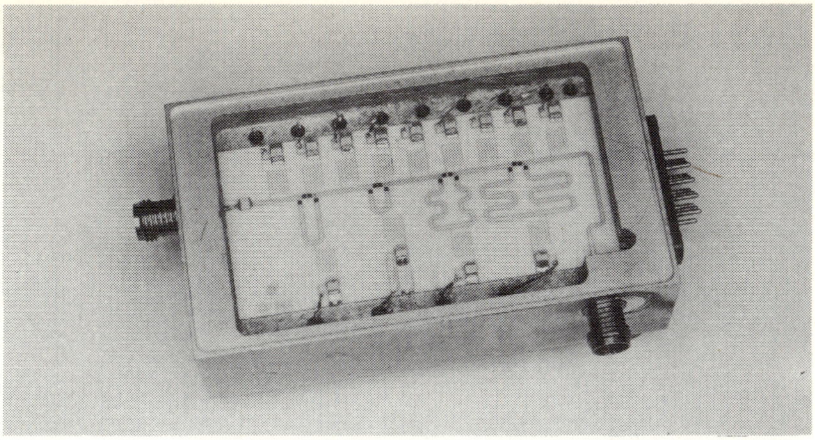

Figure IX-6 **L-Band Switched Line 4-Bit Alumina Microstrip Phase Shifter (Portions of this development were supported by the Naval Research Laboratories, Washington, DC, under Contract N00014-72-C-0213)**

2. Time Delay

Phase steering of an array antenna corresponds to setting the phase of each of the array elements of the antenna to the steady state phase which would be obtained were time delay control to have been used. Figure IX-9 shows a linear array schematic of elements uniformly separated by a distance, S, from each other. If S does not exceed a half wavelength, only one beam will be generated. The *time delay,* τ_N, required for the Nth element is determined from

$$\tau_N = NS \sin \theta \quad \text{(IX-2)}$$

For example, in a linear array of radiators having half wavelength spacing ($S = \lambda/2$) the time delay required at the Nth element in order to "steer the beam 30° off boresight" ($\theta = 30°$) is $\tau_N = 0.25\ N\lambda$. For the N = 10 element, a time delay of 2.5λ is needed. If this delay is provided, the antenna steering is independent of the RF frequency radiated, because the electrical spacing, S, and the time delay, τ_N, have the same dependence on frequency, resulting in constant scan angle

$$\theta = \sin^{-1} \frac{\tau_N}{NS} \quad \text{(IX-3)}$$

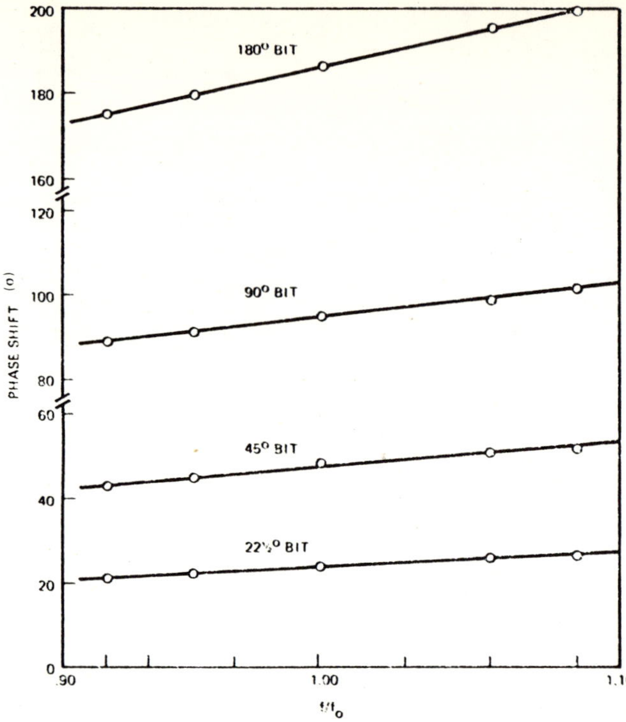

Figure IX-7 **Phase Shift vs. Frequency for L-Band Switched Line Phase Shifter**

However, at a specified frequency (f_0), if the integer number of wavelengths of required delay are omitted, the array continues to scan at the angle, θ, *at that frequency.* In the example given, the tenth element would have 0.5λ (180°) of phase delay with respect to the first element, in place of the 2.5λ time delay. But with phase shift steering, for small deviations in frequency from f_0, the beam pointing direction deviates somewhat from the θ direction obtained at f_0; for large changes in frequency, not only does scan angle change but the beam shape itself deteriorates. Pulsed modulation creates a frequency spectrum and so, for the above reasoning, an antenna used at one frequency but with short pulses functions perfectly with time delay but has deteriorated performance if phase control alone is utilized for steering. Equations (IX-4) and (IX-5) [1 (p. 83)] give the practical limits to be met for one half pulse overlap ($t - \tau/2$) between transmissions from the first and last radiating elements in a phase steering antenna.

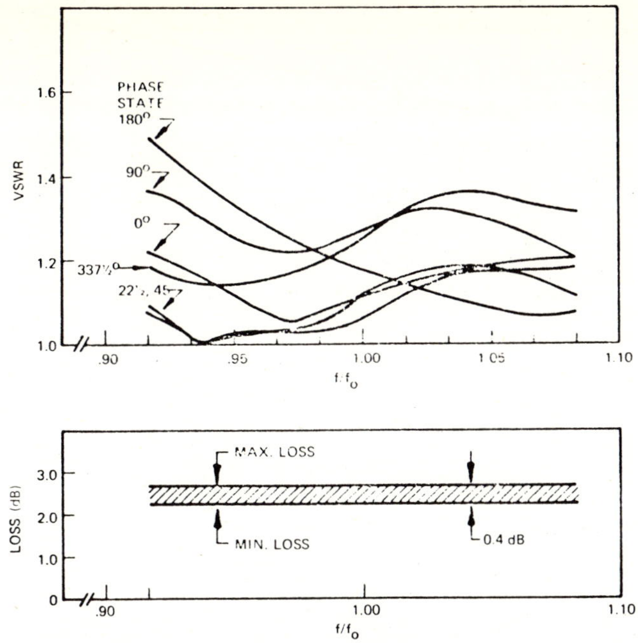

Figure IX-8 **Insertion Loss and VSWR vs. Frequency for L-Band Switched Line Phase Shifter**

$$\tau \geqslant \frac{128 \sin(\theta_M)}{\theta_0 f_0} \tag{IX-4}$$

or

$$B \leqslant \frac{\theta_0 f_0}{128 \sin(\theta_M)} \tag{IX-5}$$

where

θ_M = maximum scan angle

τ = equivalent signal pulse length (not to be confused with the time delay, τ_N, of the Nth element)

f_0 = transmission frequency in hertz

$B = \frac{1}{\tau}$ = signal bandwidth

θ_0 = beamwidth in degrees

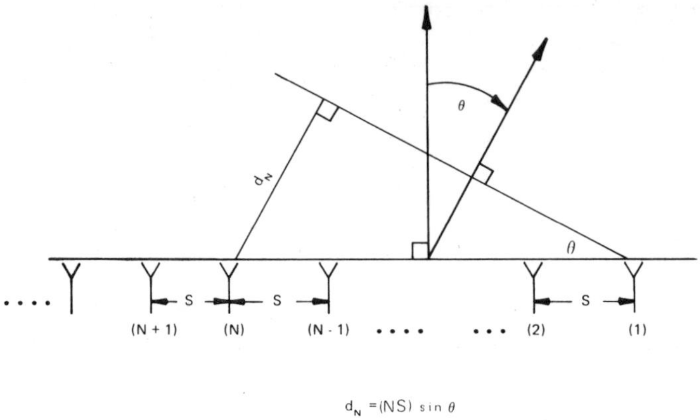

Figure IX-9 **A Linear Phased Array Steered with Time Delay**

When this criterion is applied to large radars such as the COBRA DANE Radar [1 (p. 72)], it is found that time delay for phase steered subarray groups is needed.

Figure IX-10 shows the block diagram of a 6-bit 47-wavelength time delay unit (TDU) useful for large aperture L-band radars. The device can be used at low power ahead of high power amplifiers which generate the final RF output power for various sections of the antenna array. Circulators at input and output of the time delay unit are included to minimize reflection interactions with other portions of the array control circuitry. A "dump switch" is used to interrupt RF power drive power to the RF amplifier following the TDU should a system fault (such as loss of bias to the high power phase shifters which control the high output power of each separate radiating element) be sensed by the dump switch driver circuitry. The TDU usually is required to operate over a wide bandwidth — perhaps 10 to 20 percent. With the long time delay paths switched (up to 16λ in some bits) it is impossible to select the delay paths between diodes such that resonant lengths are avoided throughout the operating bandwidth. To prevent loss resonances* in long paths, extra diodes located about λ/4 from

*An alternate method of avoiding loss resonances is to terminate the off arms in matched loads. See Reference 2 for a description of a 4-bit, 1-4 GHz time delay device on alumina substrate built using this approach.

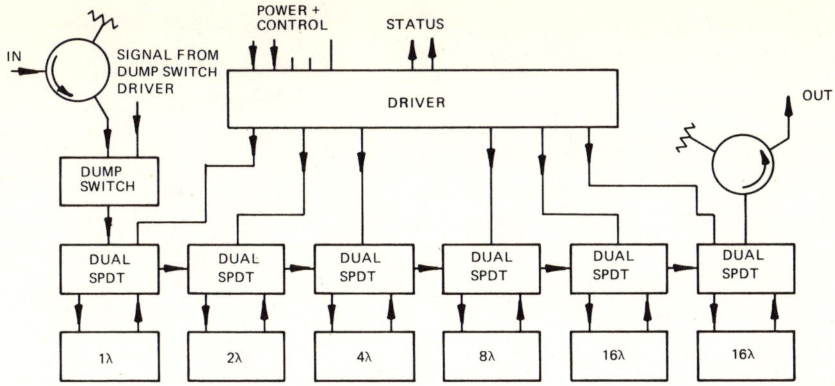

Figure IX-10 **Block Diagram of a 6-Bit 47-Wavelength Switched Time Delay Device**

the primary switching diodes at each junction are added, as shown in Figure IX-11.

Because many individually biased control bits are cascaded in the TDU, multiple reflections among them can combine for certain delay and frequency conditions which cause large phase errors and relatively high signal reflections. A means of minimizing these interactions is found in the *H-switch** configuration, also shown in Figure IX-11. Since one diode is always conducting and one is always nonconducting at the input and output of each bit, the input and output reflection coefficients of all bits can be made similar in both magnitude and phase regardless of the delay states chosen for either bit. With good design, the reflection coefficient magnitudes are small, below 0.1. If in addition the bits are quarter wavelength spaced at the center frequency, their reflections self cancel almost completely. The resulting "H" shape of the switches in adjacent bits gives this method of reflection control its name.

The photograph of a practical 6-bit time delay is shown in Figure IX-12. The switches are built into a stripline sandwich and the long delay lines are externally connected. A summary of the measured data (over 64 possible bit state combinations) is given in Figure IX-13. The total average insertion loss for the device, which switches up to 47 wavelengths of delay, is about 9 dB.

*The author was introduced to this "H-switch" method by H. Stinehelfer of Microwave Associates, Inc.

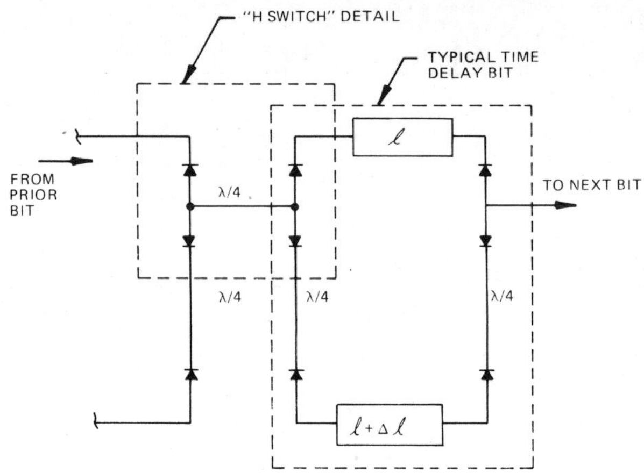

Figure IX-11 **Detail Showing Long Delay Bit with Extra Isolation Diodes. Also Shown is "H-Switch" Detail for Reducing VSWR Interactions Between Adjacent Bits**

C. Transmission Phase Shifters

1. Two Element Loaded Line

Theory

The switched path circuit described in the preceding section has some special advantages for time delay applications, but it is a brute force method of phase control. *With the switched path approach, all of the microwave power propagating on the transmission line is switched between the two paths, regardless of how much phase shift is needed.* Therefore, maximum RF voltage and current stresses are placed on the diodes in all bits, even those with small phase shift.

The diode insertion loss is equivalent to that obtained with a signal passing through a pair of SPDT switches. Therefore, not only are diode stresses high even in small bits when path switching is used, but these bits also have as much diode loss as does the largest (180°) bit.

In Chapter V, it is shown that, for reflection phase shifters*, the amount of RF stress on the diode and the resulting diode inser-

*A similar result is obtained in this section for the transmission phase shifter (Equation (IX-13)).

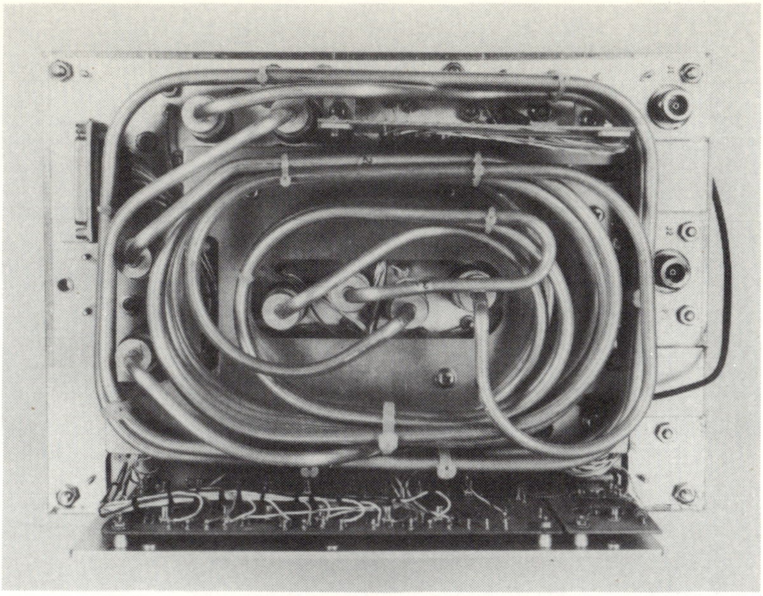

Figure IX-12 **Photograph of Stripline 6-Bit 47-Wavelength Switched Time-Delay Unit Showing RF Ports, Driver Circuit, and Coaxial Lines used to Obtain the Switched Delay**

tion loss can be decreased as the amount of phase shift obtained is reduced. Thus, for small bits or small phase shift sections used to form larger bits, a circuit which does not subject the diodes to maximum stresses is desirable. To build a very high power phase shifter, it is necessary to design a circuit which uses many diodes with only small phase shift per diode. When many diodes are used it is important that the circuit which couples them to the task does so without itself introducing undue amounts of insertion loss or reflections. One of the simplest circuits for achieving these objectives is a length of transmission line which is perturbed only slightly by each of many diode circuits coupled to the line. With this configuration, the desired small phase shift per diode is obtained while circuit complexity, losses, and reflections are held to a minimum. This phase shifter circuit, the subject of our examination in this section, commonly is called by several names — the *transmission, iterative,* or *loaded line* phase shifter. Diode phase shifters using this small perturbation [3-6] approach* made

*Previously, a low power varactor diode transmission phase shifter had been explored [7] producing large phase shift per section; however, correction for the mismatch of the two phase shifting diodes by a third diode was difficult, and the approach did not prove superior to the reflection phase shifter circuit usually used for large bits and described in the next section.

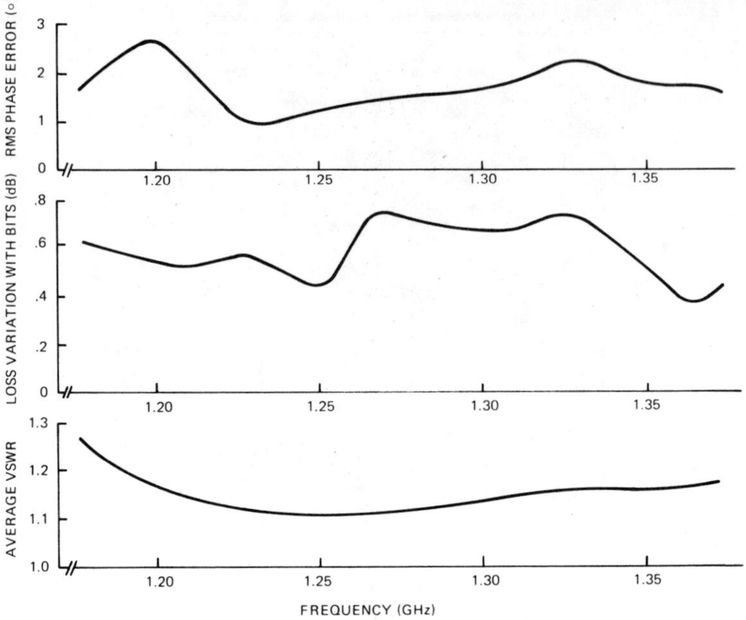

Figure IX-13 **Summary of Measured Performance of 6-Bit 47-Wavelength Time Delay Unit**

practical the diode steered array antennas used in the MSR [1 (p. 68)] and PAR [1 (p. 73)] radars of the US Safeguard System. These radars were the first to employ diode phase shifters for direct control of elements which individually radiated kilowatts of peak power for long pulse lengths.

Before examining the transmission phase shifter, consisting of a spaced diode pair, let us consider the effect of a single shunt disturbance on the transmission line between matched generator and load shown in Figure IV-14(a). With our sign convention, the line shunting susceptance, jB, is positive for capacitive susceptances and negative for inductive susceptances. The transmitted voltage, V_T, at jB can be expressed as the superposition of the incident and reflective voltages, V_I and V_R. The incident voltage V_I is that which would have existed at jB were this susceptance not in place. The reflected voltage, V_R, is defined such that the actual voltage, V_T (the resulting voltage transmitted toward the load), in the presence of the discontinuity is the sum of the incident and reflected voltages. That is

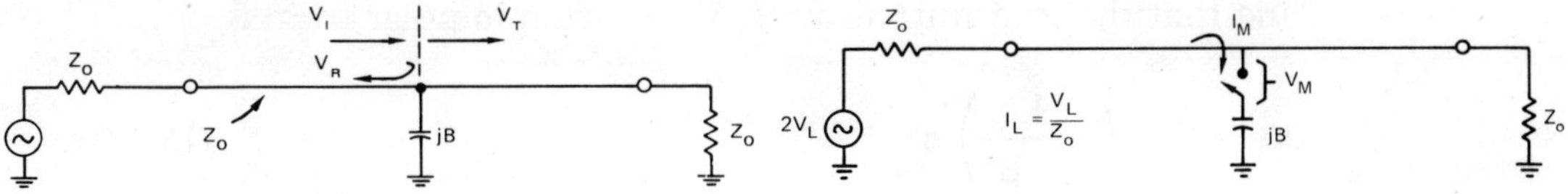

a) The Superposition of Incident and Reflected Steady State Voltages to Represent the Transmitted Voltage at an Obtacle in a Line

b) Diode Represented by an Ideal Switch to Determine Voltage V_M and Current I_M Stresses Required by the Phase Shifting Function

Figure IX-14

$$V_T = V_I + V_R \tag{IX-6}$$

The transmitted and reflected voltages are by definition related to the incident voltage by the *transmission coefficient* T, and the *reflection coefficient* Γ, respectively.

$$V_T = TV_I \tag{IX-7}$$

$$V_R = \Gamma V_I \tag{IX-8}$$

With these definitions for Γ and T, it follows from Equation (IX-6) that

$$T = 1 + \Gamma \tag{IX-9}$$

Using the normalized susceptance, jb, (defined as $jB/Y_0 = jBZ_0$), the reflection coefficient can be written in terms of the total normalized admittance ($y = 1 + jb$), which includes the load conductance, as

$$\Gamma = \frac{1 - y}{1 + y}$$

This expression reduces, for a simple line shunting normalized susceptance of jb, to

$$\Gamma = \frac{-jb}{2 + jb}$$

From Equation (IX-9) it follows that for this case

$$T = \frac{2}{2 + jb}$$

and that the transmitted wave, V_T, written in polar form is

$$V_T = V_I \left(\frac{4}{4 + b^2}\right)^{1/2} e^{-j \tan^{-1}(b/2)} \qquad \text{(IX-10)}$$

From this expression we see that the transmitted wave is reduced from the incident wave in amplitude by the factor $2/\sqrt{4 + b^2}$ and changed in transmission phase by

$$\Delta\phi = -\tan^{-1}\left(\frac{b}{2}\right) \qquad \text{(IX-11)}$$

For example, if the shunt susceptance has a normalized value of +j0.2 (corresponding to a shunt capacitor), the propagating wave is *delayed* by about 0.1 rad or 5.7°. The wave magnitude is also reduced by the factor 0.995, representing an insertion loss due to reflection of 0.04 dB. Similarly, a shunt inductor with normalized susceptance of -j0.2 produces a phase *advance* in the *steady state* transmission phase of the propagating wave. An analogous analysis for series capacitive and inductive reactances shows that these values produce *delays and advances* which are respectively also calculable from Equation (IX-11) simply by replacing the susceptance, b, with the corresponding normalized reactance, x.

The peak power, P_M, and minimum insertion loss can be estimated* for the transmission phase shifter in a manner similar to that used in Chapter V for the reflection phase shifter. Referring to Figure IX-14(b), suppose that the susceptance, jB, is switched across the line by a diode the reactances of which are tuned resonately at the desired frequency and the resistances of which are negligibly small compared to Z_0 and 1/B. The stresses, V_M and I_M, can be set to any ratio for a given phase shift by appropriate selection of Z_0 and jB. The maximum switchable power, P_M, at which V_M and I_M are incurred is calculated by noting that

$$V_M \approx V_L \quad \text{(switch open)} \qquad \text{(IX-12(a))}$$

$$I_M \approx I_L b \quad \text{(switch closed)} \qquad \text{(IX-12(b))}$$

$$V_M I_M = V_L I_L b = P_M (2 \tan \Delta\phi)$$

and thus the maximum power sustainable per diode is

*This derivation was suggested to the author by William Rushforth, Raytheon Co., Wayland, Massachusetts.

$$P_M = \frac{V_M I_M}{2 \tan(\Delta\phi)} \tag{IX-13}$$

Equation (IX-13) has the same significance for a transmission phase shifter as does Equation (V-55) for a reflection phase shifter. The approximations in Equation (IX-12) become exact expressions (and Equation (IX-13) is then also exact) if a *pair* of elements are properly spaced to form the two-element loaded line circuit described next. Notice that the $\tan(\Delta\phi)$ in the denominator implies that a single diode is limited to 90° of phase shift (at which $P_M \rightarrow 0$) in a transmission phase shifter. We will see that, in fact, much smaller phase shift is practical (only about 25° per diode) for reasonable transmission match over a useful bandwidth.

Using a small perturbational approach, the equalized insertion loss is proportional to the $V_M I_M$ product, and can be scaled directly from the corresponding expression (Equation (V-59)) for the reflection phase shifter. Thus

$$\text{Loss}_E \approx 2 \frac{f}{f_{CS}} \tan(\Delta\phi) \tag{IX-14}$$

From this analysis it is clear that phase shifters can be made simply by introducing a controllable reactive element which is placed either in shunt or in series with the transmission line; the problem with a single element phase shifter, however, is that, while it produces phase shift, it also introduces an appreciable reflected wave magnitude. From Equation (IX-11) the reflection coefficient for the above example has a magnitude of 0.10. The VSWR is related to the magnitude of the reflection coefficient by

$$\text{VSWR} = \frac{1 + |\Gamma|}{1 - |\Gamma|} \tag{IX-15}$$

Thus, in this case, with $|\Gamma| = 0.1$, the VSWR is 1.22. If one attempts to obtain a large amount of phase shift, say 360°, without paying heed to how the reflected waves from each of the individual elements might combine, a very mismatched circuit with high VSWR, high insertion loss, and unpredictable phase shift is likely to result.

Consideration of reflections leads to the design of a complete phase shifter with a number of *sections*. Each section consists of a *sym-*

metric pair of small magnitude reactive control elements spaced about a quarter wavelength apart.* The control elements of a section may be either in series or in shunt with the line. The quarter wavelength spacing causes their symmetric reflections to be almost completely mutually cancelling, resulting in low VSWR. The transmitted wave then encounters little net reflection but its transmission phase is additively increased or decreased by the sum of the phase perturbations caused by the individual control elements. This discussion defines in general terms the transmission phase shifter circuit; we now analyze it in detail in order to demonstrate two important properties of this spaced element pair circuit.

1) *The maximum phase shift introduced by any pair of spaced reflections on a transmission line, whether they are symmetric or not, can be related to their individual reflection coefficients.* This fact is useful not only in estimating the phase shift of the transmission phase shifter, but also in *estimating the phase error* due to VSWR interaction of phase shift bits, circuit components, imperfect connectors, and the common circuit interconnections.

2) With the transmission phase shifter, using symmetric transmission line perturbations, *the reflections from any number of individual phase shift sections do not combine any more unfavorably than do those from a single section,* provided that they follow one another directly on the transmission line without intervening Z_0 line lengths. It is this property which makes it practical to build a complete 0 to 360° phase shifter that may consist of ten or more individual sections without concern for the overall VSWR of the cascade of sections being too high. This important result is proved using an equivalent unloaded line model for the transmission phase shifter which is derived using the ABCD matrix.

The first result uses the transmission and reflection coefficients already described. Figure IX-14 shows a pair of shunt perturbations spaced an electrical distance, θ, apart on a lossless transmission line. Although the analysis is to be applied in the steady state, the various interactions of the incident wave with the two obstacles can be visualized as though they occur sequentially in time, the steady state condition being the summation of all of the possible interactions of the waves which are transmitted through and

**Symmetric* applies to the two spaced elements, not to their two states, which may or may not have the same reactance magnitudes.

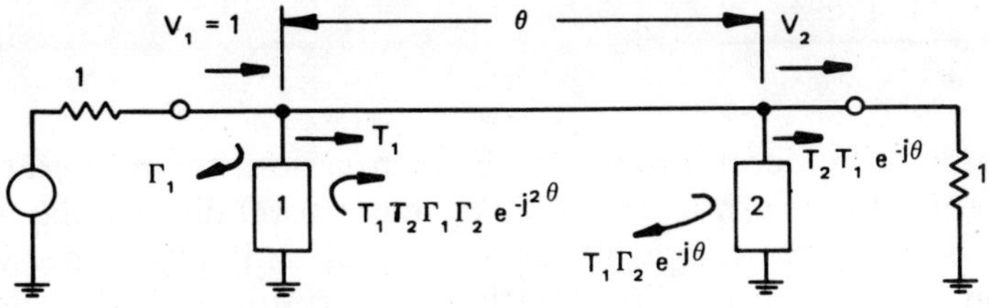

Figure IX-15 **Some of the Steady State Voltages of a Spaced Pair of Transmission Line Perturbations**

reflected from these obstacles on the line. This summation is easy to perform, as is shown below.

Suppose there is a voltage wave of unit amplitude incident to the left of obstacle 1 shown in Figure IX-15. This arrangement sets up both a reflection, Γ_1, back towards the generator, as well as a transmitted wave, T_1. The transmitted wave, after traversing θ from the first obstacle, passes through obstacle 2 while moving toward the load. This first "pass" corresponds to a propagating voltage toward the load of $T_2T_1e^{-j\theta}$. There also occurs a reflected wave with voltage $T_1\Gamma_2e^{-j\theta}$ which travels back toward obstacle 1, with re-reflection from obstacle 1; thereafter, an infinite number of re-reflections occur between the two obstacles. For each round trip between the obstacles an additional wave passes through the second obstacle toward the load. The final wave, V_2, reaching the load has a voltage given by

$$V_2 = V_1(T_1T_2e^{-j\theta} + \Gamma_1\Gamma_2T_1T_2e^{-j3\theta} + (\Gamma_1\Gamma_2)^2T_1T_2e^{-j5\theta} + \ldots)$$

or, simplifying

$$\frac{V_2}{V_1} = T_1T_2e^{-j\theta}(1 + \Gamma_1\Gamma_2e^{-j2\theta} + (\Gamma_1\Gamma_2e^{-j2\theta})^2 + \ldots) \qquad \text{(IX-16)}$$

The infinite series within the parenthesis can be recognized as the form

$$\frac{1}{1-X} = 1 + X + X^2 + X^3 + \ldots \qquad \text{(IX-17)}$$

which converges for $|X| < 1$. This sum permits the output to input voltage ratio to be written in closed form as

$$\frac{V_2}{V_1} = \frac{T_1 T_2 e^{-j\theta}}{1 - \Gamma_1 \Gamma_2 e^{-j2\theta}} \tag{IX-18}$$

Neglecting the denominator of this expression, the numerator suggests that the transmitted wave, V_2, has a total phase delay equal to the sum of the delays introduced by T_1 and T_2 together with the phase delay of the line length θ. This result indicates that the phase perturbations of a pair of obstacles (whether or not the obstacles are symmetrical) are additive to the phase delay of the line between them. A *phase error* term, the value of which is the argument of the denominator in Equation (IX-18), is added to this sum.

For the case when the two obstacles consist of susceptances of the same sign the complex product, $\Gamma_1\Gamma_2$, has a resultant phase of $\pm 180°$. With the spacing between the obstacles near 90°, the $-j2\theta$ factor gives approximately an additional 180°, and thus the total denominator phase is nearly 360° (equivalent to 0°), allowing the denominator phase to be neglected. This situation is the basis for the design of the transmission phase shifter.

Equation (IX-18) also gives a general result useful for analyzing the transmission phase interaction of *any two* sources of reflection spaced along a transmission line. If a pair of obstacles are separated by an electrical distance, θ, on a lossless transmission line, the net transmission phase of a wave passing by both of them has a mean value delayed by θ plus the separate transmission phase delays of the two obstacles. The total delay is this mean value plus a variable amount, $\Delta\phi_E$, given by

$$\Delta\phi_E = \tan^{-1}[1 - \Gamma_1\Gamma_2 e^{-j2\theta}] \tag{IX-19}$$

The quantity $\Delta\phi_E$ is called the *VSWR interaction phase error* because its value is the amount of the error which results when the total phase through cascaded devices is estimated as the sum of the individual phase delays obtained by measuring each between perfectly matched generator and load. The variable amount, $\Delta\phi_E$, can be recognized as arising from the denominator argument in Equation (IX-18). This interactive effect often is encountered in mismatch phase error analyses for microwave circuits. Thus, Equation (IX-19) can be used to estimate the range of transmission phase variation occuring due to the random spacing of any pair of mismatches having respective reflection coefficients, Γ_1 and Γ_2. If

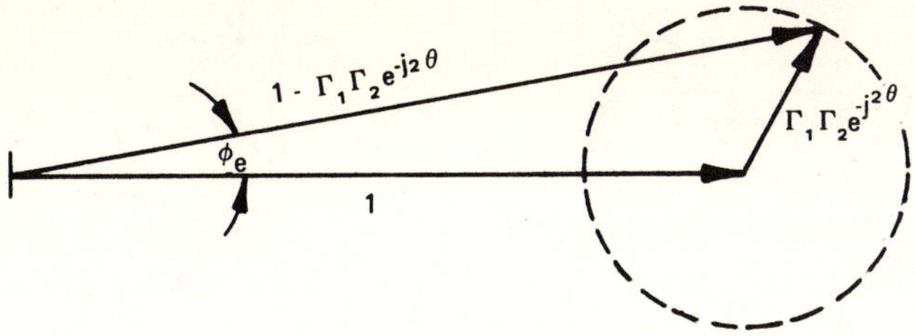

$$-\mathrm{TAN}^{-1}\ |\Gamma_1\Gamma_2| < \phi_e < +\mathrm{TAN}^{-1}\ |\Gamma_1\Gamma_2|$$

Figure IX-16 **The Transmission Phase Change Introduced by Interaction Between Spaced Mismatches**

Γ_1 and Γ_2 have small magnitudes, the range $\pm\Delta\phi_E$ is easily estimated using the diagram shown in Figure IX-16. When the VSWR of an obstacle is known instead of $|\Gamma|$, the reflection coefficient magnitude is calculated from

$$|\Gamma| = \frac{\text{VSWR} - 1}{\text{VSWR} + 1} \tag{IX-20}$$

Suppose, for example, that two obstacles have VSWR values of 1.2 and 1.3, respectively. The corresponding reflection coefficient magnitudes are 0.09 and 0.13, resulting in a maximum transmission phase variation due to variations in their spacing of (0.09) (0.13) = 0.01 rad or 0.6°. This variation in transmission phase, $\Delta\phi_E$, is the VSWR interaction phase error because, if the transmission phases of two separate two-port networks are measured separately (each using perfectly matched source and load) and the two networks then are cascaded (again each between perfectly matched generator and load), the resulting transmission phase is the sum of their separately measured transmission phases with an "error" within $\pm\Delta\phi_E$. The specific value depends upon their phase separation, θ, according to Equation (IX-19). Actually, *perfectly matched devices — generator and load — do not exist. Therefore the phase error,* $\pm\Delta\phi_E$, *is also introduced during the measurement of any practical (imperfectly matched)* two-port network at both the generator and the load side of the network. Interaction *phase error allowances must be made for both pairs of reflection interactions.*

The analysis method just presented is useful for providing insight into how spaced discontinuities can produce transmission phase shift and how variable spacing between them causes the transmis-

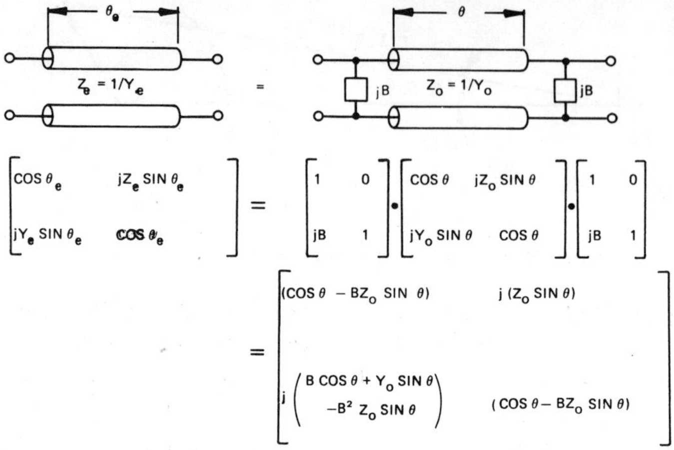

Figure IX-17 **Uniform Line Equivalent Circuit for a Transmission Phase Shifter Section**

sion phase variation as large as $\pm\Delta\phi_E$. However, this method is not convenient for describing how a cascade of two symmetric element sections can be arranged to have minimal VSWR interaction with one another and thereby produce a well-matched complete multi-bit phase shifter. For this evaluation we develop the unloaded line equivalent circuit for the symmetric element transmission circuit, as shown in Figure IX-17.

Uniform Line Equivalent Circuit

Figure VI-5 demonstrates how to determine a lumped element equivalent circuit for a uniform length of transmission line. Using the same approach, a uniform unloaded line equivalent circuit can be derived for the two element phase shift section shown in Figure IX-17 consisting of a spaced pair of susceptances. To determine the equivalence conditions, the A terms of the ABCD representations of the two circuits first are equated. This action gives

$$\cos\theta_E = \cos\theta - BZ_0 \sin\theta \qquad \text{(IX-21)}$$

Next, by equating the matrix term ratio, B/C, for the uniform line to that for the loaded line section, an expression for Y_E of the uniform line is obtained.

$$Y_E = Y_0[1 - (BZ_0)^2 + 2BZ_0 \cot\theta]^{1/2} \qquad \text{(IX-22)}$$

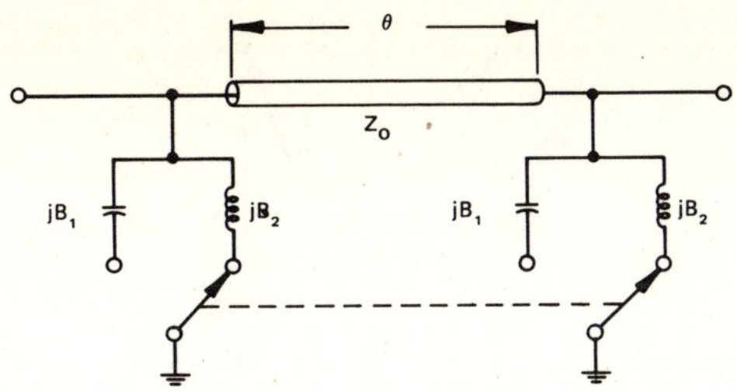

Figure IX-18 **Switched Transmission Phase Shifter Section**

Equation (IX-21) can be used to calculate the phase shift; Equation (IX-22) can be used to evaluate the degree of mismatch introduced by the susceptances. First, examine the phase shift. Consider the circuit in Figure IX-18 when $\theta = 90°$, $B_1 = +j0.2$, and $B_2 = -j0.2$. Since $\theta = 90°$, $\cos\theta = 0$; when the switches are set to state 1, $\cos\theta_E$ equals -0.2. When the switches select state 2, $\cos\theta_E$ is +0.2.

The net electrical length of the loaded-line circuit for these two conditions is described by the vector diagram in Figure IX-19. The electrical length of the line itself, $\theta = 90°$, is represented by the vertical vector with origin at the center of the unit radius circle. This vector has an angle of *90° measured counter clockwise* from the horizontal right axis. When the 90° length of transmission line is loaded by a pair of shunt capacitive susceptances, Equation (IX-21) indicates that the cosine of its angle is negative. From this fact it follows that the total electrical length, θ_{E2}, must be somewhat longer than 90° since its projection onto the horizontal axis is *negative and therefore falls to the left of the vertical, 90° vector.* From Equation (IX-21) the length of this projection is $-B_2Z_0$.

Similarly, when the line length is loaded inductively, the cosine of its equivalent electrical length is positive and the net electrical length is less than 90°, as is shown for the vector with angle θ_{E1}. The phase shift, $\Delta\phi$, obtained is equal to the difference in the equivalent electrical lengths ($\Delta\phi = \theta_{E2} - \theta_{E1}$). For the present example, the values for θ_{E1} and θ_{E2} are 101.6° and 78.5°, respectively, providing a net phase shift of 23°.

It can also be seen from Figure IX-19 that the total phase shift, $\Delta\phi$, is equal to the sum of the individual phase shifts, $\Delta\phi_1$ and $\Delta\phi_2$. These phase shifts represent the perturbations of the original 90° line length introduced alternately by the pair of capacitors,

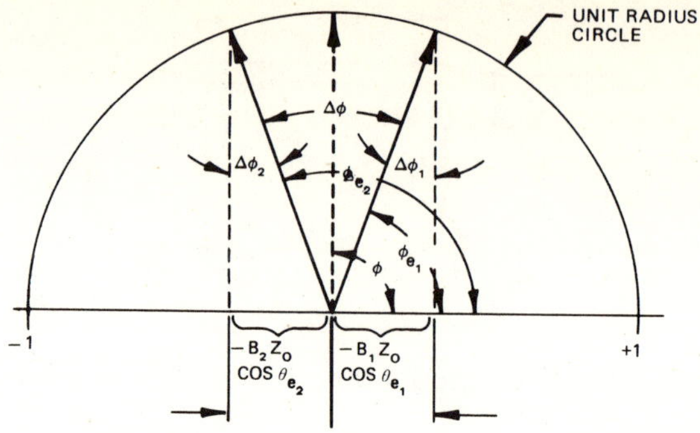

Figure IX-19 **Graphical Representation of Transmission Phase Shifter Phase Length Change**

B_1, and the pair of inductors, B_2. From Figure IX-19 it can be seen that

$$\Delta\phi = \Delta\phi_1 + \Delta\phi_2 \tag{IX-23}$$

$$\sin(\Delta\phi_1) = B_1 Z_0 \tag{IX-24}$$

$$\sin(\Delta\phi_2) = B_2 Z_0 \tag{IX-25}$$

If $\Delta\phi_1$ and $\Delta\phi_2 \ll 90°$, $\Delta\phi_1 \approx B_1 Z_0$ and $\Delta\phi_2 \approx B_2 Z_0$, the total phase shift can be estimated from

$$\Delta\phi \approx (B_2 - B_1) Z_0 \tag{IX-26}$$

This approximate expression for phase shift applies even when $B_1 Z_0 \neq B_2 Z_0$ (provided both $B_1 Z_0$ and $B_2 Z_0$ are much less than unity) and even when $\theta \neq 90°$ (provided it is within ±20° of 90°). It is, therefore, a very useful expression for designing transmission phase shifter sections. A literal statement of this result is that *for a quarter wave spaced pair of symmetrically switched susceptances, the phase shift produced is numerically equal in radians to the difference in normalized susceptance switched by one of them.* This same result could have been inferred from Equation (IX-18); however, deriving the uniform line equivalent circuit for the transmission phase shifter permits a more direct estimate of what effects the finite VSWR contributions from each of the sections have on one another when a long cascade forms a complete phase shifter.

To analyze reflection interactions among the sections, consider the characteristic admittance, Y_E, of the uniform line equivalent circuit given by Equation (IX-22). For the previous example, with $\theta = 90°$, the last term, $2BZ_0 \cot \theta$, equals 0. Moreover, since the line shunting susceptance appears in a squared term, its contribution to the quantity in brackets is always negative. Thus the unloaded-line equivalent circuit characteristic admittance, Y_E, is always less than the admittance, Y_0, of the actual loaded line. This statement means that, near the design center frequency, the transmission phase shifter circuit appears as if it were a length of higher characteristic impedance transmission line. The admittance ratio, Y_E/Y_0, is easily estimated from Equation (IX-22). For the present example, with $|BZ_0| = 0.2$, Y_E equals $0.98\ Y_0$. Thus, if the loaded circuit were built with Z_0 of 50 Ω, the equivalent uniform line model (Figure IX-19) would have a characteristic impedance of 51 Ω. The maximum VSWR that could be experienced between Y_0 generator and load when separated by an arbitrarily long cascade of Y_E admittance line sections (provided, of course, that there are no Y_0 admittance line lengths between the sections) is given by

$$\begin{aligned} \text{VSWR} &\leqslant (Y_E/Y_0)^2 \text{ if } Y_E \geqslant Y_0 \\ \text{VSWR} &\leqslant (Y_0/Y_E)^2 \text{ if } Y_E \leqslant Y_0 \end{aligned} \qquad \text{(IX-26)}$$

For the 23° phase shift section example, the maximum VSWR therefore is no more than 1.02 for any length cascade of such sections. Even this small mismatch can be avoided by either

1) Designing a quarter wave matching transformer for the input and output ends of the cascade,
2) Choosing the starting value of characteristic impedance of the transmission line used in the phase shifter sections to be somewhat lower than 50 Ω, so that with the susceptance loading, the equivalent unloaded-line impedance is 50 Ω, or
3) Arranging the line loading so that the susceptance in one state is zero and selecting the spacing between susceptances B in the other state to satisfy

$$\theta = \tan^{-1}\left(\frac{BZ_0}{2}\right) \qquad \text{(IX-27)}$$

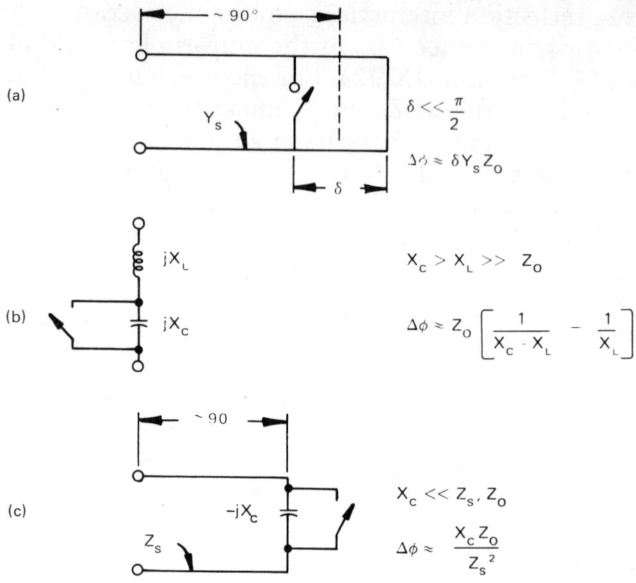

Figure IX-20 **Practical Diode Circuited used to Implement the Suitable Susceptances, B_1 and B_2, Required in the Transmission Phase Shifter Shown in Figure IX-17. The Approximate Values of Phase Shift ($\Delta\phi$) in Radians) are Obtained when a Pair of Elements of Small Normalized Susceptance are Spaced About 90° Apart on a Z_0 Impedance Line.**

for which, at one frequency, the match is perfect. In practice, however, for sections having 23° or less of phase shift, the theoretical VSWR is usually low enough that such measures need not be taken.

While the match of this transmission circuit is very good for small phase shift values, it deteriorates rapidly as one attempts to design for more than about 45° of phase shift per section, in which case even the match improving techniques described above do not give satisfactory results over the practical bandwidths of 10-20% commonly used for phased array antennas. This limitation, however, is not a problem. The principal application for the transmission circuit is when small phase shift per diode is to be obtained because either a small phase shift bit is required or large bits are to be composed of many small phase shift sections.

2. Switched Susceptance Methods

Switched Stub Length

Three practical circuit methods for obtaining diode susceptance switching for the transmission phase shifter are shown in Figure IX-20, with approximate expressions given for the phase shift *obtained from a two element section.* While these units typify the transmission phase shifters built to date, there is an unlimited number of possible means of implementing switchable line loading reactive elements. These three are representative, not exhaustive, coverage of this approach.

The first, shown in Figure IX-20(a), utilizes a diode to vary the length of a shunt stub about a mean of 90°. This method is appropriate at low frequencies for which the diode's reactances are small enough to be neglected. The average length of the stub can be made a quarter wavelength at the center frequency of the phase shifter's operating band. Then, forward bias on the diode results in a total length less than 90°; reverse bias, a length greater than 90°. If both lengths are near 90°, the difference in electrical lengths of the stubs, δ, is proportional to the susceptance change experienced on the main line and, hence, to the phase shift obtained.

Of course, the average stub length need not be 90°. Another effective choice is to make one state 90°, producing zero susceptance on the main line and, accordingly, a perfect transmission match at f_0. Then the spacing between stubs can be chosen to satisfy Equation (IX-27), producing a perfect match at f_0 for the second state as well.

An advantage of this switched stub length circuit is that the phase shift and transmission match can be made experimentally adjustable by designing the stub in coaxial line with sliding short circuit terminations. Both stubs and main line can be made with impedances near 50 Ω, and very high power handling thus can be achieved.

Figure IX-21 shows schematically how the switched stub has been realized with a coaxial L-band model [5]. A photograph of the eight-section (16-diode) phase shifter is shown in Figure IX-22; the measured phase shift and insertion loss are shown in Figure IX-23. For all cases, the VSWR of the section cascade was below 1.2.

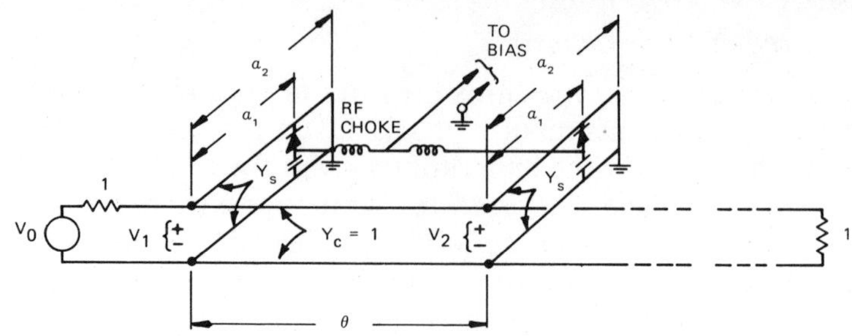

Figure IX-21 **Equivalent Circuit of Switchable Stub Phase Shifter for use at L-Band**

The electrical lengths, α_1 and α_2, were first made adjustable by moving both the diode and the short beyond it about a center value which would yield about 22½° of phase shift at 1300 MHz/s. The 800 V breakdown voltage diodes in each stub were mounted in series with bypass capacitors of about 25 pF, as shown in Figure IX-21. This arrangement served as a bias isolation dc block as well as an approximate series resonator of the diode and mount inductance (about 0.8 nH) short circuiting the stub near the diode position when the diode was forward biased. The diode equivalent circuit parameters are shown in Figure IX-24. Characteristic impedance levels of 50 Ω were used in both the main transmission line and the stubs. The data shown in Figure IX-23 were observed at low RF power (a few milliwatts); no change in phase shift was observed under high power excitation, but the increase of insertion loss with all diodes reverse biased is shown in the data. This result typifies the behavior of all PIN diode circuits with RF power.

For each value of $\alpha_2 - \alpha_1$, the average stub length was made approximately equal to 90°; all stubs were adjusted mechanically to the same dimensions. As may be seen from the measured data, a relatively good experimental reproducibility of the phase shift from section to section was obtained. The incremental phase shift per section varied between 21° and 24°; the average value was 23° for a total of 184° phase shift with the eight-section cascade at the center frequency, 1300 MHz, as shown in Figure IX-25. The calculated phase shift agreed fairly closely with that measured over the 1250-1350 MHz bandwidth. Calculations used the approximation of Equation (IX-26), taking into account:

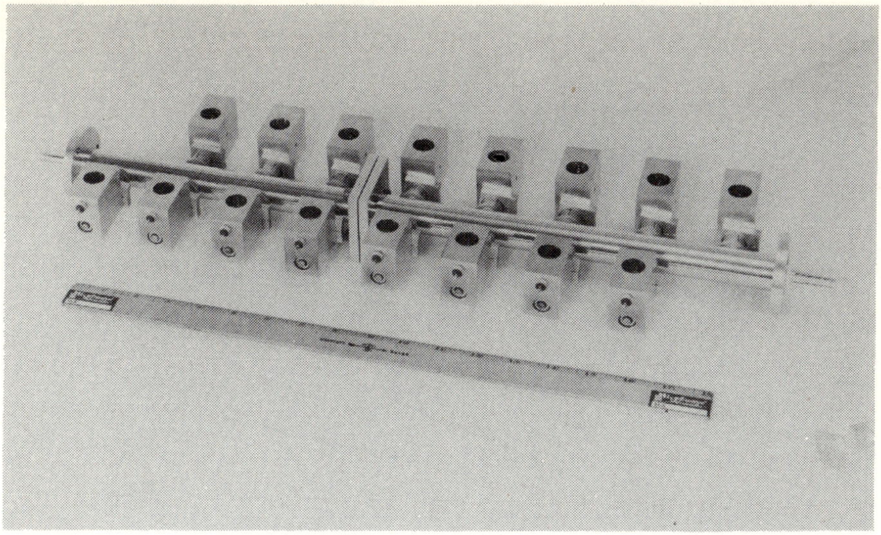

Figure IX-22 **Photograph of Eight-Section (16-Diode) L-Band Transmission Phase Shifter Operable to over 100 kW Peak Power**

1) The difference $\alpha_2 - \alpha_1$ increases with frequency and with it the phase shift.
2) With a change in frequency, the average value $(\alpha_2 + \alpha_1)/2$ departs from 90°; the difference, $B_2 - B_1$, then increases due to the cotangent expression, $B_1, B_2 = Y_S \cot \alpha_{1,2}$.

The switched stub model demonstrates how very high RF peak power can be controlled even with diodes having limited breakdown voltage. A phase shifter section was tested while the reverse bias leakage current was monitored. The RF peak power at which the average dc current (with -100 V bias) increased to 10 μA per diode was defined as the maximum survivable power. RF pulses of 5 μs were used with 0.001 duty cycle at 1300 MHz. Figure IX-26 shows how the sustainable power increased as phase shift per section decreased. The calculated curve was based on the power at 22½° phase shift and the simplified equivalent circuit shown in Figure IX-21. It becomes less accurate at small phase shift settings at which parasitic reactances contribute a larger fraction of the susceptance change, a situation which may account for the wider disparity at small phase shift values between calculated and measured power handling. In any event, however, the practicality of the approach for surviving peak powers in excess of 100 kW is apparent.

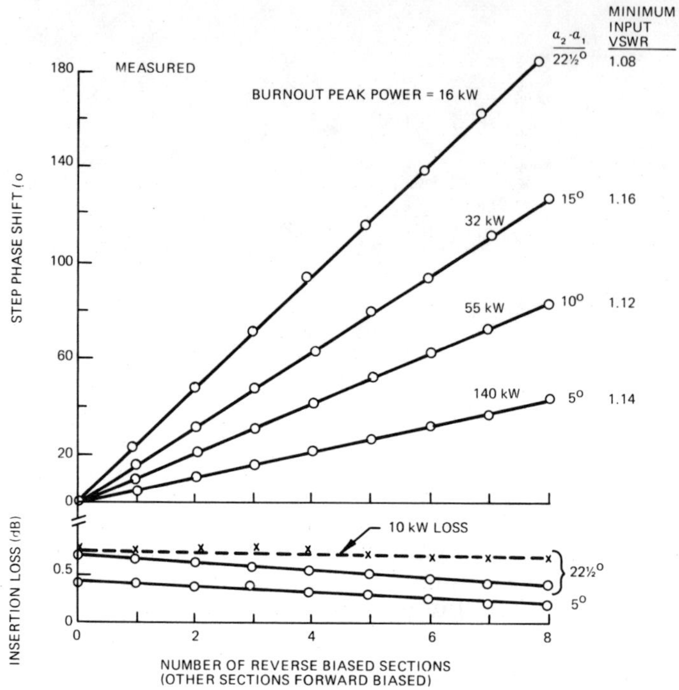

(o) DENOTES LOW RF POWER MEASUREMENT

(x) DENOTES 10 kW PEAK RF POWER MEASUREMENT (AT 5 μs PULSE WIDTH, AND .001 DUTY CYCLE AT 1300 MHz)

Figure IX-23 **L-Band Measurements of Phase Shift, Insertion Loss and Ultimate (Burnout under Reverse Bias) Peak Power Capability**

Switched LC

The second method (Figure IX-20(b)) consists of a series LC circuit mounted in shunt with the main line. The L and C values are chosen so that at the operating center frequency the inductive reactance, jX_L, has a magnitude which is about half that of jX_C, the capacitive reactance of the diode itself. The resulting total reactance then switches between $\pm jX_L$, giving inductive and susceptive line loading in the two bias states. With this method, the forward biased diode state provides inductive susceptance, resulting in shortening of the main line; the reverse bias state gives capacitive loading for line lengthening. For 22½° per section, the reactance, X_L, must have a magnitude five times the value of Z_0, the main line characteristic impedance.

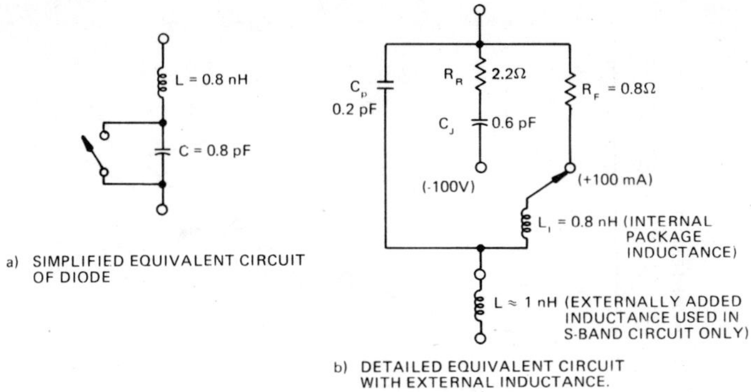

Figure IX-24 **Equivalent Circuit Models used for L- and S-Band Phase Shifter Diodes in Pill Package (Package Type -30, Table III-2)**

Accordingly, to use this approach, either very small capacitance diodes with large series inductance are required, or the main line impedance, Z_0, must be made a low value. The advantage of the circuit is that it permits diode capacitance and inductance to be absorbed into the necessary circuit elements, an arrangement which is especially desirable at higher frequencies. A disadvantage is that the capacitive reactance (which usually represents the diode junction at reverse bias) is subjected to a voltage equal to twice the line voltage, because the LC circuit approaches series resonance when $|X_L| = |X_C/2|$.

The diode described in Figure IX-24 can be used for a very compact phase shifter implementation at S-band. In this frequency range, the inductance and capacitance parameters of the diode and its mount are significant and can be employed to form the LC elements almost directly. The simplified diode equivalent circuit shown in Figure IX-24(a) was used for the design. At 3 GHz, a switchable reactance of $\pm j33\ \Omega$ can be achieved by placing about 1 nH of external inductance in series with the diode as shown in the detailed equivalent circuit in Figure IX-24(b). Phase shift almost invariant with frequency can be expected, as indicated by the susceptance plot in Figure IX-27. For 22½° of phase shift per section, a normalized susceptance of $jB_1, B_2 = \pm j0.2$ is achieved by mounting the resulting LC combination across a $Z_C = 6.25\ \Omega$ characteristic impedance transmission line.

An eight-section phase shifter model was constructed, shown in Figure IX-28. Four-section, quarter-wavelength transformers us-

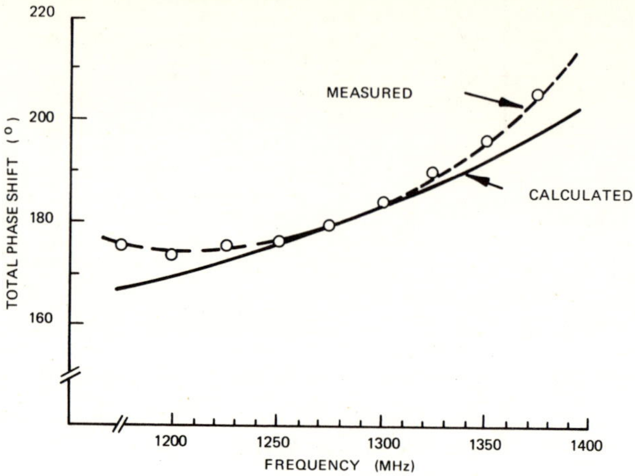

Figure IX-25 **Measured and Calculated Phase Shift vs. Frequency Characteristic of L-Band Model.**

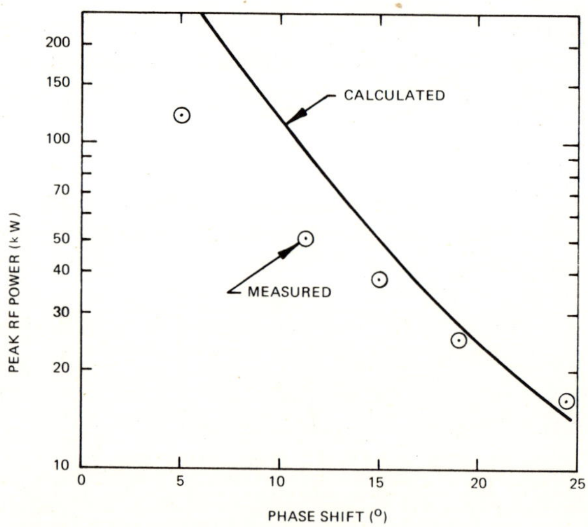

Figure IX-26 **Peak Power Survivability vs. Phase Shift of the L-Band Switched Stub Circuit.**

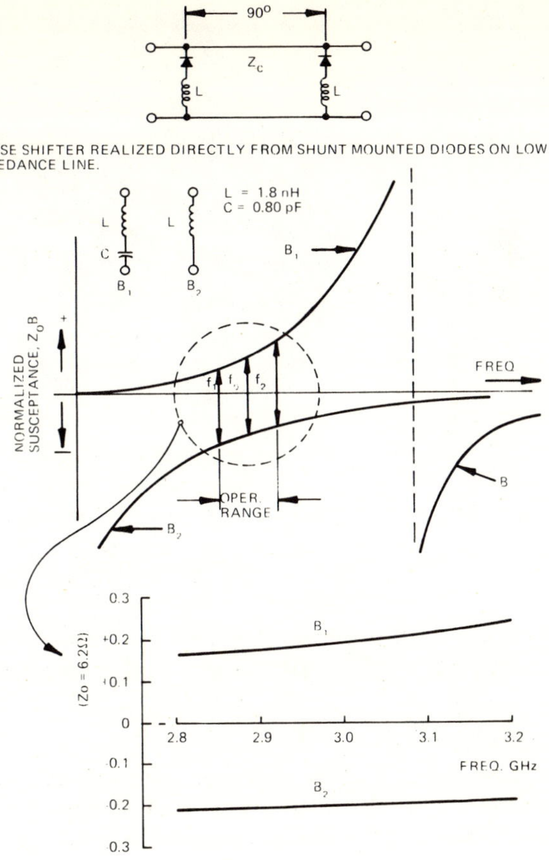

a) PHASE SHIFTER REALIZED DIRECTLY FROM SHUNT MOUNTED DIODES ON LOW IMPEDANCE LINE.

b) PHASE SHIFT CAN BE MADE NEARLY INVARIANT WITH FREQUENCY

Figure IX-27 **The Switched LC Phase Shifter**

ing a Tchebyschev reflection coefficient distribution were used to match the 50 Ω impedance levels at input and output ports to the 6.25 Ω characteristic impedance line of the phase shifter circuit. The 6.25 Ω value, one eighth of 50 Ω, was chosen in order to apply tabulated transformer design values [8,9]. The added series inductance of 1 nH was effected by a small post of adjustable height in series with each diode mount. The post lengths of a single section were adjusted until the measured phase shift of that section was about 22½° at the design center frequency, 3 GHz; then all the remaining posts of the 16-diode circuit were made mechanically identical. The phase shift versus bias state is plotted in Fig-

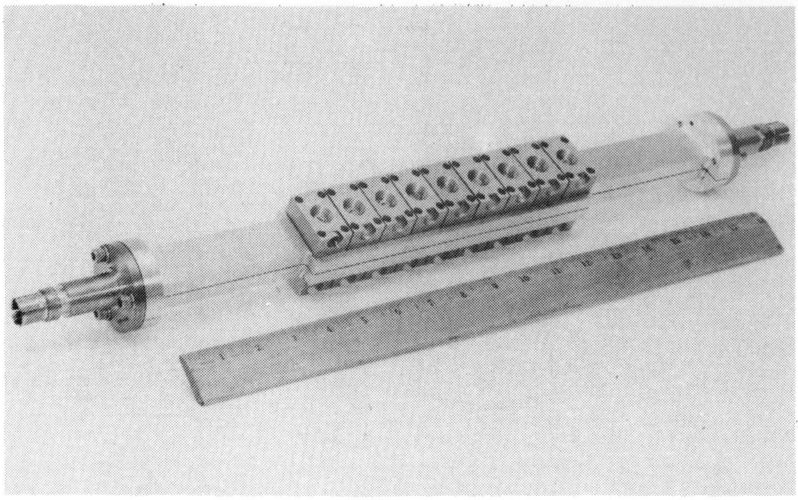

Figure IX-28 **S-Band Eight-Section (16-Diode) Transmission Phase Shifter using the Series LC Method with Low Impedance (6.2 Ω) Main Line**

ure IX-29 along with insertion loss, all measured at low power. The mechanical reproducibility of mounts and approximate similarity of diodes produced very nearly the same phase shift for all of the sections, 20° to 23° at 3 GHz.

The maximum input VSWR and insertion loss of the circuit with diodes removed were 1.25 dB and 0.3 dB, respectively, over the 2.8-3.2 GHz bandwidth. The maximum, low RF power level, insertion loss with diodes installed in the circuit was 0.9 dB and occurred when all were forward biased. Using the mean values for the detailed equivalent circuit reactance parameters of Figure IX-24(b), the effective value for L consistent with 22½° of phase shift per section is 0.5 nH. Taking circuit losses into account, the loss per forward biased diode was 0.037 dB, corresponding to a value of $R_F = 0.8\ \Omega$; similarly, R_R is 2.2 Ω (the values noted in Figure IX-24). These resistances include the loss effects of the mountings.

The measured phase shift variation with frequency shown in Figure IX-29 was very nearly the same (+4%) for either a 200 MHz increase or decrease in RF frequency.

Generally, for phased array radar applications, the bandwidth is limited by antenna considerations to 20% or less. However, transmission phase shifters can be designed with much larger theoretical bandwidths if required [10] .

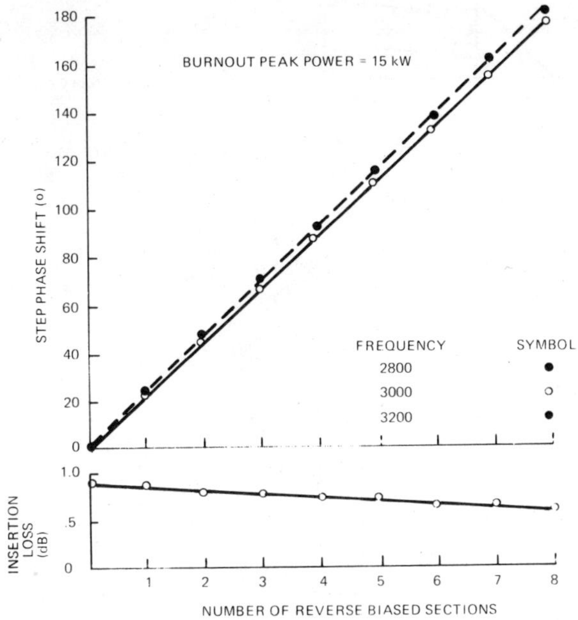

Figure IX-29 **Measured Performance of Eight-Section S-Band Phase Shifter (Figure IX-28)**

Using the detailed circuit model in Figure IX-24(b) with an external inductance, L equal to 0.5 nH, the RF voltage (at reverse bias) across the junction capacity, C_J, is calculated to be about 1.6 times the line voltage. Using the voltage limited peak power argument advanced previously with 900 V breakdown diodes, a peak input power survivability of 16 kW at 3 GHz is calculated. The model was tested, and first diode failures were found to occur at 15 kW. The maximum insertion loss per diode was about 0.04 dB; therefore, continuous RF power levels up to 500 W could be sustained.

Transformed C

The LC circuit just described has the advantage of simplicity, but since the diode junction experiences an RF voltage stress greater than the line voltage, lower line impedance must be used to increase peak power sustaining ability. The 6 Ω line impedance described is nearly the lowest characteristic impedance practical.* A means for operating with higher line impedance at frequencies for which the peak power diode reactances are appreciable is available

*For a long section of transmission line

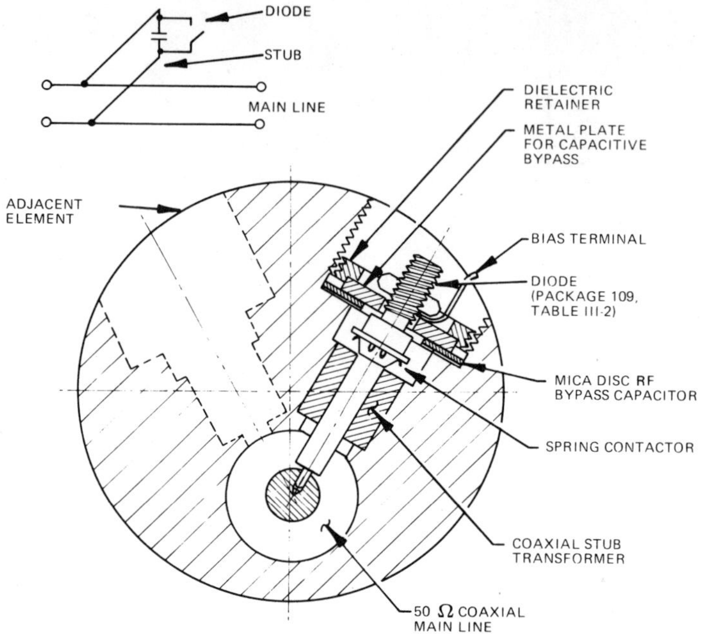

Figure IX-30 **Schematic and Plan Diagram of Transformed Capacitance Transmission Phase Shifter Element**

through the transformed capacitance circuit* shown in Figure IX-20(c).

A sketch showing how this method can be realized at S-band using a large capacitance diode is given in Figure IX-30. Adjacent stubs can be fitted within a cross-section less than a half wavelength in diameter, as is necessary for close spacing in a planar phased array. A convenient main line impedance of 50 Ω can be used since the large diode capacitance (low reactance) is transformed** to a high (inductive) reactance by the shunt stub, decoupling the diode as necessary from the main line energy.

An accurate estimate of performance requires a detailed equivalent circuit model which includes not only the diode package parasitic

*The transformed capacitance circuit to achieve high power handling was suggested to the author by G. DiPiazza, Bell Telephone Laboratories, Whippany, New Jersey.

**Actually, a quarter-wave line length causes impedance *inversion* rather than transformation (multiplication by a constant number). That is, a capacitor has its complex reactance (normalized to the line impedance) inverted by the 90° stub. This inversion results in an *inductive* susceptance on the main line. The change of reactive sign affects which bias state produces the shortest electrical length but not the phase shift magnitude itself.

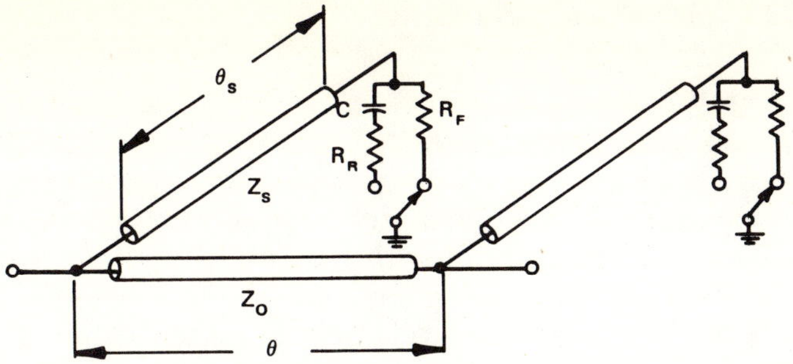

Figure IX-31 **Transformed Capacitance Transmission Phase Shifter**

reactances but also the reactances formed by the transformer stub and its junction with the diode package and the main line. A good appreciation of the operation, however, can be gained using the simplified circuit shown in Figure IX-31.

The design proceeds as follows. Suppose, for high power handling, the 2 pF, 100 μ (4 mil) I region diode described in Table III-1 is selected for use at 3 GHz in a type 109 ceramic package (Table III-2), which has about 0.7 pF package capacitance. Let us neglect package and mount inductance and approximate the diode equivalent circuit as a 3 pF capacitor with R_F and R_R resistances equal to 0.3 Ω (from Table III-1) under forward and reverse bias* as shown in Figure IX-31.

At 3 GHz, the reactance of 3 pF is $jX_C = -j17.7$ Ω. If we wish to obtain 0.4 rad (22½°) of phase shift per section, this reactance must produce (from Equation (IX-26)) a normalized susceptance change of 0.4 on the 50 Ω main line. Using a quarter-wave transformer (impedance inverter, to be precise), a zero susceptance is presented to the main line under forward bias. Therefore, the transformed value, jX_T, of $-jX_C$ must produce the 0.4 normalized susceptance, since the Z_S stub has a length $\theta_S = 90°$. For pure reactances, X_T and X_C are related to Z_S by

$$Z_S^2 = -X_T X_C \qquad \text{(IX-28)}$$

and the phase shift *magnitude* (from Equation (IX-26) with one susceptance zero) is

$$\Delta\phi = \frac{Z_0}{|X_T|} \qquad \text{(IX-29)}$$

*This assumption is conservative since the real circuit has these resistances in series with only the junction capacitance of $C_J = 2$ pF.

Hence, the proper value for Z_S is

$$Z_S = \sqrt{\frac{Z_0 X_C}{\Delta\phi}} \qquad \text{(IX-30)}$$

For this example, $Z_S = 47\ \Omega$, which produces an *inductive* reactance $jX_T = +j125\ \Omega$ in shunt with the $50\ \Omega$ main line. Since the main line susceptance loading under forward bias is zero, the stub spacing θ can be selected to give a perfect match at $f_0 = 3$ GHz. Using Equation (IX-27) and noting that the inductive line loading has *negative* susceptance, the appropriate spacing is

$$\theta = \tan^{-1}\left(\frac{2}{-0.4}\right) = 101^\circ \qquad \text{(IX-31)}$$

The circuit described in Figure IX-31 now is defined completely. It can be analyzed in its two separate bias states for the lossless case ($R_F = R_R = 0$) using Program WHITE (Chapter VI) to determine the phase shift and VSWR performance as a function of frequency. However, when there is a need for many phase shifter evaluations of this kind it is more practical to write a special computer program which can:

1) Calculate the transmission phase in both states and subtract to determine phase shift,
2) Include diode resistance to determine diode dissipative losses, and
3) Perform a calculation of VSWR in both states.

With a suitable dedicated computer program, rapid evaluation of various circuit designs is possible. The performance shown in Figure IX-32 for the S-band example circuit was calculated in this way.

Coincidentally, equal dissipative loss occurs in both bias states. Notice, however, that phase shift variation and VSWR have minimal values over a 10% bandwidth centered somewhat higher than 3 GHz. It is not surprising that these parameters are not centered ideally about 3 GHz, as the design criterion did not take frequency variations into account. As this example shows, computer analysis provides a convenient means for determining what changes can improve bandwidth or other performance parameters.

In addition, with the computer analysis, a unique method of calculating the diode voltage stress from insertion loss under reverse

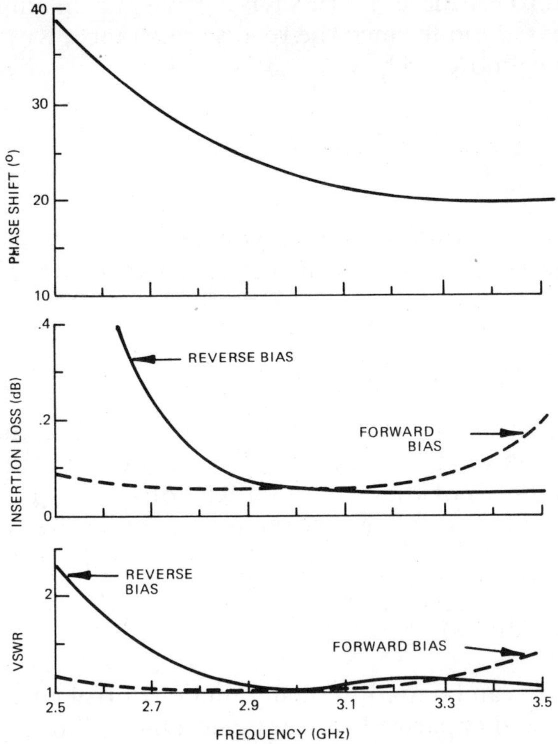

Figure IX-32 **Calculated Performance for Transformed C Phase Shifter (Figure IX-31) with C = 3 pF, $R_F = R_R$ = 0.3 Ω, Z_T = 47 Ω, Z_0 = 50 Ω, and θ = 101° @ 3.0 GHz**

bias is practical, as demonstrated below. Notice that under reverse bias the insertion loss at 3 GHz is 0.06 dB, or 1.4% of the incident power. This figure must be the result of diode dissipation, since all other circuit elements are presumed lossless and the VSWR at this frequency is unity*. Then the dissipative power loss per diode (Loss/diode) as a fraction of the incident power P is half the dissipative loss, since two diodes share this loss equally**. Thus, in this example, Loss/diode = 0.007. But the individual diode

*If the VSWR were not unity, the dissipative loss would be obtained by subtracting the return loss (Equation (VII-49)) from insertion loss.

**To be precise, the diode in the stub nearer the load experiences a slightly lower incident power, the power incident on it having been diminished by losses in the first stub; in practice, though, this difference is negligible.

can be used to estimate the rms RF current, I_R, in amperes through the back biased diode since the reverse resistance, R_R, used in the calcualtion is known. Thus

$$(\text{Loss/diode}) = \frac{P_D}{P} = |I_R|^2 R_R/P \qquad \text{(IX-32)}$$

Finally, the magnitude of the RF voltage stress, V_D, in volts rms is just the product of I_R and the capacitive reactance magnitude, X_C, or

$$|V_D| = |X_C| \cdot I_R \qquad \text{(IX-33)}$$

Combining, we can write an expression for the maximum incident power, P_{MR} (in watts), sustainable by the circuit in the reverse bias state in terms of the maximum RF voltage, V_{MR} (in volts rms), which the diode can tolerate in the reverse bias state

$$P_{MR} = \frac{R_R}{(\text{Loss/diode})} \left| \frac{V_{MR}}{X_C} \right|^2 \qquad \text{(IX-34)}$$

This equation can be used with any diode control circuit, provided that the diode dissipative loss can be determined in the circuit of interest.

For the present example, the diodes can survive RF voltages up to 1000 V peak (707 V rms) for a pulse length of about 100 μs and duty cycle of 5% when biased at about -200 V. Therefore, the phase shifter example described can survive up to

$$P_{MR} = \left(\frac{0.3 \text{ ohm}}{0.007}\right) \left(\frac{707 \text{ rms volts}}{17.7 \text{ ohms}}\right)^2$$

$$= 70 \text{ kilowatts}$$

The power dissipated under forward bias with this circuit happens to be the same as the power dissipated under reverse bias; since, by assumption, $R_F = R_R$, it follows that the RF current through the diode is the same in both bias states. The magnitude of this current, from Equation (IX-33), is

$$I = I_R = \frac{707 \text{ volts}}{17.7 \text{ ohms}} = 40 \text{ rms amperes} \qquad \text{(IX-35)}$$

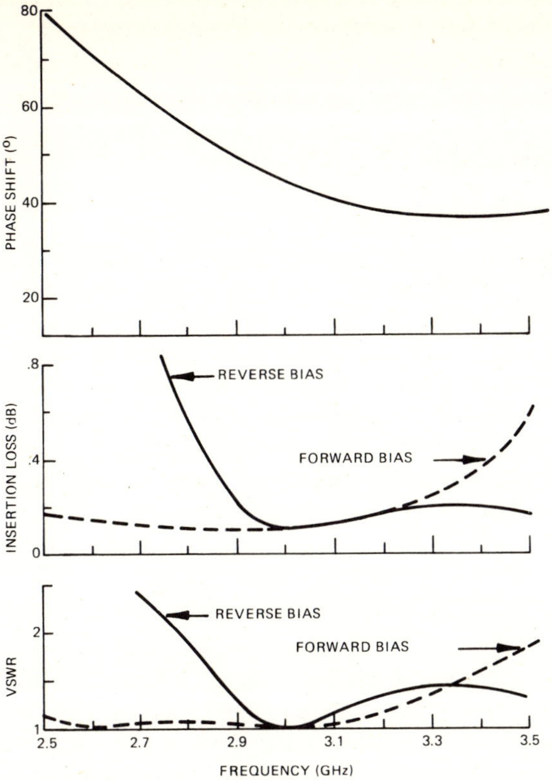

Figure IX-33 **Calculated Performance for Transformed C Phase Shifter (Figure IX-31) with C = 3 pF, $R_F = R_R$ = 0.3 Ω, Z_T = 33 Ω, Z_0 = 50 Ω, θ_S = 90°, and θ = 112° @ 3 GHz**

It is interesting to note that if the reverse RF voltage (707 V) and the forward current (40 A) are substituted into Equation (IX-13) along with the phase shift obtained per diode (22½°/2), the peak power calculated is about 70 kW, verifying this power limit expression.

If the same transformed C circuit is designed such that a phase shift of 0.8 rad (45°) is obtained per section, one finds by using the foregoing procedure that 90° stubs each of 33 Ω characteristic impedance and separated from one another by 112° are required. The resulting calculated performance for this 45° bit is shown in Figure IX-33.

The phase shift, loss, and VSWR contours with frequency are similar to the 22½° bit except that, of course, loss is greater as more phase shift is obtained, and, for a given maximum VSWR, the bandwidth is smaller. Since the dissipative loss is about double that of the 22½° bit, and the other diode parameters are similar, the peak sustainable RF power is about 35 kW.

3. *Three Element Loaded Line*

The two element loaded line circuit just described is only practical for up to 45° of phase shift; two such sections would be needed for 90°. Since both sections are switched together, the overall equivalent circuit appears as a *three element loaded line* phase shifter. It follows that some performance improvement may be possible if the center element susceptance is chosen to be somewhat different than twice the end element susceptance. We analytically explore the three element transmission phase shifter for this reason. The analysis further demonstrates that the ABCD matrix analysis is useful not only for computer solution of circuit cascades of several elements, as applied in Chapter VI, but for *manual manipulation* of several cascade circuit elements as well.

Figure IX-34 describes the circuit to be analyzed. The generator, load, and line impedances have been made unity; thus, the values b_1 and b_2 represent normalized susceptances. For illustration, the circuit is analyzed for the situation when the elements are spaced by a quarter wavelength. The five individual ABCD matrices (see Figures VI-3 and VI-4 for their definitions) are shown along with the sequence of their multiplication *beginning at the load end to* obtain the resulting overall matrix representation, $ABCD_{1,2}$. The complete five matrix multiplication can be performed easily using only the steps shown in Figure IX-34.

For the case in which the load impedance is unity ($V_2/I_2 = 1$), it follows from the defining equations for the ABCD matrix that

$$V_1 = AV_2 + BI_2$$

$$V_1 = (A + B)V_2 \qquad \text{(IX-36)}$$

Neglecting ohmic dissipation, for the circuit to be matched it is necessary that

$$V_1 = V_2 = |A + B| \qquad \text{(IX-37)}$$

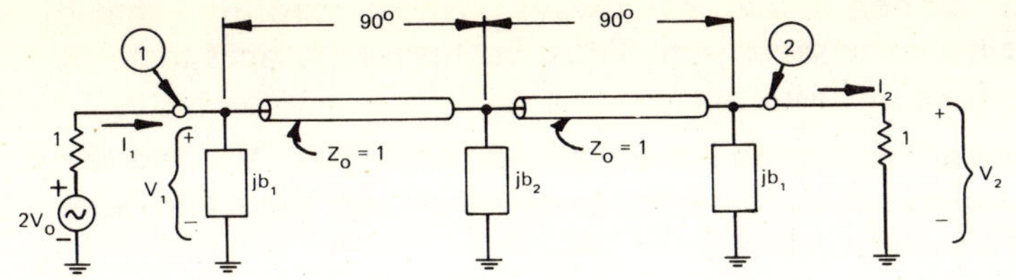

$$\begin{bmatrix} A & B \\ C & D \end{bmatrix}_{1,2} = \begin{bmatrix} 1 & 0 \\ jb_1 & 1 \end{bmatrix} \bullet \begin{bmatrix} 0 & j \\ j & 0 \end{bmatrix} \bullet \begin{bmatrix} 1 & 0 \\ jb_2 & 1 \end{bmatrix} \bullet \begin{bmatrix} 0 & j \\ j & 0 \end{bmatrix} \bullet \begin{bmatrix} 1 & 0 \\ jb_1 & 1 \end{bmatrix}$$

$$\begin{bmatrix} -b_1 & j \\ j & 0 \end{bmatrix}$$

$$\begin{bmatrix} -b_1 & j \\ -jb_1 b_2 + j & -b_2 \end{bmatrix}$$

$$\begin{bmatrix} b_1 b_2 - 1 & -jb_2 \\ -jb_1 & -1 \end{bmatrix}$$

$$= \begin{bmatrix} b_1 b_2 - 1 & -jb_2 \\ jb_1 (b_1 b_2 - 1) - jb_1 & b_1 b_2 - 1 \end{bmatrix}$$

Figure IX-34 **The Three Element Loaded Line Phase Shifter and its ABCD Matrix Evaluation When the Elements are Quarter Wavelength Spaced**

which implies that

$$|A + B|^2 = 1 \qquad \text{(IX-38)}$$

substituting the values for A and B into Equation (IX-38) gives as the condition for a transmission match

$$b_2 = \frac{2b_1}{b_1^2 + 1} \qquad \text{(IX-39)}$$

Since the denominator is always positive, it follows that b_1 and b_2 must have the same sign. Thus, both susceptances must be either capacitive or inductive.

When the input is matched, the input voltage, V_1, and the generator voltage, $2V_0$, have the same phase and the transmission phase, ϕ, of the network is given by

$$\phi = \arg\left(\frac{V_2}{V_0}\right) = \arg\left(\frac{V_2}{V_1}\right) \tag{IX-40}$$

$$= -\arg(A + B)$$

For the matched three element loaded line phase shifter

$$\phi = -\tan^{-1}\left(\frac{b_2}{b_1 b_2 - 1}\right) \tag{IX-41}$$

Substituting the value for b_2 from Equation (IX-39)

$$\phi = \tan^{-1}\left(\frac{2b_1}{1 - b_1^2}\right) \tag{IX-42}$$

The basic electrical length of the network is -180°, corresponding to the phase delay of the two 90° line sections. The susceptances b_1 and b_2 serve to perturb this length; if they are made switchable, the resulting phase shift is the difference in transmission phase perturbations of the half wavelength of line. For example, if $b_1 = +1$, from Equation (IX-39), it follows that, for a match, $b_2 = +1$ as well, and

$$\phi(b_1 = +1) = -270^\circ \tag{IX-43}$$

and switching all susceptances between ±1 gives a phase shift

$$\Delta\phi = \phi\,(b_1 = +1) - \phi\,(b_1 = -1) = -180^\circ$$

While Equation (IX-44) shows that 180° of phase shift is theoretically possible at the center frequency, the match deteriorates too rapidly with frequency deviations from f_0 to make this design a practical 180° bit for most applications. However, it can be practically used for up to 90°. Figure IX-35 shows both the phase perturbation, ϕ_p, to the 180° line length obtained with positive

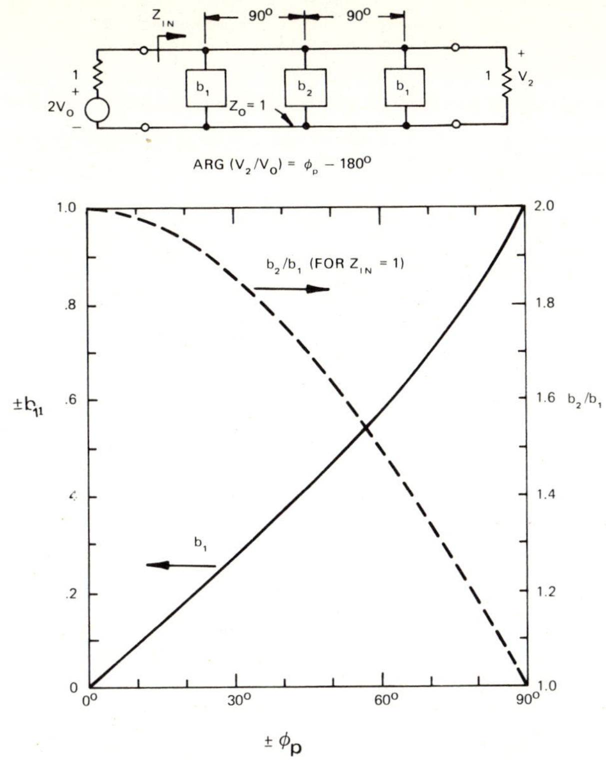

Figure IX-35 **Match Condition and Line Length Perturbation of the Three Element Loaded Line Phase Shifter (Courtesy of H. Griffin, Microwave Associates, Inc.)**

ing susceptances and the ratio, b_2/b_1, that is required for a match. Negative susceptances produce the same magnitude of phase perturbation but the opposite sign. Thus, switching between susceptances of equal magnitude but opposite sign gives phase shift of twice the perturbation shown in Figure IX-35.

4. *Lumped Element Phase Shifters*

All of the phase shifters described so far require some transmission line sections which have electrical lengths that are appreciable portions of the operating wavelength. For high frequency circuits this requirement is not a disadvantage; in fact, it is usually difficult to realize circuits the dimensions of which are not comparable to a wavelength.

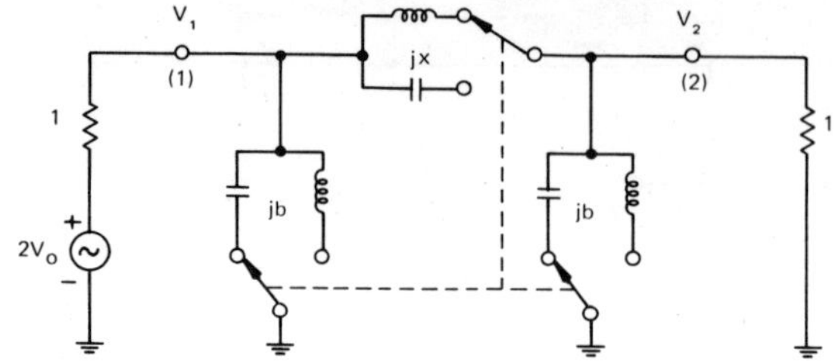

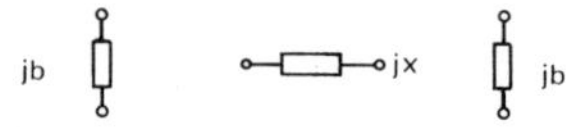

$$\begin{bmatrix} A & B \\ C & D \end{bmatrix}_{1,2} = \begin{bmatrix} 1 & 0 \\ jb & 1 \end{bmatrix} \cdot \begin{bmatrix} 1 & jx \\ 0 & 1 \end{bmatrix} \cdot \begin{bmatrix} 1 & 0 \\ jb & 1 \end{bmatrix}$$

$$= \begin{bmatrix} (1-bx) & jx \\ jb\,(2-bx) & (1-bx) \end{bmatrix}$$

FOR A MATCH, $|A+B|^2 = 1$

$$x = \frac{2b}{1+b^2} \qquad \text{(IX-44)}$$

AND UNDER THIS CONDITION,

$$\phi = \text{ARG}\,(V_2/V_o) = -\text{ARG}\,(A+B)$$

$$\phi = \text{TAN}^{-1}\left(\frac{2b}{b^2-1}\right) \qquad \text{(IX-45)}$$

Figure IX-36 **Lumped Element π Configuration Phase Shifter**

However, at relatively low RF frequencies, especially in the 100-500 MHz range, quarter wavelengths of line are both bulky and costly to fabricate. Interestingly, they also introduce greater loss than at high frequencies, because "copper loss" increases at lower frequencies for a given transmission line type, cross-section, and required electrical length. At first, this occurrence might seem strange since skin effect is reduced with lower frequency, but, although skin effect decreases as the square root of frequency, the *physical length of line required for a given electrical length increases with the reciprocal of frequency*. Hence, the overall effect is that distributed circuits tend to be lossier at lower frequencies.

Two lumped element phase shifter circuits*, requiring no line lengths for their operation, are shown in Figures IX-36 and IX-37.

*The practicability of the lumped element phase shifter was first described to the author by G. DiPiazza, Bell Telephone Laboratories, Whippany, New Jersey in August 1968. However, experimental results [11] and bandwidth analysis [10] were not published until May 1971.

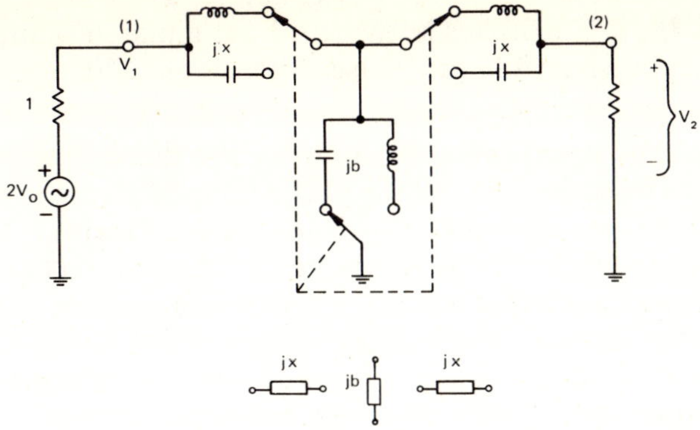

$$\begin{bmatrix} A & B \\ C & D \end{bmatrix}_{1,2} = \begin{bmatrix} 1 & jx \\ 0 & 1 \end{bmatrix} \bullet \begin{bmatrix} 1 & 0 \\ jb & 1 \end{bmatrix} \bullet \begin{bmatrix} 1 & jx \\ 0 & 1 \end{bmatrix}$$

$$= \begin{bmatrix} (1-bx) & jx\,(2-bx) \\ jb & (1-bx) \end{bmatrix}$$

FOR A MATCH, $|A+B|^2 = 1$

$$b = \frac{2x}{1+x^2} \qquad \text{(IX-46)}$$

AND UNDER THIS CONDITION

$$\phi = \text{ARG}\,(V_2/V_o) = -\text{ARG}\,(A+B)$$

$$\phi = \text{TAN}^{-1}\left(\frac{2x}{x^2-1}\right) \qquad \text{(IX-47)}$$

Figure IX-37 **Lumped Element Tee Configuration Phase Shifter**

Their ABCD matrix representations and performance equations are derived in the same manner used for the three element transmission phase shifter.

Because the circuits alternately resemble high- and low-pass filters, they are sometimes called "high-pass/low-pass" phase shifters. While this observation is valid, it does not emphasize that *it is their lumped element aspect which distinguishes them from other circuits**. Hence, the name *lumped element phase shifter* is more descriptive.

The phase shift obtained is the difference in transmission phase ϕ resulting from switching the x and b values appropriately in each circuit. Since ϕ has the same magnitude but opposite sign when switching is performed between $\pm$x and $\pm$b elements (similar to the three element loaded line circuit), a phase shift equal to twice the

*Actually, the two element transmission phase shifter (Figure IX-17) resembles "high-pass/low-pass" filter circuitry equally well in its two respective bias states.

value of ϕ can be so obtained. For example, using the π circuit in Figure IX-36, if b switches between ± 1 for a match, x must likewise switch between ± 1; the phase shift, $\Delta\phi$ is $-270^\circ - (-90^\circ)$ or 180°.

Notice that the equations for match for the three element loaded line circuit (Equation (IX-39)) and the lumped element π circuit (Figure IX-36) are identical if b_2 is interchanged with x and b_1 is interchanged with b. Furthermore, under these interchanges, the transmission phase expressions are also similar except that the argument for the $\tan^{-1}$ expression in Equation (IX-45) which describes the π lumped element circuit differs in sign from that given in Equation (IX-42). This sign difference corresponds to a difference of 180° in the transmission phase of the two circuits, the additional -180° of phase delay being due to the two 90° line lengths in the loaded line circuit.

Thus, in Equation (IX-45), we have for the π circuit

$$\phi_\pi = \tan^{-1}\left(\frac{2b}{b^2 - 1}\right)$$

If we rewrite Equation (IX-42), interchanging b_1 with b, we get

$$\phi = \tan^{-1}\left(\frac{2b}{1 - b^2}\right) = \tan^{-1}\left(\frac{2b}{b^2 - 1}\right) - 180^\circ = \phi_\pi - 180^\circ$$

Accordingly, with these substitutions ($b_2 \rightarrow x$; $b_1 \rightarrow b$), the design and graphs in Figure IX-35 can be used for the lumped π circuit in Figure IX-36.

Similarly, because of the duality of the π and tee lumped circuits, with the substitutions $b_2 \rightarrow b$ and $b_1 \rightarrow x$ the graphs in Figure IX-35 can be used for the tee circuit in Figure IX-37.

For example to design for a total phase shift of 90°, select $\phi_P = \pm 45^\circ$ from Figure IX-35 and, from the graphs, determine that $b_1 = 0.41$ and $b_2/b_1 = 1.71$. For the three different networks the appropriate normalized values are then

1) *Three Element Loaded Line*
   $b_1 = 0.41$, $b_2 = 0.70$

2) *Lumped π Circuit* ($b_2 \rightarrow x$; $b_1 \rightarrow b$)
   $b = 0.41$, $x = 0.70$

3) *Lumped Tee Circuit* ($b_2 \rightarrow b$, $b_1 \rightarrow x$)
   $b = 0.70$, $x = 0.41$

From these equivalences, it can be seen that the three element loaded line, the lumped element π, and the lumped element tee circuits are alternate means for realizing a three element phase shifter.

The physical basis for the similarity in performance of such seemingly different configurations perhaps can be appreciated by recalling that in the loaded line circuit a quarter wavelength line section *inverts* admittance. Therefore, in the three element loaded line circuit the output *admittance,* $1 + jb_1$, is inverted to an *impedance* of magnitude $1 + jb_1$ to the right of the center stub. Interchanging b_1 and x_1, this value is the same impedance seen in the tee circuit to the right of the center susceptance, b. By symmetry, the generator impedance and the input b element of the loaded line circuit likewise are transformed to a normalized impedance having the value $1 + jx_1$ to the left of the center susceptance. Thus, the three element loaded line circuit is, at the center frequency, electrically identical to the lumped element tee circuit.

D. Reflection Phase Shifters

1. *Hybrid Couplers*

By far, the most versatile diode phase shifter circuit is one built with a 3 dB hybrid coupler and symmetric phase-controllable, reflective terminations [12-15], as shown in Figure IX-38. The term "hybrid" has come to mean a coupler which has two outputs which are 90° out of phase with each other to facilitate the transmission-from-reflection operation of the phase shifter.

The same reference port numbers have been used for the coupler in Figure IX-38 as were used in Chapter VI to describe the backward wave hybrid coupler. Thus, power incident at (1) exits the coupler at 3 dB ports (2) and (4), at which it is reflected back into the coupler by the low loss terminations. Due to the symmetry of the terminations and to the quadrature (90°) phase of the two signals, the hybrid redirects the power to the normally decoupled port (3). If a backward wave hybrid coupler is used, very broad bandwidth, in terms of input VSWR at (1), is obtained, as demonstrated in Figure VI-16. Thus, through the use of the 3 dB hybrid, a transmission phase shifter can be built using any reflection networks with variable reflection phase angle.

A circulator could be used with a single reflective termination to provide matched transmission performance; however, the combined effect of cost, insertion loss, size, magnetic field requirements, circulator nonreciprocity, and so forth are such that this

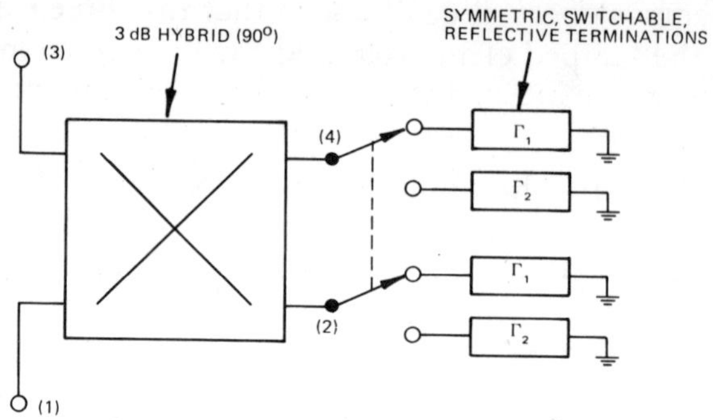

Figure IX-38 **The Reflection Phase Shifter**

approach is rarely taken in favor of the two terminations and hybrid coupler. However, the theory to be described applies to both the hybrid and circulator circuit approaches.

The advantages of the circuit shown in Figure IX-38 are:

1) It can be made using the least number of diodes (two per bit) commensurate with reciprocal operation,
2) Any phase shift increment can be obtained with proper design of the terminating circuits,
3) The transmission match of the bit is dependent only upon the design of the hybrid coupler and is separate from the design of the terminations, and
4) The terminations can be optimized with respect to phase shift-versus-frequency, insertion loss balance in the two bias states, or power handling capacity without regard to transmission match. In practice, as seen with previous control circuit examples, it is usually not possible to optimize all of these functions separately in one design; some compromises must be made.

The properties required of the hybrid coupler in Figure IX-38 are:

1) It must provide a 3 dB power split for the two output arms and
2) There must be a 90° phase difference in its output signals.

Given these properties, as was shown in demonstration of the scattering matrix in Figures VI-15 and VI-16, reflections from symmetric terminations on the 3 dB arms exit the fourth, normally decoupled, port of the hybrid; the reflective nature of the control termination is converted to matched transmission operation for the phase shifter bit. Except in the case of reflective array antennas, this matched transmission is always desired. The 3 dB, 90° properties of the coupler can be realized in TEM transmission line

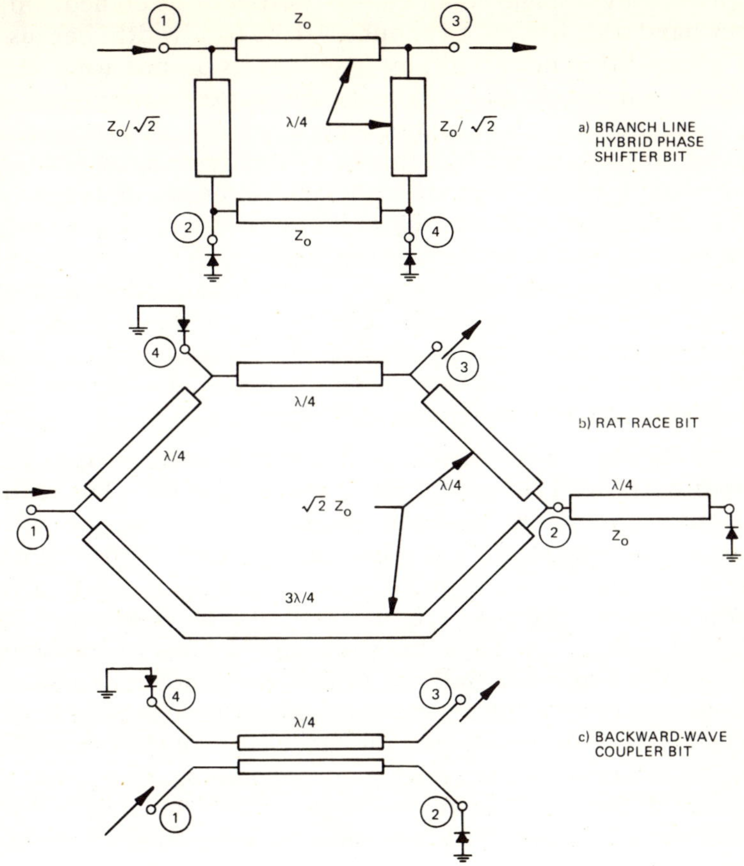

Figure IX-39 Methods of Achieving Hybrid Coupler Properties

with at least three different circuit types (shown schematically in Figure IX-39):

1) The branch line hybrid coupler.
2) The rat race coupler with a 90° line section added to one of its output ports.
3) The backward wave, proximity-coupled hybrid.

A 3 dB 90° hybrid is also obtained in waveguide using a short slot between narrow walls of adjacent waveguides, but this approach is less commonly used because of its size and the difficulty of cascading multiple phase shift bits.

1) The *Branch Line Hybrid Coupler* has the advantage that it can be formed in one plane — when it is built either in microstrip or

stripline, only a single conductor pattern need be etched. On the other hand, it is limited to about a 5-10% bandwidth, because the 90° phase difference, 3 dB power split at its output arms, input transmission match, and directivity are realized perfectly at only the frequency for which all line lengths are 90°. A further disadvantage stems from the fact that in a 50 Ω system, two of the arms of the coupler must be made with a characteristic impedance of 35 Ω ($Z_0/\sqrt{2}$ to be exact). This requirement results in a fairly wide center conductor. At high frequencies, the width of the line becomes comparable to its length. When this situation applies, the intersections of the lines can not be represented as a simple connection of transmission lines; rather, they must be modeled as complex networks. This fact makes the essential computer analysis of the phase shifter circuit more difficult.

2) The *Rat Race* is not strictly a "hybrid" because its two 3 dB output ports are 180° out of phase rather than 90°. The required phases are obtainable, however, simply by extending the reference plane of one port by 90°, as shown in Figure IX-39(b). An advantage of the rat race for high frequency phase shifters is that the characteristic impedances required to build the coupler are 50 and 70.7 Ω ($Z_0\sqrt{2}$). Thus, the wide line problem of the branch line coupler is avoided and, practically, the bandwidth achieved with the rat race coupler is greater than with the branch line. This difference occurs despite the fact that, like the branch line coupler, the rat race device achieves its coupling, phase, and directive properties only at its design center frequency. Its net electrical path length is 1 ½ wavelengths (even without the 90° reference plane movement required for a phase shifter), whereas the branch line coupler's net path is only 1 wavelength.

3) The *Backward Wave Hybrid Coupler,* shown in Figure IX-39(c), gives the broad bandwidth shown in Figure VI-16. This result occurs because, although the 3 dB power split is realized only at the center frequency, the 90° phase difference between the output arms of the coupler, the input match, and the directivity theoretically are frequency independent. A practical coupler, though, must have transmission line connections, and it is at such junctions that the theoretical independence properties are compromised. Nevertheless, this circuit gives bandwidth approaching an octave with reasonable VSWR — a much greater figure than is usually achievable in phased array antennas.

2. Terminations

Broad Bandwidth LC

Virtually any pair of symmetric reflections can be used as terminations; one of the simplest means is to install two PIN diodes directly at the 3 dB output ports of the coupler, a simplified model for which is shown in Figure IX-40. To the diode model consisting simply of a capacitance, C, a series inductance, L, has been added, the need for which is apparent in the following description.

Suppose it is desired to design a 180° phase shift bit. To achieve 180°, we wish the reactance of L to fall diametrically opposite that of C on the Smith Chart. Such a situation occurs if the normalized impedances in the two states are reciprocals of one another. In general, for 180° we wish

$$\frac{j\omega L}{Z_0} = j\omega C \cdot Z_0$$

$$Z_0 = \sqrt{\frac{L}{C}} \qquad \text{(IX-48)}$$

Equation (IX-48) is frequency independent [9, 14], as is the 180° phase shift, provided that the inductive reactance is small compared to the capacitive reactance (i.e., the total reactance in the reverse bias state of the diode is represented adequately as $-j/\omega C$).

Suppose we wish to build a 180° bit using a packageless, 0.1 pF junction capacitance PIN diode (such as the MA-47897 in Table III-1). For $Z_0 = 50\ \Omega$, the required series inductance is L = 0.25 nH.

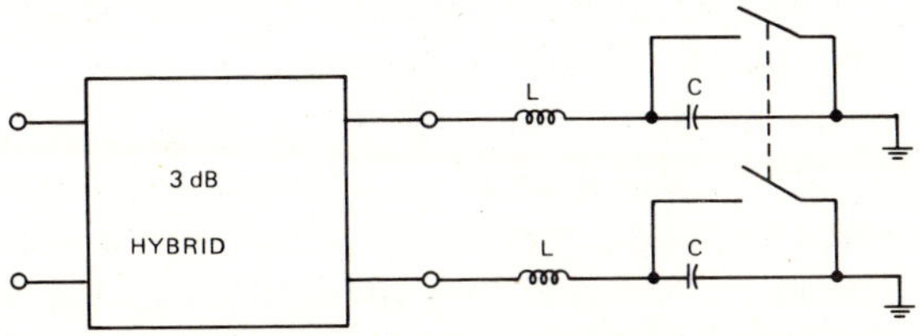

Figure IX-40 **Reflection Phase Shifter with Diodes Attached Directly to the 3 dB Ports**

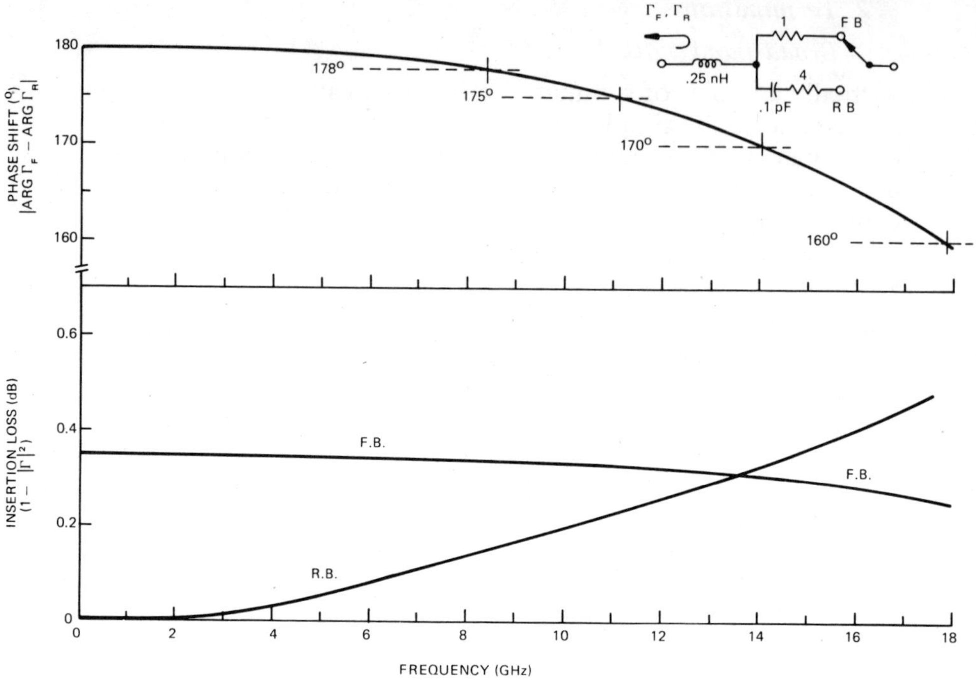

Figure IX-41 **Phase Shift and Loss Theoretically Obtainable with a 0.1 pF Chip Diode (MA-47897) using Tuning Described by Equation (IX-48)**

Figure IX-41 shows that the reflection phase shift is within 2° of 180° up to 8 GHz, and within 20° to 18 GHz. The absorptive loss, equal to $(1 - |\Gamma|^2)$, also is shown for each bias state. Although the absorptive loss in the two bias states is not equalized except in the vicinity of 14 GHz, the design would appear to be nearly ideal for much of the commonly used microwave band. In actual practice, however, it is unlikely that such broadband performance ever would be obtained, because:

1) Couplers do not operate from 1-18 GHz. Single quarter wave couplers are limited to about an octave bandwidth (Figure VI-16).
2) Mismatches at the 3 dB ports of a practical coupler produce changes in phase shift, which are particularly large for certain values of the argument of the termination reflection coefficient Γ [16].

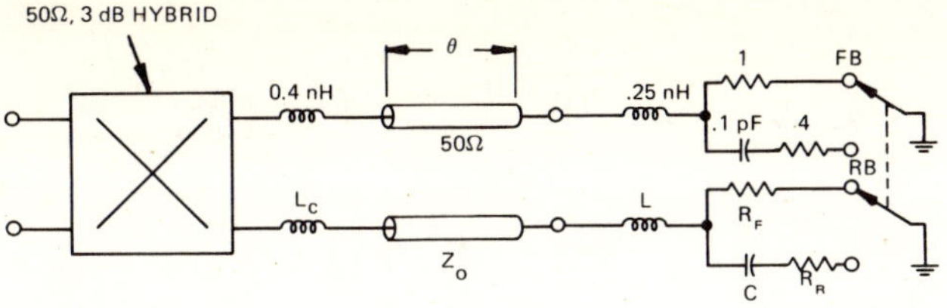

Figure IX-42 **Reflection Phase Shifter with Mismatched 3 dB Coupler Ports (VSWR = 1.1 at 2 GHz) Represented by Series Inductance L_C**

Phase Shift Error from Coupler Mismatch

The second effect, producing phase shift error as a result of coupler mismatch, can be visualized with the following example. Suppose a 3 dB backward wave hybrid has mismatches at its 3 dB ports which can be represented as series inductances*, as shown in Figure IX-42. *The phase shift error from coupler mismatch occurs because the argument of Γ is affected by the mismatch in one state more than it is in the other.* This variation is shown diagrammatically for the present example in the Smith Chart in Figure IX-43. The reverse bias state has its reflection coefficient, Γ_R, affected only slightly by the series reactance of L_C. However, in the forward state at 2 GHz, Γ_F corresponds to an inductive reactance of only +j3 Ω (+j0.06 normalized to 50 Ω). The addition of the reactance of L_C = 0.4 nH is +j5 Ω, for a total of +j0.16 normalized reactance. The new reflection coefficient, Γ_F, has been shifted 11°, as shown in Figure IX-42, resulting in a corresponding phase shift error. In this example, with series inductive coupler mismatch, the termination state with 180° reflection coefficient argument is affected more, since this situation corresponds to nearly zero normalized reactance at the coupler terminals where a perturbing series reactance has a very pronounced effect.

The termination for this example (consisting of the 180° LC model described in Figure IX-41) was combined with an ideal coupler with L_C = 0.4 nH series inductance (1.1 VSWR inductive mismatch at f_0 = 2 GHz) and series 50 Ω transmission lines of

*Similar results are obtained with other forms of reactive mismatches, shunt capacitors, shunt inductors, and so forth, but the maximum phase shift error may occur with different spacings to the termination. If the mismatch is partially resistive, the maximum mismatch phase error is smaller; thus, the totally reactive mismatch model used here is conservative.

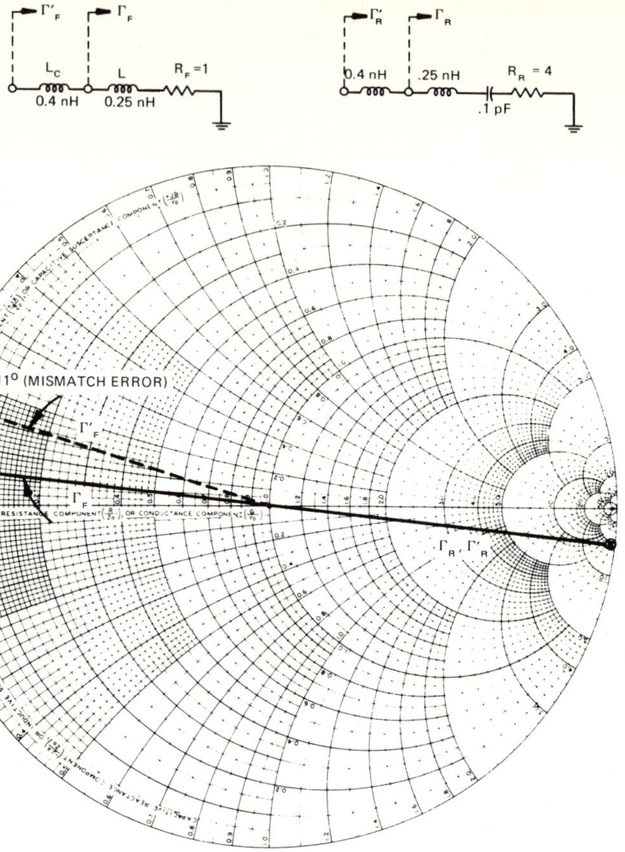

Figure IX-43 **Reflections Coefficient Diagram Showing Unequal Angular Movement Phase Shift Error of Γ_F and Γ_B Caused by Series Inductance L_C**

electrical length θ at output ports 2 and 4. The combined circuit to be analyzed is shown in Figure IX-42.

The reflection coefficient of the termination in each bias state is calculated using

$$\Gamma' = \frac{Z_L - Z_0}{Z_L + Z_0} \tag{IX-49}$$

where Z_0 is the coupler characteristic impedance and Z_L is the total impedance of the termination in the bias state and frequency of interest. In this case

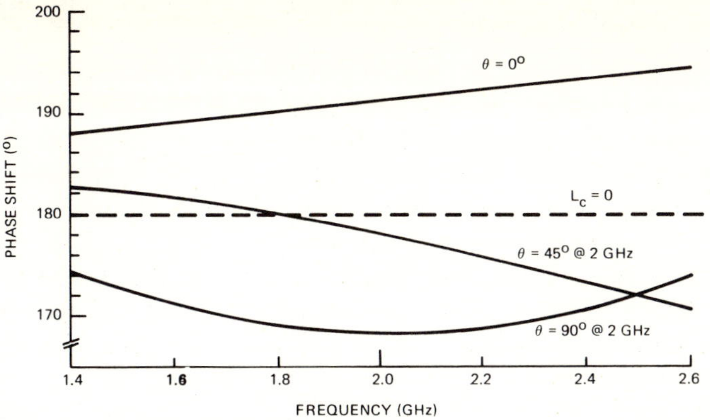

Figure IX-44 **Phase Shift Error Caused by Coupler Mismatch for Various Spacings Between Coupler and Control Termination**

$$Z_L \text{ (F state)} = R_F + j\,[2\pi f(L_C + L)]$$

$$Z_L \text{ (R state)} = R_R + j\left[2\pi f(L_C + L) - \frac{1}{2\pi f C}\right] \qquad \text{(IX-51)}$$

An electrical spacing, θ, between termination and mismatched coupler can be taken into account by transforming the diode impedance down the θ linelength using the Smith Chart subroutine (Figures VI-1 and VI-21) before adding it to $j(2\pi f L_C)$.

The computed results are shown plotted in Figure IX-44. The phase shift error introduced (the dashed curve for $L_C = 0$ gives 180° phase shift over the full band considered) is sensitive not only to the mismatch but to the electrical spacing, θ, between the control termination and the control termination. The phase error magnitude is large, about ±12° at 2 GHz, even for this modest coupler mismatch* with VSWR = 1.1.

While such mismatches can always be tuned out over a small bandwidth, perhaps 20%, matching a coupler to a VSWR < 1.02 over an octave band (as would be necessary to hold the introduced

*Theoretically, the backward wave coupler is perfectly matched for all frequencies at both ports 2 and 4, with matched load and generator at ports 1 and 3. However, in practice, some discontinuity reactance (usually a series inductance) is introduced in connecting output lines to the coupler; consequently, real couplers always have some mismatch. A VSWR less than or equal to 1.1 over an octave bandwidth is representative of excellent performance.

phase shift error to below a few degrees) is generally so difficult as to be impractical. Accordingly, in considering broadband reflection phase shifters, a much greater effort may be required to achieve the coupler match than to minimize the phase shift variation of the terminations themselves.

The broadband 180° termination (Figure IX-41) offers broad bandwidth; however, with practical diodes, it generally gives considerable insertion loss imbalance in its two bias states, the forward state having the higher insertion loss at low frequencies (as shown by the data in Figure IX-41). When only modest bandwidth (less than 20%) is required, as is usually the case for phased array control, it is more desirable to use a design which yields near equalized insertion loss in both states. With diodes having larger junction capacitance, R_F is more nearly equal to R_R, in which case the circuit shown in Figure IX-45 gives equalized loss at the design center frequency. Power handling usually is increased compared with the broadband termination, as this method permits use of a larger capacitance diode. Furthermore, phase shift error due to coupler mismatch is reduced because the impedance presented to the coupler is of the same order of magnitude in both bias states; consequently, a small series reactance tends to shift both states equally on the Smith Chart, introducing little phase shift error.

To obtain the 180° phase shift and the equalized loss, it is necessary that the reactances $\pm jX_L$ switched by the termination have the values $\pm j1$ when normalized to the impedance, Z_0, of the coupler. Generally, 50 Ω couplers are used; thus, a capacitive reactance of -j100 Ω in series with an inductive reactance of +j50 Ω is required. Usually it is not practical to make diodes with the capacitance required for -j100 Ω for each particular frequency of interest. A "transformer" consisting of a Z_T characteristic impedance line of θ electrical length is used to convert the actual reactances to those desired, as the following example demonstrates.

Suppose a 180° bit is to be designed using a 1 pF diode with $R_F = R_R = 0.5\ \Omega$, parameters similar to the MA-47892 in Table III-1, at 3.18 GHz (the frequency at which a 1 pF capacitive reactance is j50 Ω), using the termination circuit in Figure IX-45. If jX_L (3.18 GHz) = +j25 Ω, (L = 1.25 nH), the total impedance at point S switches between $Z_S = (0.5 \pm j25)\ \Omega$. If this termination could be used directly at the output of a $Z_0 = 25\ \Omega$ coupler, the normalized reactance would switch ± j1, shown as points 1F and 1R respectively in Figure IX-46. The resulting calculated phase

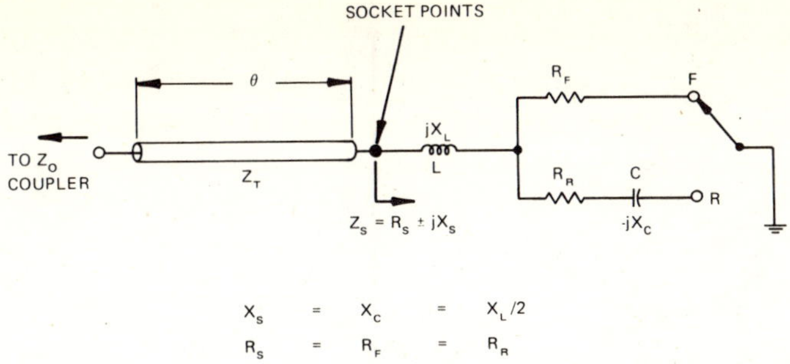

Figure IX-45 **Loss Equalized Termination**

shift and insertion loss with frequency is shown in Figure IX-47. To "transform" a $Z_0 = 50\ \Omega$ coupler to the required "socket impedance" of $Z_S = 25\ \Omega$, a quarter-wave transformer can be used. Its impedance is

$$Z_T = \sqrt{Z_0 Z_S} \qquad \text{(IX-52)}$$

The corresponding phase shift performance is shown as the dotted curve in Figure IX-47. A more slowly varying phase shift with frequency often can be obtained by selecting a value for θ which is less than 90° for 180° bits; however, perfect loss equalization may not be obtained.

The transformer permits achievement of 90° and 45° phase shift if Z_T is chosen to transform the coupler Z_0 to higher socket impedance (Z_S) values. From Figure IX-46, 90° and 45° phase shift are obtained by switching

$$X_S/Z_S = \pm\ 0.41\ (90°) \qquad \text{(IX-53)}$$

$$X_S/Z_S = \pm\ 0.20\ (45°) \qquad \text{(IX-54)}$$

In this example, with $X_S = 25\ \Omega$ and $Z_S = 61\ \Omega$ and $125\ \Omega$ for 90° and 45° of phase shift, (from Equation (IX-52)) Z_T must be $55.2\ \Omega$ and $79.0\ \Omega$, respectively. Figure IX-48 shows the calculated performance with these Z_T values and $\theta = 90°$ at $f_0 = 3.18$ GHz. As with the 180° bit, equalized loss is obtained at f_0.

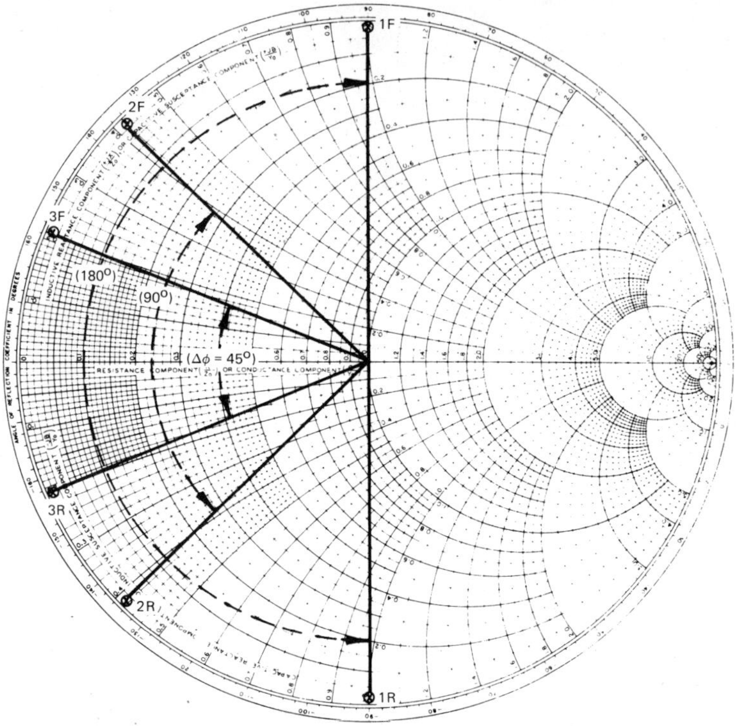

Figure IX-46 **Reflection Coefficients of Loss Equalized Terminations for 180°, 90°, and 45° of Phase Shift**

Series and Parallel Resonant Termination

Both equalized loss and a flat phase shift versus frequency characteristic at f_0 can be obtained using the method of Burns and Stark [17, 18]. This procedure consists of adding tuning elements to obtain alternately series and parallel resonances in the two bias states. Since a 180° phase shift is obtained switching between an open and a short on any impedance line, the specific socket impedance, Z_S, can be chosen as that which gives equal losses. The method is a little more complicated to implement than the transformed LC circuit, but the advantage of added phase shift flatness may offset the additional complexity.

Either diode state may be tuned for series resonance, with the other tuned for parallel resonance. When the forward bias state is series resonated, the circuit is called the *forward mode;* series resonance in the reverse biased state is called *reverse mode,* as shown in Figure IX-49.

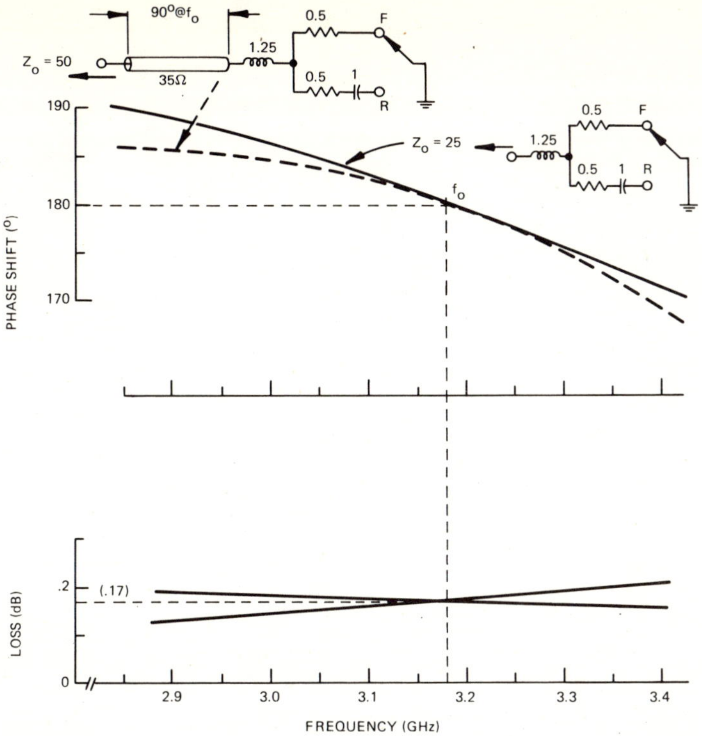

Figure IX-47 **Performance of 180° Loss Equalized Terminations**

When the elements L_T and C_T are tuned to give their respective resonances, the net impedance is real, as is now demonstrated. Consider the reverse mode case. The VSWR on a Z_0 line with the resonated diode termination is equal to the magnitude of the total normalized conductance or resistance, whichever is greater. Thus

$$\text{VSWR R} = \frac{Z_0}{R_R} \qquad \text{(IX-55)}$$

$$\text{VSWR F} = \frac{{X_C}^2}{R_F Z_0} \qquad \text{(IX-56)}$$

For equal loss in both states, Z_0 must be chosen to satisfy

$$Z_0 = X_C \sqrt{\frac{R_R}{R_F}} \qquad \text{(IX-57)}$$

2. Stark[18] gives these & also gives
$\text{Loss}(180° \text{ bit}) = 17.27\, \omega C_j \sqrt{R_R R_F}$ when IX-57 is met.
$\text{Loss}(\Delta\phi \text{ bit}) = \text{Loss}(180° \text{ bit}) \times \sin\frac{\Delta\phi}{2}$

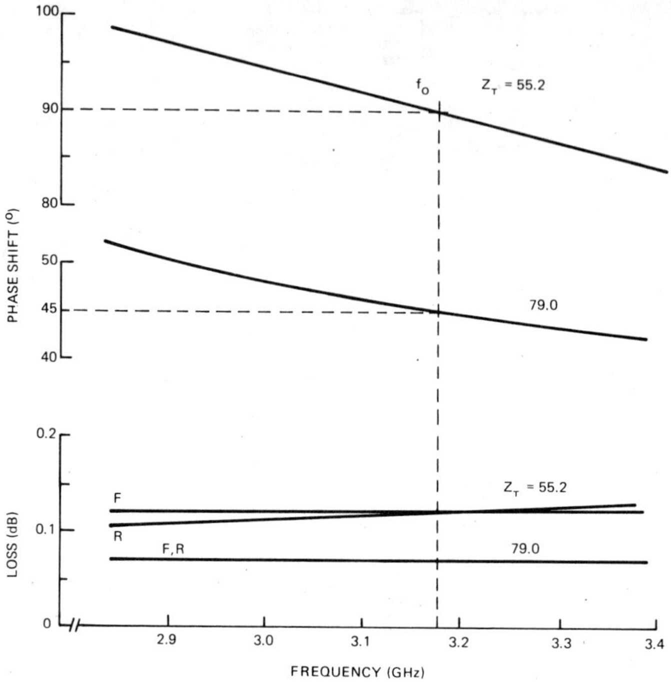

Figure IX-48 **Performance of 90° and 45° Loss Equalized Terminations (Figure IX-45)**

in which case the insertion loss has the same loss as does the transformed LC circuit at f_0, the minimum value for a 180° bit (Equation (V-60), derived in Chapter V).

$$\text{Loss (180° bit)} \approx 18(f/f_{CS}) \text{ (decibels)} \tag{IX-58}$$

In the event that $R_F = R_R$, the appropriate value for Z_0 (from Equation (IX-57)) is X_C, the magnitude of the capacitive reactance of the diode. This particular choice also gives nearly invariant phase shift over a bandwidth of about 10%. This fact is demonstrated by noting that, near f_0, the phase shift can be written approximately as

$$\Delta\phi \approx 180° + 110° \left[\frac{x}{Z_0} - \frac{b}{Y_0}\right] \tag{IX-59}$$

$$\text{where } x \approx 2\left(\frac{\Delta f}{f_0}\right) X_C \tag{IX-60}$$

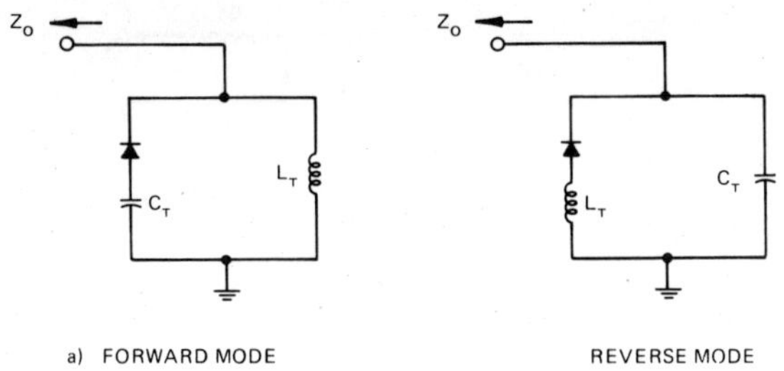

Figure IX-49 **Series and Parallel Resonant Circuits to Give Equalized Loss and "Flat" Phase Shift (After Burns and Stark [13]. The Diode Equivalent Circuit is Similar to that in Figure IX-45.**

and $$b \approx 2 \left(\frac{\Delta f}{f_0}\right) \frac{1}{X_C} \tag{IX-61}$$

The constant multiplier, 110°, is determined graphically by examining the Smith Chart near the z = 0 and z = ∞ points. Equations (IX-60) and (IX-61) are expressions for the net reactance and susceptance, respectively, resulting from a slight frequency deviation, Δf, from series resonance and from parallel resonance. (This approximation is derived in Chapter VII as Equation (VII-34).) Substituting Equations (IX-60) and (IX-61) into Equation (IX-59) gives

$$\Delta\phi \sim 180^\circ + 2 \left(\frac{\Delta f}{f_0}\right) 110^\circ \left[\frac{X_C}{Z_0} - \frac{Z_0}{X_C}\right] \tag{IX-62}$$

$$\approx 180^\circ \quad (\text{for } \Delta f \ll f_0,\ X_C = Z_0)$$

which proves that when Z_0 is chosen equal to X_C, the phase shift, $\Delta\phi$, is nearly constant for small frequency deviations about f_0.

Figure IX-50 shows the calculated performance for the reverse mode circuit using the same diode example chosen for the transformed LC example in Figure IX-47. The phase shift variation is less than that of the transformed LC, and the flat slope response at f_0 is obtained as predicted by Equation (IX-62).

In practice, a 45° length of 50 Ω line should be added to the circuitry shown in Figure IX-50 between the termination and coupler to minimize coupler mismatch phase error; the 45° line length

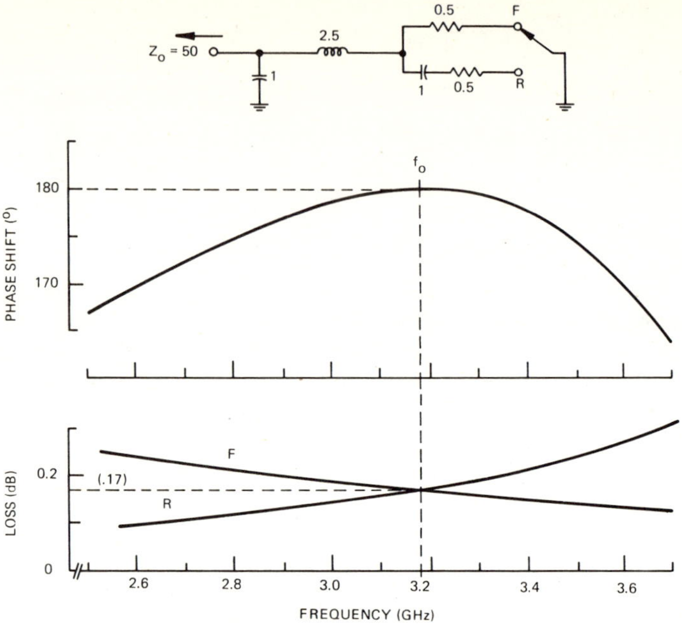

Figure IX-50 **Performance of the Series and Parallel Resonant 180° Termination**

gives the least error as described in Figure IX-44. The series and parallel resonant terminations, however, may not provide superior performance when applied to terminations with diodes having appreciable package and mount parasitic reactances. For this reason it is appropriate to analyze both the transformed LC and the series and parallel resonant terminations to determine which can better accommodate the junction and package reactances of the particular diode used.

The addition of a Z_0, 45° line ahead of the termination produces $\pm jZ_0$ reactive switching, just as does the transformed LC circuit. Therefore, adding a quarter wave transformer permits achievement of any other desired values of phase shift with the series and parallel resonant terminations, as was shown for the transformed LC circuit. The results obtained in this way for 90° and 45° phase shift are shown in Figure IX-51. In these cases, however, the phase shift variation with frequency (except for the immediate vicinity of f_0) is not significantly better or worse than is the performance obtained with the transformed LC circuit for these bits shown in Figure IX-48.

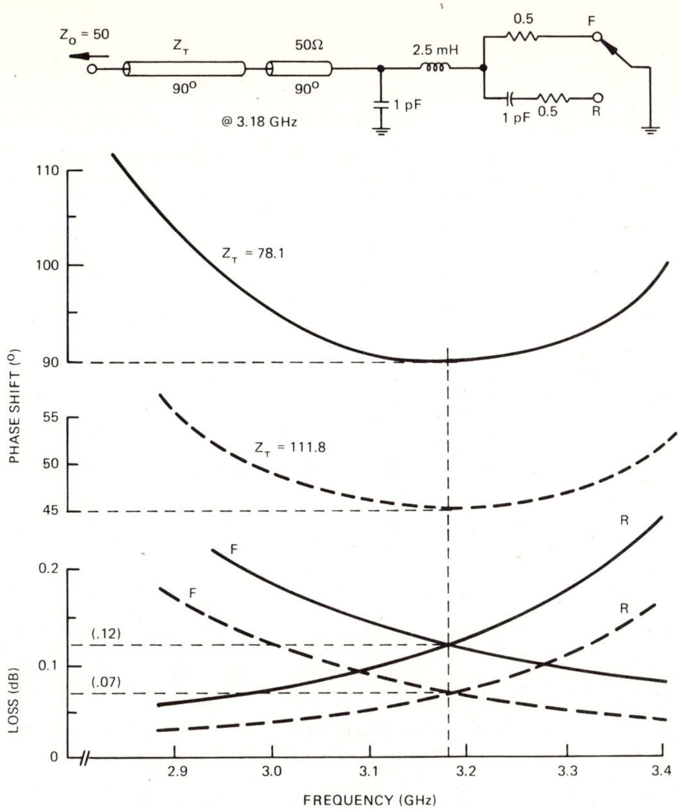

Figure IX-51 **The Series and Parallel Resonant Terminations Transformed to Give 90° and 45° Phase Shift**

It should be noted that the minimum phase state for the series and parallel resonant 45° and 90° bits is obtained with reverse bias while, for the transformed LC circuit, it is obtained under forward bias. Either state of a 180° bit may be defined as a reference; for the data in Figure IX-50, the forward state was defined as the minimum phase state. Had the reverse bias state been chosen instead, the phase shift versus frequency contour in Figure IX-50 would have been concave rather than convex; i.e, 180° would have been the minimum phase shift rather than the maximum value as shown. There are sometimes practical advantages to selecting one or the other error characteristic as, for example, to compensate for the phase error of other bits in cascade.

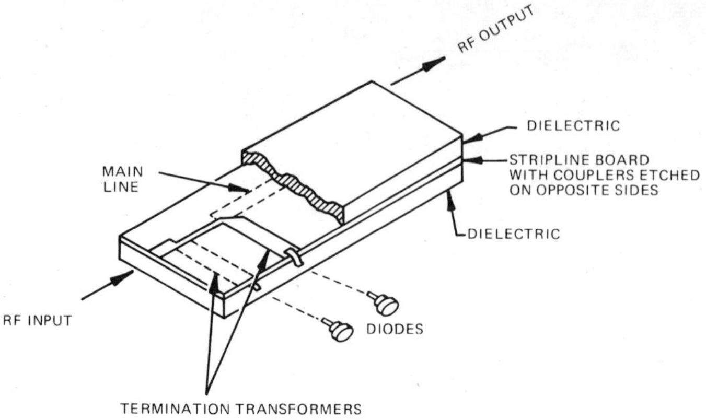

Figure IX-52 **Plan View of Stripline (Dielectric Sandwich) Reflection Phase Shifter Bit using Backward Wave 3 dB Coupler**

Peak Power

Both the transformed LC and the series and parallel resonant terminations have exactly the same center frequency dissipative loss at reverse bias, because only the diode is assumed to have resistance, and the same diode is used for both examples. Therefore, it follows that identical RF currents are established at f_0 in the diodes in both circuits. From the diode equivalent circuit and the insertion loss, the RF voltage stress across the diode's reactance, X_C, can be calculated from Equation (IX-34). Alternatively, recognizing that the diode appears either as an open or short circuit at Z_S (socket impedance — the transformed impedance of the coupler at the short and open circuit of the termination), the maximum power, P_{MR}, sustainable by the 180° bit in terms of the maximum rms RF voltage, V_{MR}, sustainable by the diode in state R can be expressed [18] in terms of the circuit parameters. The RF current, I_L, incident on the termination is related to the power on the whole bit by $(P_{MR}/2) = I_L{}^2 Z_0$. In the reverse mode termination, in the reverse bias state, the diode shorts the line and the current through the diode is $2I_L$. The line impedance, Z_0, is given by Equation (IX-57). Finally, the diode voltage magnitude, in volts rms, is $V_{MR} = 2I_L X_C$. Combining

$$P_{MR} = \frac{V_{MR}{}^2}{2} \cdot \frac{1}{X_C} \sqrt{\frac{R_R}{R_F}} \qquad \text{(IX-63)}$$

For the 1 pF diode example, $X_C = 50\ \Omega$ and $R_R = R_F = 0.5\ \Omega$.

Assuming a bulk breakdown voltage of 1200 V with 200 V bias, rms RF voltages up to $V_{MR} = 707$ V (corresponding to 1000 V peak) might be sustained (before diode failure); thus, P_{MR} equal to 5 kW can be sustained at the input to the two-diode, 180° bit.

Using Equation (IX-34) and noting that the fraction of incident power lost *per diode* is 0.02 in the 180° bit gives

$$P_{MR} = \frac{0.5 \text{ ohm}}{.02} \left(\frac{707 \text{ volts}}{50 \text{ ohms}}\right)^2 = 5 \text{ kilowatts} \qquad \text{(IX-64)}$$

While both methods for estimating voltage limited power are equally applicable at f_0, note that it is only proper to apply Equation (IX-63) *when the terminations represent* open or short circuit impedances *directly on a* Z_0 coupler. However, Equation (IX-34) is based upon the diode dissipative loss and can be applied at any frequency for which the diode dissipative loss, X_C, and R_R are known.

For example, Equation (VI-34) can be applied to calculate the voltage limited survival power for the 90° and 45° bits by noting that the dissipative losses at 3.18 GHz are, from Figure IX-51, 0.12 dB and 0.07 dB, respectively. The P_{MR} for these bits can be obtained by scaling of the 180° bit performance; thus

$$P_{MR}\ (90^\circ \text{ bit}) = \frac{0.17 \text{ decibel}}{0.12 \text{ decibel}}\ 5 \text{ kilowatts} = 7.1 \text{ kilowatts} \qquad \text{(IX 61)}$$

$$P_{MR}\ (45^\circ \text{ bit}) = \frac{0.17 \text{ decibel}}{0.07 \text{ decibel}}\ 5 \text{ kilowatts} = 12.1 \text{ kilowatts} \qquad \text{(IX-65)}$$

One must, however, be sure to use Equation (VI-34) with the proper assignment of values to its variables. Important points to keep in mind are

1) Loss/diode must be the *dissipative* loss of *only* the diode the voltage, V_{MR}, of which determines the power limit, P_{MR}. Reflection loss of the circuit must be excluded. (With ideal 3 dB couplers, the VSWR < 1.02 for $0.8\ f_0 \leqslant f \leqslant 1.2\ f_0$, and corresponding reflection loss is less than 0.0004 dB.)
2) The reactance, X_C, must be evaluated at the frequency for which the Loss/diode is known. For example, if the variation of

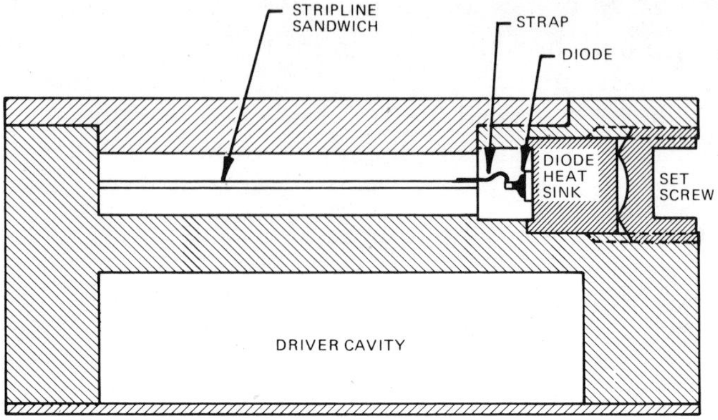

Figure IX-53 **Section View of Typical Stripline Phase Shifter Assembly**

X_C with frequency is overlooked, it could appear that the 180° bit has higher V_{MR} stress at 3.6 GHz than at 3.18 GHz since the loss is seen to increase with f in Figure IX-50. Actually, the voltage stress is about constant over the band, the extra loss being attributable to decreasing X_C with frequency.

3. Practical Stripline Phase Shifters

Stripline Circuitry

One of the most practical methods to realize multi-bit phase shifters is with a stripline dielectric sandwich structure, as shown in Figures IX-52 and IX-53. This method allows the 3 dB backward wave couplers, the bias blocks and returns, and the termination elements to be etched directly on the stripline circuitry, as shown in Figures IX-54 and IX-55. The only components which must be "installed" in the circuit are the diodes themselves. Open circuited 3 dB hybrid couplers can serve as low loss "bias blocks" between bits. Not only is low loss obtained, but low VSWR and higher reliability are gained in comparison with lumped capacitors, because no separate connections are needed to install them. The LC low-pass filter to introduce bias can be realized as a "printed flag" distributed circuit, consisting of an open-circuited low impedance 90° line length and a high impedance 90° line length, as illustrated in Figure IX-55.

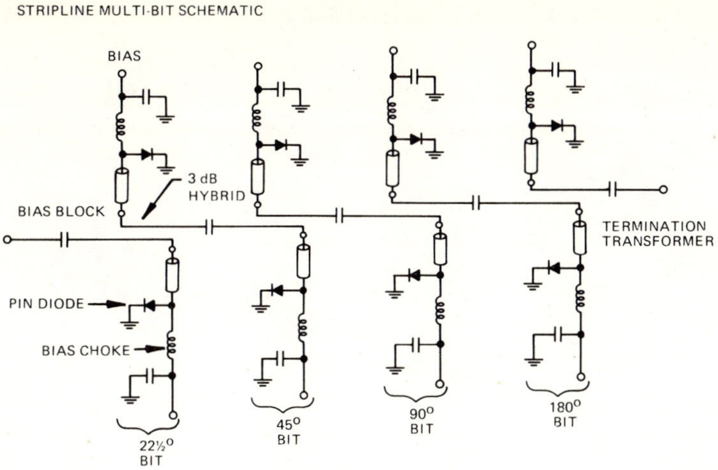

Figure IX-54 **Schematic for 4-Bit Stripline Hybrid Coupler Shifter Showing Method of Biasing**

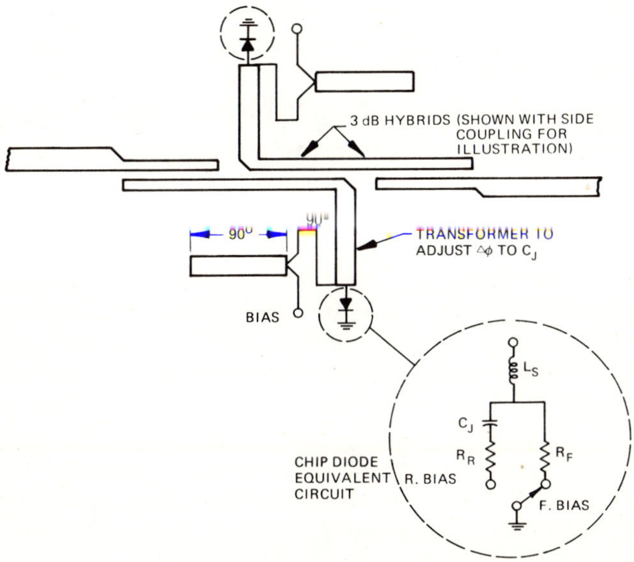

Figure IX-55 **Detailed Single Bit Stripline Schematic Showing Printed Circuit Elements**

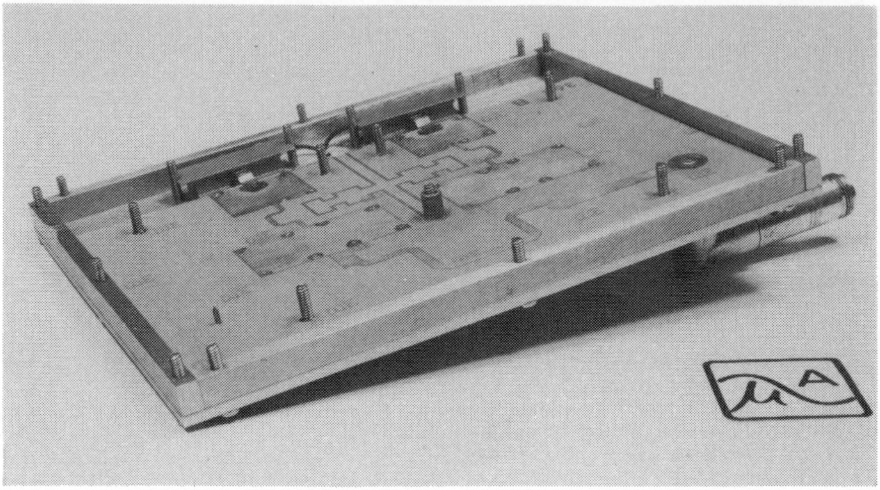

Figure IX-56 **Photograph of High Power UHF (400-450 MHz) 180° Stripline Phase Shifter**

UHF

A photograph of a 180° stripline bit designed for the 400-450 MHz bandwidth is shown in Figure IX-56. The design is similar to that described except, for space limitation purposes, lumped capacitors are used for the diode bias blocks, and the center conductor is dc grounded. The terminations use lumped element transformers* which also reduce the physical size of the circuit.

Measured low power performance for this 180° bit is shown in Figure IX-57. Diodes similar to the MA-47890 (Table III-1) were operated at -200 V reverse bias and +200 mA per diode forward bias. When high power testing was performed using 1500 μs RF pulses and a 0.06 duty cycle, the first diode failures occurred in the reverse bias state at 10 kW of incident power. Thus, for phased array operation, with short-circuit loads of any phase, the device could be operated safely with up to 2500 W of input power.

L-Band

A three bit L-band phase shifter is shown in Figure IX-58. As with the UHF model, the diode bias is introduced through lumped element capacitors, and similar diodes are used; however, distributed line transformers are employed to adjust the phase shift of each bit. The measured data for this model are shown in Figure IX-59.

*The π lumped equivalent circuit of a 90° line length is described in Figures VI-5 and VI-31.

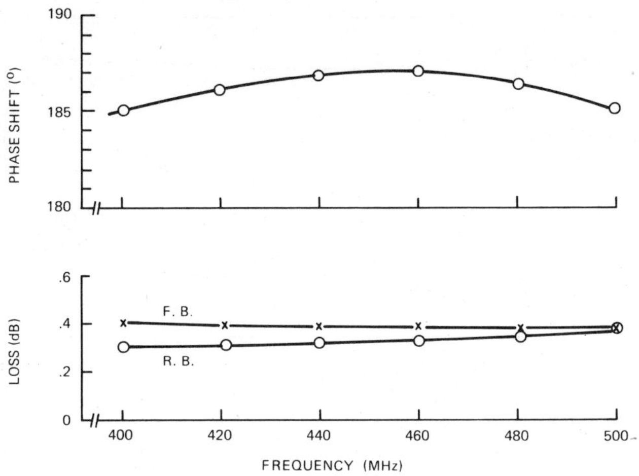

Figure IX-57 **UHF Phase Shifter Performance**

Figure IX-58 **Three-Bit L-Band High Power Phase Shifter**

This phase shifter has been tested to failure at power levels of 4.1-4.8 kW peak, using 2 ms pulses and a 0.06 duty cycle. The diode failures occurred under reverse bias (-200 V) in the 180° bit. Under these conditions, the maximum rated input power is 1 kW peak with a totally reflecting load of any phase. Calculated

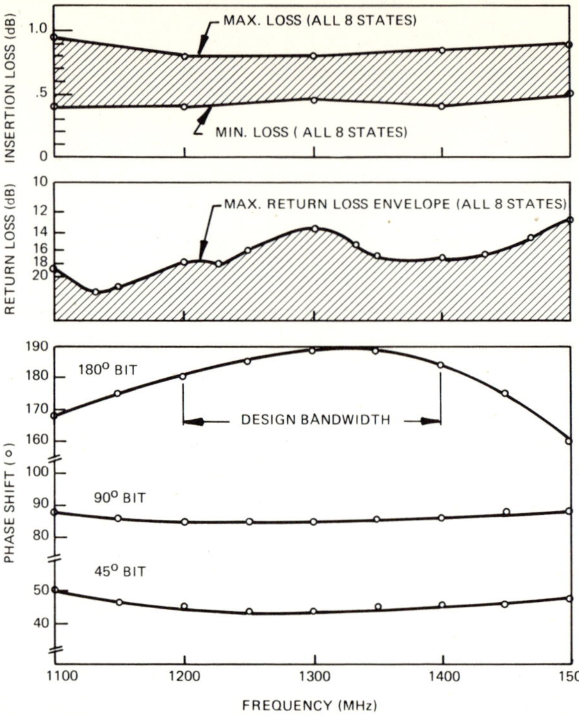

Figure IX-59 **Measured Performance for L-Band Stripline Phase Shifter**

RF voltage stress at 4.1 kW with a matched load is approximately 1,000 V rms.

Each diode, biased with either +200 mA of forward bias or -200 V reverse bias, is mounted on an 8-32 threaded stub (type 150 in Table III-2) which serves both as a mechanical mount and thermal path for heat dissipated in the diode. The upper (mesa) contact of the diode has a metal cover with a flexible strap for soldering directly to the center conductor of the stripline board. The use of the strap eliminates the transmission of mechanical stresses to the diode chip from differential thermal expansion of circuit board and housing.

To afford protection of the I region, a vitreous glass of approximately 50 μm (2 mils) was fired directly onto the silicon (see Figure IX-60) in wafer form. Thereafter the scribed chips are essentially "self-packaged" without need for a separate hermetic

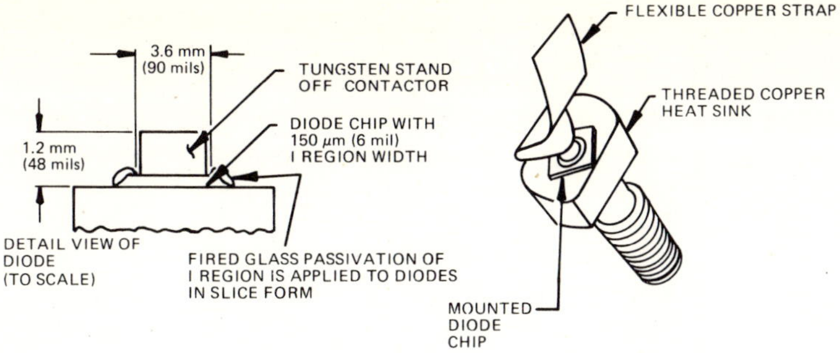

Figure IX-60 **The High Voltage PIN Chip Mounted on a Heat Sink**

enclosure, resulting in both lower cost and reduced parasitic capacitance. The typical measured parameters of this diode, of which more than 90,000 were made, are given for the MA-47890 in Table III-1.

This design* is used for phase steering at high RF peak power in the COBRA DANE radar [1] antenna. This project is the largest diode phase shifter effort reported in the literature [19], encompassing a 2 J pulsed RF energy control capability *(1000 W peak power for 2000 μs).* Measured data were obtained for over 15,000 separate three-bit phase shifters.

Four phase shifters were combined in a single housing; in this way, a portion of the array antenna power division is associated directly with the phase shifters, and the losses of a connector interface are eliminated. Figure IX-61 shows the microwave diagrams for the 4-to-1 divider and for the individual phase shifter channels; Figure IX-62 is a photograph of the complete, four phase shifter assembly.

In production quantities it was found that about a ±10% tolerance on capacitance distribution could be achieved. Prior to installation in the phase shifters, diodes were screened only with respect to capacitance (measured at -200 V and 1 MHz) and dc breakdown (a screen value of 1000 V minimum at 10 μA was used). The phase shift tolerance depends mainly on the precision with which the diode capacitance is reproduced. Figure IX-63 shows the

*These phase shifters were designed and built by Microwave Associates, Inc., of Burlington, Massachusetts and used by Raytheon Company of Wayland, Massachusetts in the construction of the COBRA DANE radar for the United States Air Force. Raytheon is the System Contractor for this space surveillance radar installed at Shemya Air Force Base on Shemya Island in the Aleutians.

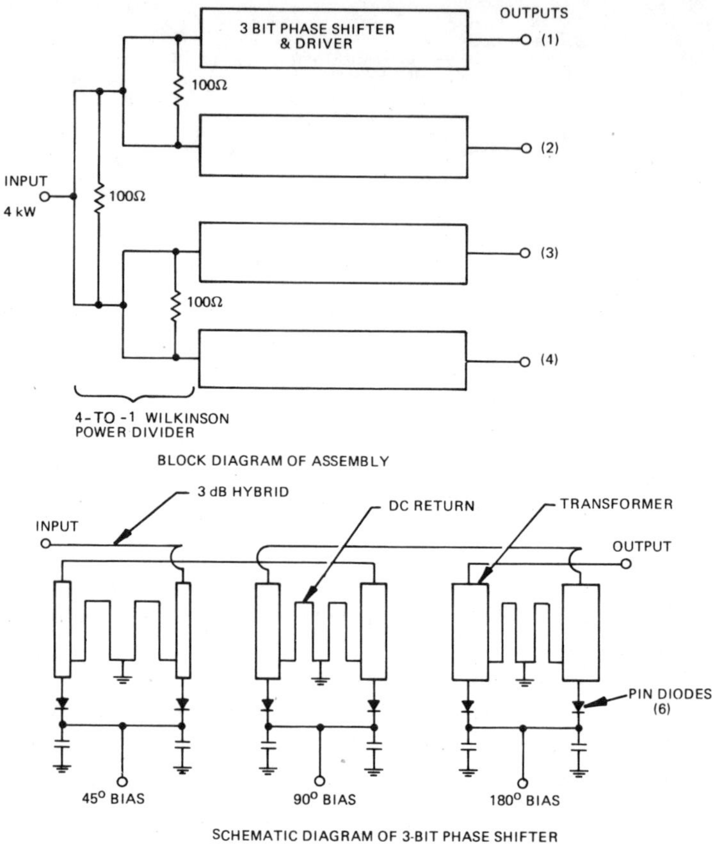

Figure IX-61 **The 4-to-1 Divider and Phaser Assembly**

computer calculated phase shift for the 180° bit for various C_J values. From this figure it is apparent that only about ±3% variation in C_J is tolerable in the 180° bit, while progressively more variation is acceptable in the 90° and 45° bits using the C_J range shown in Table IX-1. An overall phase shift error of about ±10° rms could be held with no need for tuning.

However, insertion phase variations of ±10° would, when combined with the phase shift error, cause the 13° rms overall phase error specification to be exceeded. This problem was solved through the design of an output connector with changeable phase length increments of ±10°.

Bit	C_J Range	Tolerance
45°	2.7-3.3	±10%
90°	2.8-3.2	± 7%
180°	2.9-3.1	± 3%

Table IX-1 Capacitance Tolerance

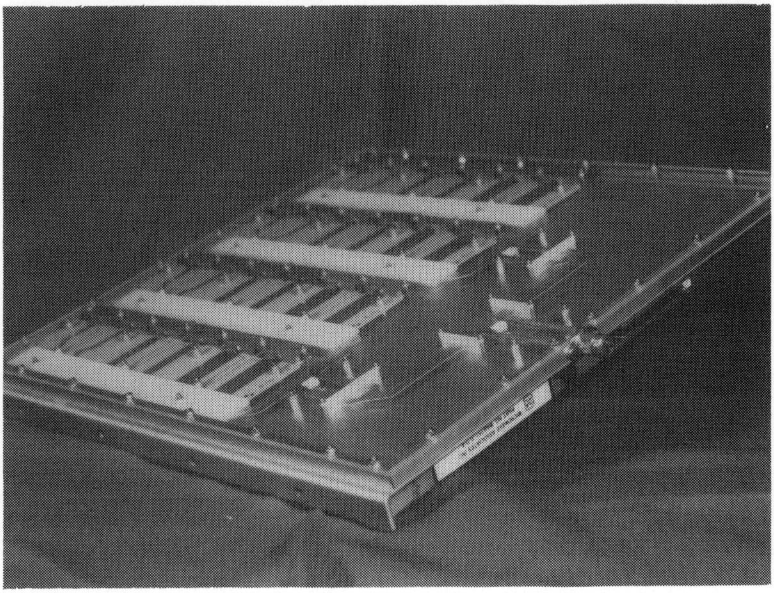

Figure IX-62 Photograph of COBRA DANE 4-to-1 Driver and High Power Phase Shifter Assembly

The phase shifters are equipped with a printed circuit driver actuated by a 5 mW (5 V at 1 mA) balanced signal input (Figure IV-4). Each driver circuit contains three independently controlled sections for the 45°, 90°, and 180° bits. The driver signal input is balanced with respect to ground and does not trigger with induced noise voltages *(common mode voltage)* as high as ±25 V on the twisted pair input wires, although a *difference* in potential of 5 V *(differential mode voltage)* produces the desired switching.

Total switching time is 25 μs including delay; the actual transition time, however, is usually below 5 μs.

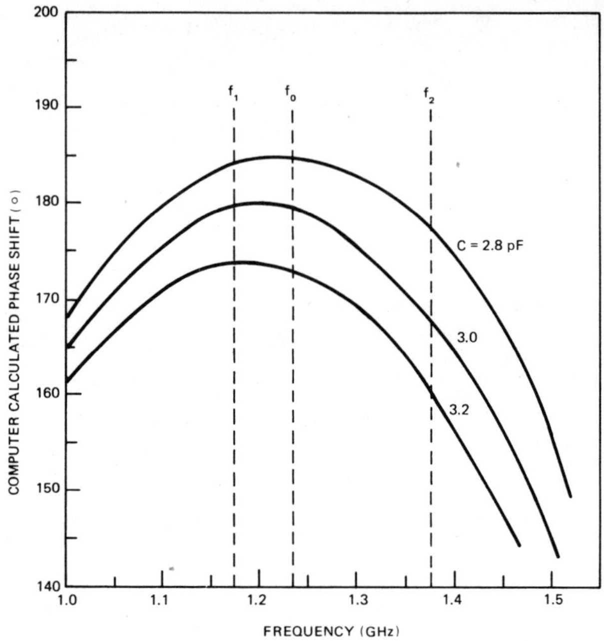

Figure IX-63 **Computer Calculated Phase Shift Performance with Diode Capacitance Variation**

All phase shifters were tested using a Hewlett Packard computer-controlled network analyzer programmed to calculate the phase shift error in every bias state, compare it with a phase standard, and calculate the rms error. In the event 13° rms error is exceeded (11° rms at band center), the computer calculates the effect of a +10° and -10° insertion phase offset and, if it predicts that either change would bring the performance within requirements, the computer prints an instruction to the operator to change the insertion phase trim. In this way, over 98% of the phase shifters met specifications with no other reworking. Measured data is shown in Figure IX-64 for the first 12,000 units. Except for insertion phase trimming, *no production tuning* of the circuits was performed. For operating reliability, all phase shifters were required to survive full input power into a short-circuited load of any phase. Accordingly, *all 15,000* devices were so tested, stressing the diodes to the same levels that would be incurred with matched loads at 3-4 kW of input power per channel.

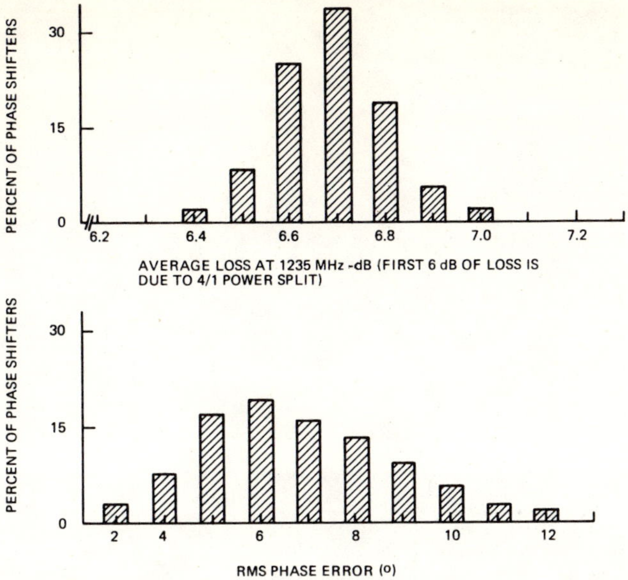

Figure IX-64 **Performance Distribution for the 15,000 Phase Shifters Measured at 1235 MHz**

S-Band

S-band phased arrays are used mainly for ground and ship search and tracking radars. Low cost as well as low insertion loss are of primary importance for these applications. Figure IX-65 shows a complete three-bit stripline phase shifter. The built-in, TTL-compatible driver provides +100 mA forward bias per diode and -100 V of reverse bias. Unpackaged diodes were used in this model, the PIN diode chips being fitted with top heat sinks and connector straps and then mounted on base heat sinks per the drawing in Figure IX-53. A photograph of this diode chip subassembly is shown in Figure IX-66.

Measured S-band performance is shown in Figures IX-67 and IX-68. The "phase shift error" plotted in Figure IX-68 is the departure in measured phase shift from an ideal characteristic (0°, 45°, 90°, 135°, . . . 315°). The phase errors shown include

1) The fundamental variation of phase shift of the termination.
2) Any interactions of the terminations with coupler mismatches.
3) Phase shift interactions among bits.
4) Measurement errors.

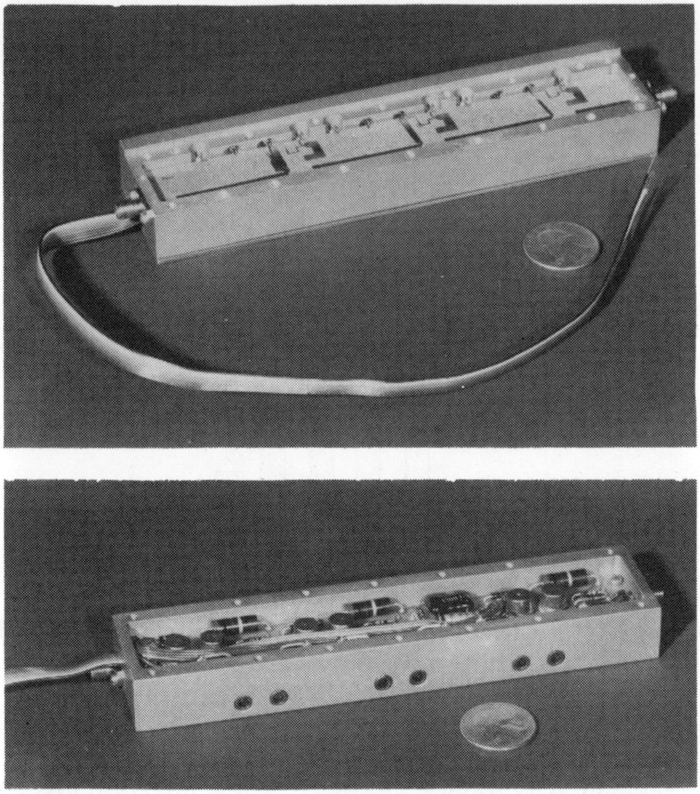

Figure IX-65 **S-Band, Three-Bit Stripline Phase Shifter and TTL Driver. Plan View Sketches IX-52 Through IX-55.**

In view of the number of phase shift error contributing causes, the average error of only 3° over a 20% bandwidth is small. Not included, however, is the error which would be introduced by production tolerances on diode capacitance, since this model was designed about the specific diodes available, similar to the MA-47892 described in Table III-1. Thus, the overall phase error in production would be comparable to that of the L-band high power model.

The phase shifter was tested to high power burnout, which occurred at 4 kW peak using 0.1 ms pulses and 0.05 duty cycle. The high power limit is set by the maximum RF peak voltage sustainable by the diodes in the 180° bit in the reverse biased condition. At the 4 kW level, the diodes in this bit experience an RF voltage of about 550 V rms. (The diodes were biased with +150 V and -150 mA for this high power test.) The minimum average diode

Figure IX-66 **Unpackaged PIN Diode Subassembly**

loss, found using Hines' theory, is about 0.3 dB. The remaining 0.5 dB of average loss measured at low power, shown in Figure IX-69, can be attributed to dissipative and reflection losses in the circuit. The RF losses near the burnout power with all diodes reverse biased can increase by up to 0.5 dB, but the loss increase near rated power of 1 kW peak would be only about 0.1 dB.

C-Band

One of the most promising commercial applications of the phased array is the electronically steered C-band microwave landing system (MLS). With the phase-scanned MLS approach, linear array antennas generate rapidly swept fan beams to provide elevation and azimuth angle data to aircraft prior to landing. The precision of this data is sufficient to permit a fully automated (Category III) landing with zero visibility, even down to the runway surface

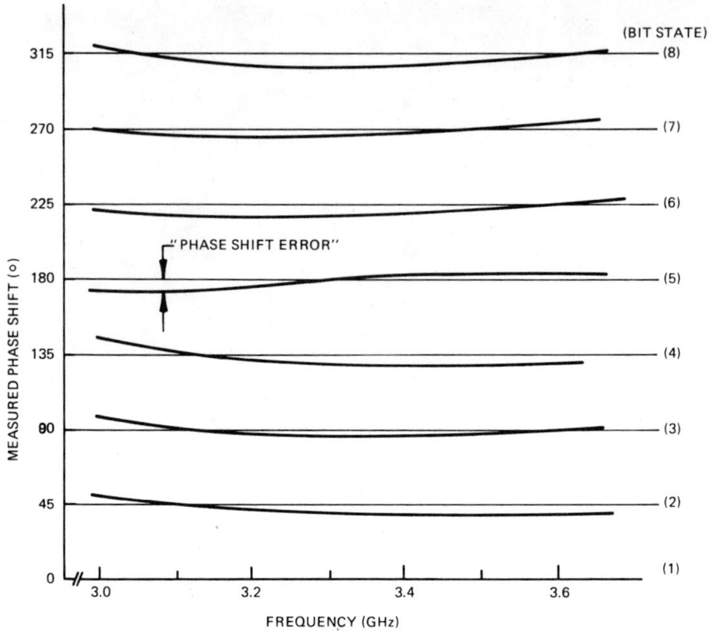

Figure IX-67 **Measured Phase Shift of the S-Band Stripline Phase Shifter vs. Frequency and Bias State**

itself [20]. The PIN diode phase shifter with inherently fast switching speed, discrete phase switching, and temperature insensitivity is well suited as the control element for this rapidly scanned antenna beam.

Figure IX-70 shows a four-bit phase shifter designed using techniques similar to those described for the L and S band models. A complete driver with diode fault sensing is mounted on a common baseplate but outside of the RF enclosure to facilitate driver checking and repair. The large area layout is permissible since the phase shifters are not used in a two-dimensional array structure. Diodes with 0.2-0.7 pF junction capacitance (similar to the MA-47895 and MA-47896 in Table III-1) were used. The smaller (0.2 pF) capacitance diodes were employed for the larger (90° and 180°) phase shift bits. Measured RF results are shown in Figures IX-71 and IX-72.

The power requirements of the driver are -40 V at -40 mA and +5 V at +300 mA maximum. The device switches in about 400 ns and there is about an additional 400 ns of fixed delay. Thus,

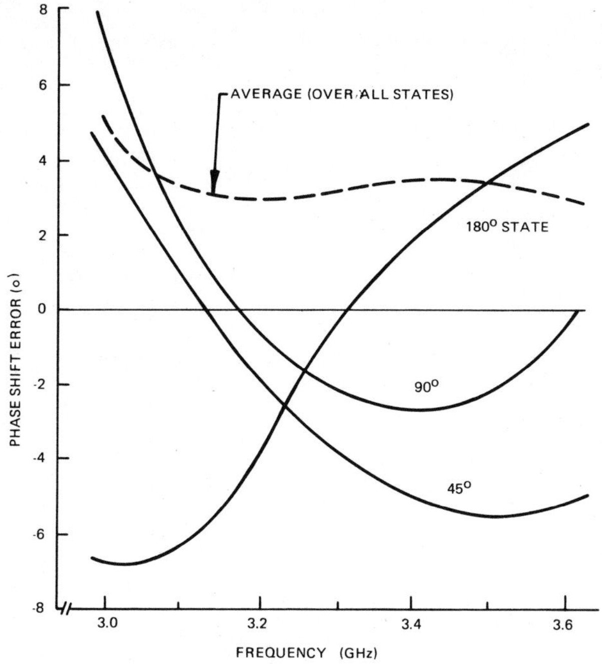

Figure IX-68 **Detail of Measured Phase Shift Error of the S-Band Phase Shifter**

complete switching occurs in less than 1 μs. The maximum currents are drawn from both supplies when all four bits are in the forward biased (longest electrical length) state. In the all-reverse biased state, the current magnitudes drawn from the -40 V and +5 V supplies are less than 1 mA each. Proportionate current levels are drawn for intermediate bias conditions. For example, with any one bit forward biased (with the remaining bits reverse biased), the bias drawn would be -40 V at -10 mA and +5 V at +75 mA. On the average, there are only two bits forward biased; thus, the average bias is -40 V at -20 mA and +5 V at +150 mA. Considerably less bias current would be adequate if increased insertion loss and switching time were acceptable. For example, the driver design could be modified to cut the currents in half with a 0.3-0.5 dB increase in insertion loss and a switching time increase of 2-3 μs.

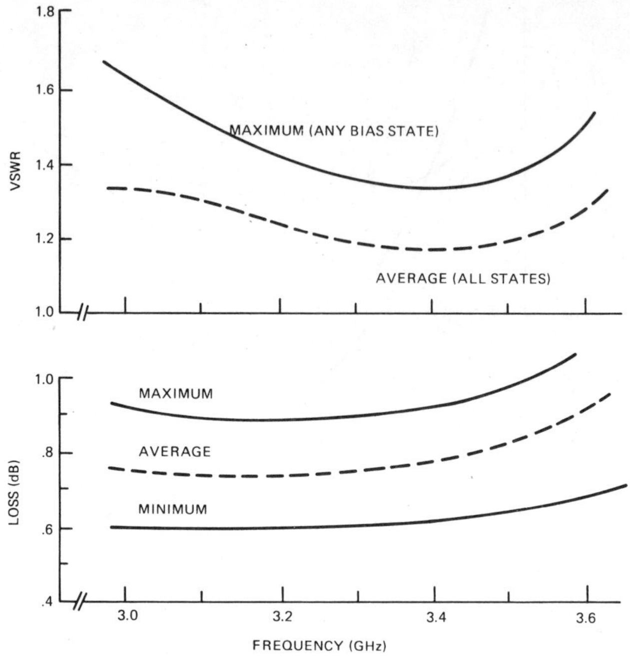

Figure IX-69 **Measured VSWR and Loss of the S-Band Phase Shifter**

X-Band

A primary application for X-band phase shifters is in multi-function phased array antenna steering, in which low weight and cost are especially important. Figure IX-73 shows a four-bit stripline X-band phase shifter. The outside dimensions of the housing shown are 0.5 x 1.0 x 4.25 in (13 x 25 x 108 mm) which can also contain an integrated circuit driver. (A discrete element driver requires an additional housing of the same size.) Diodes similar to the MA-47896 (see Table III-1) but having 0.2-0.4 pF junction capacitance are used, with lower capacitance diodes used for the 90° and 180° bits, as with the C-band circuit described previously. Typical performance is shown in Figures IX-74 and IX-75.

Projected high power phase shifter requirements of airborne arrays are usually below 100 W peak. A model was tested to more than 1 kW peak without failure using 1 μs pulse length and 0.001 duty cycle. The reverse bias loss versus power data are shown in Chapter II (Figure II-23) as a demonstration of the power loss increase under different reverse bias levels.

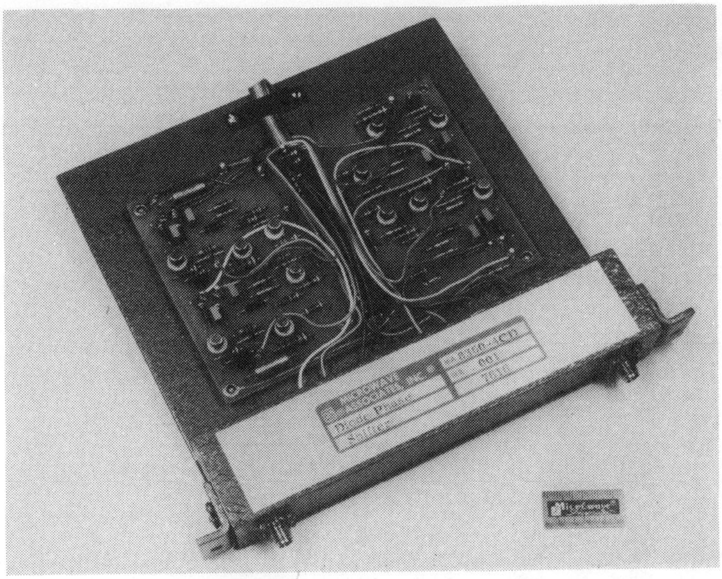

Figure IX-70 Four-Bit, Stripline, C-Band Phase Shifter with Fault Sensing Driver Designed for the Microwave Landing System (MLS)

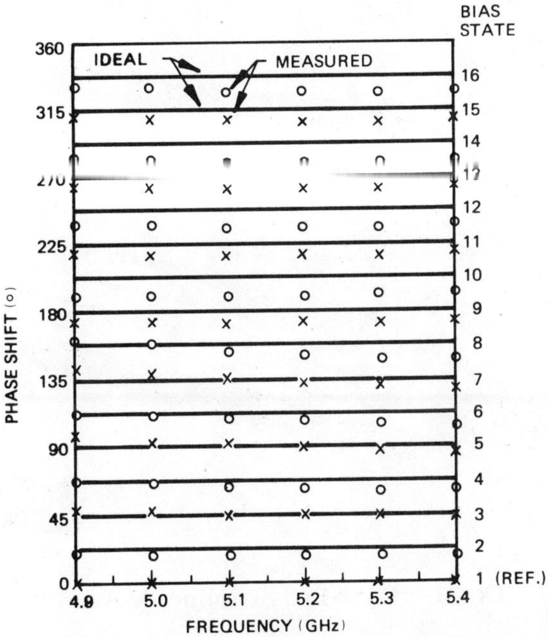

Figure IX-71 Measured Phase Shift of C-Band Phase Shifter

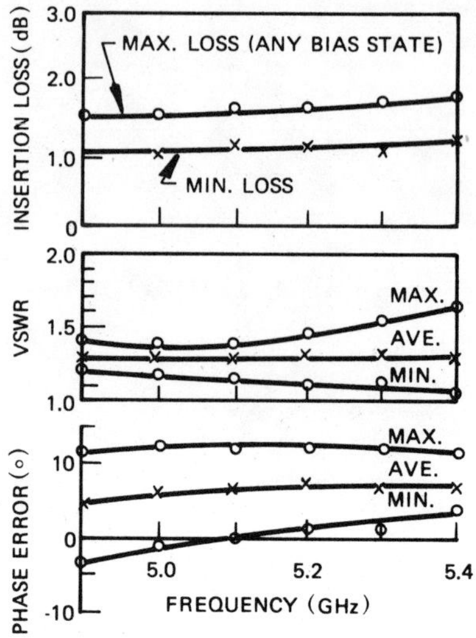

Figure IX-72 Measured Loss, VSWR, and Phase Shift Error of the C-Band Phase Shifter.

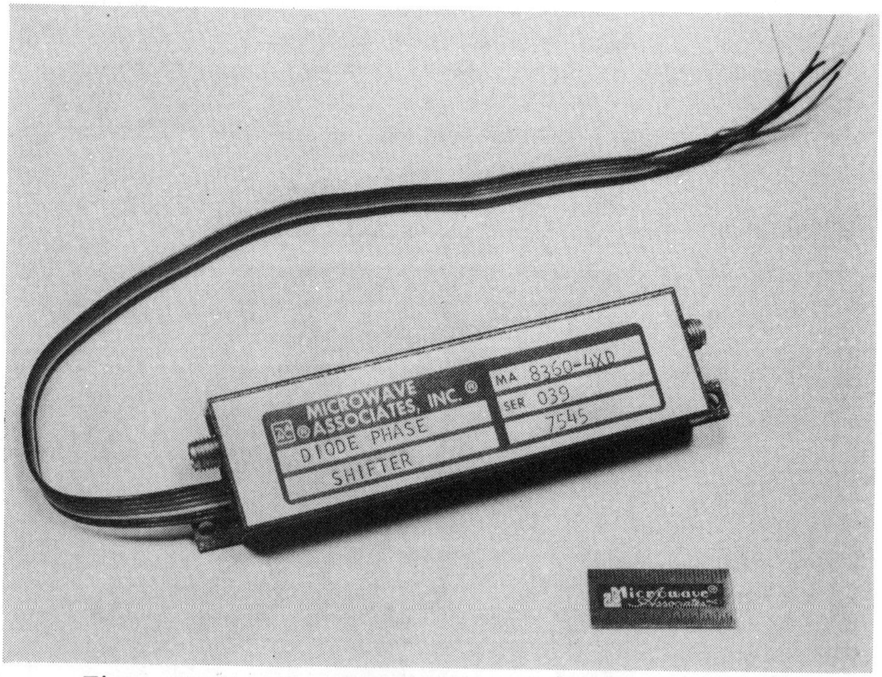

Figure IX-73 Four-Bit, Stripline, X-Band Phase Shifter.

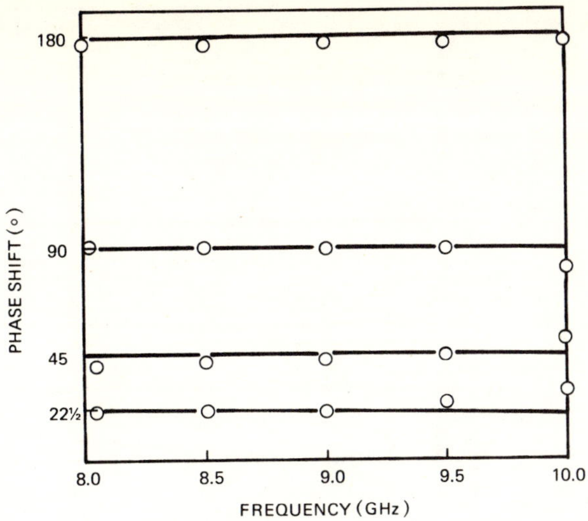

Figure IX-74 **Typical Phase Shift with Frequency for the X-Band Stripline Phase Shifter.**

This housing is useful for engineering evaluation but weighs 2 oz, a prohibitively high weight for an airborne antenna containing 3000-5000 elements. Figure IX-76 shows a skeletal housing with the coaxial connectors eliminated and a trough for direct stripline interconnection to the feed and radiating elements. This housing, complete with three-bit phase shifter circuit and flatpack driver, weighs less than 0.6 oz. However, the direct stripline interconnection is not suitable for rapid connection and disconnection. This aspect is yet another which must be addressed in the design of the airborne array antenna.

Ku-Band

In order to permit the realization of portable microwave landing systems, some exploration has been made of diode phase shifters at 15.5-15.6 GHz. These devices consist of a microstrip 180° bit using a branch-line 3 dB hybrid coupler. At Ku-band, the wavelength in Teflon substrate microstrip is short, about 0.5 in (13 mm). To minimize junction discontinuities in the coupler, the ground plane spacing is only 0.005 in (0.13 mm). Figure IX-77 shows a sketch of the center conductor circuit pattern; the dc return, connecting center conductor to ground, is omitted in the sketch.

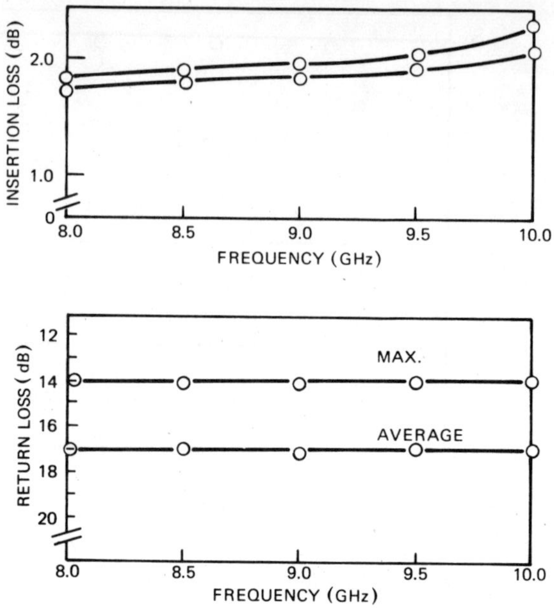

Figure IX-75 **Typical Insertion Loss and Return Loss for the X-Band, Four-Bit Stripline Phase Shifter**

The chip diodes are mounted directly on the TFG substrate since the operating power (a few hundred milliwatts) is low. The measured performance is shown in Figure IX-78. The high insertion loss obtained, about 3 dB, is composed mainly of circuit loss. Lower loss probably could be obtained with the backward wave, stripline overlay coupler used in the lower frequency models described, but added development effort is needed to permit the design of suitable couplers in this band.

E. Schiffman Phase Shifters

1. Match and Phase Relations

In their classic paper on coupled lines Jones and Bolljahn [21] described the properties of proximity coupled lines obtained when they are interconnected in different ways. One of these, with symmetric shorts on ports 2 and 4 (shown in Figure VI-16) is the basis for the reflection phase shifter. Another connection, shown in Figure IX-79, results in a two-port network as ports 2 and 3 are connected to each other, ideally by a zero length line. This action produces an "all pass" network with input impedance

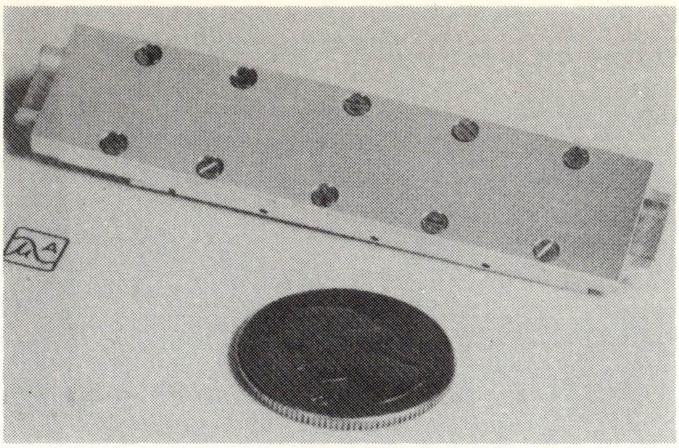

a) Upper view showing through for stripline overlap joint RF connection

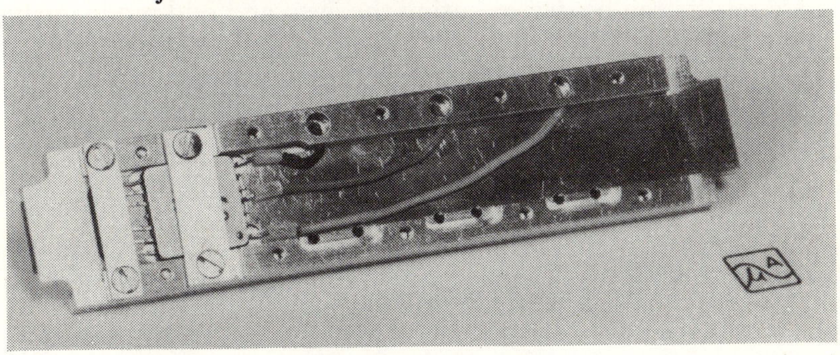

b) Lower view showing flatpack integrated circuit driver and wiring to 3-bits

Figure IX-76 Light Weight (0.6 oz) X-Band Three-Bit Phase Shifter Using Stripline Interconnections (Portions of this engineering were supported by the Air Force Avionics Laboratory, Wright Patterson Air Force Base, Ohio, under Contract F33615-72-C-1967)

$$Z_0 = \sqrt{Z_{OO} Z_{OE}} \tag{IX-66}$$

and transmission phase length, ϕ, determined from

$$\cos \phi = \frac{(Z_{OE}/Z_{OO}) - \tan^2 \theta}{(Z_{OE}/Z_{OO}) + \tan^2 \theta} \tag{IX-67}$$

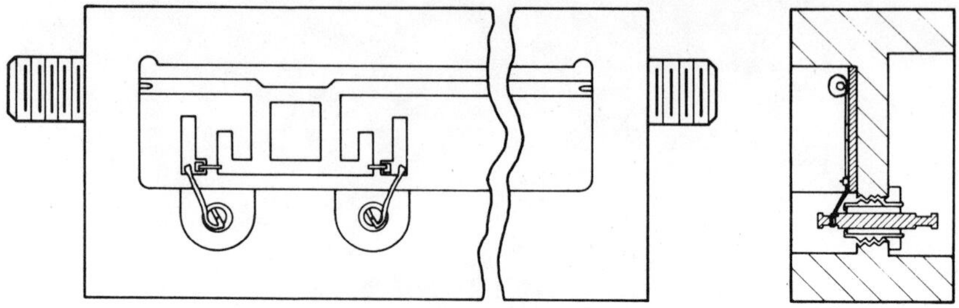

Figure IX-77 **Sketch of Ku-Band Phase Shifter Bit**

where θ is the electrical length of the coupling section and Z_{OE} and Z_{OO} are the even and odd mode impedances, respectively*. These results can be derived using the even and odd mode analysis shown in Figure IX-79.

This coupled line interconnection results in an "all pass" network that theoretically is perfectly matched in a Z_0 system at all frequencies. As a reference, when used as a four-port coupler, the coupled arm coupling (in decibels) is given by

$$\text{Coupling} = -20 \log_{10} \left(\frac{Z_{OE} - Z_{OO}}{Z_{OE} + Z_{OO}}\right) = -20 \log_{10} \left(\frac{\rho - 1}{\rho + 1}\right) = -20 \log_{10} k \tag{IX-68}$$

and plotted in Figure IX-80.

2. Constant Phase Shift

Schiffman observed that such coupling also produces a dispersive phase response**, as shown in Figure IX-81, which might be used to provide a nearly constant phase shift compared to a uniform nondispersive line of suitably chosen length [22]. This fact is illustrated in Figure IX-82, which shows ϕ_1 versus θ for a Schiffman section with $\rho = 3.01$ (6 dB coupling). Also shown is the phase $\phi = K\theta$ of a uniform line. It can be seen that the choice of $K = 3$ results in a phase shift of $\Delta\phi = \phi_0 - \phi_1$, which is equal to 90° within ±4.8° for $55° < \theta < 125°$ — a frequency ratio of 2.27:1. Other

*For a detailed discussion of these coupled line properties, see Chapter VI, Backward Wave Coupler Example.

**Networks which have an electrical length, θ, proportional to frequency (i.e., θ = Kf) produce uniform delay to all signals. *A dispersive network is one which has an electrical length which is not directly proportional to frequency.* It is called dispersive because it distorts (or "disperses") pulses which, being made up of multiple frequency components, do not all pass through the network with the same time delay (*group delay*).

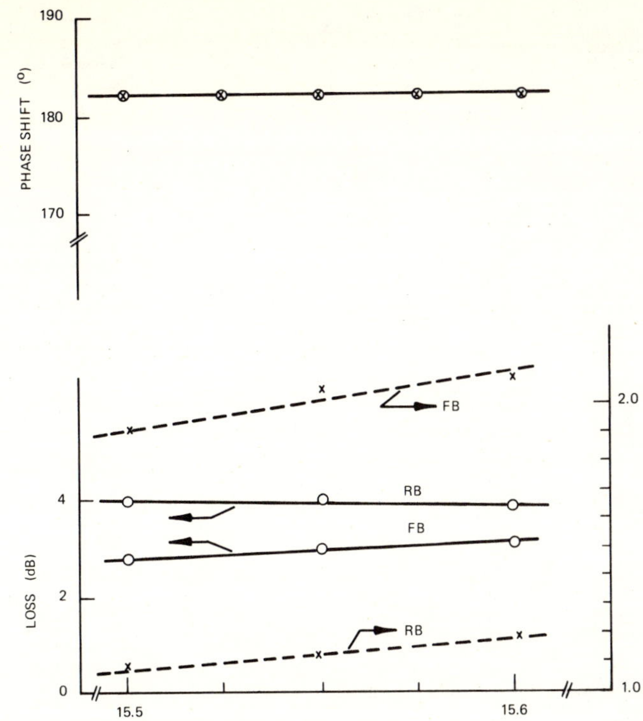

Figure IX-78 **Measured Performance of Ku-Band 180° Phase Shifter Bit**

phase shift and bandwidth performance combinations can be obtained by appropriate selection of K and ρ. A graphical approach, similar to that shown in Figure IX-82, provides the initial approximate values for K and ρ. A computer program to evaluate $\Delta\phi$ versus θ can then be used for the final selection of K and θ, from

$$\Delta\phi = k\theta - \cos^{-1}\left(\frac{\rho - \tan^2\theta}{\rho + \tan^2\theta}\right) \tag{IX-69}$$

An electronically controlled phase shifter is obtained by using diode switches to select between transmission through the longer, uniform line (nondispersive) path and the electrically shorter (dispersive) path. Figure IX-83 shows the stripline circuit method used by Grauling and Geller [23] to build a 1.9 GHz phase shifter with 16% bandwidth for a seradyne* application. Four bits having 22 ½°, 45°, 90°, and 180° phase shift were built. A Schiffman

*Frequency shift by linear phase modulation.

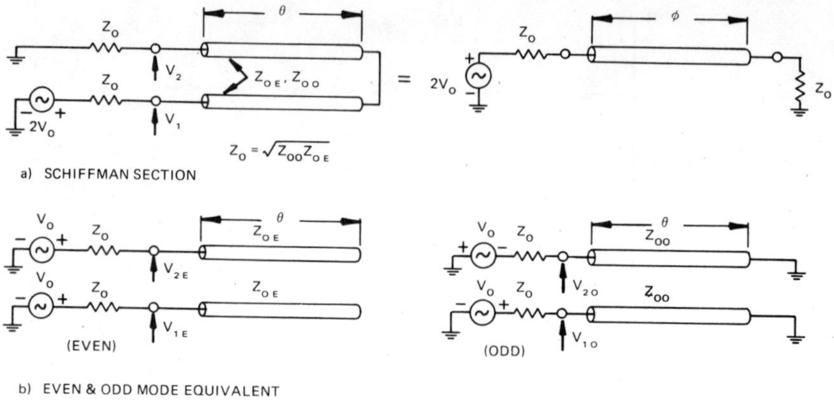

$$V_{1E} = V_0 \left(\frac{-jZ_{OE} \cot\theta}{Z_0 - jZ_{OE} \cot\theta} \right) \qquad V_{1O} = V_0 \left(\frac{jZ_{OO} \tan\theta}{Z_0 + jZ_{OO} \tan\theta} \right)$$

$$V_1 = V_{1E} + V_{1O} = V_0$$

$$V_0/V_2 = \frac{2Z_0 + j(Z_{OO} \tan\theta - Z_{OE} \cot\theta)}{-j(Z_{OO} \tan\theta + Z_{OE} \cot\theta)}$$

$$\phi = \mathrm{ARG}\,(V_0/V_2) = -\tan^{-1}\left(\frac{Z_{OO} \tan\theta - Z_{OE} \cot\theta}{2Z_0} \right) - 90^\circ = \cos^{-1}\left(\frac{\dfrac{Z_{OE}}{Z_{OO}} - \tan^2\theta}{\dfrac{Z_{OE}}{Z_{OO}} + \tan^2\theta} \right)$$

c) MATCH AND PHASE RELATIONSHIPS

Figure IX-79 **Schiffman Phase Shift Section with Even and Odd Mode Analysis**

section for more than 90° of phase shift requires coupling tighter than 3 dB, and is not practical using the stripline edge coupling shown in Figure IX-83; therefore, the 180° bit was realized by the cascade combination of two 90° circuits.

The Schiffman phase shifter using switched paths is not the lowest loss approach to phase shifting. All of the comments made concerning minimum insertion loss and potential insertion loss resonances for switched path and time delay phase shifters apply to the circuit in Figure IX-83. Furthermore, phase shift errors due to VSWR interactions between the Schiffman paths and the SPDT switches within a bit, as well as VSWR interactions between the bits themselves, make the realization of practical octave band phase shifters unlikely despite the theoretical expectations implicit in Figure IX-82. Nonetheless, the *Schiffman phase shifter approach is a unique method for obtaining nearly constant phase shift of any magnitude over considerable bandwidths,* and should

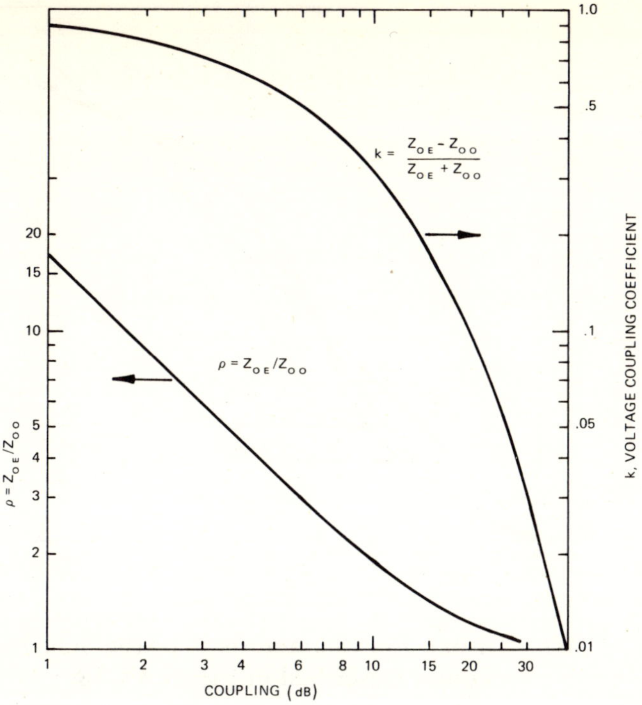

Figure IX-80 **Coupling Coefficient, k, and Impedance Ratio ($\rho = Z_{OE}/Z_{OO}$) versus Power Coupling (dB) for the Backward Wave Hybrid Coupler**

be considered for those applications where this characteristic is of prime importance.

F. Continuous Phase Shifters

1. Uses

A varactor diode can be used to provide continuously variable phase modulation of a microwave signal. As shown in Figure I-12, the diode is a capacitor the value of which depends upon the reverse voltage impressed across its terminals. Since the change of capacity can occur in less than a nanosecond, very high speed microwave signal processing is possible.

In fact, the response is so rapid that the microwave voltages used must be kept small compared to the bias voltage in order to avoid nonlinear operation resulting from the varactor's response to the

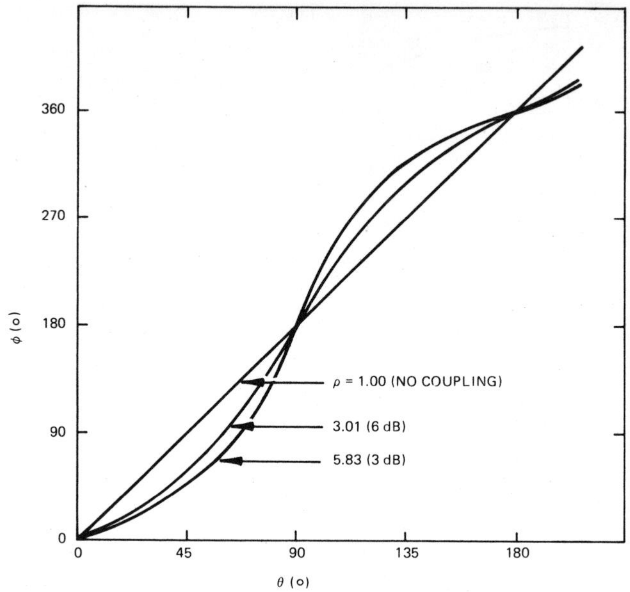

Figure IX-81 **Dispension Characteristic of the Schiffman [22] Section Described in Figure IX-78**

RF signal itself. As a result, continuous phase shifters are limited to less than one watt input power. This single characteristic has prevented the diode continuous phase shifter from enjoying the intense development efforts afforded the PIN diode and the ferrite-controlled phase shifters as steering elements for microwave phased array antennas.

Nonetheless, the varactor's speed, high circuit Q at microwave frequencies, voltage variable capacity, small size, and infinitesimal bias power requirement (the diodes are always operated in reverse bias) make it ideally suited to various low power microwave phase control applications, such as:

1) Seradyne frequency modulation, in which the linear phase modulation of an RF signal with time produces discrete FM modulation sidebands.
2) A phase trimmer in the drive circuit of a high power RF amplifier to control high power output phase. This action is needed both where high power tubes are "paralleled" and in microwave linear particle accelerators in which output RF phase must be controlled to achieve specific charged particle acceleration.

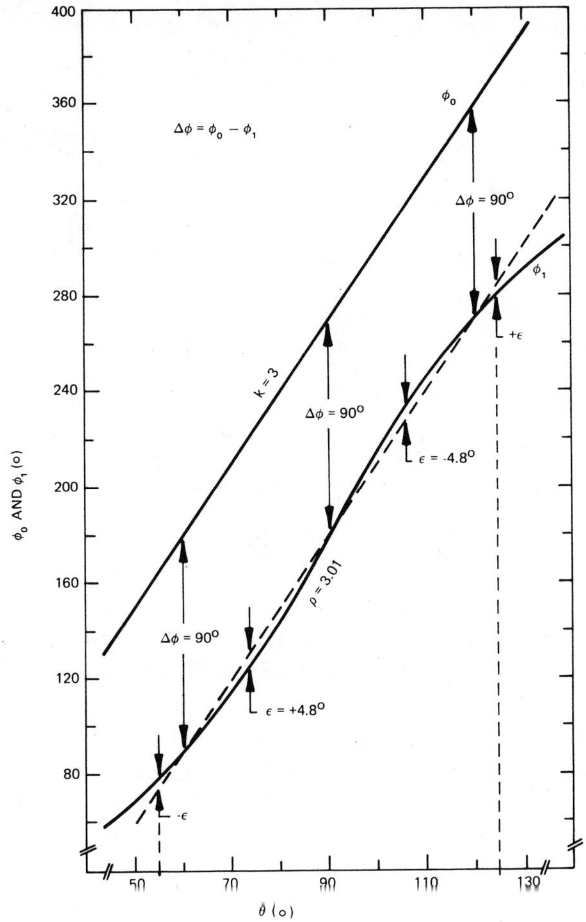

Figure IX-82 **An Octave Bandwidth Schiffman [22] Phase Shifter Formed by a ρ = 3.01 Coupled Line Pair θ Long and a Uniform Line Path 3θ Long, Phase Shift is 90° ±4.8° for $55° \leqslant \theta \leqslant 125°$, a 2.27:1 Frequency Ratio**

3) Microwave countermeasure circuits in which deceptive phase information is applied to received signals before they are retransmitted to their source.
4) A "line stretcher" in the reference arm of nulling circuit for a microwave bridge network.

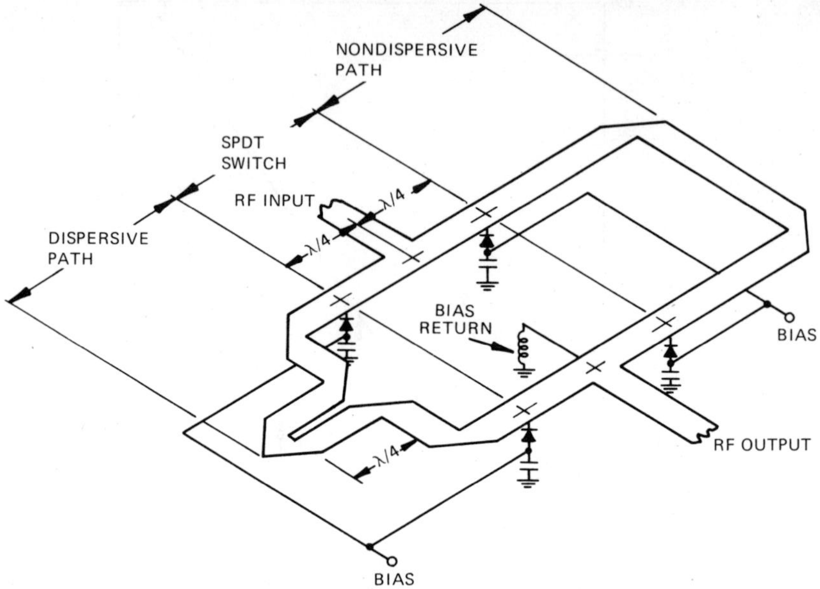

Figure IX-83 **A Stripline Schiffman Phase Shifter using Diode Path Switching (After Grauling and Geller [23])**

2. Figure of Merit

Usually the first question about the applicability of a varactor phase shifter is – "How much phase shift can be obtained per decibel of insertion loss [3]?" This question can be answered by considering the design of a varactor termination for a reflection phase shifter. Transmission circuits are not usually advantageous because of the limit on how much phase shift can be obtained with good VSWR. Thus, the reflection circuit shown in Figure IX-38 providing 0-180° or more of phase shift with only two diodes is almost always the best network choice.

Figure IX-84 shows the equivalent circuit of the diode chip embedded within the most general [24] two-port network realizable from lossless, reciprocal, linear passive ("brassware") circuitry. From the figure it is evident that, after specification of the diode, only the choice of the impedance level of operation (selected by the turns' ratio, n) and the series reactance, jX_P, provide meaningful design leverage; the line length, $\beta\ell$, serves only to add a fixed transmission path between the termination and the coupler.

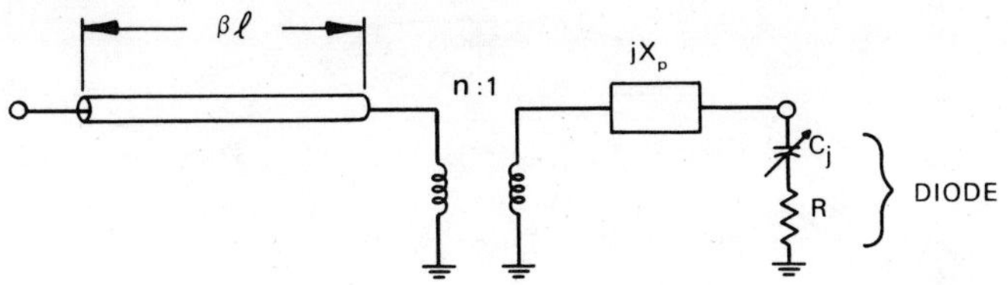

Figure IX-84 **General Equivalent Circuit of Varactor Diode and Phase Shifter Termination Circuit**

To formulate the diode selection it is desirable to put the diode reactance change and resistance into a convenient format. Thus, if we define

$$X_{Cj} = \frac{1}{2\pi fC} \qquad \text{(IX-71)}$$

$$M \equiv \frac{C_{MAX}}{C_{MIN}} \qquad \text{(IX-72)}$$

$$f_C \equiv \frac{1}{2\pi C_{MIN} R} \qquad \text{(IX-73)}$$

$$\Delta X = \frac{1}{\omega C_{MIN}} - \frac{1}{\omega C_{MAX}} \qquad \text{(IX-74)}$$

It follows that

$$\frac{\Delta X}{R} = \frac{f_C}{f}(1 - 1/M) \qquad \text{(IX-75)}$$

For any given varactor diode, the capacitance change factor, M, is usually equal to or greater than 5; thus, the quantity (1 – 1/M) approaches unity and the resulting $\Delta X/R$ is approximately equal to f_C/f. The designer who wishes to minimize loss should seek the highest ratio of f_C/f, paying only small heed to the C change (M) available which, it can be seen, only weakly affects the obtainable $\Delta X/R$ ratio once a value of M of 5 or more is achieved.

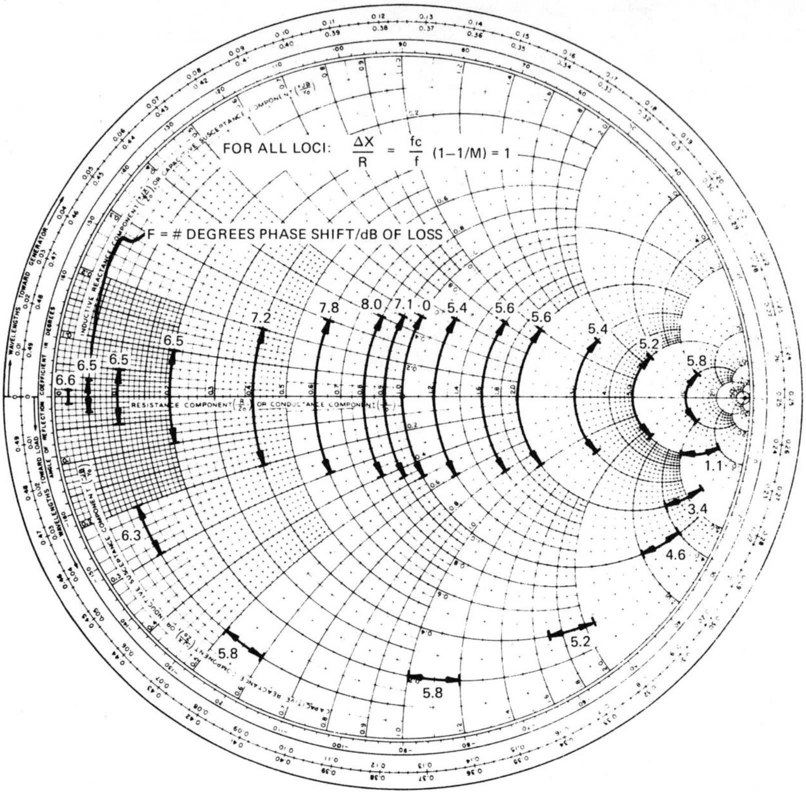

Figure IX-85 **Impedance Loci Traversed by Various Varactor Terminations Showing How F Varies with Location on the Smith Chart**

With the diode chosen, the only variables left are n and jX_P. To establish how these should be chosen most advantageously, consider the Smith Chart plot shown in Figure IX-85. The numerous loci drawn on this coordinate set have been determined in the following manner. All plots were made for the case in which the ratio $(\Delta X/R)\ (1 - 1/M)$ equals unity. Two distinct, general curve families can be seen in this plot.

First, consider the set of loci of termination impedance changes which are symmetric to the horizontal (zero reactance) line of the chart. In these cases, it is assumed that just enough series reactance jX_P has been used to make the resulting reactance change of the termination vary between positive and negative values of equal magnitude. The different loci of this set are obtained by varying n (changing the normalization factor) or, equivalently, the socket impedance of the termination. For each locus, the ratio was found of the number of degrees phase shift obtained (which is just equal to the change in reflection coefficient angle measured at the periphery of the chart) to the magnitude of the largest absorptive loss, $1 - |\Gamma|^2$, on the locus. The resulting ratio is a *loss figure of merit,* F, for the varactor phase shifter. In general, the insertion loss is not constant on a locus and the largest value (obtained at the point on the locus nearest to the center of the Smith Chart) is used in the specification of Γ. The second family of loci is obtained by adjusting both n and jX_P to produce impedance variation contours in the lower half of the Smith Chart (an exactly symmetric family could be plotted in the upper half of the chart, but the same F would be obtained).

It can be seen that the largest value for F is achieved near the center of the chart, at which F = 8; however, operation this close to the chart's center would produce unduly large absolute values of loss (10-20 dB). Therefore, although F has somewhat lower values near the chart's periphery — the absolute loss is less, and is equal to zero on the periphery itself. Of all the contours examined, those symmetrically disposed about the X = 0 axis of the Smith Chart with $R/Z_0 \ll 1$ give nearly the best figure of merit F found anywhere on the chart and, additionally, the lowest absolute value of insertion loss. From Figure IX-85, the value for F as R/Z_0 approaches zero is found graphically to be 6.6°/dB.

To express the phase shift per decibel of loss analytically for a general $\Delta X/R$, note that at $X = 0$ and $R/Z_0 \ll 1$

$$\text{Insertion Loss} = \frac{P_{\text{INCIDENT}}}{P_{\text{REFLECTED}}} = 20 \log_{10} \left(\frac{1 + R/Z_0}{1 - R/Z_0} \right)$$

$$\approx 17.4 \, R/Z_0 \quad \text{(decibels)} \qquad \text{(IX-76)}$$

Also, for small $\Delta X/Z_0$ excursions about zero, the phase shift magnitude, $|\Delta\phi|$, is

$$|\Delta\phi| = 2 \tan^{-1} \left[\frac{\Delta X/Z_0}{1 - \left(\dfrac{\Delta X/Z_0}{2} \right)^2} \right]$$

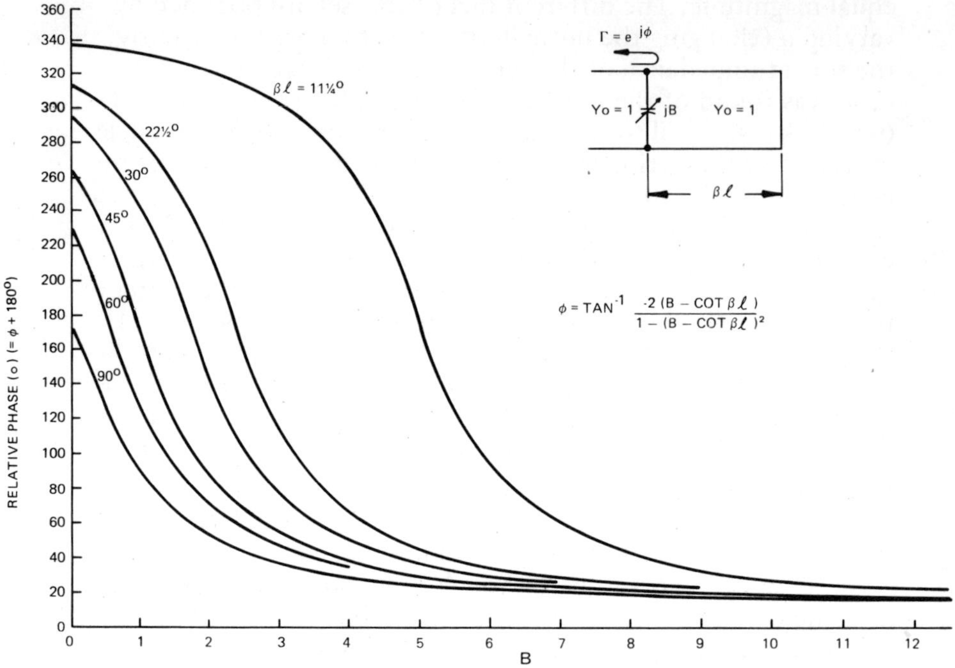

Figure IX-86 **Continuously Phase Variable, Single Varactor Termination**

$$\approx (2)\,(\Delta X/Z_0)\ \text{(radians)}$$
$$\approx (115°)\,(\Delta X/Z_0) \qquad \text{(IX-77)}$$

and the figure of merit, F, for the continuous phase shifter is defined as the number of degrees of phase shift per decibel of loss:

$$F = \frac{\Delta\phi\ \text{(degrees)}}{\text{Loss (decibels)}} = \frac{(115°)\,(\Delta X/Z_0)}{17.4\ R/Z_0} = 6.6\ \frac{\Delta X}{R}\ \text{(degrees/decibel)} \qquad \text{(IX-78)}$$

Finally, substituting the result from Equation (IX-75) gives

$$F = \left(\frac{f_C}{f}\right)\left(1 - \frac{1}{M}\right)\left(\frac{6.6°}{\text{decibel}}\right) \qquad \text{(IX-79)}$$

as the limit for the phase shift per decibel of loss in a continuous phase shifter in which the absolute loss is low. Although Equation

(IX-79) is derived under small phase shift approximations, well-designed practical phase shifters with up to 180° phase shift typically achieve about 50-75% of this ideal figure of merit performance.

3. Termination Designs and Loss Limitations

While the termination circuit described in Figure IX-84 is completely general, it does not lend itself to simple adjustment. A more versatile circuit, which gives nearly optimum performance in most cases, is shown in Figure IX-86. It consists of a varactor diode mounted directly across the transmission line and followed by a sliding short circuit, the distance of which from the varactor determines the phase shift. As may be seen from Figure IX-86, 360° of phase shift can be approached asymptotically by a diode the capacitive susceptance of which is variable from zero to infinity, followed by a short-circuit line length of practically zero length. This situation can be appreciated by examining the Smith Chart. When the diode is biased so that its susceptance is zero, the very short length of line produces an inductive susceptance of large magnitude and negative sign, and the reflection coefficient corresponds to that of a very small inductance. As the value of capacitive susceptance increases from zero, it adds to the negative susceptance value, causing the reflection coefficient to sweep clockwise along the periphery of the Smith Chart, eventually passing through zero, all positive susceptances, and, finally, returning to a short circuit. The net angular rotation approaches 360° in this manner.

In practice, however, continuous phase shift ranges greater than 180° produce high insertion loss, since at some point the diode capacitance and the shunt inductance of the shorted line transform the relatively small diode resistance to a value near Z_0, creating high absorption. Figure IX-87 illustrates this situation by an example. A varactor diode with only 1 Ω series resistance has a parallel equivalent circuit resistance of $R(1 + Q^2) = 50\ \Omega$ when the diode $Q = X_C/R = 7$. If a short stub having an input impedance near $+j7\ \Omega$ is used, the net shunt reactance is nearly infinite (parallel resonance) and nearly total absorption (infinite insertion loss) occurs. Loss peaks in the loss versus phase shift characteristic generally occur at the parallel resonance of the stub and the shunt diode, their magnitudes depending on the ratio of transformed resistance to the coupler characteristic impedance, $[R(1 + Q^2)/Z_0]$. This loss is not unique to the termination circuit in Figure IX-87.

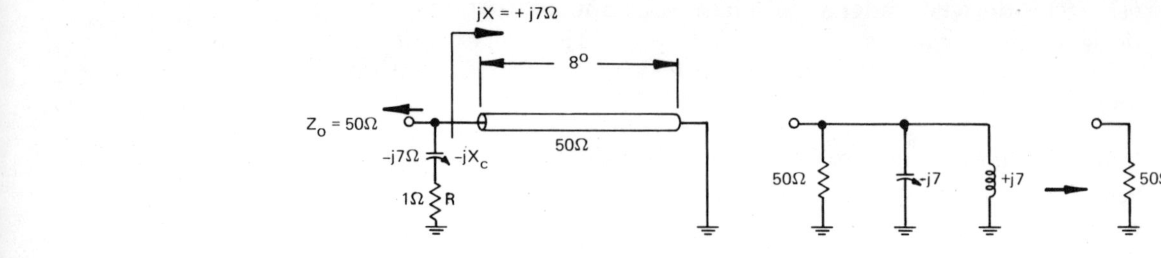

a) CONTINUOUS TERMINATION NEAR RESONANCE b) PARALLEL EQUIVALENT CIRCUIT

Figure IX-87 **Termination Resonance Condition Demonstrating How a 1Ω Diode Resistance is Transformed to 50Ω, Resulting in Total Power Absorption**

A similar peak occurs in the general circuit in Figure IX-84, at the series resonance when $jX_P = -jX_C$, the peak loss magnitude depending upon the ratio N^2R/Z_0. Moreover, a large value of N is usually needed as phase shift greater than 180° is sought.

4. Frequency Variability

Figure IX-88 shows a phase shifter termination; Figure IX-89 is its admittance versus reverse bias contour on the Smith Chart. Figure IX-90 shows the measured and calculated phase shift and loss. For the diode used, C_{MAX} = 10 pF, C_{MIN} = 2 pF, and R = 2.6 Ω, giving f_C = 31 GHz. From Equation (IX-79), at f = 1 GHz, F = 165°/dB. The actual results, from Figure IX-89, give 240° with 1.8 dB maximum loss (or 133°/dB), and correspond to better than 80% of the ideal performance estimated for the small phase shift circuits.

As frequency is varied, both X_C and the line length, θ, change; the net result is a reduction of both phase shift and peak loss, as shown in Figure IX-91. Special circuitry to reduce frequency variability is sometimes practical. Phase shift variation over a 20% bandwidth centered at 2 GHz was reduced to ±20° out of 360° by Henoch and Tamon using a series resonant circuit in the termination [25].

For some applications, such as the seradyne frequency translator, it is desirable to have a linear phase shift versus applied voltage characteristic. Garver has shown that a 360° phase shift range can be linearized within ±3% using a two varactor termination circuit in which the varactor C(V) law is matched to a portion of the tangent function reactance characteristic. This function is suggested for linear phase shift because it matches the reactance at the input of a transmission line with a sliding short circuit termination [26].

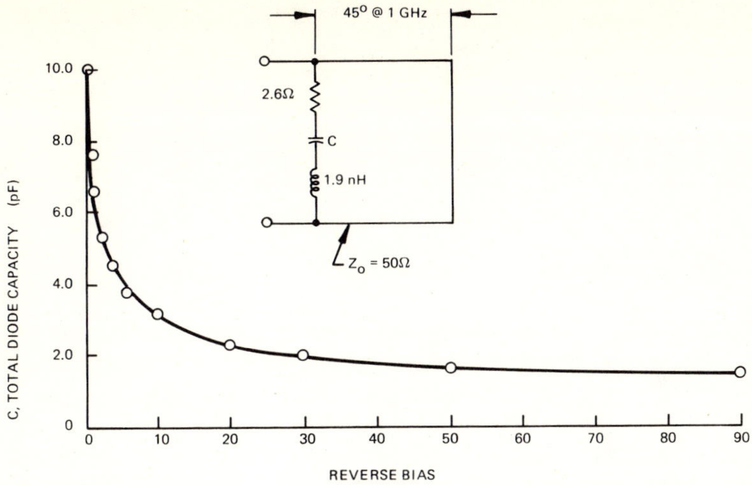

Figure IX-88 **Varactor Termination Parameters**

Garver has also shown that resistance can be added to the termination to obtain nearly constant, but a somewhat higher minimum, insertion loss for all phase shift values.

5. *Temperature Variability*

The varactor diode capacitance changes very little with temperature; consequently, the phase shift is likewise only slightly temperature dependent. In Figure IX-92, the phase shift variations shown is the total due to both diode capacitance change and circuit dimensional changes.

6. *Power Variations*

As the RF power level increases, the RF waveform makes excursions into the forward bias region of the varactor (when the applied bias is near zero) and into the breakdown region (with applied bias near breakdown). The result is that harmonic signals are generated, loss increases, and the phase shift observed at the fundamental RF frequency changes. With the circuitry discussed, designed about $Z_0 = 50\ \Omega$, such nonlinear power behavior is experienced near zero bias at 10-100 mW of RF power; at a watt or more, a pronounced change in the phase shift-bias characteristic occurs, as shown in Figure IX-93.

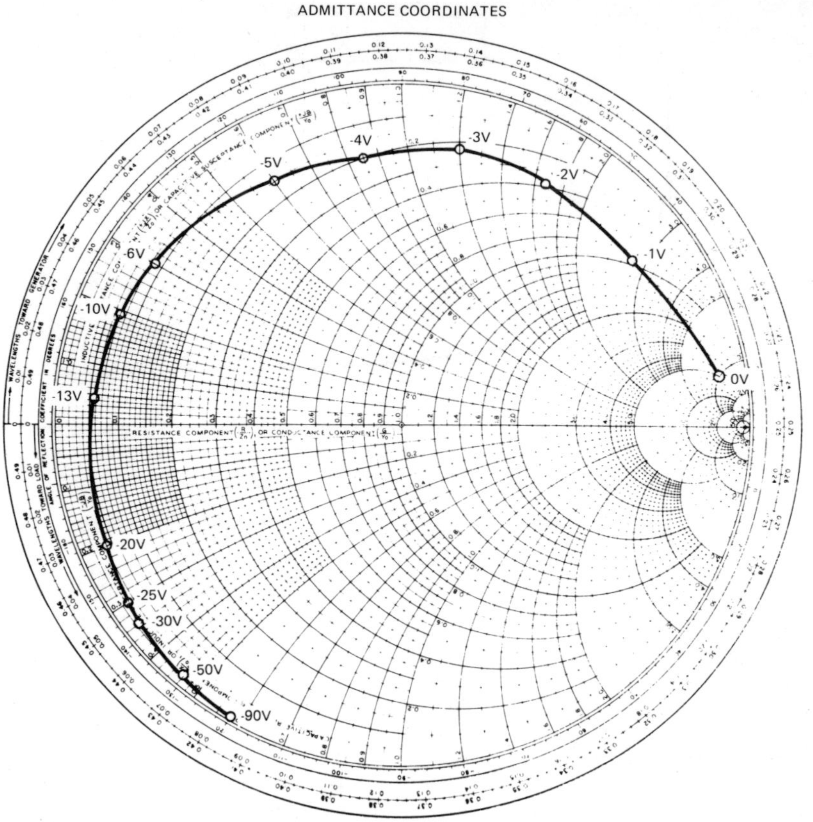

Figure IX-89 **Phase Shift and Loss of Termination Circuit in Figure IX-87**

Notice that the phase shift range increases with increasing RF power as the RF excursions into the forward bias diode state produce a larger effective capacitance than is obtained at low power with zero bias. Despite the fact that phase shift increases, such phase shift variability with power is usually undesirable, and the RF power level must be limited to a few milliwatts for linear operation (with power).

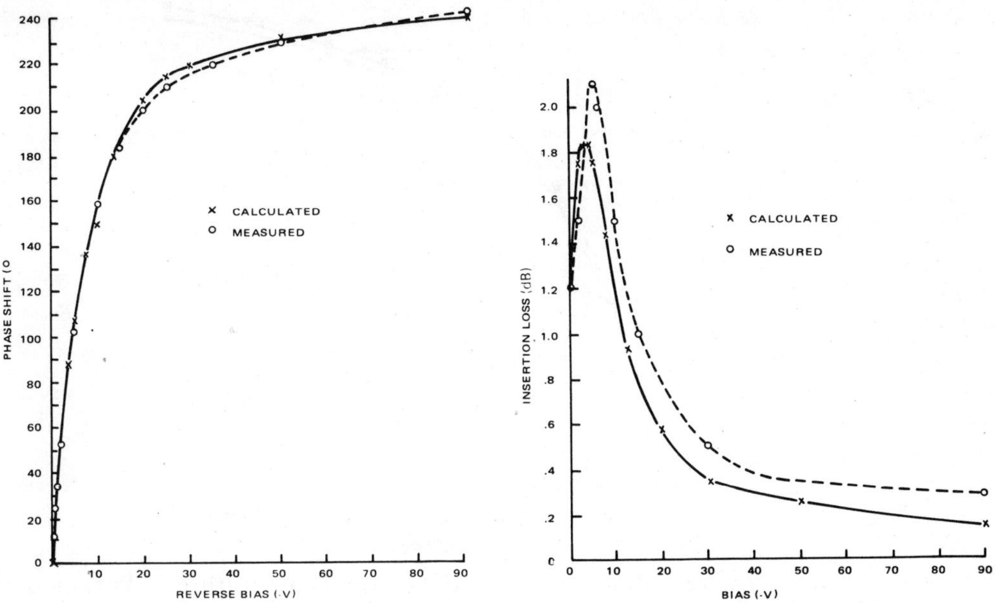

Figure IX-90 Calculated Admittance at 1 GHz for Termination in Figure IX-5

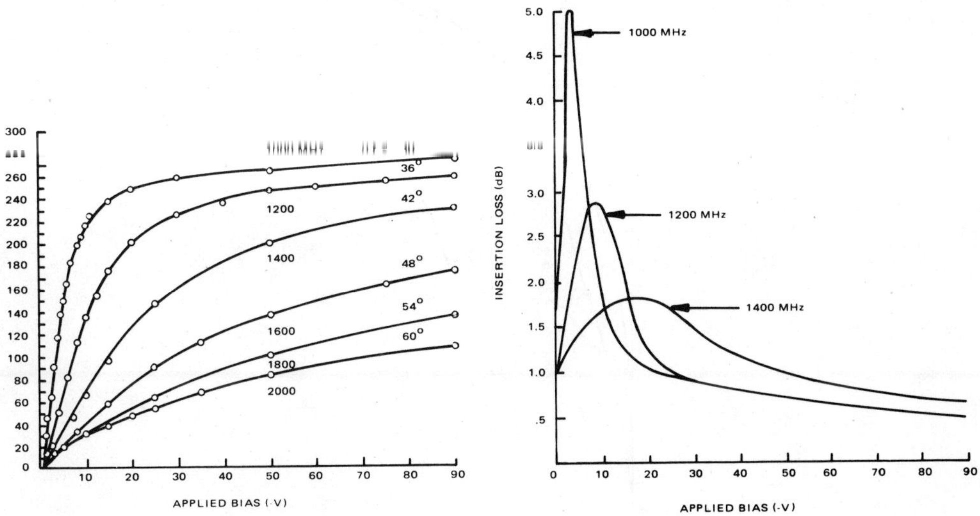

Figure IX-91 Typical Phase Shift vs. Frequency of Continuous Phase Shifter

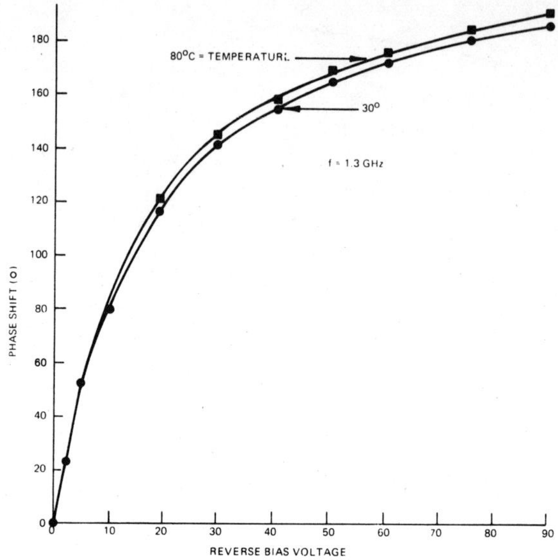

Figure IX-92 **Phase Shift vs. Bias with Temperature as Parameter**

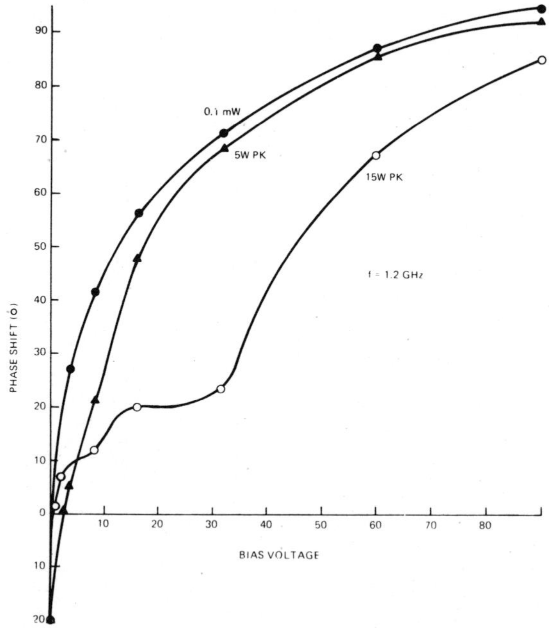

Figure IX-93 **Typical Variation of Phase Shift with RF Input Power**

Acknowledgments

The phase shifter theory, techniques, and results described in this chapter were collected over a fifteen year period at Microwave Associates. Many friends within and outside M/A contributed, and it is unlikely all will be included no matter how carefully prepared an acknowledgment is made. I would like to thank especially Charles Buntschuh, Victor Camilleri, Carlos Dantas, Harry Esterly, David Fryklund, Dan Gallagher, George Garas, Carmen Genzabella, John Gottlander, Henry Griffin, William Hoelzer, Vincent Mistretta, Vern Philbrook, Gordon Simpson, Charles Ward, and Richard Ziller, who helped design and build the phase shifters. For helpful discussions I am indebted to Frank Brand, Norman Brown, Seymour Cohn, Dan Conlon, Gerald DiPiazza, Don Temme, Ronald Gutmann, Marion Hines, Peter Ledger, Kenneth Mortenson, Frank O'Hara, William Rushforth, Bob Smith, Robert Tennenholtz, and Richard Walker.

References

[1] Kahrilas, Peter J.: *Electronic Scanning Radar Systems (ESRS) Design Handbook,* Artech House, Inc., Dedham, Massachusetts, 1976.

[2] Lynes, Guy D.; Johnson, Gerald E.; Huckleberry, B.E.; and Forrest, Neil H.: "Design of a Broad-Band Four-Bit Loaded Switched-Line Phase Shifter," *IEEE Transactions on Microwave Theory and Techniques, Vol. MTT-22, No. 6,* June 1974, pp. 693-697.

[3] White, J.F.: *Semiconductor Microwave Phase Control,* Northeast Research and Engineering Meeting (NEREM), 5 November 1963, pp. 106-107.

[4] White, J.F.: *A Diode Phase Shifter for Array Antennas,* IEEE International Symposium, 21 May 1964.

[5] White, J.F.: "High Power PIN Diode Controlled, Microwave Transmission Phase Shifter," *IEEE Transactions on Microwave Theory and Techniques, Vol. MTT-13, No. 2,* March 1965, pp. 233-242.

[6] White, J.F.: "Phase Shifter Having Means to Simultaneously Switch First and Second Reactive Means Between a State of Capacitive and Inductive Reactance," *US Patent #2, 491,314; United Kingdom Patent #1,101,843;* and *French Patent #1,476,448.*

[7] Dawirs, H.N.; and Swarner, W.G.: "A Very Fast Voltage Controlled Microwave Phase Shifter," *Microwave Journal, Vol. 5,* June 1962, pp. 99-106.

[8] Young, Leo: Tables for Cascaded Homogeneous Quarter-Wave Transformers, *IRE Transactions on Microwave Theory and Techniques, Vol. MTT-7,* April 1964, pp. 233-237. (Also see Reference 8 below)

[9] Matthei, G.L.; Young, Leo; and Jones, E.M.T.: *Microwave Filters, Impedance Matching Networks, and Coupling Structures,* McGraw-Hill, Inc., New York, 1964.

[10] Garver, Robert V.: "Broadband Diode Phase Shifters," *1971 IEEE-GMTI International Symposium Digest,* pp. 178-179. Also published as a full paper with the same title in *IEEE Transactions on Microwave Theory and Techniques, Vol. MTT-20, No. 5,* May 1972, pp. 314-323.

[11] Onno, Peter; and Plilkins, Andrew: "Miniature Multi-Kilowatt PIN Diode MIC Digital Phase Shifters," *1971 IEEE-GMTI International Symposium Digest,* pp. 22-23.

[12] Hardin, R.H.; Downey, E.J.; and Munushian, J.: "Electronically Variable Phase Shifter Utilizing Variable Capacitance Diodes" (Letters), *Proceedings of the IRE, Vol. 48,* May 1960, pp. 944-945.

[13] White, J.F.: "Review of Microwave Phase Shifters" (Inivted Paper), *Proceedings of the IEEE, Vol. 56, No. 11,* November 1968, pp. 1924-1931.

[14] White, J.F.: "Diode Phase Shifters for Array Antennas" (Invited Paper), *IEEE Transactions on Microwave Theory and Techniques, Vol. MTT-22, No. 6,* June 1974, pp. 658-674.

[15] Barber, M.R.; Sodomsky, K.F.; and Zacharias, A.: "Microwave Switches, Limiters, and Phase Shifters," Chapter 10 of *Microwave Semiconductor Devices and Their Circuit Applications* (H.A. Watson, ed.), McGraw-Hill, Inc., New York, 1969.

[16] Garver, R.V.; Bergfried, D.; Raff, S.; and Weinschel, B.O.: "Errors in S_{11} Measurements Due to Residual SWR of the Measurement Equipment," *IEEE Transactions on Microwave Theory and Techniques, Vol. MTT-20, No. 2,* January 1972, pp. 61-69. Also covered in Garver, R.V.: *Microwave Diode Control Devices* (Chapter 10), Artech House, Inc., Dedham, Massachusetts, 1976.

[17] Burns, R.W.; and Stark, L.: "PIN Diodes Advance High-Power Phase Shifting," *Microwaves, Vol. 11,* November 1965, pp. 38-48.

[18] Stark, L.: "Microwave Theory of Phased-Array Antennas – A Review," *Proceedings of the IEEE, Vol. 62, No. 12,* December 1974, pp. 1661-1701. *This is a particularly comprehensive and clearly written paper about phased array antennas and their components (including ferrite and diode phase shifters).*

[19] White, J.F.; Genzabella, C.; Fryklund, D.; and Ziller, R.: "PIN Diode and Circuit Performance for 15,000 High Power L-Band Phase Shifters," *European Microwave Conference Digest,* September 1975.

[20] Cox, R.M.; and Sebring, J.R.: "MLS-A Practical Application of Microwave Technology," *IEEE MTT Symposium Digest,* Session 19, 19 June 1976.

[21] Jones, E.M.T.; and Bolljahn, J.T.: "Coupled-Strip-Transmission-Line Filters and Directional Couplers," *IRE Transactions on Microwave Theory and Techniques, Vol. MTT-4,* April 1956, pp. 75-81.

[22] Schiffman, B.M.: "A New Class of Broadband Microwave 90° Phase Shifters," *IRE Transactions on Microwave Theory and Techniques, Vol. MTT-6,* April 1958, pp. 232-237.

[23] Grauling, C.H., Jr.; and Geller, B.D.: "A Broadband Frequency Translator with 30 dB Suppression of Spurious Sidebands," *IEEE Transactions on Microwave Theory and Techniques, Vol. MTT-18, No. 9,* September 1970, pp. 651-652.

[24] Montgomery, C.G.; Dicke, R.H.; and Purcell, E.M.: *Principles of Microwave Circuits,* Vol. 8 of MIT Radiation Laboratory Series, Cambridge, Massachusetts. For general two-port network analysis, see pp. 104-110.

[25] Henoch, Bengt T.; and Tamm, Peter: "A 360° Reflection-Type Diode Phase Modulator," *IEEE Transactions on Microwave Theory and Techniques, Vol. MTT-19, No. 1,* January 1971, pp. 103-104.

[26] Garver, R.V.: "360° Varactor Linear Phase Modulator," *IEEE Transactions on Microwave Theory and Techniques, Vol. MTT-17, No. 3,* March 1969, pp. 137-147.

Answers

Chapter I

1. See derivation in Chapter II, Equations (II-23) to (II-25).
2. About 0.8 Ω-cm.
3. Because each acceptor captures a donated electron, and no free carriers exist.
4. With acceptor backgrounds of 0, 10, 10^2, 10^3, 10^4, and 10^5. The approximate total impurity concentrations are 10^{14}, 10^{15}, 10^{16}, 10^{17}, 10^{18}, and 10^{19}, and the corresponding $\mu_N \approx 1450$, 1400, 1100, 600, 400, and 200 cm^2/V-s giving resistivities with 10^{14} net electrons/cm^3 of approximately 430, 450, 570, 1050, 1560, and 3125 Ω-cm.
5. a) (R = 0) V = 0.54, 0.64, 0.66 V @ I = 0.1, 1.0, 10A
   P = 0.05, 0.64, 6.6 W @ I = 0.1, 1.0, 10A

   b) (R = 0.1 Ω) V = 0.54, 0.74, 1.66 V @ I = 0.1, 1.0, 10A
   P = 0.05, 0.74, 16.6 W @ I = 0.1, 1.0, 10A

Chapter II

1. About 20 V (from Figure II-5).
2. About 500 MHz (from Figure II-4).
3. About 1 μs (from Figure II-9).
4. 0.2 Ω (See sample calculation in Equation (II-35)).
5. $R_I \propto 1/Q \quad Q = Q_o e^{-t/\tau}$
   @ 1 μs, $Q/Q_o = 0.61$, R = 0.2 Ω/0.61 = 0.33 Ω
   @ 10 μs, $Q/Q_o = 0.00674$, R = 29.7 Ω
   @ 15 μs, $Q/Q_o = 5.53 \times 10^{-4}$, R = 361 Ω
6. Isolation $= 10 \log_{10} (1 + Z_o/(R_C + R_I))^2$
   35.9 dB, 33.5 dB, 5.3 dB, 0.57 dB @ 0, 1, 10, 15 μs

Chapter III

1. $\Delta T = P_D \theta (1 - e^{-t/\tau_T})$

$$\approx P_D \theta \left[\frac{t}{\tau_T} - \frac{1}{2}\left(\frac{t}{\tau_T}\right)^2 + \ldots \right]$$

   Error is less than first neglected term, hence $E < (50\%)\, P_D \theta t/\tau_T$
2. $HC = 200\ \mu J/^\circ C$, $\tau_T = 1000\ \mu s$
3. $P_D = (\Delta T/\theta)/(1 - e^{-t/\tau_T})$ = 32 W for 1 ms = 210 W for 100 μs
   = 20 W for CW = 51 W for 500 μs = 2010 W for 10 μs
   = 23 W for 2 ms = 110 W for 200 μs = 20010 W for 1 μs

4. τ_T = 5 ms

P_D = 4 W for CW = 102 W for 200 μs
= 12 W for 2 ms = 202 W for 100 μs
= 22 W for 1 ms = 2002 W for 10 μs
= 42 W for 500 μs = 20002 W for 1 μs

5. Long pulses are most affected by an increase in θ. No, because ΔT increases linearly with θ, and decreases more slowly with t/τ_T.

Chapter IV

1. $q = Q_0 e^{-t/\tau} + I_D t$

where $Q_o = I_F \tau$, $I_D = -I_R$, and, when switching occurs, $t = T_S$, $q = 0$; thus

$0 = I_F \tau e^{-T_S/\tau} - I_R T_S$ $\qquad T_S = \left(\frac{\tau I_F}{I_R}\right) e^{-T_S/\tau}$ q.e.d.

2. Rewrite the answer to Question 1 as

$\frac{T_S}{\tau} = 2.5\, e^{(-T_S/\tau)}$ since $\frac{I_F}{I_R} = 2.5$

By trial and error evaluate

T_S/τ	1.0	1.1	0.95	0.96
$2.5\, e^{-T_S/\tau}$	0.92	1.02	0.97	0.96

thus $T_S = 0.96\ \tau = 0.96\ \mu$s.

3. $T_S/\tau = 0.57$

Chapter V

1. f_{CS} = 300 GHz.

2.

t	P_D(W)	I_M (A rms)	V_M (V rms)
CW	20	7.1	838
100 μs	210	22.9	2,718
10 μs	2,010	70.1	8,409
1 μs	20,010	224	26,532

3. V_M (V peak) = (1200 − 100) V/2 = 550 V peak = 389 V rms

4. $P_M = V_M I_M/2$, $Z_0 = V_M^2/P_M$

t	V_M (V rms)	I_M (A rms)	P_M(W)	$Z_0(\Omega)$
CW	389	7.1	1,381	110
100 μs	389	22.9	4,454	34
10 μs	389	70.1	13,634	11
1 μs	389	224.	43,568	3.5

5. $P_M = 7566$ W, $Z_0 = 880\ \Omega$
6. $P_M = 2250$ kW, $Z_0 = 880\ \Omega$
7. $f_{CS} \approx 50$ GHz, $P_M \approx 62$ kW, $Z_0 \approx 50\ \Omega$

Chapter VI

1. For $Z_L = 0$, $Z_{IN} = Z_C \tanh(\gamma\ell)$

$\gamma\ell = \alpha\ell + j\beta\ell$

For $\ell = \lambda/4$, $\gamma\ell = \alpha\ell + j90^\circ$

$$Z_{IN} = Z_C \left[\frac{e^{\alpha\ell}e^{j90^\circ} - e^{-\alpha\ell}e^{-j90^\circ}}{e^{\alpha\ell}e^{j90^\circ} + e^{\alpha\ell}e^{-j90^\circ}}\right] = Z_C \left[\frac{e^{\alpha\ell} + e^{-\alpha\ell}}{e^{\alpha\ell} - e^{-\alpha\ell}}\right]$$

Since $e^{\alpha\ell} \approx 1 + \alpha\ell$, $Z_{IN} \approx \dfrac{Z_C}{\alpha\ell}$

Converting α in nepers to decibels,

$$Z_{IN} \approx \frac{Z_C}{[\alpha(\text{dB}/\lambda)/8.68588] \cdot (\lambda/4)} \approx 34.7\ Z_C/\alpha\ (\text{dB}/\lambda)$$

q.e.d.

2. $$Z_{IN} = Z_T \frac{Z_L/Z_T + j\tan\theta}{1 + j(Z_L/Z_T)\tan\theta} \rightarrow Z_T \cdot \frac{1}{(Z_L/Z_T)} \quad \text{as } \tan\theta \rightarrow \infty$$

$$Z_{IN} = \frac{Z_T}{RZ_0/Z_T} = Z_0 \text{ if } Z_T^2 = RZ_0^2\text{; i.e., if } Z_T = \sqrt{R}\ Z_0$$

3. Insertion Loss (dB) $\approx 10 \log_{10}(1 + GZ_0)$

$$= 10 \log_{10}\left[1 + \frac{Z_0\,\alpha}{34.7\ Z_C}\right]$$

but $\log_{10}(1 + x) \approx 0.434x \qquad (x \ll 1)$

Thus $\text{Loss(dB)} \approx 0.0125 \dfrac{Z_0}{Z_C}\,\alpha\ (\text{dB}/\lambda)$

For $Z_C = Z_0$, $\alpha = 1$ dB/λ, Loss ≈ 0.01 dB.

Appendices

A. Constants and Formulas

Constants

Symbol	Description	Value
π	Pi	3.1415926
ϵ_0	Free Space Permittivity	0.08854186 picofarad/centimeter $\approx 10^{-9}/36\pi$ farad/meter
ϵ_R	Relative Dielectric Constant of air (at sea level, 23°C, 50% RH)	1.00065
μ_0	Free Space Permeability	$4\pi \times 10^{-7}$ henry/meter (exactly) 12.566370 nanohenries/centimeter
$\sqrt{\mu_0/\epsilon_0}$	Free Space Wave Impedance	376.7303 ohms $\approx 120\pi$ ohms
	Radian	$360/2\pi$ degrees (exactly) 57.295780 degrees
e	Charge of an electron	1.60206×10^{-19} coulomb
m	Mass of an electron	9.1083×10^{-34} gram
c	Velocity of light in free space	2.99793×10^{10} centimeters/second
k	Boltzmann's Constant	1.38044×10^{-23} joule/Kelvin 8.63×10^{-5} electron-volt/Kelvin
e/kT	Diode Law exponent coefficient at room temperature	38.6 volts^{-1} at T = 300 Kelvin
h	Planck's Constant	6.62517×10^{-34} joule-seconds
J	joule	1 watt for 1 second (watt-second) 10^7 ergs
e	Base of Natural Logarithm	2.7182818
Base Changes	logarithm (base e) =	2.3025850 log (base 10)
	logarithm (base 10) =	0.43429448 ln (base e)

Formulas and Conversions

Reactances of L (nH) and C (pF) @ f (GHz)

+j6.2831853 ohms → 1 nanohenry @ 1 gigahertz

$-j159.15494$ ohms → 1 picofarad @ 1 gigahertz

@ f

$jX_L = +j6.2831853 \cdot f \cdot L$ ohms

$-jX_C = -j159.15494/(f \cdot C)$ ohms

Parallel Plate Capacitor (area, A, and spacing, d)

$C = \epsilon_0 \epsilon_R A/d$

$= \epsilon_R (0.0885)\ A/d$ picofarads/centimeter

$= \epsilon_R (0.225)\ A/d$ picofarads/inch

Photon Energy

1 electron-volt = 1.602×10^{-19} joule

= 1.602×10^{-12} erg

E = hf = Energy of 1 photon of electromagnetic-radiation at frequency f

= $(4.136 \times 10^{-6}$ electron-volt) f (gigahertz)

Metric and English Equivalences

1 inch = 2.540005 centimeters

1 centimeter = 0.39370 inch

1 square inch = 6.4516254 square centimeters

1 square centimeter = 0.15499970 square inch

1 cubic centimeter = 0.061023 cubic inch

1 cubic inch = 16.387162 cubic centimeters

$1\ \mu$ = 1 micron = 10^{-4} centimeter *exactly*

1 Å = 1 angstrom = $10^{-4} \mu = 10^{-8}$ centimeter *exactly*

1 mil = 10^{-3} inch = 25.4 microns

1 meter = 39.3700 inches

1 kilogram = 1000 grams = 2.2046223 pounds

1 gram = 0.0352739 ounces

1 ounce (avoirdupois) = 28.349527 grams

1 ounce (apothecary or Troy) = 31.103481 grams

1 pound = 453.59243 grams

1 liter = 1000 cubic centimeters *exactly*

1 liter = 1.056710 quarts (US)

1 quart (US) = 0.9463334 liter

1 (gram) calorie = 4.1868 joules

1 British Thermal Unit (BTU) = 0.2520 kilogram-calories

1 horsepower = 550 foot-pounds/second = 746 watts

1 kilogram/square millimeter = 1,422.3397 pounds/square inch

Wavelength (free space)

$\lambda = c/f$

$$= \frac{29.9793 \text{ centimeters}}{f \text{ (gigahertz)}} \approx \frac{30 \text{ centimeters}}{f \text{ (gigahertz)}}$$

$$= \frac{11.80285 \text{ inches}}{f \text{ (gigahertz)}} \approx \frac{11.8 \text{ inches}}{f \text{ (gigahertz)}}$$

Propagation Loss

$$\alpha \text{ (nepers)} = \frac{\text{Insertion Loss (decibels)}}{8.685889638}$$

Loss (decibels) = 0.11512925 · Loss (nepers)

Temperature Conversions

K = °C + 273

°F = 1.8(°C) + 32

°C = (°F − 32)5/9

K	*°C*	*°F*	*K*	*°C*	*°F*
213	-60	-76	300	+23	+73.4
223	-50	-58	303	+30	+86
233	-40	-40 *exactly*	313	+40	+104
243	-30	-22	323	+50	+122
253	-20	-4	333	+60	+140
263	-10	+14	343	+70	+158
273	0	+32 *exactly*	353	+80	+176
283	+10	+50	363	+90	+194
293	+20	+68	373	+100	+212 *exactly*

Common Mathematical Equivalences [20, 21]

1. $$\frac{1}{1-x} = 1 + x + x^2 + x^3 + \ldots \infty \quad (x < 1)$$

2. $$\frac{1}{(1-x)^2} = 1 + 2x + 3x^2 + 4x^3 + \ldots \infty \quad (x < 1)$$

3. $$e^x = 1 + x + \frac{x^2}{2!} + \frac{x^3}{3!} + \ldots \infty$$

4. $$\ln(1 + x) = x - \frac{x^2}{2} + \frac{x^3}{3} - \frac{x^4}{4} + \ldots \infty \quad (x < 1)$$

5. $$\log_{10}(1 + x) = k\left[x - \frac{x^2}{2} + \frac{x^3}{3} - \frac{x^4}{4} + \ldots \infty\right] \quad (x < 1)$$

   where $k = \log_{10}e = 0.43429448 \ldots$

6. $$\sin x = x - \frac{x^3}{3!} + \frac{x^5}{5!} - \frac{x^7}{7!} + \ldots \infty \quad (x < \infty)$$

7. $$\cos x = 1 - \frac{x^2}{2!} + \frac{x^4}{4!} - \frac{x^6}{6!} + \ldots \infty \quad (x < \infty)$$

8. $$\tan^{-1} x = x - \frac{x^3}{3} + \frac{x^5}{5} - \frac{x^7}{7} + \ldots \infty$$

9. $$\sinh x = x + \frac{x^3}{3!} + \frac{x^5}{5!} + \frac{x^7}{7!} + \ldots \infty \quad (x^2 < \infty)$$

10. $$\cosh x = 1 + \frac{x^2}{2!} + \frac{x^4}{4!} + \frac{x^6}{6!} + \ldots \infty \quad (x^2 < \infty)$$

Maclaurin's Theorem (Expansion about x = 0)

11. $f(x) = f(0) + x \frac{df(0)}{dx} + \frac{x^2}{2!} \frac{d^2f(0)}{dx^2} + \ldots \frac{x^n}{n!} \frac{d^nf(0)}{dx^n} + \ldots \infty$

Taylor's Theorem (Expansion about x = a)

12. $f(x) = f(a) + (x-a) \frac{df(a)}{d^x} + (x-a)^2 \frac{d^2 f(a)}{dx^2} + (x-a)^n \frac{d^nf(a)}{dx^n} + \ldots \infty$

Quadratic Formula

13. $ax^2 + bx + c = 0; \quad x = \frac{-b \pm \sqrt{b^2 - 4ac}}{2a}$

B. Properties of Materials******* [2-10, 27]

Symbol	Name	Density @20°C (gm/cm³)	Melting Point (°C)	Relative Hardness	Specific Heat (cal/gm-°C)	Thermal Resistivity (°C-cm/W)
	Air (still)					4200
Al	Aluminum	2.70	660	2.9	0.226	0.46
Al_2O_3	Alumina (aluminum oxide)					5.41
Sb	Antimony	6.62	630	3	0.049	5.26
Be	Beryllium	1.82	1878	3	0.425	0.61
BeO	Beryllia (99.5%)					0.43
BeO	Beryllia (100%)		2550			1.2
B	Boron	2.3	2300	9.5	0.307	
BN	Boron Nitride	2.20	3000			3.15
	Brass (66 Cu, 34 Zn)					0.8-1.0
C	Carbon (diamond)	2.22	>3500	10	0.165	0.15
C	Carbon (graphite)					4-15
Cu	Copper (annealed)	8.96	1083	3	0.0921	0.25
	Epoxy (high conductivity)					60
	Ferrite	4-5****				16
Ga	Gallium	5.91	29.8	1.5	0.079	
Ge	Germanium	5.36	937.4	6.2	0.073	1.7
	Glass (crown-window)					95
	Glass (iron-sealing)					88.4
Au	Gold	19.3	1063	2.5	0.031	0.34
He	Helium	0.1664*	-268.94**		1.25	719
H	Hydrogen	0.08375*	-252.8**	—	3.415	588
In	Indium	7.31	156	1.2	0.057	4.2
Fe	Iron	7.87	1535	4	0.108	1.3
PB	Lead	11.34	327.4	1.5	0.030	2.9
—	Lucite™ (Plexiglas™)					
—	Kovar™ (29 NI, 17 Co, 0.3 Mn, 53.7 Fe)					5.94
Mg	Magnesium	1.74	651	2	0.249	0.65
Hg	Mercury	13.55	-38.9		0.033	12
—	Mica (ruby)					130-210
Mo	Molybdenum	10.2	2610	6	0.065	0.68
—	Mylar™					1040
—	Nichrome (65 Ni, 12 Cr, 23 Fe)					7.62
Ni	Nickel	8.9	1453	5	0.112	1
N	Nitrogen	1.1649*	-195.8**	—	0.247	—
—	Nylon™					482
—	Oil (transil)					
—	Paper (royal grey)					800
Pd	Palladium	12	1552	4.8	0.059	1.4

Linear Thermal Expansion ($\times 10^{-6}/^{\circ}C$)	Elasticity Modulus (kg/mm^2)	Tensile Strength (kg/mm^2)	Dielectric Constant 1 MHz	Dielectric Constant 3 GHz	Electrical Resistivity (Ω-cm)	Symbol	Name
			1.0	1.0			Air (still)
22.9	7250	6.3	—	—	2.1×10^{-5}	Al	Aluminum
			9.8	9.8	—	Al_2O_3	Alumina (aluminum oxide)
8.5-10.8	7900	1.05	—	—	3.9×10^{-5}	Sb	Antimony
	3200					Be	Beryllium
6.0			6.4		—	BeO	Beryllia (99.5%)
7.5			7.1		—	BeO	Beryllia (100%)
2			—	—	—	B	Boron
					—	BN	Boron Nitride
					3.9×10^{-6}		Brass (66 Cu, 34 Zn)
	500		5.5 @ 100 MHz		—	C	Carbon (diamond)
			—	—	2.0×10^{-5}	C	Carbon (graphite)
16.5	11000	22.5	—	—	1.72×10^{-6}	Cu	Copper (annealed)
			—	—			Epoxy (high conductivity)
			—	10-11***			Ferrite
18			—	—		Ga	Gallium
6.1			—	16	—	Ge	Germanium
			—	5	—		Glass (crown-window)
			—	8	—		Glass (iron-sealing)
14.2	7300	11.5	—	—	2.4×10^{-6}	Au	Gold
			1.0	1.0	—	He	Helium
—			1.0	1.0	—	H	Hydrogen
33		0.3	—	—	9×10^{-6}	In	Indium
11.7	20000	20.5	—	—	9.7×10^{-6}	Fe	Iron
28.7	1800	1.33	—	—	2.2×10^{-5}	PB	Lead
					—	—	Lucite™ (Plexiglas™)
			—	—	$(4.5\text{-}8.5) \times 10^{-5}$	—	Kovar™ (29 Ni, 17 Co, 0.3 Mn, 53.7 Fe)
25.2	4600	9.15	—	—	4.5×10^{-6}	Mg	Magnesium
			—	—	9.6×10^{-5}	Hg	Mercury
			5.4	5.4	—	—	Mica (ruby)
4.9	35000	120.0	—	—	4.8×10^{-6}	Mo	Molybdenum
			3			—	Mylar™
			—	—	1×10^{-4}	—	Nichrome (65 Ni, 12 Cr, 23 Fe)
13.3	21000	32.3			6.9×10^{-6}	Ni	Nickel
—	—	—	1.0	1.0		N	Nitrogen
			3.14	2.84	—	—	Nylon™
			2.2	2.2		—	Oil (transil)
			3.0	2.7	—	—	Paper (royal grey)
11.8	12000	14.0	—	—	1.1×10^{-5}	Pd	Palladium

Symbol	Name	Density @20°C (gm/cm^3)	Melting Point (°C)	Relative Hardness	Specific Heat (cal/gm-°C)	Thermal Resistivity (°C-cm/W)
P	Phosphorus	1.8	44.1		0.177	
Pt	Platinum	21.45	1769	4.3	0.032	1.45
—	Polyethylene					304
	Polystyrene					965
SiO_2	Quartz					
	(fused SiO_2)					7-15
	(sapphire-single crystal)					
—	Rexolite™ (1422)					
	Rexolite™ (2200)					
Rh	Rhodium	12.44	1966	6	0.060	1.1
—	Rubber (neoprene)					
—	Rubber (silicone)					482
	Shellac (natural)					482
Si	Silicon (single crystal)	2.4	1410	7	0.176	1.2
—	Silicon Thermal Grease					117
	Silicon Carbide					5.8
Ag	Silver	10.49	960.8	2.7	0.056	0.25
NaCl	Sodium Chloride (dry salt)					
PbSn	Solder (37% 63%, eutectic)		183			2.0
AuSn	Solder (80% 20%, eutectic)		280			
Na	Sodium	0.97	97	0.4	0.295	0.74
	Steel, Stainless (0.1 C, 18 Cr, 8 Ni, 73.9 Fe)					
	Steel, Carbon (1 C, 0.5 Mn, 98.5 Fe)					
Ta	Tantalum	16.6	299.6	7	0.036	1.8
—	Teflon™ (pure)					482
—	TFG (Teflon™-fiberglass) woven microfibre					
Te	Tellurium	6.24	449.5	2.3	0.047	16.7
—	Tellurium Copper					
Sn	Tin	7.3	231.9	1.8	0.054	1.6
Ti	Titanium	4.5	1675	4	0.142	
W	Tungsten	19.3	3410	7	0.034	0.50
H_2O	Water (distilled)	1.00	0.0	—	1.00	160
Zn	Zinc	7.14	419.4	2.5	0.09	0.9

*Per liter, material is gaseous at normal temperatures

**Boiling Point

***Varies with type, batch, etc.

****Varies with purity, composition

*****Material exhibits plastic (i.e., inelastic) flow with time and/or temperature.

******In plane of weave

Linear Thermal Expansion ($\times 10^{-6}/^{\circ}C$)	Elasticity Modulus (kg/mm^2)	Tensile Strength (kg/mm^2)	Dielectric Constant 1 MHz	Dielectric Constant 3 GHz	Electrical Resistivity (Ω-cm)	Symbol	Name
125			–	–	–	P	Phosphorus
8.9	15000	16	–	–	1.1×10^{-5}	Pt	Platinum
			2.26	2.26	–	–	Polyethylene
70		5	2.55	2.55	–		Polystyrene
					–	SiO_2	Quartz
			3.78	3.78			(fused SiO_2)
			3.78	3.78			(sapphire-single crystal)
			2.53	2.53	–	–	Rexolite™ (1422)
			2.62	2.62			Rexolite™ (2200)
8.1	3000		–	–	5.1×10^{-6}	Rh	Rhodium
			6.26	4.00	–	–	Rubber (neoprene)
			3.20	3.13	–	–	Rubber (silicone)
			3.47	2.86	–		Shellac (natural)
4.2	11000		11.8	11.8	8.5×10^{4}	Si	Silicon (single crystal)
					–	–	Silicon Thermal Grease
							Silicon Carbide
18.9	7200	15.1	–	–	1.62×10^{-6}	Ag	Silver
			5.90	5.90		NaCl	Sodium Chloride (dry salt)
			–	–		PbSn	Solder (37% 63%, eutectic)
			–	–		AuSn	Solder (80% 20%, eutectic)
71			–	–	4.6×10^{-6}	Na	Sodium
			–	–	9×10^{-5}		Steel, Stainless (0.1 C, 18 Cr, 8 Ni, 73.9 Fe)
			–	–	1.3-2.2 x 10[illegible]		Steel, Carbon (1 C, 0.5 Mn, 98.5 Fe)
6.6	19000	50	–	–	1.3×10^{-5}	Ta	Tantalum
100*****				2.03		–	Teflon™ (pure)
						–	TFG (Teflon™-fiberglass)
18.5******		11			2.4-2.6***		woven
16-100		5			2.2-2.4***		microfibre
16.8	2100	1.12	–	–		Te	Tellurium
						–	Tellurium Copper
23	41100	1.4	–	–	1.14×10^{-5}	Sn	Tin
8.5	8500		–	–	4.8×10^{-5}	Ti	Titanium
4.3	35000	270	–	–	5.48×10^{-6}	W	Tungsten
–	–	–	78.2	76.7	1×10^{6}	H_2O	Water (distilled)
17-39	8400	10.5	–	–	6×10^{-6}	Zn	Zinc

*******This listing of material properties is, of necessity, only partially complete. Each user has special requirements for material property information. For this reason, space has been left for one to write in additional material types and properties.

C. Thermal Resistance Calculations

The temperature rise, ΔT, due to a steady heat dissipation (power flow), P_D, between parallel faces of area A and separation d of a cylindrical object (Figure C-1) of homogeneous material can be calculated from the thermal resistance, θ, using

$$\Delta T = \frac{P_D}{\theta} \tag{C-1}$$

which is recognized as the thermal equivalent of Ohm's law with temperature rise (ΔT) replacing voltage, heat flow (P_D) replacing current, and thermal resistance (θ) replacing electrical resistance.

The thermal resistance, θ, can, in turn, be calculated from thermal resistivity, ρ_T, (equal to the reciprocal of thermal conductivity) in a manner exactly analogous to the calculation of electrical resistance. Thus

$$\theta = \rho_T(d/A) \tag{C-2}$$

The calculations, though simple, are complicated by the variety of units used for

1) *Heat Flow*

   Watts
   Calories (grams/second)
   Calories (kilograms/second)
   BTUs (British Thermal Units/hour)

2) *Temperature Difference*

   °C (Celsius)
   °F (Fahrenheit)

3) *Length*

   cm (centimeter)
   in (inch)
   ft (foot)

The problem of various dimensional systems is addressed here by calculating θ in degrees Celsius per watt using centimeters as a length standard. Conversion factors for the other common systems of units are used to remove dimensional inconsistencies in either the calculation or application of the results. Thus, if thermal re-

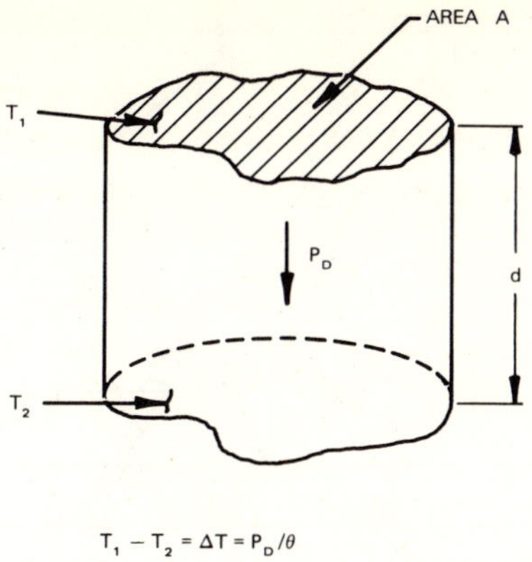

Figure C-1 **Cylindrical Heat Flow Model**

sistivity is taken from Appendix B, its units are degrees Celsius-centimeter per watt and the corresponding thermal resistance θ is

$$\theta\left(\frac{^\circ\text{Celsius}}{\text{watt}}\right) = \rho_T\left(\frac{^\circ\text{Celsius-centimeters}}{\text{watt}}\right)\ \frac{\text{d(centimeters)}}{\text{A(centimeter}^2)} \tag{C-3}$$

Thermal resistance can be calculated and/or expressed in the other common units using the following conversions:

1 calorie (gram)	=	0.001 calorie (kilogram)
	=	4.1868 joules
	=	4.1868 watt-seconds
	=	0.003968 British Thermal Unit
1 centimeter	=	0.39370 inch
	=	0.032808 foot
1°Celsius	=	1.8°Fahrenheit (*exactly*)

Table C-1 **Conversions for Thermal Calculations**

Definitions [4]

One *calorie* is the amount of heat required to raise the temperature of one gram of water from 3.5°C to 4.5°C.

One *BTU (British Thermal Unit)* is the amount of heat required to raise the temperature of one pound of water at its maximum density one degree Fahrenheit.

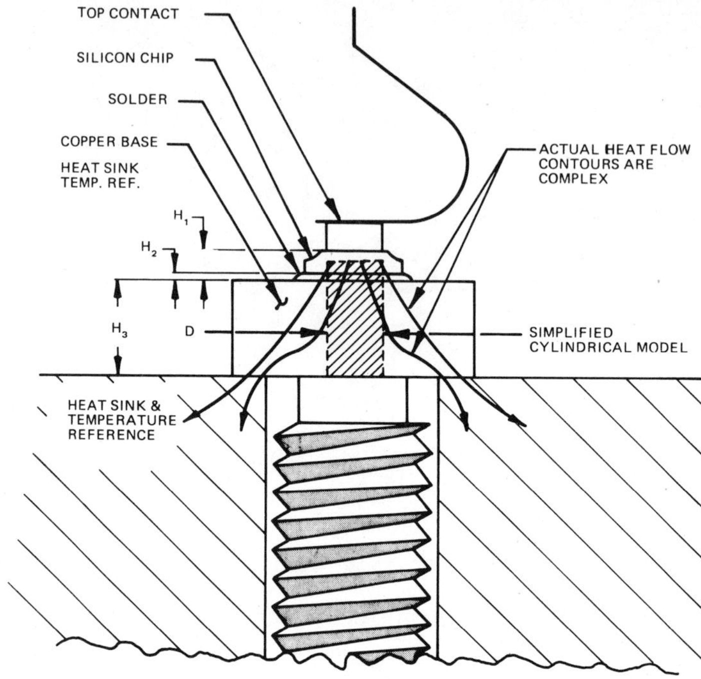

Figure C-2 **Illustrating the Simplified Uniform Diameter Cylindrical Heat Flow Model for Estimating Diode Thermal Resistance**

Example Calculation of θ

Calculate the thermal resistance from the junction of a PIN diode to the mounting heat sink using the simplified cylindrical heat path model shown in Figure C-2. Before doing so, note that, compared to the simple model path, the actual heat flow contours to the heat sink are longer, but diffuse out in the copper base beyond the diameter D of the model; these two considerations contribute counteracting errors to the simplified thermal resistance estimate. The total thermal resistance ($\theta = \theta_1 + \theta_2 + \theta_3$) is the sum of the separate thermal resistances incurred as the I region heat traverses:

1) an average silicon path of half the thickness of the silicon chip height, H_1;
2) the thickness, H_2, of the solder used to attach the chip to its copper base; and
3) the copper thickness, H_3, between the solder and the heat sink, the temperature of which the overall thermal resistance, θ, is referenced.

The separate resistances are calculated using the parameters in Table C-2.

$$
\begin{aligned}
D &= 0.092 \text{ inch} \\
H_1 &\approx 0.010 \text{ inch} \\
H_2 &\approx 0.001 \text{ inch} \\
H_3 &= 0.100 \text{ inch}
\end{aligned}
$$

Table C-2 **Parameters for Thermal Resistance Example**

ρ (Silicon) = 1.2 (°Celsius-centimeters/watt)
ρ (SnPd solder) = 2 (°Celsius-centimeters/watt)
ρ (Copper) = 0.25 (°Celsius-centimeter/watt)

Thermal resistivities of the various materials are from Appendix B. Notice that the mechanical dimensions are in inches while the material resistivities are in centimeters. This disparity is handled easily in the equations by *carrying through the dimensions in the equations and using appropriate multipliers from Table C-1 to cancel unwanted units;* thus

$$\theta_1 = \rho \text{ (silicon) } \frac{H_1/2}{(\pi D^2/4)}$$

$$= \left(\frac{1.2°\text{Celsius-centimeters}}{\text{watt}}\right)\left(\frac{0.005 \text{ inch}}{0.0065 \text{ square inch}}\right)\left(\frac{0.3970 \text{ inch}}{1 \text{ centimeter}}\right)$$

$$= 0.37°\text{Celsius/watt}$$

$$\theta_2 = \rho \text{ (SnPb solder) } \frac{H_2}{(\pi D^2/4)}$$

$$= \left(\frac{2°\text{Celsius-centimeters}}{\text{watt}}\right)\left(\frac{0.001 \text{ inch}}{0.0065 \text{ square inch}}\right)\left(\frac{0.3970 \text{ inch}}{1 \text{ centimeter}}\right)$$

$$= 0.12°\text{Celsius/watt}$$

$$\theta_3 = \rho \text{ (copper) } \frac{H_3}{(\pi D^2/4)}$$

$$= \left(\frac{0.25°\text{Celsius-centimeter}}{\text{watt}}\right)\left(\frac{0.100 \text{ inch}}{0.0065 \text{ square inch}}\right)\left(\frac{0.3970 \text{ inch}}{1 \text{ centimeter}}\right)$$

$= 1.53°\text{Celsius/watt}$

$\theta = \theta_1 + \theta_2 + \theta_3$

$\theta = 2°\text{Celsius/watt}$

In practice, the geometry shown in Figure C-2 is comparable to that used for the MA-47890 PIN diode (Table III-1) with the type 150 packaging (Table III-2) for which a measured θ of 1.5°C/W is obtained; thus, the simplified model produces, in this case, a conservative result. However, for most engineering purposes the accuracy is sufficient and the effort saved is considerable, compared with performing a more rigorous estimate taking into account the actual diffused heat flow contours, as sketched in Figure C-2.

The calculated thermal resistance, θ, can be combined with the minimum thermal heat sinking capacity, HC, (Equation (III-4)) to estimate minimum thermal time constant, τ_T, as shown in Chapter III (Equation (III-6)).

D. Coaxial Lines

Coaxial lines, consisting of one conductor located within and on the center line of another hollow conductor, completely contain the propagating energy. In addition, when coaxial lines are operated in their dominant (TEM) mode, their characteristic impedances (neglecting losses) do not vary with frequency. Consequently, they have wide application in microwave systems.

There are numerous shapes which may be used for "coax" lines; the most common are shown in Figure D-1, which have a round center conductor axially located within either a round or square outer conductor. Also shown are two common unshielded configurations, *trough line* and *slab line.*

For any uniform, lossless, reciprocal, TEM transmission line, the ratio of the incident voltage to the incident current at any frequency is a constant, defined as the *characteristic impedance,* equal to

$$Z_0 = \sqrt{\frac{L}{C}} \qquad \text{(D-1)}$$

where L and C are respectively the distributed inductance and capacitance per unit length.

For general TEM line cross sections the analytic solution for Z_0 involves finding C, since both the distributed capacitance and inductance for any TEM line are related directly to the characteristic impedance by [11 (p. 84)]

$$Z_0 = \frac{\sqrt{\mu_0 \epsilon_0 \epsilon_R}}{C} \qquad \text{(D-2)}$$

$$Z_0 = \frac{L}{\sqrt{\mu_0 \epsilon_0 \epsilon_R}} \qquad \text{(D-3)}$$

1) *Round Coax*

The round inner and outer conductor coax gives the lowest possible Z_0 for given conductor spacing and cross sectional size and an exact analytic expression for Z_0.

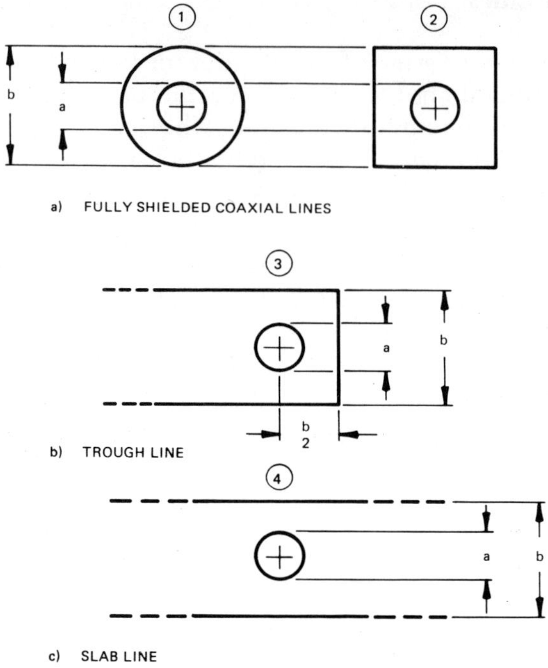

Figure D-1 **TEM Transmission Line Configurations**

Thus, round coax should be used whenever low impedance (below about 20 Ω) or precise estimates of the impedance value must be made.

For coaxial line with round center and outer conductors, the capacitance and inductance per unit length is given by [1]

$$C = \frac{2\pi\epsilon_0\epsilon_R}{\ln(b/a)} \tag{D-4}$$

$$L = \frac{\mu_0\mu_R}{2\pi} \ln(b/a) \tag{D-5}$$

where

$$\epsilon_0 = 8.854186 \times 10^{-14} \text{ farad/meter} \tag{D-6}$$

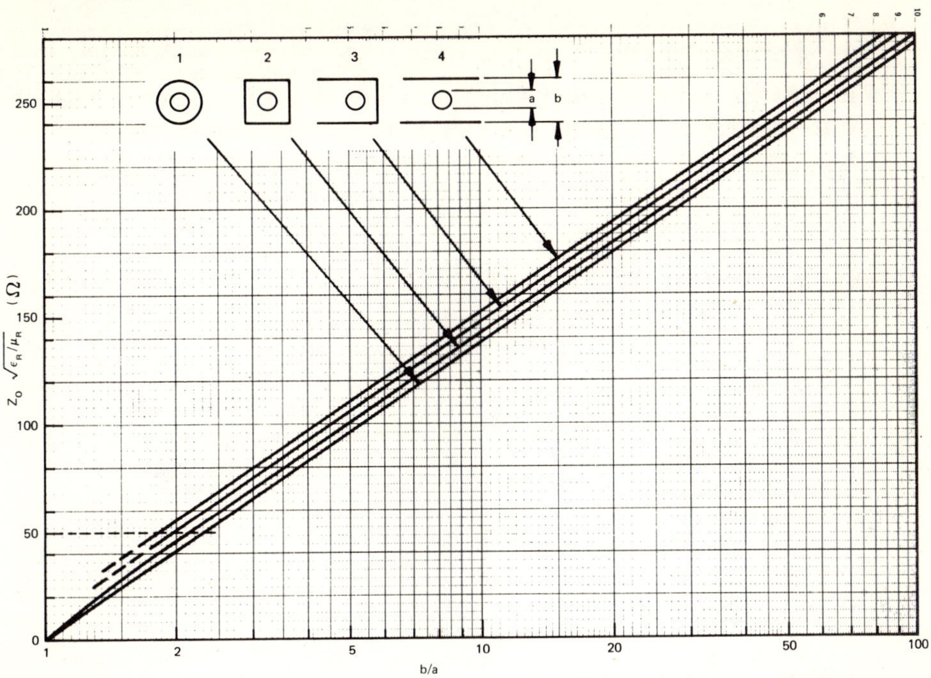

Figure D-2 **Characteristic Impedance of Various Coaxial Lines**

$$\mu_0 = 4\pi \times 10^{-7} \text{ henry/meter} \tag{D-7}$$

$$\sqrt{\frac{\mu_0}{\epsilon_0}} = 376.7303 \text{ ohms (in a vacuum)} \tag{D-8}$$

The corresponding characteristic impedance of round coaxial line is given precisely by

$$Z_{01} = 59.96 \sqrt{\frac{\mu_R}{\epsilon_R}} \ln\left(\frac{b}{a}\right) \tag{D-9}$$

$$Z_{01} = 138.06 \sqrt{\frac{\mu_R}{\epsilon_R}} \log_{10}\left(\frac{b}{a}\right) \tag{D-10}$$

Figure D-2 shows graphically the characteristic impedance of round coax as a function of the cross sectional ratio, b/a.

The characteristic impedance functions for the remaining three TEM line geometries shown in Figure D-1 are given by the approximate expressions below. This formula is taken from Cohn [28], but References 29 through 31 should be consulted for the original derivations and other work.

2) *Square Coax*

$$Z_{02}\sqrt{\epsilon_R/\mu_R} \approx 138 \log_{10}\left(1.0787\,\frac{b}{a}\right) \tag{D-11}$$

This formula is accurate to within 1.5% at 17 Ω; accuracy improves above 17 Ω.

3) *Trough Line*

$$Z_{03}\sqrt{\epsilon_R/\mu_R} \approx 138 \log_{10}\left(1.17\,\frac{b}{a}\right) \tag{D-12}$$

```
C   CØAX:  GIVES ZØ ØF VARIØUS CØAX LINE GEØMETRIES
C
           WRITE(16,100)
100        FØRMAT(' B/A RANGE (START,STEP,STØP)= '$)
           READ(16,300)BA1,BASTEP,BA2
           WRITE(16,200)
200        FØRMAT(' RELATIVE DIELECTRIC AND PERMEABILITY CØNSTANTS
         + ',/,' (IF DIFFERENT FRØM UNITY) = '$)
           READ(16,300) DIELC,PERMC
           IF(DIELC.LT.1.0) DIELC=1.0
           IF(PERMC.LT.1.0) PERMC=1.0
300        FØRMAT(4G)
           K=SQRT(PERMC/DIELC)
           BA=BA1
350        ZØ1=K*60*ALØG(BA)
           ZØ2=K*(138*ALØG10(1.0787*BA))
           ZØ3=K*(135*ALØG10(1.17*BA))
           ZØ4=K*(135*ALØG10(1.2732395*BA))
           IF(BA.LT.(BA1+BASTEP/2)) WRITE(16,400)
400        FØRMAT(' B/A            ZØ1       ZØ2       ZØ3       ZØ4',/)
           WRITE(16,500)BA,ZØ1,ZØ2,ZØ3,ZØ4
500        FØRMAT(F7.3,2X,4F8.2)
           IF(BA.GT.(BA2+BASTEP/2)) GØ TØ 600
           BA=BA+BASTEP
           GØ TØ 350
600        CØNTINUE
           END

SYSTEM?..
.
```

Figure D-3 Program COAX, a FORTRAN IV Program used to Calculate the Characteristic Impedances in Figure D-2

```
RUN COAX

B/A RANGE (START,STEP,STOP)= 1 .25 10

RELATIVE DIELECTRIC AND PERMEABILITY CONSTANTS
(IF DIFFERENT FROM UNITY) =

B/A           Z01        Z02        Z03        Z04

 1.000         0.00       4.54       9.41      14.48
 1.250        13.39      17.91      22.78      27.85
 1.500        24.33      28.84      33.71      38.78
 1.750        33.58      38.08      42.95      48.02
 2.000        41.59      46.08      50.95      56.02
 2.250        48.66      53.14      58.01      63.08
 2.500        54.98      59.46      64.33      69.39
 2.750        60.70      65.17      70.04      75.11
 3.000        65.92      70.38      75.25      80.32
 3.250        70.72      75.18      80.05      85.12
 3.500        75.17      79.62      84.49      89.56
 3.750        79.31      83.76      88.63      93.69
 4.000        83.18      87.62      92.49      97.56
 4.250        86.82      91.26      96.13     101.20
 4.500        90.24      94.68      99.55     104.62
 4.750        93.49      97.92     102.79     107.86
 5.000        96.57     101.00     105.87     110.94
 5.250        99.49     103.92     108.79     113.86
 5.500       102.28     106.71     111.58     116.65
 5.750       104.95     109.37     114.24     119.31
 6.000       107.51     111.93     116.79     121.86
 6.250       109.95     114.37     119.24     124.31
 6.500       112.31     116.72     121.59     126.66
 6.750       114.57     118.98     123.85     128.92
 7.000       116.75     121.16     126.03     131.10
 7.250       118.86     123.27     128.14     133.20
 7.500       120.89     125.30     130.17     135.24
 7.750       122.86     127.26     132.13     137.20
 8.000       124.77     129.17     134.04     139.10
 8.250       126.61     131.01     135.88     140.95
 8.500       128.40     132.80     137.67     142.74
 8.750       130.14     134.54     139.41     144.47
 9.000       131.83     136.23     141.10     146.16
 9.250       133.48     137.87     142.74     147.81
 9.500       135.08     139.47     144.34     149.40
 9.750       136.64     141.02     145.89     150.96
10.000       138.16     142.54     147.41     152.48
10.250       139.64     144.02     148.89     153.96
CP UNITS      2

EXIT

SYSTEM?..
.
```

Figure D-4 Sample Execution of Program COAX Shown in Figure D-3

4) *Slab Line*

An approximate formula for slab line characteristic impedance which gives a value 1% high at 50 Ω and increasing accuracy above 50 Ω is [12 (Volume 1)]

$$Z_{04} \sqrt{\epsilon_R / \mu_R} \approx 138 \log_{10} \left(\frac{4}{\pi} \cdot \frac{b}{a} \right) \quad (Z_0 \geqslant 50 \text{ ohms}) \qquad \text{(D-13)}$$

Much better accuracy can be obtained using Wheeler's result [32] which gives the ratio b/a in terms of Z_{04}.

$$\frac{b}{a} = \frac{4}{\pi} \exp\left\{ -\frac{Z_{04}}{60} \left[1 - \frac{2}{9} \exp\left(-\frac{4Z_{04}}{60} \right) - 0.00736 \exp\left(-\frac{8Z_{04}}{60} \right) + 0.01932 \exp\left(-\frac{12Z_{04}}{60} \right) - 0.0043 \exp\left(-\frac{16Z_{04}}{60} \right) \pm \ldots \right] \right\} \qquad \text{(D-14)}$$

The characteristic impedance expressions in Equations (D-10) through (D-13) are shown graphically in Figure D-2. For greater computational accuracy, a computer program is used to calculate all four impedances given μ_R, ϵ_R, and (b/a) and is shown in Figure D-3. A sample execution is given in Figure D-4, from which it can be seen that the approximate formulas for trough line and slab line give invalid results for small b/a values (i.e., finite impedances for b/a = 1.0). These transmission lines should not be used for very low impedances, so this limitation is not usually a serious one. For more coaxial line impedance detail (including rectangular center conductor coax, etc.), Volume 1 of Reference 12 and References 28 through 32 should be consulted.

Distributed Capacitance and Inductance

Occasionally, the actual values of the distributed capacitance and/or inductance of a coaxial line are desired. These figures can be obtained for any TEM line for which Z_0 is known, from Equations (D-2) and (D-3), respectively.

E. Microstrip

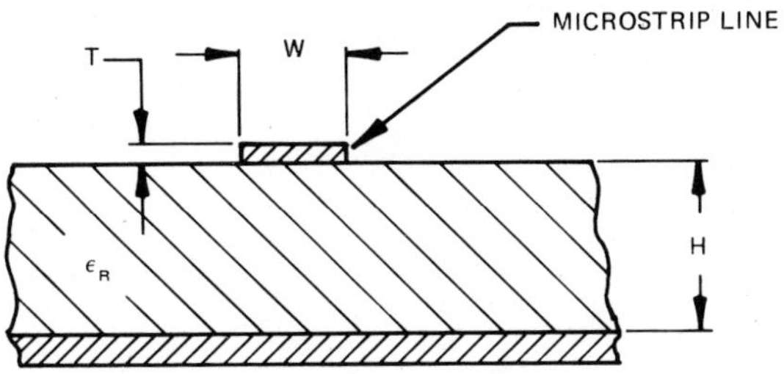

Figure E-1 **Microstrip Transmission Line Cross Section**

Microstrip transmission line consists of a single thin "center conductor" above an infinite ground plane. For support, a dielectric substrate separates the two conductors, but the dielectric is omitted above the center conductor for ease of access to the circuit. This omission results in an *inhomogeneous* transmission medium, with RF energy propagating partly in air and partly in the dielectric constant; $\epsilon_{EFFECTIVE}$ for such transmission lines is somewhere between 1.0 and ϵ_R, the substrate dielectric constant. The effective dielectric constant is found from the *filling factor,* q, the ratio of dielectric area to total area in the area of a rectangle to which the microstrip geometry is mapped. The value for $\epsilon_{EFFECTIVE}$ can be found from [14]

$$\epsilon_{EFFECTIVE} = 1 + q(\epsilon_R - 1) \qquad \text{(E-1)}$$

Alternately, $\epsilon_{EFFECTIVE}$ can be found for a given W/H geometry from

$$Z_0(\epsilon_R = 1) = Z_0(\epsilon_R = \epsilon_R)\sqrt{\epsilon_{EFFECTIVE}} \qquad \text{(E-2)}$$

by determining Z_0 for both $\epsilon_R = 1$ and $\epsilon_R = \epsilon_R$.

A summary of the early work in microstrip lines by Wheeler [14], Presser [15], and Schneider [16] was made by Kwon [13]. The resulting characteristic impedance for narrow and wide strip geometries is given for synthesis calculations, in which W/H can be determined for given Z_0 by

Wide strip ($Z_0 \sqrt{\epsilon_R} < 125$ ohms)

$$W/H = \frac{2}{\pi}\left[d_K - 1 - \ln(2\,d_K - 1) + \frac{\epsilon_R - 1}{2\epsilon_R}\left(\ln(d_K - 1) + 0.293 - \frac{0.517}{\epsilon_R}\right)\right] \quad \text{(E-3)}$$

where

$$d_K = \frac{188.5\,\pi}{Z_0\sqrt{\epsilon_R}}$$

Narrow strip ($Z_0 \sqrt{\epsilon_R} > 125$ ohms)

$$W/H = \frac{4}{(e^K/2) - e^{-K}} \quad \text{(E-4)}$$

where

$$K = \frac{\pi Z_0(\sqrt{\epsilon_R + 1}/\sqrt{2})}{188.5} + \frac{\epsilon_R - 1}{2(\epsilon_R + 1)}\left[\ln\left(\frac{\pi}{2}\right) + \frac{1}{\epsilon_R}\ln\left(\frac{4}{\pi}\right)\right]$$

For analysis, when the geometry is specified, Z_0 can be calculated from

Wide strip ($W/H > 1$)

$$Z_0 = \frac{188.5}{\sqrt{\epsilon_R}}\left[\frac{W/H}{2} + \frac{\ln(4)}{\pi} + \frac{(\epsilon_R - 1)\ln[(e\pi^2)/16]}{2\pi\,\epsilon_R^2} + \frac{\epsilon_R + 1}{2\pi\,\epsilon_R}\ln\left(\frac{\pi e[(W/H)/2 + 0.94]}{2}\right)\right] \quad \text{(E-5)}$$

where e = 2.7182818 (base of the natural logarithm)

Narrow strip ($W/H < 2$)

$$Z_0 = \frac{188.5}{\pi}\cdot\sqrt{\frac{2}{\epsilon_R + 1}}\left[\ln\left(\frac{8}{W/H}\right) + \frac{1}{8}\left(\frac{W/H}{2}\right)^2 - \frac{\epsilon_R - 1}{2(\epsilon_R + 1)}\left(\ln\left(\frac{\pi}{2}\right) - \frac{1}{\epsilon_R}\ln\left(\frac{4}{\pi}\right)\right)\right] \quad \text{(E-6)}$$

```
C  PRØGRAM  "MICRØ"  (FØRTRAN IV)  WRITTEN BY A. H. KWØN
C
C  CALCULATES PARAMETERS ØF MICRØSTRIP
C
           PI=3.141592654
50         WRITE(16,510)
           READ(16,505)Z1,Z2,Z3,ER,H,T
           IF(Z1.EQ.0) GØ TØ 450
           WRITE(16,520)ER,H,T
           Z=Z1
100        EZ=Z*SQRT(ER)
           IF(EZ.GT.125) GØ TØ 130
           DK=(188.5*PI)/EZ
           CC=(ER-1)*(ALØG(DK-1)+.293-.517/ER)/(2*PI*ER)
           W=2*((DK-1)/PI-ALØG(2*DK-1)/PI+CC)
           GØ TØ 150
130        HA=PI*Z*SQRT((ER+1)/2)/188.5
           H1=HA+(ER-1)*(ALØG(PI/2)+ALØG(4/PI)/ER)/(2*(ER+1))
           W=4/(EXP(H1)/2-EXP(-H1))
150        WID=W*H
           CALL WIDTH(PI,WID,H,T,ER)
           ER1=(ER+1)/2+(ER-1)/(2*SQRT(1+10/W))
           Q=(ER1-1)/(ER-1)
           WRITE(16,500)Z,W,WID,ER1,Q
350        CØNTINUE
           Z=Z+Z3
           IF(Z.GT.Z2) GØ  TØ 400
           GØ TØ 100
500        FØRMAT(5F10.3)
505        FØRMAT(10G)
510        FØRMAT(/,/,' MICRØSTRIP LINE WIDTH CALCULATIØN',/,
       +   ' PLEASE PRØVIDE FØLLØWING INFØ:',/,
       +   ' IMPEDANCE START, STØP & STEP IN ØHMS (Z1,Z2,Z3)',/,
       +   ' DIELECTRIC CØNSTANT ØF SUBSTRATE (ER)',/,
       +   ' SUBSTRATE THICKNESS IN MILS (H)',/,
       +   ' STRIP THICKNESS IN MILS (T)',/,/,
       +   ' INPUT IN ØRDER: Z1,Z2,Z3,ER,H,T = '$)
520        FØRMAT(/,/,' SUBSTRATE DIELECTRIC CØNSTANT = ',F10.4,/,
       +   ' SUBSTRATE THICKNESS (MILS)     =',F11.4,/,
       +   ' CØNDUCTØR THICKNESS (MILS)     =',F11.4,/,
       +   /,/,6X,'Z0',8X,'W/H',8X,'W',8X,'ER1',7X,'Q',/)
530        FØRMAT(/,/,' FØR ANØTHER RUN, TYPE "1"',/)
400        WRITE(16,530)
           READ(16,505)II
           IF(II.EQ.1) GØ TØ 50
450        STØP
           END
C
C  WIDTH CØRRECTIØN CALCULATIØN
           SUBRØUTINE WIDTH(PI,WID,H,T,ER)
           AA=H/(2*PI)
           BB=2*T
           IF(WID.LE.AA) GØ TØ 20
           IF(AA.LE.BB) GØ TØ 50
           DW=T/PI*(1+ALØG(2*H/T))
           GØ TØ 30
20         IF(WID.LE.BB) GØ TØ 50
           DW=T/PI*(1+ALØG(4*PI*WID/T))
30         IF(DW.LE.(1.33*T)) GØ TØ 50
           IF(ER.GE.5)DW=DW/ER
           WID=WID-DW
50         RETURN
           END
```

Figure E-2 **FORTRAN Program MICRO [17] to Calculate Microstrip Z_0 and $\epsilon_{EFFECTIVE}$**

Effective Dielectric Constant

The effective dielectric constant for all of the above cases is given by Schneider's approximation

$$\epsilon_{EFFECTIVE} = \frac{\epsilon_R + 1}{2} + \frac{\epsilon_R - 1}{2}\left[1 + \frac{10}{W/H}\right]^{-1/2} \qquad \text{(E-7)}$$

Conductor Thickness Correction

These equations are for conductors of zero thickness ($T \approx 0$); the width correction, recommended by Wheeler, can be used to correct for the fringing fields associated with the finite thickness of the center conductor strip. This effective added width, ΔW, caused by strip thickness is given by

Wide strip $(W/H > 1/(2\pi) > 2\,T/H)$

$$\Delta W = \frac{T}{\pi}\left[1 + \ln\left(\frac{2H}{T}\right)\right] \qquad \text{(E-8)}$$

```
MICRØSTRIP LINE WIDTH CALCULATIØN
PLEASE PRØVIDE FØLLØWING INFØ:
IMPEDANCE START, STØP & STEP IN ØHMS (Z1,Z2,Z3):
DIELECTRIC CØNSTANT ØF SUBSTRATE (ER)
SUBSTRATE THICKNESS IN MILS (H)
STRIP THICKNESS IN MILS (T)

INPUT IN ØRDER: Z1,Z2,Z3,ER,H,T = 5 200 5 2.32 7 1

SUBSTRATE DIELECTRIC CØNSTANT =     2.3200
SUBSTRATE THICKNESS (MILS)    =     7.0000
CØNDUCTØR THICKNESS (MILS)    =     1.0000
```

ZO	W/H	W	ER1	Q
5.000	46.456	325.192	2.259	0.954
10.000	22.022	154.155	2.207	0.915
15.000	13.958	97.708	2.164	0.882
20.000	9.966	69.761	2.126	0.853
25.000	7.594	53.159	2.094	0.828
30.000	6.029	42.202	2.065	0.807
35.000	4.922	34.455	2.039	0.787
40.000	4.101	28.705	2.016	0.770
45.000	3.469	24.280	1.995	0.754
50.000	2.968	20.777	1.976	0.739
55.000	2.563	17.943	1.958	0.726
60.000	2.230	15.607	1.942	0.713
65.000	1.950	13.653	1.927	0.702
70.000	1.714	11.997	1.912	0.691
75.000	1.511	10.578	1.899	0.681
80.000	1.336	9.352	1.887	0.672
85.000	1.205	8.434	1.876	0.664
90.000	1.073	7.512	1.865	0.656
95.000	0.957	6.702	1.855	0.648
100.000	0.855	5.988	1.845	0.640
105.000	0.765	5.355	1.836	0.633
110.000	0.685	4.793	1.827	0.627
115.000	0.613	4.293	1.819	0.620
120.000	0.550	3.843	1.811	0.614
125.000	0.493	3.450	1.803	0.608
130.000	0.442	3.094	1.796	0.603
135.000	0.397	2.776	1.789	0.598
140.000	0.356	2.491	1.782	0.593
145.000	0.319	2.236	1.776	0.588
150.000	0.287	2.007	1.770	0.583
155.000	0.257	1.802	1.765	0.579
160.000	0.231	1.617	1.759	0.575
165.000	0.207	1.452	1.754	0.571
170.000	0.186	1.304	1.749	0.568
175.000	0.167	1.171	1.745	0.564
180.000	0.150	1.052	1.740	0.561
185.000	0.135	0.945	1.736	0.558
190.000	0.121	0.848	1.732	0.555
195.000	0.109	0.762	1.728	0.552
200.000	0.098	0.684	1.725	0.549

Figure E-3 **Sample Execution of Program MICRO (Given in Figure E-2)**

Narrow strip $(1/(2\pi) > W/H > 2\,T/H)$

$$\Delta W = \frac{T}{\pi}\left[1 + \ln\left(\frac{4\pi W}{T}\right)\right] \tag{E-9}$$

To make the original analysis theoretically tractable, these width corrections were based on the assumption that $\epsilon_R = 1$, which clearly is not the case for dielectrically supported microstrip line. The width correction, ΔW, takes into account the change in distributed inductance as well as the change in distributed capacitance, both caused by the finite thickness of the strip. The inductance change resulting from finite strip thickness is not dependent upon ϵ_R, but as ϵ_R becomes larger, the fringing capacitance from the edges of the strip (which are in air) becomes smaller compared with the distributed capacitance of the line beneath the strip (which is in ϵ_R dielectric). Accordingly, fringing capacitance becomes negligible as ϵ_R approaches infinity. Cohn* recommends using the full correction indicated by Equations (E-8) and (E-9) for $\epsilon_R \approx 1$, and one-half the indicated correction (associating equal contributions to ΔW from inductive and capacitive effects) for $\epsilon_R \approx 10$, with proportionate adjustments for intermediate values of ϵ_R. This adjustment, though approximate, should be adequate since ΔW itself is usually a small correction term.

The effective dielectric constant of inhomogeneous microstrip is also somewhat dependent on frequency; thus, microstrip has dispersion (i.e., an insertion phase which is not linearly proportional to frequency). For many applications dispersion can be neglected; however, Getsinger's results [33] can be consulted if dispersion effects are thought to be significant.

Sample graphs of Z_0 and $\epsilon_{EFFECTIVE}/\epsilon_R$ versus W/H are given in Figures VII-27 and VII-28 for ϵ_R values of 1.0, 2.5, and 10. Equations (E-3) through (E-7) have an accuracy estimated at better than 3 percent when used within the ranges specified. Because of the variety of ϵ_R and conductor thickness combinations, it is not practical for a single chart to cover all design applications. For this reason Equations (E-3), (E-4), (E-7), (E-8), and (E-9) have been used to generate the computer program MICRO [17] shown in Figure E-2. An execution of this program, as shown in Figure E-3, usually suffices to define the various Z_0 versus W/H values needed for a specific design, once ϵ_R and T have been assigned.

*Private communication with Seymour B. Cohn.

F. Stripline

A conveniently realized TEM transmission line is made by chemically etching thin copper bonded to Teflon fiberglass (TFG) dielectric sheets, assembled in the configuration shown in Figure F-1. Exact solution for the characteristic impedance is usually unnecessary since the approximate analysis by Cohn is accurate within 1.3% [8, 18].

Wide strip (W/B $\geq$ 0.35)

The characteristic impedance is given by

$$Z_0 \sqrt{\epsilon_R} = \frac{94.15 \text{ ohms}}{[(W/B)/(1 - T/B)] + [C_F/0.0885\epsilon_R)]} \tag{F-1}$$

where

$$C_F = \frac{0.0885\ \epsilon_R}{\pi}\left[2K \ln(K+1) - (K-1)\ln(K^2 - 1)\right] \tag{F-2}$$

$$K = \frac{1}{1 - T/B} \tag{F-3}$$

Narrow strip (W/B $<$ 0.35)

The fringing fields on each side of the center strip conductor cannot be treated independently; the approximate formulation is that used for slab line

$$Z_0 \sqrt{\epsilon_R} = 60 \ln\left(\frac{4B}{\pi D}\right) \tag{F-4}$$

where the strip with width W and thickness T is related to a round center conductor of diameter D by

$$D \approx \frac{W}{2}\left[1 + \frac{T}{\pi W}\left(1 + \ln\left(\frac{4\pi W}{T}\right) + 0.51\ \pi\left(\frac{T}{W}\right)^2\right)\right] \tag{F-5}$$

provided T/W $\leq$ 0.11

A graph of $Z_0\sqrt{\epsilon_R}$ versus W/B for various thickness ratios, T/B, is given in Figure F-1. For more accurate evaluation, computer pro-

Narrow strip I use $W = 2D - \frac{T}{\pi}\left[4.22417 + \ln\left(\frac{D}{T} - .1275\right)\right]$ excellent for $0 \leq \frac{T}{W} \leq 1$, usable up to $\frac{T}{W} = 5$

where $D = \frac{4}{\pi} B \exp(-Z_0\sqrt{\epsilon_r}/59.96)$

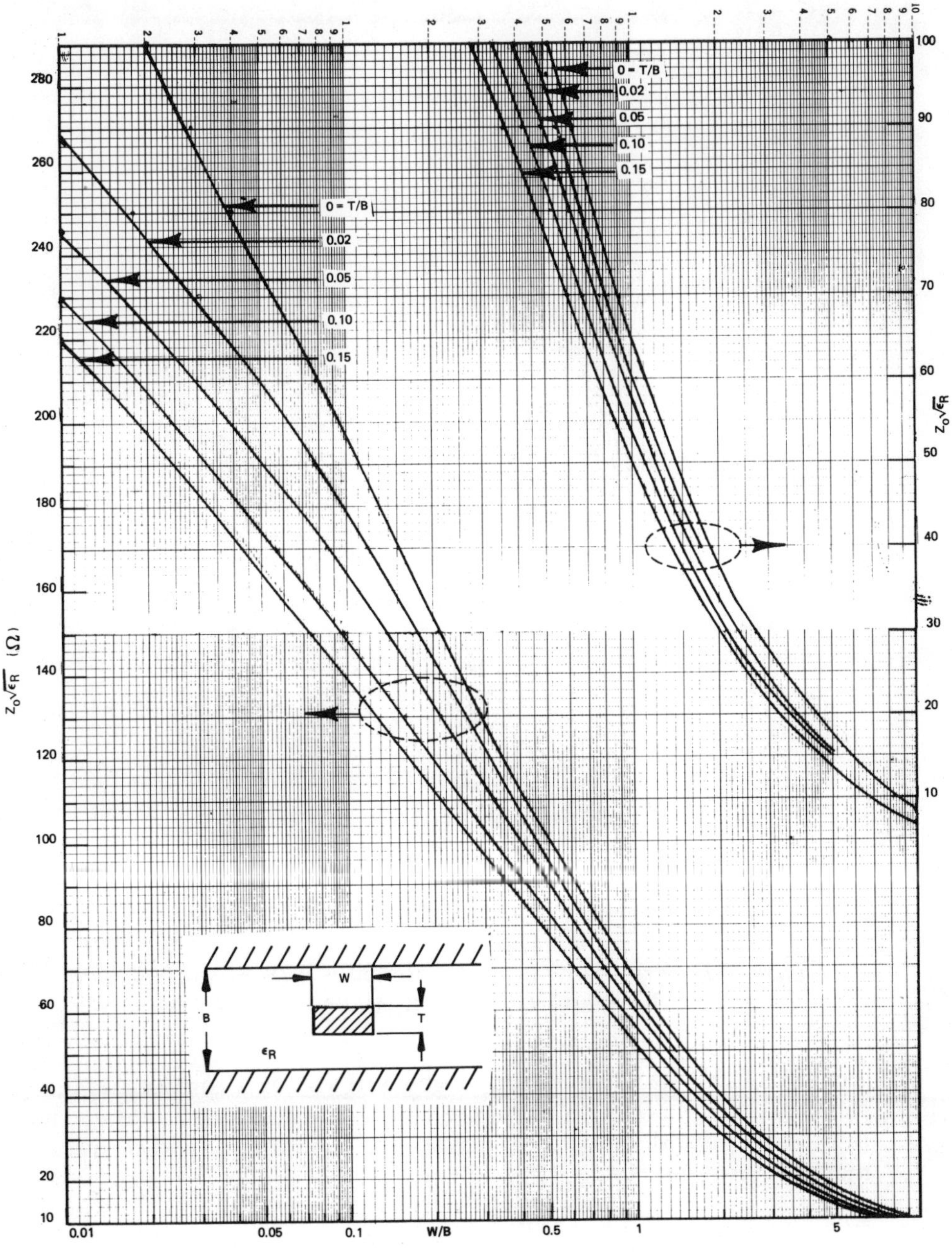

Figure F-1 Stripline Characteristic Impedance

```
C   ZØ.F4:   STRIPLINE CHARACTERISTIC IMPEDANCE - BY H. HØWE
            WRITE(16,10)
10          FØRMAT(' STRIPLINE CHARACTERISTIC IMPEDANCE',/)
            WRITE(16,20)
20          FØRMAT(' GRØUND PLANE SPACING, B, =   '$)
            READ(16,25)B
25          FØRMAT(3G)
            WRITE(16,30)
30          FØRMAT(' CØNDUCTØR THICKNESS, T, =   '$)
            READ(16,25) T
            WRITE(16,40)
40          FØRMAT(' REL. DIELECTRIC CØNSTANT, ER,  =   '$)
            READ(16,25) E
45          WRITE(16,50)
50          FØRMAT(' ZØ (IN ØHMS) RANGE (START,STØP,STEP)=   '$)
            READ(16,25) D1,D2,D3
            IF(D1.LT.0.0000001) GØ TØ 500
            WRITE(16,60)
60          FØRMAT('        ZØ       WIDTH',/)
            PI=3.1415926535
            ZØ=D1
70          IF(ZØ.GT.D2) GØ TØ 320
            R=.01
            W=.001
            W1=.010
            T1=(2./(1.0-T/B))*ALØG((1.0/(1.0-T/B))+1.0)
            T2=((1.0/(1.0-T/B))-1.0)*ALØG(1.0/((1.0-T/B)**2.0)-1.0)
            C=1.0/PI*(T1-T2)
100         D=.5*W*(1.0+T/(PI*W)*(1.0+ALØG((4.0*PI*W)/T)))
            Z=60.0/SQRT(E)*ALØG(4.0*B/(PI*D))
            IF(ABS(Z-ZØ).LT.R) GØ TØ 200
            IF((Z-ZØ).LT.0.0) GØ TØ 130
            IF(W.GT.B*0.35) GØ TØ 140
            W=W+W1
            GØ TØ 100
130         W=W-W1
            W1=W1/10.0
            W=W+W1
            GØ TØ 100
140         W1=0.010
150         Z=94.15/(SQRT(E)*(((W/B)/(1.0-T/B))+C))
            IF(ABS(Z-ZØ).LT.R) GØ TØ 220
            IF((Z-ZØ).LT.0.0) GØ TØ 170
            W=W+W1
            GØ TØ 150
170         W=W-W1
            W1=W1/10.0
            W=W+W1
            GØ TØ 150
200         Z=(100.0*Z+0.5)/100.0
220         WRITE(16,300) ZØ,W
300         FØRMAT(F10.1,F10.5)
            ZØ=ZØ+D3
            GØ TØ 70
320         WRITE(16,330)
330         FØRMAT(/,/,/)
            GØ TØ 45
500         STØP
            END
```

Figure F-2 **Program Z0 [19] Calculates W/B Stripline Given the Required Z_0, T/B, and ϵ_R**

gram ZO [19], which was used to generate the data for the graph, can be used and is shown in Figure F-2; a sample execution is given in Figure F-3. A detailed treatment of stripline theory and techniques is given by Howe [8].

```
STRIPLINE CHARACTERISTIC IMPEDANCE

GRØUND PLANE SPACING, B, =  .250

CØNDUCTØR THICKNESS, T, =  .0014

REL. DIELECTRIC CØNSTANT, ER,  =  2.32

ZØ (IN ØHMS) RANGE (START,STØP,STEP)=  10 100 5

      ZØ       WIDTH

     10.0    1.42300
     15.0    0.91100
     20.0    0.65520
     25.0    0.50200
     30.0    0.39930
     35.0    0.32620
     40.0    0.27140
     45.0    0.22870
     50.0    0.19460
     55.0    0.16660
     60.0    0.14331
     65.0    0.12361
     70.0    0.10673
     75.0    0.09210
     80.0    0.08014
     85.0    0.07025
     90.0    0.06154
     95.0    0.05387
    100.0    0.04713


ZØ (IN ØHMS) RANGE (START,STØP,STEP)=
```

Figure F-3 **Sample Execution of Program Z0 (Given in Figure F-2)**

G. Waveguide

Designation: IEC R( )	Designation: EIA WR( )	Recommended Operating Frequency Range For TE_{01} Mode: IEC (GHz) 1.25fc - 1.9fc	Recommended Operating Frequency Range For TE_{01} Mode: EIA (GHz)	Cut-off For TE_{01} Mode: Frequency (GHz)	Cut-off For TE_{01} Mode: Wavelength (cm)	Power Rating (megawatts) (see note 1) 1.25fc - 1.9fc	Theoretical attenuation lowest to highest frequency (dB/100ft.)	JAN WG RG( )	Material Alloy	JAN FLANGE: Choke UG( )/U	JAN FLANGE: Cover UG( )/U	EIA WG WR( )	DIMENSIONS (Inches): Inside	Tol. (±)	Outside	Tol. (±)	Wall Thickness (nom.)
3	2300	0.32-0.49	0.32-0.49	0.256	116.84	246-348	.040-.027	290	Alum.			2300	23.000 11.500	0.020	23.376 11.876	.020	0.188
4	2100	0.35-0.53	0.35-0.53	0.281	106.68	205-290	.046-.031	291	Alum.			2100	21.000 10.500	0.020	21.376 10.876	.020	0.188
5	1800	0.41-0.62	0.41-0.62	0.328	91.44	150-213	.058-.039	201	Alum.			1800	18.000 9.000	0.020	18.250 9250	.020	0.125
6	1500	0.49-0.75	0.49-0.75	0.393	76.20	104-148	.076-.051	202	Alum.			1500	15.000 7.500	0.015	15.350 7.750	.015	0.125
8	1150	0.64-0.98	0.64-0.96	0.513	58.40	61.5-87.1	.113-.076	203	Alum.			1150	11.500 5.750	0.015	11.750 6.000	.015	0.125
9	975	0.76-1.15	0.75-1.12	0.605	49.53	44.2-62.6	.145-.098	204	Alum.			975	9.750 4.875	0.010	10.000 5.125	.010	0.125
12	770	0.96-1.46	0.96-1.45	0.766	39.12	27.6-39.1	.206-.140	205	Alum.			770	7.700 3.850	0.010	7.950 4.100	.010	0.125
14	650	1.14-1.73	1.12-1.70	0.908	33.02	19.6-27.8	.317-.214 .266-.180	69 103	Brass Alum.		417A* 418A*	650	6.500 3.250	0.010	6.660 3.410	.010	0.080
18	510	1.45-2.20	1.45-2.20	1.157	25.91	12.09-17.1	.456-.309 .382-.259	337 338	Brass Alum.			510	5.100 2.550	0.010	5.260 2.710	.010	0.080
22	430	1.72-2.61	1.70-2.60	1.372	21.84	8.6-12.2	.588-.399 .494-.334	104 105	Brass Alum.		435A* 437A*	430	4.300 2.150	0.008	4.460 2.310	.008	0.080
26	340	2.17-3.30	2.20-3.30	1.736	17.27	5.4-7.6	.837-.567 .702-.475	112 113	Brass Alum.		553* 554*	340	3.400 1.700	0.005	3.560 1.860	.005	0.080
32	284	2.60-3.95	2.60-3.95	2.078	14.43	3.5-5.0	1.136-.777 .953-.652	48 75	Brass Alum.	54B 585A	53 584	284	2.840 1.340	0.005	3.000 1.500	.005	0.080
40	229	3.22-4.90	3.30-4.90	2.577	11.63	2.44-3.46	1.514-1.026 1.270-.860	340 341	Brass Alum.			229	2.290 1.145	0.005	2.418 1.273	.005	0.064
48	187	3.94-5.99	3.95-5.85	3.152	9.510	1.52-2.15	2.140-1.467 1.795-1.231	49 95	Brass Alum.	148C 406B	149A 407	187	1.872 0.872	0.005	2.000 1.000	.005	0.064
58	159	4.64-7.05	4.90-7.05	3.711	8.078	1.17-1.66	2.617-1.773 2.195-1.487	343 344	Brass Alum.			159	1.590 0.795	0.004	1.718 0.923	.004	0.064
70	137	5.38-8.17	5.85-8.20	4.301	6.970	0.79-1.12	3.470-2.390 2.910-2.004	50 106	Brass Alum.	343B 440B	344 441	137	1.372 0.622	0.004	1.500 0.750	.004	0.064
84	112	6.58-10.00	7.05-10.00	5.259	5.700	0.52-0.73	4.761-3.292 3.993-2.761	51 68	Brass Alum.	522B 137B	51 138	112	1.122 0.497	0.004	1.250 0.625	.004	0.064
	102	(7.23)-(11.0)	7.00-11.0	5.785	5.182	0.48-0.68	5.093-3.450 4.272-2.894	320	Brass Alum.	1494	1493	102	1.020 0.510	0.003	1.148 0.638	.003	0.064
100	90	8.20-12.5	8.20-12.40	6.557	4.572	0.33-0.47	6.614-4.570 5.547-3.833	52 67	Brass Alum.	40B 136B	39 135	90	0.900 0.400	0.003	1.000 0.500	0.003	0.050
120	75	9.84-15.0	10.00-15.00	7.868	3.810	0.26-0.34	8.078-5.472 6.775-4.590	346 347	Brass Alum.			75	0.750 0.375	0.0003	0.850 0.475	0.003	0.050
140	62	11.9-18.0	12.4-18.0	9.486	3.160	0.18-0.25	10.696-7.246 8.971-6.077 6.762-4.581	91 349 107	Brass Alum. Silver	541A	419	62	0.622 0.311	0.002	0.702 0.391	0.003	0.040
180	51	14.5-22.0	15.0-22.0	11.574	2.590	0.12-0.17	14.406-9.759 12.082-8.185	352 351	Brass Alum.			51	0.510 0.255	0.0025	0.590 0335	0.003	0.040
220	42	17.6-26.7	18.0-26.5	14.047	2.137	0.066-0.094	22.042-15.464 18.487-12.970 13.936-9.778	53 121 66	Brass Alum. Silver	596A 598A	595 597	42	0.420 0.170	0.0020	0.500 0.250	0.003	0.040
260	34	21.7-33.0	22.0-33.0	17.328	1.730	0.053-0.076	26.465 17.928 22.197-15.036	354 355	Brass Alum.			34	0.340 0.170	0.0020	0.420 0.250	0.003	0.040
320	28	26.4-40.1	26.5-40.0	21.08	1.422	0.036-0.051	35.413-23.989 29.701-20.120 22.391-15.168	271 96	Brass Alum. Silver	600A	599	28	0.280 0.140	0.0015	0.360 0.220	0.002	0.040
400	22	33.0-50.1	33.0-50.0	26.34	1.138	0.023-0.033	49.491-33.526 41.508-28.119 31.292-21.198	272 97	Brass Alum. Silver		383	22	0.224 0.112	0.0010	0.304 0.192	0.002	0.040
500	19	39.3-59.7	40.0-60;0	31.36	0.956	0.016-0.023	64.367-43.603 40.697-27.569	358	Brass Silver		1529*	19	0.188 0.094	0.0010	0.268 0.174	0.002	0.040
620	15	49.9-75.8	50.0-75.0	39.86	0.752	0.010-0.144	92.152-62.425 58.265-39.470	273 98	Brass Silver		385	15	0.148 0.074	0.0010	0.228 0.154	0.002	0.040
740	12	60.5-92.0	60.0-90.0	48.35	0.620	0.0069-0.0098	123.128-83.409 77.851-52.737	274 99	Brass Silver		387	12	0.122 0.061	0.0005	0.202 0.141	0.002	0.040
900	10	73.8-112	75.0-110.0	59.01	0.508	0.0046-0.0066	165.920-112.397 104.906-71.065	359 -	Brass Silver		1528*	10	0.100 0.050	0.0005	0.180 0.130	0.002	0.040
1200	8	92.3-140	90.0-140.0	73.6	0.406	0.0030-0.0042	146.611-99.317	278	Silver		1527*	8	0.0800 0.0400	0.0003	0.120 0.080	0.001	0.020
1400	7	113-173	110.0-170.0	90.9	0.330	0.0019-0.0028	200.185-135.609	276	Silver		1525*	7	0.0650 0.0325	0.00025	0.105 0.073	0.001	0.020
1800	5	145-220	140.00-220.0	115.7	0.259	0.0012-0.0017	288.036-195.120	275	Silver		1524*	5	0.0510 0.0255	0.00025	0.091 0.066	0.001	0.020
2200	4	172-261	170.0-260.0	137.3	0.218	0.00086-0.00122	372.048-252.032	277	Silver		1526*	4	0.0430 0.0215	0.00020	0.083 0.062	0.001	0.020
2600	3	220-335	220.0-325.0	176.2	0.170	0.00054-0.00076	529.155-358.459		Silver			3	0.0340 0.0170	0.00020	0.156 dia	0.001	

Note 1: True theoretical values at 1 atmos. dry air at 20 C, no safety factor included.
*Contact Flange

H. Stripline Backward Wave Hybrid Coupler

Stripline backward wave* hybrid couplers can be produced conveniently by etching parallel strips on opposite sides of a stripline pair centered between ground planes. A complete analysis of this structure is given by Shelton [26] who gives design equations for both tight coupling (about 10 dB or less) and loose coupling for strips which are completely overlapping, as shown in the cross section in Figure H-1, or offset from one another.

For control circuits including reflection phase shifters, balanced switches, and duplexers, typically the most useful coupling value is 3 dB with full overlap of the conductor pattern to achieve phase symmetry of the output ports. Shelton's design procedure for this full overlap, tight coupling condition then can be reduced to

$$S = \frac{1}{\rho}\left[1 - \frac{Z_0 \epsilon_R \ln 4}{60\,\pi^2}\right] \tag{H-1}$$

where

$$\rho = \frac{Z_{OE}}{Z_{OO}} \qquad Z_{OE} = \rho Z_0 \qquad Z_{OO} = Z_0/\rho$$

$$\rho = \frac{1+k}{1-k} \tag{H-2}$$

$$k = \text{Voltage Coupling Coefficient} = 10^{-[\text{Coupling dB}/20]} \tag{H-3}$$

$$W = \frac{S(1-S)\,(C_O - C_{FO})}{2} \tag{H-4}$$

where

$$C_O = \frac{120\,\pi\sqrt{\rho}}{Z_0\sqrt{\epsilon_R}} \tag{H-5}$$

$$C_{FO} = -\frac{2}{\pi}\left[\frac{1}{1-S}\ln S + \frac{1}{S}\ln(1-S)\right] \tag{H-6}$$

*See Chapter VI for the circuit performance analysis of the backward wave hybrid coupler.

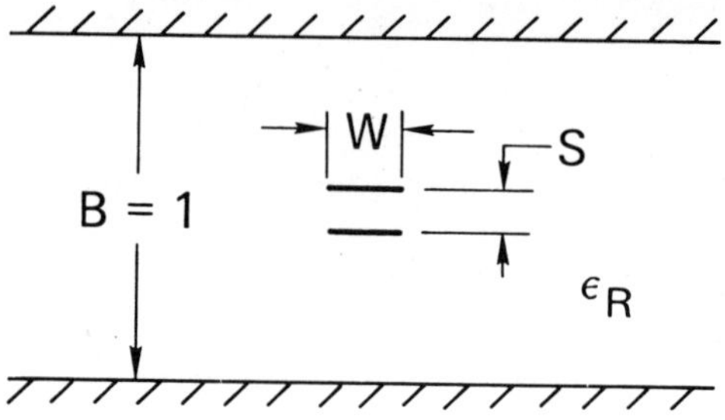

Figure H-1 **Stripline Backward Wave Coupler Cross Section**

Equations (H-1) and (H-4) are valid under the conditions that

$\rho > 2$ (i.e., coupling is 9.5 dB or tighter) (H-7)

and

$$\frac{\rho \epsilon_R Z_0 \ln 4}{60\, \pi^2} < 0.5 \qquad \text{(H-8)}$$

The condition of Equation (H-8) in a 50 Ω system requires that

$\rho \epsilon_R < 18.25$ (H-9)

For example, the tightest coupling for which the equations are valid with various ϵ_R values are:

Tightest Coupling (dB)	0.95	1.9	2.9	3.9
ϵ_R	1	2	3	4

If the conditions in Equations (H-7) and (H-8) are not met, or if an offset geometry is desired (to permit use of given S or B), the coupler design may be possible but Shelton's complete formulation should be consulted.

Coupling, S/B, and W/B are shown graphically in Figure H-2 as derived from Equations (H-1) and (H-4).

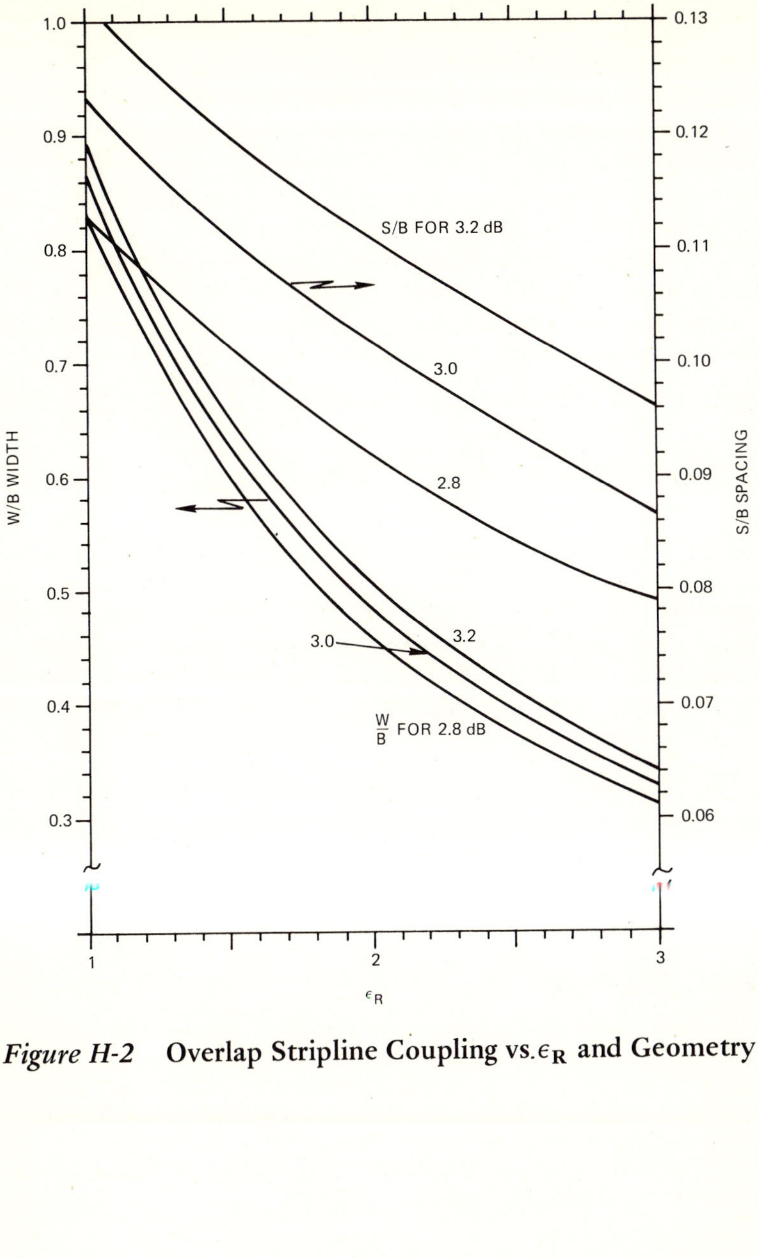

Figure H-2 **Overlap Stripline Coupling vs. ϵ_R and Geometry**

I. Bias Blocks and Returns

Analyzing Bias Elements

PIN and varactor diodes require low frequency bias which must be separated from the RF circuit, yet be introduced across the same diode terminals as is the RF. Bias circuit configurations are limitless yet the bias separating elements and their general behavior can be described in terms of how the bias *blocks* and *returns* affect RF performance, usually VSWR and loss. In this appendix we examine the VSWR of *lumped, distributed,* and *coupled* bias circuits, for 50 Ω characteristic impedance. The lumped and distributed results shown in Figure IA-1 and IA-2 were obtained directly using Program WHITE, Figure VI-30. Inclusion of loss or more complex element equivalent circuits (distributed capacitance of coils, series inductance in capacitors, resistive and line loss effects, etc.) is usually possible with the same program.

Lumped Elements

Figure IA-1 shows the VSWR obtained with either a shunt coil, L, or a series capacitor, C. The curves were generated starting with a coil of +j500 Ω at 1 GHz across a 50 Ω line. This point corresponds to L = 80 nH ($500/2\pi$ to be exact), and gives a VSWR ≈ 1.1 at 1 GHz, as shown. The normalized admittance of the coil is

$$y = 1 - j\left(\frac{Z_0}{\omega L}\right) \tag{IA-1}$$

Similarly, a series capacitance as shown in Figure IA-1, has normalized impedance

$$z = 1 - j\left(\frac{Y_0}{\omega C}\right) \tag{IA-2}$$

If $(Z_0/\omega L) = (Y_0/\omega C)$, y = z at all frequencies and both elements have the same VSWR performance, because both are the same radial distance from the Smith Chart center (see Appendix J) for any ω. The shunt L and series C circuits are then exact mathematical *duals* of each other. If L = 80 nH, the corresponding C is found from

$$C = \frac{L}{{Z_0}^2} = 32 \text{ picofarads} \tag{IA-3}$$

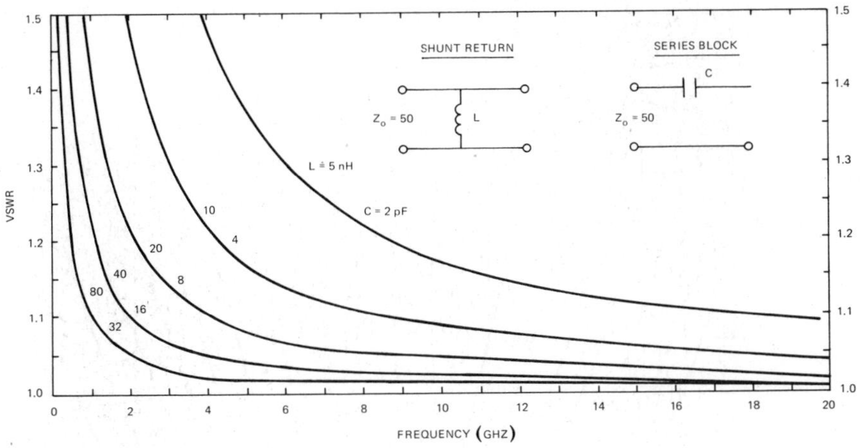

Figure IA-1 **VSWR of Lumped Element Returns and Blocks**

Other choices for L and C giving the same (10) normalized impedance or admittance value at 2, 4, 8, and 16 GHz also are shown in Figure IA-1. It must be emphasized that these contours apply for pure L and C elements. Actual elements, however, have additional "parasitic" reactances as well as resistance which must be modeled specifically for each situation.

Distributed Elements

A short-circuited transmission line has an input reactance which approaches infinity (for a lossless line) as the length approaches 90°; its reciprocal (the susceptance) approaches zero, as can be verified from Equation (J-29) or the Smith Chart in Figure J-5, The susceptance as a function of electrical length, θ, and stub admittance, $1/Z_A$, is

$$jB = -j\,(1/Z_A)\cot\theta \tag{IA-4}$$

and the normalized admittance with the load is

$$y = 1 - j\left(\frac{Z_0}{Z_A}\right)\cot\theta \tag{IA-5}$$

The resulting VSWR in a 50 Ω system is shown in Figure IA-2. The normalized impedance of an open-circuited stub in shunt with a matched load has the same format.

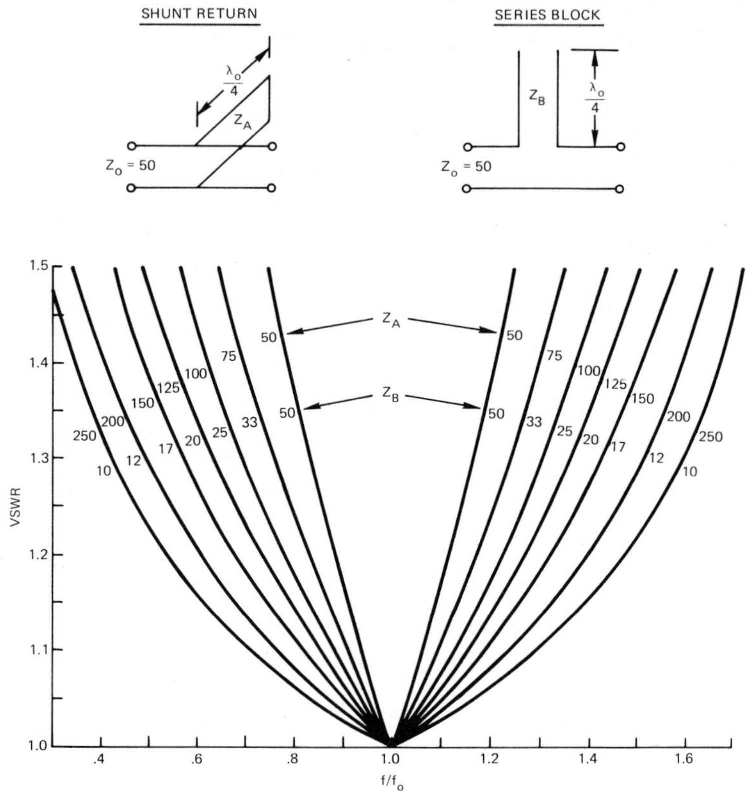

Figure IA-2 **VSWR of Quarterwave Stubs**

$$z = 1 - j\left(\frac{Z_B}{Z_0}\right)\cot\theta \qquad \text{(IA-6)}$$

Thus, for $(Z_0/Z_A) = (Z_B/Z_0)$, the two circuits are duals of one another, and the same VSWR vs. f/f_0 contours apply for both, as shown in Figure IA-2.

Coupled Element Block

A 3 dB hybrid coupler with open circuit terminations can be used instead of a series capacitor as a series block. This action is especially convenient in phase shifter circuits which already have need for the couplers. The VSWR performance of this circuit is shown in Figure VI-16, and a sample application in Figure IX-55.

J. The Smith Chart

Introduction

When a uniform transmission line is terminated in a resistive load equal to its characteristic impedance, Z_0, energy flows between generator and load without reflection, the impedance seen by the generator is equal to Z_0, and this situation is called the *matched line condition.* For any other impedance termination of the line, however, reflection occurs and some of the incident power returns to the generator – where re-reflection can take place if the generator is not matched to the line.

Mismatched loads give rise to a *standing wave* on the line – even if the generator is matched to the line – as the incident and reflected wave magnitudes alternately cancel and reinforce each other between generator and load. In addition, with a mismatched load, the input impedance to the line depends not only upon the load impedance but upon the electrical separation (*line length*) between generator and load as well. In the next section, we develop the basic transmission line equations which describe this phenomenon. This discussion is followed by the *Smith Chart*, a graphical computation method for solving these equations.

Transmission Line Equations [1, 11, 22, 23]

A uniform length of transmission line can be represented as an infinite cascade of series impedance Z and shunt admittance Y sections as shown in Figure J-1. The voltage drop, ΔV, along the line is

$$\Delta V = V(x + \Delta x) - V(x) = - IZ(\Delta x) \qquad \text{(J-1)}$$

Similarly

$$\Delta I = - VY(\Delta x) \qquad \text{(J-2)}$$

Taking the limit as $\Delta x \rightarrow 0$, we obtain two coupled differential equations for V and I.

$$\frac{dV}{dx} = - ZI \qquad \frac{d^2V}{dx^2} = - Z\frac{dI}{dx} \qquad \text{(J-3)}$$

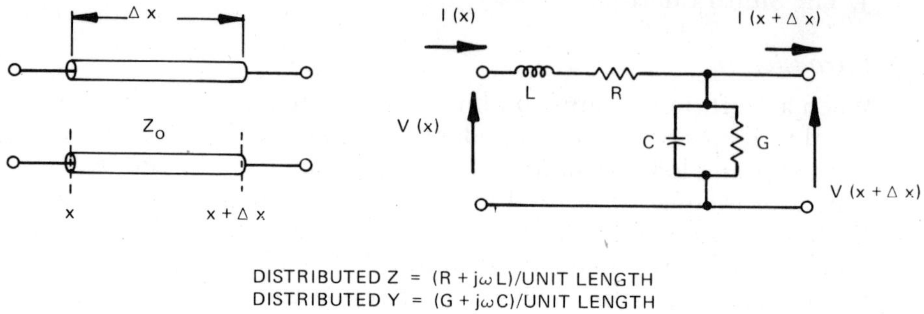

Figure J-1 **Equivalent Circuit of a Short Section of Transmission Line**

$$\frac{dI}{dx} = -YV \qquad \frac{d^2I}{dx^2} = -Y\frac{dV}{dx} \tag{J-4}$$

The expressions containing both V and I in Equations (J-3) and (J-4) can be manipulated to give

$$\frac{d^2V}{dx^2} = ZYV \tag{J-5}$$

$$\frac{d^2I}{dx^2} = ZYI \tag{J-6}$$

The exponential function is unchanged after being differentiated twice; thus, the solutions to the second order differential equations in Equation (J-5) and (J-6) are

$$V(x) = V_I e^{-\gamma x} + V_R e^{+\gamma x} \tag{J-7}$$

$$I(x) = I_I e^{-\gamma x} + I_R e^{+\gamma x} \tag{J-8}$$

in which the constants of integration have been chosen to relate to the physical situation; namely

$$\gamma = \sqrt{(R + j\omega L)(G + j\omega C)} = \sqrt{ZY}$$

$$\gamma = \alpha + j\beta \tag{J-9}$$

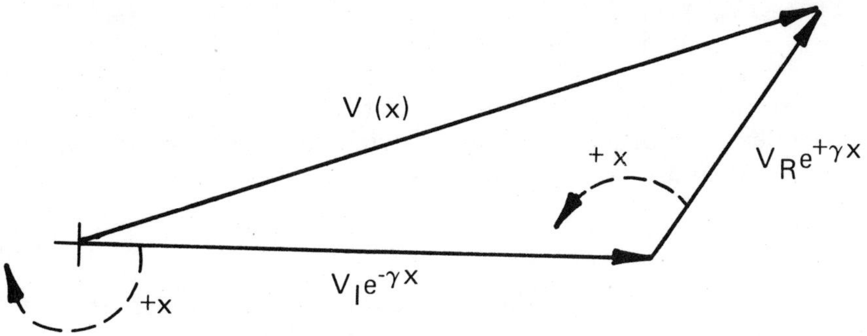

Figure J-2 **The Total Voltage on the Line is the Sum of Two Counter-Rotating Vectors**

V_I and I_I are incident wave amplitudes traveling in the +x direction toward the load*, V_R and I_R are reflected wave amplitudes returning to the generator.

The constant γ is called the *propagation constant* for the line. Usually line losses can be neglected, in which case, $\alpha \approx 0$ and

$$\gamma = j\beta = j\omega\sqrt{LC} = \frac{j2\pi f}{v_P} = \frac{j2\pi}{\lambda_G} \qquad \text{(J-10)}$$

where v_P is the phase velocity and λ_G is the wavelength. The wavelength would be measured between successive traveling wave peaks were one able to freeze time and examine the voltage along the line, the velocity of travel of the wave peak is v_P. In a TEM line $v_P = v_G$ = the velocity of propagation.

For a given load the complex ratio between incident and reflected waves has a constant magnitude but variable phase, since $V_I e^{-\gamma x}$ and $V_R e^{+\gamma x}$ travel in opposite directions. The total voltage in the line, V(x), is thus the sum of two counter-rotating vectors, the resulting amplitude of which varies with distance x as shown in Figure J-2. Passive loads cannot generate power; thus, a *reflection coefficient,* Γ, with magnitude equal to or less than unity can be defined as

$$\Gamma(x) = V_R e^{+\gamma x}/V_I e^{-\gamma x} = \rho e^{j\theta} \qquad \text{(J-11)}$$

*The terms with negative x exponents produce waves traveling in the +x direction. This situation occurs because, when combined with the time variation, $e^{j\omega t}$ (normally not written), the exponent becomes $e^{j(\omega t - \gamma x)}$, which has a constant value for increasing time, t, as one travels in the +x direction.

$$\Gamma(x) = (V_R/V_I)\Big|_{x=0} e^{+2\gamma x}$$

Equation (J-11), stated simply, says that, if Γ is evaluated as the complex ratio of incident and reflected voltages at some position, say x = 0, at any other position, x

$$\Gamma(x) = \Gamma(x = 0)\, e^{+2\gamma x} \tag{J-12}$$

from which we see that the argument of Γ goes through 360° as x goes through a half-wavelength*. It is for this reason that *standing wave maxima (or minima) are separated by one half wavelength.* This point is an important one and it is the reason why, as is shown later, a 360° rotation on the Smith Chart corresponds to a movement of only one half wavelength along the transmission line.

In a microwave *slotted line,* an electric probe is slid along a x-directed slot in the transmission line outer conductor. A rectifying crystal or heat sensitive bolometer gives an output proportional to the RF voltage amplitude on the line. The maximum and minimum amplitudes sensed by the probe are

$$|V_{MAX}| = |V_I(1 + \rho)| \tag{J-13}$$

$$|V_{MIN}| = |V_I(1 - \rho)| \tag{J-14}$$

The ratio $|V_{MAX}/V_{MIN}|$ is called the *voltage standing wave ratio* or *VSWR;* it is related to reflection coefficient magnitude by

$$\text{VSWR} = \frac{1+\rho}{1-\rho} \quad \text{where } \rho \text{ is the magnitude of } \Gamma. \tag{J-15}$$

If VSWR is measured, the relative amplitude of the reflected wave (i.e., the reflection coefficient magnitude) is

$$\rho = \frac{\text{VSWR} - 1}{\text{VSWR} + 1} \tag{J-16}$$

and the fraction of power reflected, R_R, to power incident, P_I, is

$$\frac{P_R}{P_I} = \rho^2 \tag{J-17}$$

*Physically, this fact can be appreciated by recognizing that, for example, for a movement of 180 electrical degrees on the line, both the incident wave gains 180° and the reflected wave losses 180°, producing a net relative change between them of 360°.

To determine the input impedance Z_{IN} (x) an expression for I(x) in terms common to V(x) is needed. This expression is obtained by operating on Equation (J-7) with Equation (J-3).

$$I(x) = -\frac{1}{Z}\left(\frac{dV}{dx}\right) = \frac{\gamma}{Z}\left(V_I e^{-\gamma x} - V_R e^{+\gamma x}\right) \tag{J-18}$$

then

$$Z_{IN}(x) = \frac{V(x)}{I(x)} = \sqrt{\frac{Z}{Y}}\;\frac{V_I e^{-\gamma x} + V_R e^{+\gamma x}}{V_I e^{-\gamma x} - V_R e^{+\gamma x}} \tag{J-19}$$

When the line is match terminated, $V_R = 0$ and the input impedance obtained under this condition is called the *characteristic impedance*, Z_0, of the transmission line

$$Z_0 = \sqrt{\frac{Z}{Y}} = \sqrt{\frac{R+j\omega L}{G+j\omega C}} \tag{J-20}$$

$$Z_0 = \sqrt{\frac{L}{C}} \quad \text{(lossless line; } R = G = 0\text{)} \tag{J-21}$$

Using Γ, Equation (J-19) can be written

$$Z_{IN}(x) = Z_0 \frac{1 + \Gamma(x)}{1 - \Gamma(x)} \tag{J-22}$$

In particular if a terminating impedance, Z_L, is used and the line length is reduced to zero the input impedance is just the load impedance and

$$Z_L = Z_0 \frac{1 + \Gamma_L}{1 - \Gamma_L} \tag{J-23}$$

Since the choice of x = 0 is arbitrary Equation (J-23) holds for any x; thus, the simultaneous values for z(x) and Γ(x) are related by

$$z = \frac{1 + \Gamma}{1 - \Gamma} \tag{J-24}$$

Equivalently

$$\Gamma = \frac{z - 1}{z + 1} \tag{J-25}$$

where $z = Z(x)/Z_0$ is called the *normalized impedance*. This variable is used with the Smith Chart to describe the variation of impedance along a mismatched line. The most common transmission line analysis requires the calculation of the input impedance, z_{IN}, given the impedance at some other point on the line, the location of which we may define to be $x = 0$. With our sign convention energy propagates from the generator to the load in the $-x$ direction; therefore, we use the $-x$ variable in Equation (J-19). The input impedance to be calculated is usually on the generator side of the specified impedance which is at $x = 0$; thus

$$z(-x) = \frac{V_I(x{=}0)e^{+\gamma x} + V_R(x{=}0)e^{-\gamma x}}{V_I(x{=}0)e^{+\gamma x} - V_R(x{=}0)e^{-\gamma x}} = \frac{e^{+\gamma x} + \Gamma\,(x{=}0)e^{-\gamma x}}{e^{+\gamma x} - \Gamma\,(x{=}0)e^{-\gamma x}} \tag{J-26}$$

Replacing $\Gamma(x{=}0)$ with $z\,(x{=}0)$ through the use of Equation (J-25) permits Equation (J-26) to be written [26]

$$z(-x) = \frac{z(x{=}0)\cdot\dfrac{e^{\gamma x} + e^{-\gamma x}}{2} + \dfrac{e^{\gamma x} - e^{-\gamma x}}{2}}{\dfrac{e^{\gamma x} + e^{-\gamma x}}{2} + z(x{=}0)\cdot\dfrac{e^{\gamma x} - e^{-\gamma x}}{2}}$$

$$z(-x) = \frac{z(x{=}0)\cosh\gamma x + \sinh\gamma x}{\cosh\gamma x + z(x{=}0)\sinh\gamma x} \tag{J-27}$$

Equation (J-27) gives the impedance at $-x$, given the impedance at $x{=}0$. Since the choice of location for $x{=}0$ is arbitrary, Equation (J-27) is valid for any pair of impedances regardless of whether one or both are at the respective ends of the transmission line. Usually, however, one desires to know how a normalized terminating (load) impedance, $z_L = Z_L/Z_0$, is transformed by a length ℓ of Z_0 line. Therefore Equation (J-27) is usually written

$$z_{IN} = \frac{z_L + \tanh(\gamma\ell)}{1 + z_L\tanh(\gamma\ell)} \tag{J-28}$$

where the substitution of $\tanh\gamma\ell$ for $\sinh\gamma\ell/\cosh\gamma\ell$ has been made. If the line has small enough dissipation to neglect loss ($\alpha \approx 0$), Equation (J-28) reduces to

$$z_{IN} = \frac{z_L + j\tan\theta}{1 + jz_L\tan\theta} \tag{J-29}$$

where $\theta = 2\pi\ell/\lambda$ is the electrical length of the line. In this text, Equations (J-28) and (J-29) are termed the *Smith Chart Transformations* because they are the basis for the Smith Chart, the graphical means for manipulating them and more widely known than the equations themselves.

The Smith Chart Construction

In principle, one could plot for various complex values of z_L and $\gamma\ell$ the corresponding values of z_{IN}. The problem is that, even for fixed $\gamma\ell$, the entire right half of the (passive) impedance z_L plane must be mapped onto a similar and also semi-infinite z_{IN} plane. Philip Smith [24] solved this graphical problem by observing that if, instead, one plots z(x) *on the* $\Gamma(x)$ *plane* using Equation (J-25), all impedance values could be mapped within a circle of unit radius since $|\Gamma| \leqslant 1$, the requirement for the *Smith Chart.* Since the variation of Γ for a movement along the transmission line represents merely a rotating vector of constant magnitude (neglecting losses) in the $\Gamma(x)$ plane *the impedance transformation* $z_L \rightarrow z_{IN}$ *(or vice versa) is determined easily from this Smith Chart by noting the impedances which correspond to the initial and final values of* Γ. Before demonstrating the method, let us examine how the Smith Chart is constructed.

Figure J-3 shows the Γ plane, representing the real (u, horizontal axis) and imaginary (jv, vertical axis) parts of Γ. Normally Γ is written in polar form as $\Gamma = \rho e^{j\theta}$, but for this development it is desirable to express it in the u-jv rectangular form as well. Thus

$$u = \rho \cos \theta$$

$$v = \rho \sin \theta$$

$$\rho^2 = u^2 + v^2 \qquad \text{(J-30)}$$

Then, if z = r + jx, Equation (J-24) can be rewritten as

$$r + jx = \frac{1 + \rho e^{j\theta}}{1 - \rho e^{j\theta}} \cdot \frac{1 - \rho e^{-j\theta}}{1 - \rho e^{-j\theta}} \qquad \text{(J-31)}$$

Carrying out the indicated multiplication in the right side (to remove the complex quantities from the denominator) and noting that $e^{j\theta} - e^{-j\theta} = 2\,j \sin \theta$ and $e^{j\theta} + e^{-j\theta} = 2 \cos \theta$ gives

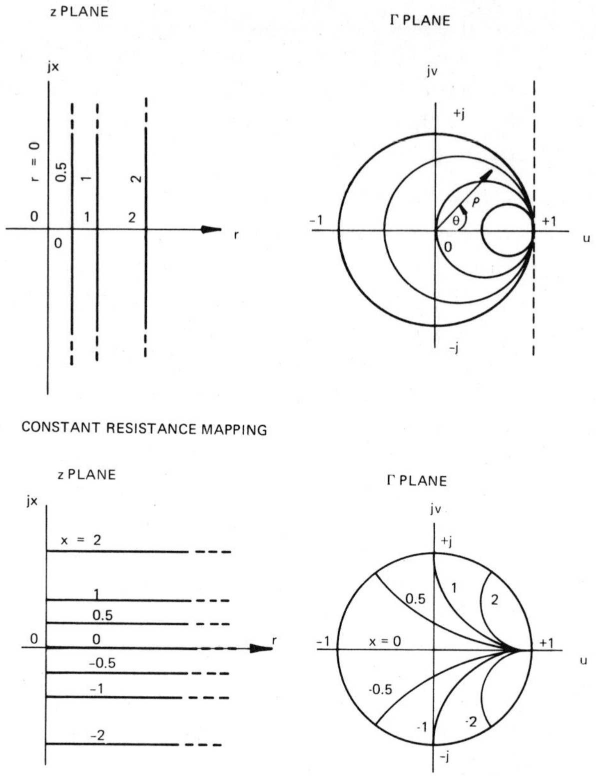

b) CONSTANT REACTANCE MAPPING

Figure J-3 **Development of the Smith Chart by Mapping z to Γ using Γ = (z − 1)/(z + 1)**

$$r + jx = \frac{1 - \rho^2 + 2\, j\rho \sin\theta}{1 + \rho^2 - 2\,\rho \cos\theta}$$

$$= \frac{1 - u^2 - v^2 + 2\, jv}{1 + u^2 + v^2 - 2\, u} \qquad \text{(J-32)}$$

In order to construct the Smith Chart we desire to know how various impedance contours in the z plane ($z = r + jx$) are mapped* onto the Γ plane ($\Gamma = \rho e^{j\theta} = u + jv$) by the mapping function of

*Because a complex function, f(z), has two parts, real and imaginary (as does its complex argument, z), it is not possible to plot f(z) versus z in a single plane. Rather, we must show how a contour in the z plane is "mapped" onto a corresponding contour in the f(z) plane by the "mapping function," which is just the function f(z) itself. For example, as is shown, the r = 0 contour (vertical axis) in the z-plane is mapped onto the $|\Gamma| = 1$ circle in the Γ plane by the complex function, Equation (J-25).

Equation (J-24), which has been written explicitly as Equation (J-32) in order to separate the real and imaginary parts of both Γ and z. Equation (J-24) is one of a class of complex functions called *bilinear transformations* [25] which have the property that circles are mapped onto circles, straight lines being considered circles of infinite radius.

Let us examine how constant resistance contours in the z plane are mapped onto the Γ plane by the mapping function, Equation (J-32). For z to have a constant resistive magnitude, the real parts of both sides of Equation (J-32) have constant magnitudes; therefore

$$\text{Constant} = r = \frac{1 - u^2 - v^2}{1 + u^2 + v^2 - 2u} \tag{J-33}$$

which, after algebraic manipulation, can be written

$$\left(u - \frac{r}{r+1}\right)^2 + v^2 = \frac{1}{(r+1)^2} \tag{J-34}$$

In the u-jv coordinate system (the Γ plane), this equation is that of a circle of radius $1/(r + 1)$ with center at $(u = r/(r + 1), jv = 0)$. Figure J-3(a) shows how constant resistance (vertical) contours in the z plane are mapped onto circles in the Γ plane.

Similarly, for constant reactance magnitude, the imaginary parts of both sides of Equation (J-32) are constant, or

$$\text{Constant} = x = \frac{2v}{1 + u^2 + v^2 - 2u} \tag{J-35}$$

This expression can be rewritten as

$$(u - 1)^2 + \left(v - \frac{1}{x}\right)^2 = \frac{1}{x^2} \tag{J-36}$$

which, for a given fixed x, is the locus of a circle in the u-jv plane with radius $1/x$ and center at $(u = 1, v = 1/x)$, as shown in Figure J-3(b).

The overlay of constant resistance and reactance contours in the Γ plane is the Smith Chart, a complete version of which is shown in Figure J-5 in connection with a sample load matching example.

Notice that the u and jv axes and the designation "Γ plane" are not stated explicitly, but they can always be assumed. As is shown, the Smith Chart can be used with either impedance or admittance parameters — but *only z relates directly to Γ on the chart.*

Using the Smith Chart

a) Impedance Inversion

An important property of the Smith Chart is that a rotation of 180° produces impedance inversion ($z_{IN} = 1/z_L = y_L$). Thus, since admittance is the reciprocal of impedance, by normalizing any impedance or admittance to a convenient value, the *complex reciprocal can be determined graphically from the chart.*

For example, given an impedance, Z_L equal to 50 + j50 Ω, the value normalized to Z_0 = 50 Ω is 1 + j1, and a 180° rotation on the chart gives 0.5 − j0.5. Un-normalizing, the actual admittance is Y = (0.5 − j0.5) (1/50) = (0.01 − j0.01) mhos. The reader may verify that the same result is obtained regardless of the value chosen for Z_0, although graphical accuracy deteriorates when the desired values produce very small or very large impedance/admittance magnitudes on the chart.

The formal proof that impedance inversion results from a 180° rotation is demonstrated using Equation (J-29) and considering the result of a transformation of a load, z_L, by a quarter-wave (180° rotation on the Smith Chart) section of lossless (no change in the radius, ρ) transmission line. Thus

$$z_{IN} = \frac{z_L + j \tan\theta}{1 + j\, z_L \tan\theta}$$

Dividing numerator and denominator by tan 90° and taking the limit as $\theta \rightarrow 90°$ gives

$$z_{IN} = \frac{j}{j z_L} = y_L$$

This inversion property is used to create *quarter wavelength impedance inverters, which have the same effect as impedance transformers when the load is purely resistive.* It also explains why a 90° short-circuit terminated stub appears as an open (parallel resonant) circuit and a 90° open-circuit terminated stub appears as a short (series resonant) circuit.

b) Admittance Notation

The Smith Chart also can be used with admittances, *except that the reflection coefficient is only obtained when the corresponding value is converted to impedance by a 180° rotation.* This fact can be proved by substituting $y_L = 1/z_L$ into Equation (J-28) or, for lossless lines, Equation (J-29). This flexibility is useful, since shunt elements added to a line have their admittances added to the transformed admittance at that point. In practice, transitions between admittance and impedance, by 180° rotations on the chart, permit necessary conversions for series or shunt element additions to the line.

It is important not to confuse the inversion of an actual normalized impedance by a 90° length of transmission line with the conversion of impedance to admittance format by the graphical procedure of a 180° rotation on the Smith Chart. *The graphical procedure yields the inverse of any complex number quite apart from any transmission line considerations.*

c) Matching Example

Suppose it is desired to match a 100 Ω resistive load to a 50 Ω transmission line by adding a short-circuited 50 Ω line stub of appropriate electrical length, θ_1, and electrical distance, θ_2, from the load as shown in Figure J-4. What are the appropriate θ_1 and θ_2 values?

At the load, the normalized impedance is $z_L = 100/50 = 2$, and the corresponding reflection coefficient is $\Gamma_L = 0.33\angle 0°$, indicating a voltage maximum at the load. Moving clockwise (toward the

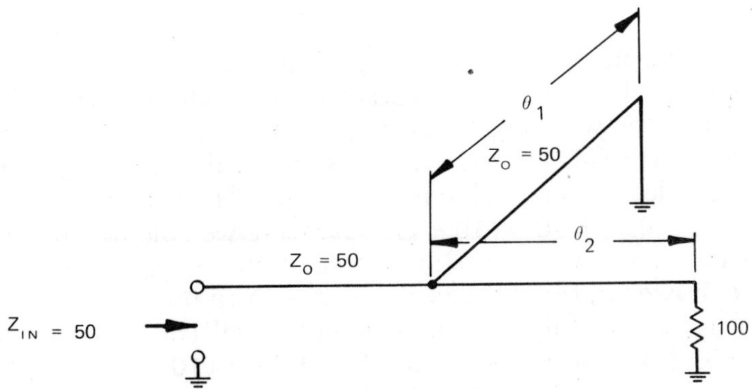

Figure J-4 **Load Matching Example**

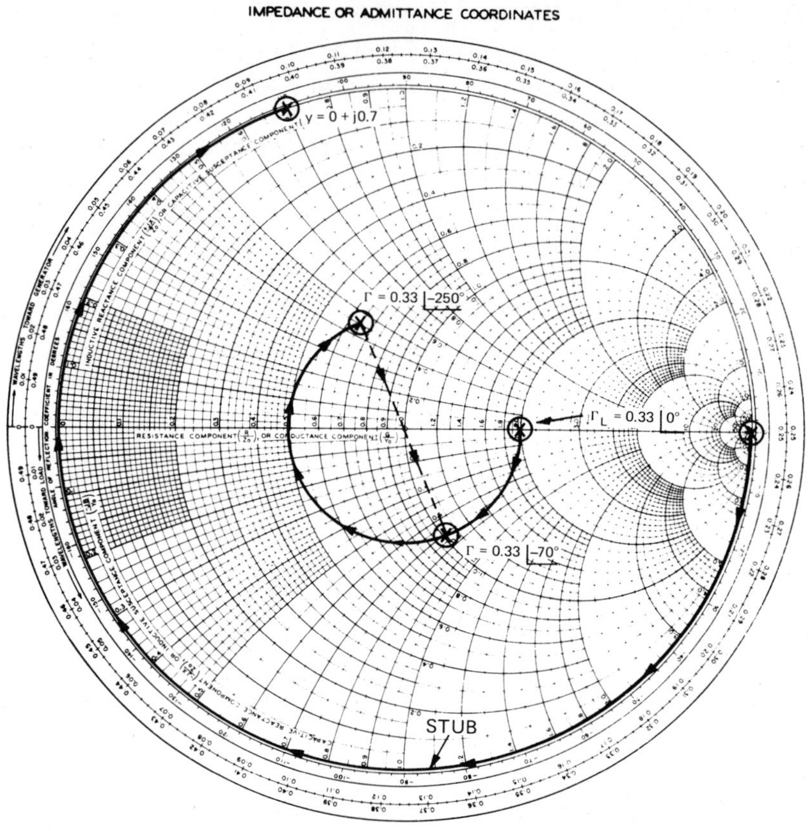

Figure J-5 **The Smith Chart with an Illustration of the Load Matching Determination for Figure J-4**

generator) we intersect the r = 1 circle at $\Gamma = 0.33\angle{-70^\circ}$, as shown in Figure J-5. The -70° angle is read from the chart's periphery. Alternately, the wavelength scale could be used. At this point (35° from the load, since movements on the line are only half the number of degrees of change in the phase of Γ) the input impedance is $(1 - j0.7) \cdot 50 = (50 - j35)\ \Omega$. Were we able to add a *series* inductive reactance of j35 Ω, the desired match would be obtained at once; however, for this example we wish to use a *shunt* stub. Therefore, we continue through an additional 180° rotation to $\Gamma = 0.33\angle{-250^\circ}$ (or, equivalently, $\Gamma = 0.33\angle{+110^\circ}$).

At this point the normalized impedance is about $z = 0.65 + j0.45$. We need not establish this figure accurately, for at this point the *admittance*, y, equals 1/z = 1 − j0.7. Addition of a normalized susceptance of +j0.7 yields y = 1, the desired match. Thus, the stub should be separated from the load by $\theta_2 = 250°/2 = 125°$.

The appropriate stub length can be found using the same chart. At the short-circuited (right side of the horizontal axis) end of the stub the admittance, y, equals $0 + j\,\infty$. We avoid reading the reflection coefficient because impedance coordinates must be determined to identify Γ (the appropriate value for z would be obtained by a 180° rotation on the chart to the position z = 0 + j0, for which $\Gamma = 1\angle 180°$).

Moving clockwise around the periphery of the chart ($|\Gamma| = \rho = 1$ since the stub has no loss, and reflected waves on it have the same magnitude as do incident waves) to the positive (upper half of chart) value for j0.7 defines a total movement toward the generator of 250° and the corresponding value for the stub length, $\theta_1 = 250°/2 = 125°$.

This matching method was chosen to demonstrate:

1) Impedance-Admittance Conversion (Inversion)
2) Shunt Susceptance Addition
3) Shorted Stub Input Susceptance

There is, however, an infinitely large number of means for matching. For example, had we chosen to use a $Z_0 = 50\sqrt{2}\ \Omega$ line between generator and load 90° long, the 50 Ω input impedance match would have been obtained without need of a shunt stub. This fact can be verified using the Smith Chart and the principles described; in this case, the matching results from the normalization, 180° rotation, and subsequent unnormalization processes.

The extent of the usefulness of the Smith Chart becomes apparent only through practice, but its basis is easy to remember through Equation (J-24) and the fact that the Smith Chart is a mapping of constant resistance and reactance *from the z plane onto the Γ plane.*

References and Bibliography

[1] Ramo, S.; and Whinnery, J.R.: *Fields and Waves in Modern Radio,* John Wiley and Sons, New York, 1944; also see the revised book by Ramo, Whinnery and Van Duzer.

[2] *ITT Reference Data for Radio Engineers – Fifth Edition,* Howard W. Sams and Co., Inc., New York, 1974.

[3] Klein, G.: "Thermal Resistivity Table Simplifies Temperature Calculations," *Microwaves,* pp. 58-59, February 1970.

[4] *Handbook of Chemistry and Physics,* Chemical Rubber Publishing Co., Cleveland, Ohio, (37th Edition, 1955, used for these appendices).

[5] *Tables of Dielectric Materials, Vols. I-IV,* Laboratory for Insulation Research, Massachusetts Institute of Technology, Cambridge, Massachusetts, January, 1953.

[6] Von Hipple, A.R. (ed.): *Dielectric Materials and Applications,* John Wiley and Sons, New York, 1954.

[7] Frados, J. (ed.): *Modern Plastics Encyclopedia,* New York, 1962.

[8] Howe, H. Jr.: *Stripline Circuit Design,* Artech House, Inc., Dedham, Massachusetts, 1974.

[9] Phillips, A.B.: *Transistor Engineering,* McGraw-Hill, Inc., New York, 1962 (references for properties of germanium and silicon).

[10] Clauser et al.: *Encyclopedia of Engineering Materials and Processes,* Reingold Publishing Corp., New York, 1963.

[11] Collin, R.E.: *Foundations of Microwave Engineering,* McGraw-Hill, Inc., New York, 1966.

[12] Saad, T.S.: *Microwave Engineers' Handbook, Vols. I and II,* Artech House, Inc., Dedham, Massachusetts, 1971.

[13] Kwon, A.H.: "Design of Microstrip Transmission Line," *Microwave Journal,* pp. 61-63, January 1976. This paper, the basis of Appendix E, summarizes and extends the earlier work of Wheeler, Presser, and Schneider in References 14, 15, and 16.

[14] Wheeler, H.A.: "Transmission Line Properties of Parallel Strips Separated by a Dielectric Sheet," *IEEE Transactions on Microwave Theory and Techniques, Vol. MTT-13, No. 3,* pp. 172-185, March 1965.

[15] Presser, A.: "RF Properties of Microstrip Line," *Microwaves,* pp. 53-58, March 1968.

[16] Schneider, M.V.: "Microstrip Lines for Microwave Integrated Circuits," *Bell System Technical Journal,* pp. 1421-1443, May-June 1969.

[17] Kwon, Andrew H.: MICRO (FORTRAN Computer Program)," *Microwave Journal Handbook,* 1976.

[18] Cohn, S.B.: "Problems in Strip Transmission Lines," *IEEE Microwave Theory and Techniques Transactions, Vol. MTT-3, No. 2,* pp. 119-126, March 1955.

[19] Howe, H. Jr.: "ZO (FORTRAN Computer Program)," *Microwave Journal Handbook,* 1974.

[20] *Standard Mathematical Tables,* Chemical Rubber Publishing Co., Cleveland, Ohio (1955 edition used for these appendices).

[21] Jolley, L.B.W.: *Summation of Series,* Dover Publications, Inc., New York, 1961. (First published by Chapman & Hall, Ltd., England, 1925; published later in the United Kingdom by Constable and Co., Ltd., London, 1961.)

[22] Altman, J.L.: *Microwave Circuits,* D. Van Nostrand Co., Inc., New York, 1964.

[23] Garver, R.V.: *Microwave Diode Control Devices,* Artech House, Inc., Dedham, Massachusetts, 1976.

[24] Smith, P.H.: "An Improved Transmission Line Calculator," *Radio Engineers Handbook* (First Edition), McGraw-Hill, Inc., New York, circa 1940, p. 111. Printed in German.

[25] Miller, K.S.: *Advanced Complex Calculus,* Harper and Brothers Publishers, New York, 1960.

[26] Shelton, J.P. Jr.: "Impedances of Offset-Coupled Strip Transmission Lines," *IEEE Transactions on Microwave Theory and Techniques, Vol. MTT-14, No. 1,* pp. 7-15, January 1966.

[27] Hansen, Max: *Constitution of Binary Alloys,* McGraw-Hill, Inc., New York, 1958. This reference is a good one for the melting point vs. composition of various alloys, especially solders and semiconductor metals.

[28] Cohn, S.B.: "Beating a Problem to Death," *Microwave Journal*, November 1969, pp. 22-24. This editorial is interesting, comparing the various formulas used for square coax line impedance.

[29] Wheeler, H.A.: "Transmission Line Impedance Curves," *Proceedings of the IRE, Vol. 38,* December 1950, pp. 1400-1403.

[30] Cristal, E.G.: "Coupled Circular Cylindrical Rods Between Parallel Ground Planes," *IEEE Transactions on Microwave Theory and Techniques, Vol. MTT-12, No. 1,* July 1964, pp. 428-439. Also see, by this author, "Characteristic Impedance of Coaxial Lines of Circular Inner and Rectangular Outer Conductors" (Correspondence), *Proceedings of the IEEE, Vol. 52,* October 1964, pp. 1265-1266.

[31] Frankel, Sidney: "Characteristic Impedance of Parallel Wires in Rectangular Trough," *Proceedings of the IRE, Vol. 30,* April 1942, pp. 182-190.

[32] Wheeler, H.A.: "The Transmission-Line Properties of a Round Wire Between Parallel Planes," *IRE Transactions on Antennas and Propagation, Vol. AP-3,* October 1955, p. 203.

[33] Getsinger, William J.: "Microstrip Dispersion Model," *IEEE Transactions on Microwave Theory and Techniques, Vol. MTT-21, No. 1,* January 1973, pp. 34-39.

Index